National Communications System
Technology & Standards Division

General Services Administration
Information Technology Section

TELECOMMUNICATIONS

Glossary of Telecommunication Terms

Federal Standard 1037C

(The use of this standard by all Federal Departments and agencies is mandatory.)

Government Institutes
Rockville, Maryland

Government Institutes, Inc., 4 Research Place, Rockville, Maryland 20850
Phone: (301) 921-2355
Fax: (301) 921-0373
Email: giinfo@govinst.com
Internet Address: http://www.govinst.com

Copyright © 1997 by Government Institutes.
Published February 1997.

01 00 99 98 5 4 3 2

This standard is issued by the General Services Administration and provides Federal departments and agencies a comprehensive source of definitions of terms used in telecommunications and directly related fields by international and U.S. Government telecommunications specialists.

Released on August 7, 1996, this Federal Standard 1037-C supersedes Federal Standard 1037-B of June 3, 1991. The standard is updated every five years and its use is mandatory for all Federal departments and agencies.

This standard was developed by a subcommittee of the Federal Telecommunication Standards Committee. The U.S. Department of Commerce, National Telecommunications and Information Administration, Institute for Telecommunication Sciences supplied the chair, secretariat and technical editorial services for the subcommittee.

Government Institutes determined that this document contained information of importance to the general telecommunications industry, so we have reproduced this material in order to serve those interested.

The publisher makes no representation of warranty, express or implied, as to the completeness, correctness or utility of the information in this publication. In addition, the publisher assumes no liability of any kind whatsoever resulting from the use of or reliance upon the contents of this book.

ISBN: 0-86587-580-4

Printed in the United States of America

FOREWORD

This standard is issued by the General Services Administration pursuant to the Federal Property and Administrative Services Act of 1949, as amended.

This document provides Federal departments and agencies a comprehensive source of definitions of terms used in telecommunications and directly related fields by international and U.S. Government telecommunications specialists.

The Paperwork Reduction Reauthorization Act of 1986, Public Law 99-500 [44 U.S.C. 3502(13)], expanded the definition of "automatic data processing equipment (ADPE)" to include . . . any equipment or interconnected systems or subsystems or equipment that is used in the automatic acquisition, storage, manipulation, management, movement, control, display, switching, interchange, transmission or reception of data or information." This expansion in the scope of ADPE is implemented in the Federal Information Resources Management Regulation (FIRMR), 41 CFR. To minimize confusion between the statutory definition of ADP and the popular meaning of that term, GSA has established the term "Federal information processing (FIP) resources" to replace the term ADPE, as defined in Public Law 99-500.

The existence of multiple definitions for the same term in this standard reflects, in most instances, differences in meaning commonly assigned to these terms by practitioners in telecommunications and other, related disciplines. In some instances, however, the differences are not the result of interdisciplinary distinctions, but rather reflect variances in American and international usage or a need for further work in the various Government, national, and international standards-development groups to reach agreement on a common definition, or as a result of legislation. Source citations for some definitions (see the *Legend* pages) reflect the tracking of specialized glossaries for consistency.

The use of this standard by all Federal departments and agencies is ***mandatory.***

Neither this nor any other glossary covering terms in an advanced-technology field such as telecommunications can be considered as complete and ageless. Periodic revisions will be made as required.

(This page is deliberately left blank.)

CONTENTS

(This page is deliberately left blank.)

FEDERAL STANDARD

Telecommunications: GLOSSARY OF TELECOMMUNICATION TERMS

1. SCOPE.

a. This glossary provides standard definitions for the fields subsumed by the umbrella discipline of telecommunications. Fields defined herein include: antenna types and measurements, codes/coding schemes, computer and data communications (computer graphics vocabulary, file transfer techniques, hardware, software), fiber optics communication, facsimile types and techniques, frequency topics (frequency modulation, interference, spectrum sharing), Internet, ISDN, LANs (MANs, WANs), modems, modulation schemes, multiplexing techniques, networking (network management, architecture/topology), NII, NS/EP, power issues, PCS/UPT/cellular mobile, radio communications, routing schemes, satellite communications, security issues, switching techniques, synchronization/timing techniques, telegraphy, telephony, TV (UHF, VHF, cable TV, HDTV), traffic issues, transmission/propagation concerns (signal loss/attenuation, transmission lines), video technology, and wave propagation/measurement terminology.

b. The terms and accompanying definitions contained in this standard are drawn from authoritative non-Government sources such as the International Telecommunication Union, the International Organization for Standardization, the Telecommunications Industry Association, and the American National Standards Institute, as well as from numerous authoritative U.S. Government publications. The FTSC Subcommittee to Revise FED-STD-1037B has rewritten many definitions as deemed necessary either to reflect technology advances or to make those definitions that were phrased in specialized terminology more understandable to a broader audience.

1.1 **Applicability.** This standard incorporates and supersedes FED-STD-1037B, June 1991. Accordingly, all Federal departments and agencies shall use it as the authoritative source of definitions for terms used in the preparation of all telecommunications documentation. The use of this standard by all Federal departments and agencies is ***mandatory.***

1.2 **Purpose.** The purpose of this standard is to improve the Federal acquisition process by providing Federal departments and agencies a comprehensive, authoritative source of definitions of terms used in telecommunications and directly related disciplines by national, international, and U.S. Government telecommunications specialists.

2. REQUIREMENTS AND APPLICABLE DOCUMENTS.

a. The terms and definitions that constitute this standard, and that are to be applied to the uses cited in paragraph 3 below, are contained on page A-1 through Z-1 of this document. There are no other documents applicable to implementation of this standard. A list of acronyms and abbreviations is presented as Appendix A. The list of abbreviations and acronyms uses **bold** font to identify those term names that are defined in this glossary. An abbreviated index of selected principal families of related term names is presented in Appendix B.

b. Within this document, symbols for units of measurement (and the font type for these symbols) are in accord with ANSI/IEEE Std. 260.1-1993, *American National Standard Letter Symbols for Units of Measurement (SI Units, Customary Inch-Pound Units, and Certain Other Units).*

3. USE.

a. All Federal departments and agencies shall use the terms and definitions contained herein. Only after determining that a term or definition is not included in this document may other sources be used. The *Legend* beginning on page *xii* is provided to assist users in determining the documentary source of the definitions.

b. Nearly all terms are listed alphabetically; a few exceptions to this rule include (1) the family of network topologies, which are grouped under the definition of "*network topology*," and (2) the family of dispersion terms, which are grouped under the definition of "*dispersion*." In all cases, ample cross references guide the reader to the location of the definition. Term names containing numerals are alphabetized as though the numbers were spelled out; thus, *"144-line weighting"* will appear in the "O" portion of the alphabet between the terms *"on-board communication station"* and *"one-way communication,"* since it is pronounced as if it were spelled "one-forty-four line. . . ." For user convenience, exceptions to the rule are taken for entries comprising numerically consecutive terms, *e.g., "digital signal 0," . . . "digital signal 4,"* which are grouped numerically following the *"digital signal"* entry.

c. An abbreviation for the term name often appears in parentheses following the term name. When both the abbreviation and the spelled-out version of a term name are commonly used to name an entity defined in this glossary, the definition resides with the more commonly used version of the term name. If the more commonly used designation is the fully spelled-out term name, then the definition resides under that name. If, however, the more common term name is the abbreviation, then the definition rests with the abbreviated spelling of that term name. For example, the definition of *"decibel"* resides under *"dB."*

d. When more than one definition is supplied for a given term name, the definitions are numbered, and the general definition is given first. Succeeding definitions are often specific to a specialized discipline, and are usually so identified.

e. Notes on definitions are **not** a mandatory part of this document; these notes are expository or tutorial in nature. When a note *follows* a source citation (such as "[JP1]"), that note is not part of the source document cited. Notes and cross references apply only to the immediately preceding definition, unless stated otherwise.

f. Three types of cross references are used: *"Contrast with," "Synonym,"* and *"See":*

(1) *"Contrast with"* is used for terms that are nearly antonyms, or when understanding one concept is aided by examining the definition of its counterpart.

(2) When term names are synonymous, the definition is placed under only one of the term names, *i.e.,* the preferred term name, which is generally the most common name. Synonyms are listed for cross-reference purposes only. The other term name entries contain only a *"Synonym"* listing; *i.e.,* the definition for synonymous term names is not repeated. Terms labeled *"Colloquial synonym"* are in occasional informal use, but may be semantically inexact or may border on slang.

(3) *"See"* is used where an undefined term name is entered as a cross reference only to direct the reader to a related term name (or term names) that is (are) defined in the glossary.

g. Term names that are semantically incorrect, that have been replaced by recent advances in technology, or that have definitions that are no longer applicable, are designated as *"deprecated."* In such case, the reader is referred to current term names, where applicable.

h. The telecommunications terms included in this glossary either are not sufficiently defined in a standard desk dictionary or are restated for clarity and convenience. Likewise, combinations of such words are included in this glossary only where the usual desk-dictionary definitions, when used in combination, are either insufficient or vague.

i. Definitions that carry the source citation "[47CFR]" (which refers to *Title 47 U.S. Code of Federal Regulations*), or "[NTIA]" (which refers to the *NTIA Manual),* or the source citation "[RR]" (which refers to the ITU *Radio Regulations*) may have a format or syntax that differs from the definitions in the remainder of FED-STD-1037C because the FTSC Subcommittee to Revise FED-STD-1037B was not authorized to make any changes whatever to the definitions in these three documents. One minor formatting change was made to definitions from NSTISSI No. 4009, *National Information Systems Security (INFOSEC) Glossary*, cited [NIS]: Often the introductory indefinite article or definite article was added at the beginning of the cited definition, and that article was added in square brackets "[ ]" to indicate that its addition was the only change made in the quoted definition.

j. Figures have been added to many definitions throughout the glossary to illustrate complex concepts or systems that are defined herein. With the exception of the figure called "**electromagnetic spectrum,**" these figures are **not** a mandatory part of this document.

k. This standard contains two appendixes, **neither** of which is mandatory.

Appendix A consists of a list of abbreviations used in this glossary. In that list, the **bold** font graces the term names that are defined in this glossary.

Appendix B consists of an abbreviated index of families of defined terms whose technologies are related. This index is provided as a tool to identify all related terms within a specific discipline so that the reader's understanding of a definition may be amplified by reading related definitions within a specific discipline. The index also provides the reader with information on the breadth and scope of disciplines addressed in the glossary.

4. **EFFECTIVE DATE.** The use of this approved standard by U.S. Government departments and agencies is mandatory, effective 180 days following the date of this standard.

5. **CHANGES.** When a Federal department or agency considers that this standard does not provide for its essential needs, a statement citing inadequacies shall be sent in duplicate to the General Services Administration (K), Washington, DC 20405, in accordance with the provisions of the Federal Information Resources Management Regulation Part 201-39.1002. The General Services Administration will determine the appropriate action to be taken and will notify the applicable agency.

6. DEVELOPMENT OF THIS STANDARD.

FED-STD-1037C was developed by a subcommittee of the Federal Telecommunication Standards Committee, the Subcommittee to Revise FED-STD-1037B. The U.S. Department of Commerce, National Telecommunications and Information Administration, Institute for Telecommunication Sciences (NTIA/ITS), 325 Broadway, Boulder, CO 80303-3328, supplied the chair, secretariat, and technical editorial services for the subcommittee. The work of the subcommittee was reviewed by the National Communications System Member Organizations, by the Federal Telecommunication Standards Committee members, and by representatives from other Federal agencies as well as representatives from industry and from the general public. The following Federal Agencies and Departments participated in the subcommittee:

Federal Aviation Administration Technical Center/Information Technology and Systems Section, ACT-142, Atlantic City International Airport, NJ
General Services Administration, Washington, DC
Joint Tactical Command, Control, and Communications Agency/ADW-S
National Security Agency, Fort Meade, MD
U.S. Department of the Army, ISEC, Ft. Huachuca, AZ
U.S. Air Force, 1842 EEG/EEMST, Scott Air Force Base, IL
U.S. Department of Commerce/NTIA/Washington, DC
U.S. Department of Commerce/NTIA/Annapolis, MD
U.S. Department of Commerce/NTIA/ITS, Boulder, CO
U.S. Department of Defense, Defense Information Systems Agency, Joint Interoperability and Engineering Organization, Center for Standards, Fort Monmouth, NJ
U.S. Department of Interior/OIRM, Washington, DC
U.S. Department of Justice/JMD/IRM/TSS, Washington, DC
U.S. Department of the Navy, NAVSEA, Arlington, VA
U.S. Marine Corps, Quantico, VA

LEGEND (for labels appended to definitions)

LEGEND	MEANING
[After...]	Definitions cited "After..." (as in "[After 2196]") are the responsibility of the FTSC Subcommittee to Revise FED-STD-1037B, and not the source-cited document, because of rewriting by the Subcommittee.
(188)	Terms and definitions in direct support of the MIL-STD-188 series of standards and their associated military handbooks. **This is *not* a source citation.**
[2196]	Terms and definitions extracted verbatim from MIL-STD-2196 (SH), *Glossary, Fiber Optics* (1989).
* **[47CFR]**	Terms and definitions extracted verbatim from *Title 47 Code of Federal Regulations, Telecommunications,* Parts 0-199 (rev. Oct. 1, 1987, Oct. 1, 1988).
[CCITT/CCIR]	Recommendations and other documents from the ITU-T (the former CCITT and CCIR). The "CCITT" and "CCIR" citations are retained in this glossary because many of the cited documents have not been reissued by the ITU-T and given a new prefix.
[FAA]	FAA (1991), *Glossary of Optical Communication Terms*, DOT/FAA/CT-TN91/9 (FAA Technical Center, Atlantic City International Airport, NJ 08405).
[FIRMR]	*Federal Information Resources Management Regulation, Title 41 CFR Chapter 201.*
[FS1045A]	Federal Standard 1045A, *Telecommunications: Automatic Link Establishment* (1994).
[ITU-T]	Recommendations and papers from the International Telecommunication Union—Telecommunication Standardization Bureau.
[JP1]	Telecommunications terms and definitions extracted from Joint Pub 1-02 (DOD Joint Staff Publication No. 1-02), 1994, *Department of Defense Dictionary of Military and Associated Terms,* and established for use by all DOD Components, which will use the terms and definitions so designated without alteration unless a distinctly different context or application is intended.
[JP1-A]	Final draft of proposed revision of Joint Pub 1-02; memo from Chief, Joint Doctrine Division, the Joint Staff, 20 October 1994.
[NATO]	NATO ADatP-2(f) *Automatic Data Processing (ADP) NATO Glossary,* June 1991.

[NIS] *National Information Systems Security (INFOSEC) Glossary*, NSTISSI No. 4009, June 5, 1992, (National Security Telecommunications and Information Systems Security Committee, NSA, Ft. Meade, MD 20755-6000). *Note:* The FTSC Subcommittee used the most recent version of the NSTISSI 4009 document that was available at the time the Subcommittee meetings were held (November 1993 through September 1994). However, the NSTISSI document may have changed asynchronously with the 1037C standard, and those NSTISSI definitions may have been amended. The NSTISSI source document is scheduled to be updated on Internet. For the latest version of the NSTISSI No. 4009 Glossary, the reader must access the document on Internet. For hard copy of the 4009 Glossary, write NSA at Ft. Meade, MD. The user may wish to enhance his/her researches into definitions by reviewing the newer version of the 4009 glossary. The changes on 4009 on Internet do not, *per se* revise FED-STD-1037C. Therefore, the user should be aware that if he/she wishes to cite official NSTISSI No. 4009 definitions in a procurement document or other official paper, the official standard glossary is FED-STD-1037C.

[NTIA] Term names and definitions extracted verbatim from the *NTIA Manual of Regulations and Procedures for Federal Radio Frequency Management.*

[OMB] OMB Circular A-130, *Management of Federal Information.*

[RR] Terms and definitions extracted verbatim from the *International Telecommunication Union Radio Regulations,* Malaga-Torremolinos (Oct. 1984, rev. 1985).

[From Weik '89] Use of this source citation acknowledges that the cited information is from *Communications Standard Dictionary*, 2nd ed., Dr. M. Weik, 1989 [Van Nostrand Reinhold Co., New York, NY], with the written permission of the holders of the copyright. These definitions are usually verbatim, but in some cases have been abbreviated or edited.

Note 1: Appreciation is extended to ASC T1A1.5 for generously providing their draft glossary of terms and definitions relating to video-quality degradation. *Note 2:* Appreciation is extended to ANSI-accredited technical committee X3K5 for use of their draft definitions, which were used as a launching point for many of the computer-related definitions herein. No source citation is given the committee's work because their definitions were edited for format consistency and for broad applicability.

** Title 47 of the Code of Federal Regulations is available from Government Institutes, 4 Research Place, Rockville, Maryland 20850; phone: 301-921-2355; fax: 301-921-0373; email: giinfo@govinst.com; Internet: http://www.govinst.com*

(This page is deliberately left blank.)

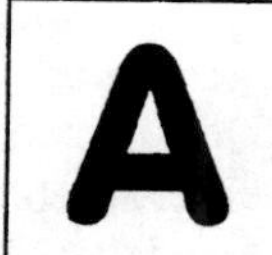

a: *Abbreviation for* **atto** (10^{-18}). *See* **metric system.**

abandoned call: A call in which the call originator disconnects or cancels the call after a connection has been made, but before the call is established.

abbreviated dialing: A telephone service feature that (a) permits the user to dial fewer digits to access a network than are required under the nominal numbering plan, and (b) is limited to a subscriber-selected set of frequently dialed numbers. *Synonym* **speed dialing.**

abort: **1.** In a computer or data transmission system, to terminate, usually in a controlled manner, a processing activity because it is impossible or undesirable for the activity to proceed. **2.** In data transmission, a function invoked by a sending station to cause the recipient to discard or ignore all bit sequences transmitted by the sender since the preceding flag sequence.

abrasive: Any of a number of hard materials, such as aluminum oxide, silicon carbide, and diamond, that are powdered and carefully graded according to particle size, and used to shape and/or finish optical elements, including the endfaces of optical fibers and connectors. *Note:* For finishing the endfaces of optical fiber connectors, abrasive particles are adhered to a substrate of plastic film, in a fashion after that of sandpaper. The film is in turn supported by a hard, flat plate. The connector is supported by a fixture that holds it securely in the proper position for finishing. The grinding motion may be performed manually or by a machine. [After FAA]

absolute delay: The time interval or phase difference between transmission and reception of a signal. (188)

absolute gain: **1.** Of an antenna, for a given direction and polarization, the ratio of (a) the power that would be required at the input of an ideal isotropic radiator to (b) the power actually supplied to the given antenna, to produce the same radiation intensity in the far-field region. *Note 1:* If no direction is given, the absolute gain of an antenna corresponds to the direction of maximum effective radiated power. *Note 2:* Absolute gain is usually expressed in dB. (188) *Synonym* **isotropic gain.** **2.** Of a device, the ratio of (a) the signal level at the output of the device to (b) that of its input under a specified set of operating conditions. *Note 1:* Examples of absolute gain are no-load gain, full-load gain, and small-signal gain. *Note 2:* Absolute gain is usually expressed in dB. (188)

absolute temperature: *See* **thermodynamic temperature.**

absorption: In the transmission of electrical, electromagnetic, or acoustic signals, the conversion of the transmitted energy into another form, usually thermal. (188) [After 2196] *Note 1:* Absorption is one cause of signal attenuation. *Note 2:* The conversion takes place as a result of interaction between the incident energy and the material medium, at the molecular or atomic level.

absorption band: A spectral region in which the absorption coefficient reaches a relative maximum, by virtue of the physical properties of the matter in which the absorption process takes place. [FAA]

absorption coefficient: A measure of the attenuation caused by absorption of energy that results from its passage through a medium. [After 2196] *Note 1:* Absorption coefficients are usually expressed in units of reciprocal distance. *Note 2:* The sum of the absorption coefficient and the scattering coefficient is the attenuation coefficient.

absorption index: **1.** A measure of the attenuation caused by absorption of energy per unit of distance that occurs in an electromagnetic wave of given wavelength propagating in a material medium of given refractive index. *Note:* The value of the absorption index K' is given by the relation

$$K' = \frac{K\lambda}{4\pi n} ,$$

where K is the absorption coefficient, λ is the wavelength in vacuum, and n is the refractive index of the absorptive material medium. (188) [After 2196] **2.** The functional relationship between the

Sun angle—at any latitude and local time—and the ionospheric absorption. (188)

absorption loss: That part of the transmission loss caused by the dissipation or conversion of electrical, electromagnetic, or acoustic energy into other forms of energy as a result of its interaction with a material medium.

absorption modulation: Amplitude modulation of the output of a radio transmitter by means of a variable-impedance circuit that is caused to absorb carrier power in accordance with the modulating wave.

abstract syntax: In open systems architecture, the specification of application-layer data or application-protocol control information by using notation rules that are independent of the encoding technique used to represent the information. (188)

abstract syntax notation one (ASN.1): A standard, flexible method that (a) describes data structures for representing, encoding, transmitting, and decoding data, (b) provides a set of formal rules for describing the structure of objects independent of machine-specific encoding techniques, (c) is a formal network-management Transmission Control Protocol/Internet Protocol (TCP/IP) language that uses human-readable notation and a compact, encoded representation of the same information used in communications protocols, and (d) is a precise, formal notation that removes ambiguities. (188)

ac: *Abbreviation for* **alternating current.**

accept: In data transmission, the condition assumed by a primary or secondary station upon correct receipt of a frame for processing.

acceptance: The condition that exists when a system or functional unit meets the specified performance and security requirements.

acceptance angle: In fiber optics, half the vertex angle of that cone within which optical power may be coupled into bound modes of an optical fiber. *Note 1:* The axis of the cone is collinear with the fiber axis, the vertex of the cone is on the fiber end-face, and the base of the cone faces the optical power source. *Note 2:* The acceptance angle is measured with respect to the fiber axis. *Note 3:* Rays entering an optical fiber at angles greater than the acceptance angle are coupled into unbound modes. (188) [After 2196]

acceptance cone: In fiber optics, the cone within which optical power may be coupled into the bound modes of an optical fiber. *Note:* The acceptance cone is derived by rotating the acceptance angle about the fiber axis. (188) [After 2196]

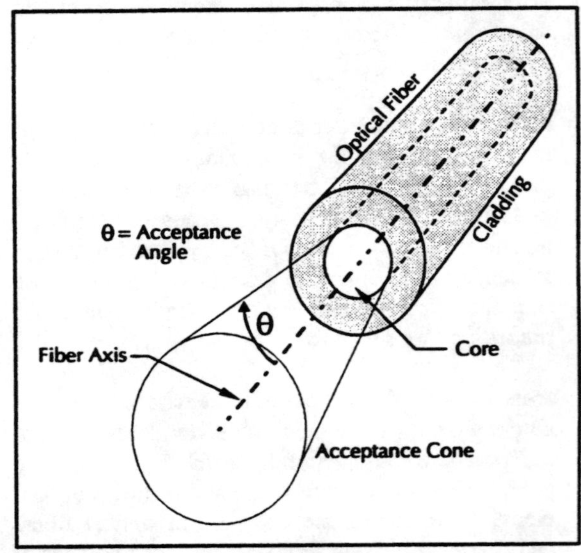

acceptance cone

acceptance pattern: **1.** Of an antenna, for a given plane, a distribution plot of the off-axis power relative to the on-axis power as a function of angle or position. (188) [After 2196] *Note:* The acceptance pattern is the equivalent of a horizontal or vertical antenna pattern. **2.** Of an optical fiber or fiber bundle, a curve of total transmitted power plotted against the launch angle.

acceptance test: A test of a system or functional unit to ensure that contractual requirements are met. *Note:* An acceptance test may be performed at the factory or user premises by the user, vendor, or a third party.

acceptance testing: Operating and testing of a communication system, subsystem, or component, to ensure that the specified performance characteristics have been met. (188)

acceptance trial: A trial carried out by nominated representatives of the eventual military users of the weapon or equipment to determine if the specified performance and characteristics have been met.

accepted interference: Interference at a higher level than that defined as permissible interference and which has been agreed upon between two or more administrations without prejudice to other administrations. [NTIA] [RR]

access: **1.** The ability and means necessary to store data in, to retrieve data from, to communicate with, or to make use of any resource of a system. (188) **2.** To obtain the use of a resource. **3.** (COMSEC) [The] capability and opportunity to gain detailed knowledge of or to alter information or material. [NIS] **4.** (AIS) [The] ability and means to communicate with (*i.e.*, input to or receive output from), or otherwise make use of any information, resource, or component in an AIS. *Note [for 3 and 4]:* An individual does not have "access" if the proper authority or a physical, technical, or procedural measure prevents him/her from obtaining knowledge or having an opportunity to alter information, material, resources, or components. [NIS] **5.** An assigned portion of system resources for one data stream of user communications or signaling. (188)

access attempt: The process by which one or more users interact with a telecommunications system to enable initiation of user information transfer. *Note:* An access attempt begins with an issuance of an access request by an access originator. An access attempt ends either in successful access or in access failure.

access category: A class to which a user, such as a person, program, process, or equipment, of a system may be assigned, based on the resources each user is authorized to use.

access charge: A charge made by a local exchange carrier for use of its local exchange facilities for a purpose such as the origination or termination of traffic that is carried to or from a distant exchange by an interexchange carrier. *Note:* Although some access charges are billed directly to local end users, a very large part of all access charges is paid by interexchange carriers.

access code: The preliminary digits that a user must dial to be connected to a particular outgoing trunk group or line. (188)

access contention: In ISDN applications, *synonymous with "contention." See* **contention (def. #2).**

access control: **1.** A service feature or technique used to permit or deny use of the components of a communication system. **2.** A technique used to define or restrict the rights of individuals or application programs to obtain data from, or place data onto, a storage device. **3.** The definition or restriction of the rights of individuals or application programs to obtain data from, or place data into, a storage device. **4.** [The] process of limiting access to the resources of an AIS to authorized users, programs, processes, or other systems. [NIS] **5.** That function performed by the resource controller that allocates system resources to satisfy user requests. (188)

access control message: A message that is a user request, a resource controller response, or a request/response between resource controllers. (188)

access coupler: *Deprecated term. See* **directional coupler.**

access denial: **1.** Access failure caused by the issuing of a system blocking signal by a communications system that does not have a call-originator camp-on feature. **2.** Access failure caused by exceeding the maximum access time and nominal system access time fraction during an access attempt. *Synonym* **system blocking.**

access-denial time: The time between the start of an access attempt and access failure caused by access denial, *i.e.*, system blocking. *Note:* Access denial times are measured only on access attempts that result in access denial.

access failure: In a communications system, an unsuccessful access that results in termination of an access attempt in any manner other than initiation of user information transfer between the intended source and destination (sink) within the specified maximum access time. *Note:* Access failure can be the result of access denial, access outage, user blocking, or incorrect access.

access group: A group of one or more stations having identical rights to use the available resources on a PBX, network or host computer.

access line: A transmission path between user terminal equipment and a switching center.

access node: In switching systems, the point where user traffic enters and exits a communications network. *Note:* Access node operations may include various operations, such as protocol conversion and code conversion.

access originator: The functional entity responsible for initiating a particular access attempt. *Note:* An access attempt can be initiated by a source user, a destination user, or the telecommunications system.

access phase: In an information-transfer transaction, the phase during which an access attempt is made. *Note:* The access phase is the first phase of an information-transfer transaction.

access point: **1.** A point where connections may be made for testing or using particular communications circuits. (188) **2.** In telephony, a junction point in outside plant consisting of a semipermanent splice at a junction between a branch feeder cable and distribution cables. (188)

access request: A control message issued by an access originator for the purpose of initiating an access attempt.

access time: **1.** In a telecommunication system, the elapsed time between the start of an access attempt and successful access. *Note:* Access time values are measured only on access attempts that result in successful access. **2.** In a computer, the time interval between the instant at which an instruction control unit initiates a call for data and the instant at which delivery of the data is completed. (188) **3.** The time interval between the instant at which storage of data is requested and the instant at which storage is started. (188) **4.** In magnetic disk devices, the time for the access arm to reach the desired track and the delay for the rotation of the disk to bring the required sector under the read-write mechanism.

accounting management: In network management, a set of functions that (a) enables network service use to be measured and the costs for such use to be determined and (b) includes all the resources consumed, the facilities used to collect accounting data, the facilities used to set billing parameters for the services used by customers, maintenance of the data bases used for billing purposes, and the preparation of resource usage and billing reports. (188) [After ANSI T1.210]

accumulator: **1.** A register in which one operand can be stored and subsequently replaced by the result of an arithmetic or logic operation. **2.** A storage register. (188) **3.** A storage battery. (188)

accuracy: The degree of conformity of a measured or calculated value to its actual or specified value. *Contrast with* **precision.**

ACD: *Abbreviation for* **automatic call distributor.**

ac-dc ringing: Telephone ringing that makes use of both ac and dc voltages and currents. *Note:* An alternating current may be used to operate a ringer and direct current to aid the relay action that stops the ringing when the called telephone is answered.

ACK: *Abbreviation for* **acknowledge character.**

acknowledge character (ACK): A transmission control character transmitted by the receiving station as an affirmative response to the sending station. *Note:* An acknowledge character may also be used as an accuracy control character.

acknowledgement: **1.** A response sent by a receiver to indicate successful receipt of a transmission. *Note:* An example of an acknowledgement is a protocol data unit, or element thereof, between peer entities, to indicate the status of data units that have been successfully received. **2.** A message from the addressee informing the originator that the originator's communication has been received and understood. [After JP1]

acknowledgement delay period: *Synonym (loosely)* **sliding window.**

A-condition: In a start-stop teletypewriter system, the significant condition of the signal element that immediately precedes a character signal or block signal and prepares the receiving equipment for the reception of the code elements. *Contrast with* **start signal.**

acoustic coupler: **1.** An interface device for coupling electrical signals by acoustical means—usually into and out of a telephone instrument. (188) **2.** A terminal device used to link data terminals and radio sets with the telephone network. *Note:* The link is achieved through acoustic (sound) signals rather than through direct electrical connection.

acoustic noise: An undesired audible disturbance in the audio frequency range. (188)

acoustic wave: A longitudinal wave that (a) consists of a sequence of pressure pulses or elastic displacements of the material, whether gas, liquid, or solid, in which the wave propagates, (b) in gases, consists of a sequence of compressions (dense gas) and rarefactions (less dense gas) that travel through the gas, (c) in liquids, consists of a sequence of combined elastic deformation and compression waves that travel though the liquid, and (d) in solids, consists of a sequence of elastic compression and expansion waves that travel though the solid. *Note 1:* The speed of an acoustic wave in a material medium is determined by the temperature, pressure, and elastic properties of the medium. In air, acoustic waves propagate at 332 m/s (1087 ft/s) at 0°C, at sea level. In air, sound-wave speed increases approximately 0.6 m/s (2 ft/s) for each kelvin above 0 °C. *Note 2:* Acoustic waves audible to the normal human ear are termed *sound waves*. [From Weik '89]

acousto-optic effect: A variation of the refractive index of a material caused by interaction with acoustic energy in the form of a wave or pulse. (188) *Note:* The acousto-optic effect is used in devices that modulate or deflect light.

acousto-optics: The discipline devoted to the interactions between acoustic waves and light waves in a material medium. (188) [After 2196] *Note:* Acoustic waves can be made to modulate, deflect, and/or focus light waves by causing a variation in the refractive index of the medium.

acquisition: **1.** In satellite communications, the process of locking tracking equipment on a signal from a communications satellite. (188) **2.** The process of achieving synchronization. **3.** In servo systems, the process of entering the boundary conditions that will allow the loop to capture the signal and achieve lock-on. (188)

acquisition and tracking orderwire: *See* **ATOW.**

acquisition time: **1.** In a communications system, the time interval required to attain synchronism. **2.** In satellite control communications, the time interval required for locking tracking equipment on a signal from a communications satellite. (188)

active device: A device that requires a source of energy for its operation and has an output that is a function of present and past input signals. *Note:* Examples of active devices include controlled power supplies, transistors, LEDs, amplifiers, and transmitters.

active laser medium: Within a laser, the material that emits coherent radiation or exhibits gain as the result of electronic or molecular transitions to a lower energy state or states, from a higher energy state or states to which it had been previously stimulated. *Note:* Examples of active laser media include certain crystals, gases, glasses, liquids, and semiconductors. *Synonym* **laser medium.**

active satellite: A satellite carrying a station intended to transmit or retransmit radio communication signals. [NTIA] [RR] (188) *Note:* An active satellite may perform signal processing functions such as amplification, regeneration, frequency translation, and link switching, to make the signals suitable for retransmission.

active sensor: **1.** A detection device that requires input energy from a source other than that which is being sensed. *Note:* An example of an active sensor is a photoconductive cell. **2.** In surveillance, a detection device that emits energy capable of being detected by itself. *Note:* An example of an active sensor is a measuring instrument that generates a signal, transmits it to a target, and receives a reflected signal from the target. Information concerning the target is obtained by comparison of the received signal with the transmitted signal. **3.** A measuring instrument in the Earth exploration-satellite service or in the space research service by means of which information is obtained by transmission and reception of radio waves. [NTIA] [RR]

active star: *See* **star coupler, multiport repeater**.

activity factor: For a communications channel during a specified time interval, such as the busy hour, the percentage of time that a signal is present in the channel in either direction. (188)

ACU: *Abbreviation for* **automatic calling unit.**

A-D: *Abbreviation for* **analog-to-digital.** *See* **analog transmission.**

Ada®: The official, high-level computer language of DOD for embedded-computer, real-time applications as defined in MIL-STD-1815. *Note:* Ada® is a registered trademark of the U.S. Government (Ada Joint Program Office).

adaptive channel allocation: In communications system traffic flow control, channel allocation in which information-handling capacities of channels are not predetermined but are assigned on demand. *Note:* Adaptive channel allocation is usually accomplished by means of a multiplexing scheme.

adaptive communications: Any communications system, or portion thereof, that automatically uses feedback information obtained from the system itself or from the signals carried by the system to modify dynamically one or more of the system operational parameters to improve system performance or to resist degradation. (188) *Note:* The modification of a system parameter may be discrete, as in hard-switched diversity reception, or may be continuous, as in a predetection combining algorithm.

adaptive differential pulse code-modulation (ADPCM): Differential pulse-code modulation in which the prediction algorithm is adjusted in accordance with specific characteristics of the input signal.

adaptive equalization: Equalization (a) that is automatically accomplished while traffic is being transmitted and (b) in which signal characteristics are dynamically adjusted to compensate for changing transmission path characteristics. (188)

adaptive predictive coding (APC): Narrowband analog-to-digital conversion that uses a one-level or multilevel sampling system in which the value of the signal at each sampling instant is predicted according to a linear function of the past values of the quantized signals. *Note:* APC is related to linear predictive coding (LPC) in that both use adaptive predictors. However, APC uses fewer prediction coefficients, thus requiring a higher sampling rate than LPC.

adaptive radio: A radio that (a) monitors its own performance, (b) monitors the path quality through sounding or polling, (c) varies operating characteristics, such as frequency, power, or data rate, and (d) uses closed-loop action to optimize its performance by automatically selecting frequencies or channels.

adaptive routing: Routing that is automatically adjusted to compensate for network changes such as traffic patterns, channel availability, or equipment failures. *Note:* The experience used for adaptation comes from the traffic being carried.

adaptive system: A system that has a means of monitoring its own performance, a means of varying its own parameters, and uses closed-loop action to improve its performance or to optimize traffic. (188)

ADC: *Abbreviation for* **analog-to-digital converter, analog-to-digital conversion.**

ADCCP: *Abbreviation for* **Advanced Data Communication Control Procedures.**

added bit: A bit delivered to the intended destination user in addition to intended user information bits and delivered overhead bits. *Synonym* **extra bit.**

added block: Any block, or other delimited bit group, delivered to the intended destination user in addition to intended user information bits and delivered overhead bits. *Synonym* **extra block.**

adder: **1.** A device whose output data are a representation of the sum of the numbers represented by its input data. *Note:* An adder may be serial or parallel, digital or analog. **2.** A device whose output data are a representation of the sum of the quantities represented by its input data. *Note:* An adder can add things other than representations of numbers. It can add voltages, *etc.* Analog adders are not limited to summing representations of numbers. An adder may operate on digital or analog data.

adder-subtracter: A device that acts as an adder or subtracter depending upon the control signal received; the adder-subtracter may be constructed so as to yield a sum and a difference at the same time. *Note:* An arithmetic adder-subtracter yields arithmetic sums and differences, whereas a logical adder-subtracter yields logical sums and differences.

additive white gaussian noise (AWGN): *Synonym* **white noise.**

add mode: In addition and subtraction operations, a mode in which the decimal marker is placed at a predetermined location with respect to the last digit entered.

add-on conference: A service feature that allows an additional party to be added to an established call without attendant assistance. *Note:* A common implementation provides a progressive method that allows a call originator or a call receiver to add at least one additional party.

address: **1.** In communications, the coded representation of the source or destination of a message. (188) **2.** In data processing, a character or group of characters that identifies a register, a particular part of storage, or some other data source or destination. (188) **3.** To assign to a device or item of data a label to identify its location. (188) **4.** The part of a selection signal that indicates the destination of a call. **5.** To refer to a device or data item by its address.

addressability: **1.** In computer graphics, the capability of a display surface or storage device to accommodate a specified number of uniquely identifiable points. **2.** In micrographics, the capability of a specified field frame to contain a specific number of uniquely identifiable points. *Note:* The addressability is usually specified as the number of identifiable horizontal points by the number of identifiable vertical points, such as 3000 by 4000.

addressable point: In computer graphics, any point of a device that can be addressed.

address field: The portion of a message that contains the source-user address and the destination-user addresses. *Note:* In a communications network, the address field is usually contained within the message header portion of the message. A message usually consists of the message header, the user data, and a trailer.

address-indicating group (AIG): A station or address designator, used to represent a set of specific and frequently recurring combinations of action or information addresses. *Note:* The identity of the message originator may also be included in the AIG. An address group is assigned to each AIG for use as an address designator.

address message: A message sent in the forward direction that contains (a) address information, (b) the signaling information required to route and connect a call to the called line, (c) service-class information, (d) information relating to user and network facilities, and (e) call-originator identity or call-receiver identity.

address message sequencing: In common-channel signaling, a procedure for ensuring that address messages are processed in the correct order when the order in which they are received is incorrect.

address part: A part of an instruction that usually contains only an address or part of an address.

address pattern: A prescribed structure of data used to represent the destination(s) of a block, message, packet, or other formalized data structure.

address resolution protocol (ARP): A Transmission Control Protocol/Internet Protocol (TCP/IP) protocol that dynamically binds a Network-Layer IP address to a Data-Link-Layer physical hardware address, *e.g.*, Ethernet address.

address separator: A character that separates the different addresses in a selection signal.

ADH: *Abbreviation for* **automatic data handling.**

adjacent-channel interference: Extraneous power from a signal in an adjacent channel. (188) *Note 1:* Adjacent channel interference may be caused by inadequate filtering, such as incomplete filtering of unwanted modulation products in frequency modulation (FM) systems, improper tuning, or poor frequency control, in either the reference channel or the interfering channel, or both. *Note 2:* Adjacent-channel interference is distinguished from crosstalk.

adjunct service point (ASP): An intelligent-network feature that resides at the intelligent peripheral equipment and responds to service logic interpreter requests for service processing.

administration: **1.** Any governmental department or service responsible for discharging the obligations undertaken in the convention of the International Telecommunication Union and the *Regulations.* [RR] **2.** Internal management of units. [JP1] **3.** The management and execution of all military matters not included in strategy and tactics. [JP1] **4.** In international telecommunications for a given country, the government agency assigned responsibility for the implementation of telecommunications standards, regulations, recommendations, practices, and procedures. **5.** In network management, network support functions that ensure that (a) services are performed, (b) the network is used efficiently, and (c) prescribed service-quality objectives are met.

administrative management complex (AMC): In network management, a complex that is controlled by a network provider, and is responsible for and performs network management functions such as network maintenance. (188) [After ANSI T1.218-1991]

ADP: *Abbreviation for* **automatic data processing.**

ADPCM: *Abbreviation for* **adaptive differential pulse-code modulation.**

ADPE: *Abbreviation for* **automatic data processing equipment.**

ADP system: *Synonym* **computer system.**

Advanced Data Communication Control Procedures (ADCCP): A bit-oriented Data-Link-Layer protocol used to provide point-to-point and point-to-multipoint transmission of data frames that contain error-control information. *Note:* ADCCP closely resembles high-level data link control (HDLC) and synchronous data link control (SDLC).

advanced intelligent network (AIN): A proposed intelligent-network (IN) architecture that includes both IN/1+ and IN/2 concepts.

advanced television (ATV): A family of television systems that is intended to be improvements over current commercial-quality television. *Note:* The ATV family includes improved-definition television (IDTV), extended-definition television (EDTV), and high-definition television (HDTV).

AECS: *Abbreviation for* **Aeronautical Emergency Communications System.** *See* **Aeronautical Emergency Communications System Plan.**

aerial cable: A communications cable designed for installation on, or suspension from, a pole or other overhead structure. (188)

aerial insert: In a direct-buried or underground cable run, a cable rise to a point above ground, followed by an overhead run, *e.g.*, on poles, followed by a drop back into the ground. *Note:* An aerial insert is used in places where it is not possible or practical to remain underground, such as might be encountered in crossing a deep ditch, canal, river, or subway line.

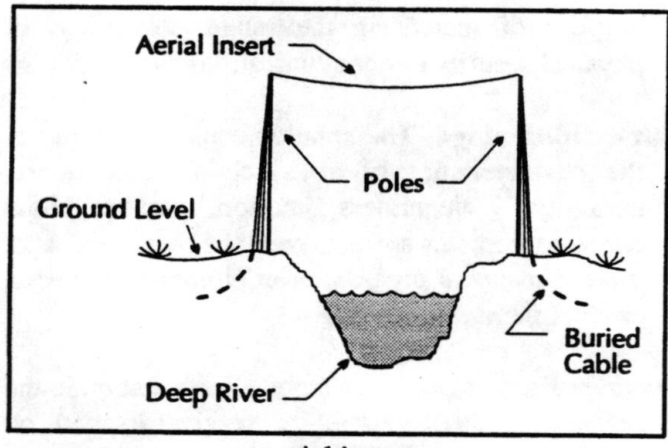

aerial insert

aeronautical advisory station: An aeronautical station used for advisory and civil defense communications primarily with private aircraft stations. *Synonym* **UNICOM station.** [NTIA]

aeronautical broadcast station: An aeronautical station which makes scheduled broadcasts of meteorological information and notices to airmen. (In certain instances, an aeronautical broadcast station may be placed on board a ship.) [NTIA]

aeronautical Earth station: An Earth station in the fixed-satellite service, or, in some cases, in the aeronautical mobile-satellite service, located at a specified fixed point on land to provide a feeder link for the aeronautical mobile-satellite service. [NTIA] [RR]

Aeronautical Emergency Communications System (AECS) Plan: The AECS Plan provides for the operation of aeronautical communications stations, on a voluntary, organized basis, to provide the President and the Federal Government, as well as heads of state and local governments, or their designated representatives, and the aeronautical industry with an expeditious means of communications during an emergency situation. [47CFR]

aeronautical fixed service: A radiocommunication service between specified fixed points provided primarily for the safety of air navigation and for the regular, efficient and economical operation of air transport. [NTIA] [RR]

aeronautical fixed station: A station in the aeronautical fixed service. [NTIA][RR]

aeronautical mobile (OR) [off-route] service: An aeronautical mobile service intended for communications, including those relating to flight coordination, primarily outside national or international civil air routes. [NTIA] [RR]

aeronautical mobile (R) [route] service: An aeronautical mobile service reserved for communications relating to safety and regularity of flight, primarily along national or international civil air routes. [NTIA] [RR]

aeronautical mobile-satellite service: A mobile-satellite service in which mobile Earth stations are located on board aircraft; survival craft stations and emergency position-indicating radiobeacon stations may also participate in this service. [NTIA] [RR]

aeronautical mobile-satellite (OR) [off-route] service: An aeronautical mobile-satellite service intended for communications, including those relating to flight coordination, primarily outside national and international civil air routes. [NTIA] [RR]

aeronautical mobile-satellite (R) [route] service: An aeronautical mobile-satellite service reserved for communications relating to safety and regularity of flight, primarily along national or international civil air routes. [NTIA] [RR]

aeronautical mobile service: A mobile service between aeronautical stations and aircraft stations, or between aircraft stations, in which survival craft stations may participate; emergency position-indicating radiobeacon stations may also participate in this service on designated distress and emergency frequencies. [NTIA]

aeronautical multicom service: A mobile service not open to public correspondence, used to provide communications essential to conduct activities being performed by or directed from private aircraft. [NTIA]

aeronautical radionavigation-satellite service: A radionavigation-satellite service in which Earth stations are located on board aircraft. [NTIA] [RR]

aeronautical radionavigation service: A radionavigation service intended for the benefit and for the safe operation of aircraft. [NTIA] [RR]

aeronautical station: A land station in the aeronautical mobile service. In certain instances, an aeronautical station may be located, for example, on board ship or on a platform at sea. [NTIA] [RR]

AF: *Abbreviation for* **audio frequency.**

AFNOR: *Acronym for* **Association Français Normal.** France's national standards-setting organization. *Note:* AFNOR provides the Secretariat for ISO TC97/SC1, Information Technology Vocabulary, which includes computers, communications information processing, and office machines.

AGC: *Abbreviation for* **automatic gain control.**

AI: *Abbreviation for* **artificial intelligence.**

AIM: *Abbreviation for* **amplitude intensity modulation.** *See* **intensity modulation.**

AIN: *Abbreviation for* **advanced intelligent network.**

AIOD: *Abbreviation for* **automatic identified outward dialing.**

AIOD leads: Terminal equipment leads used solely to transmit automatic identified outward dialing (AIOD) data from a PBX to the public switched telephone network or to switched service networks (*e.g.*, EPSCS), so that a vendor can provide a detailed monthly bill identifying long-distance usage by individual PBX stations, tie trunks, or the attendant. . . . [from 47CFR]

airborne radio relay: **1.** Airborne equipment used to relay radio transmission from selected originating transmitters. [JP1] **2.** A technique employing aircraft fitted with radio relay stations for the purpose of increasing the range, flexibility, or physical security of communications systems.

air-conditioning: The simultaneous controlling of the characteristics of air, such as temperature, humidity, cleanliness, motion, and pollutant concentration, in a space to meet the requirements of the occupants, a process, or equipment. *Synonym* **environmental control.**

aircraft Earth station: A mobile Earth station in the aeronautical mobile-satellite service located on board an aircraft. [NTIA] [RR]

aircraft emergency frequency: An international aeronautical emergency frequency, such as 121.5 MHz (civil) and 243.0 MHz (military), for aircraft stations and stations concerned with safety and regulation of flight along national or international civil air routes and maritime mobile service stations authorized to communicate for safety purposes.

aircraft station: A mobile station in the aeronautical mobile service, other than a survival craft station, located on board an aircraft. [RR]

airdrome control station: An aeronautical station providing communication between an airdrome control tower and aircraft. [NTIA] *Synonym* **airport control station.**

air-ground radiotelephone service: A public radio service between a base station and airborne mobile stations. [47CFR]

air-ground worldwide communications system: A worldwide military network of ground stations that (a) provides two-way communications links between aircraft and ground stations for navigation and control, including air route traffic control and (b) may also provide support for special functions, such as for civil aircraft providing assistance to military missions and for meeting communications requirements for aircraft flying distinguished visitors.

air portable: Denotes materiel that is suitable for transport by an aircraft loaded internally or externally, with no more than minor dismantling and reassembling within the capabilities of user units. This term must be qualified to show the extent of air portability. [JP1]

airport control station: *Synonym* **airdrome control station.**

air sounding: Measuring atmospheric phenomena or determining atmospheric conditions, especially by means of apparatus carried by balloons, rockets, or satellites.

air terminal: In grounding systems, the lightning rod or conductor placed on or above a building, structure, or external conductors for the purpose of intercepting lightning. (188)

AIS: *Abbreviation for* **automated information system.**

alarm center: A location that receives local and remote alarms. *Note:* An alarm center is usually in a technical control facility. (188)

alarm indicator: A device that responds to a signal from an alarm sensor. *Note:* Examples of alarm indicators include bells, lamps, horns, gongs, and buzzers.

alarm sensor: **1.** In communications systems, any device that (a) can sense an abnormal condition within the system and provide a signal indicating the presence or nature of the abnormality to either a local or remote alarm indicator, and (b) may detect events ranging from a simple contact opening or closure to a time-phased automatic shutdown and restart cycle. (188) **2.** In a physical security system, an approved device used to indicate a change in the physical environment of a facility or a part thereof. (188) *Note:* Alarm sensors may also be redundant or chained, such as when one alarm sensor is used to protect the housing, cabling, or power protected by another alarm sensor.

a-law: *See* **a-law algorithm.**

a-law algorithm: A standard compression algorithm, used in digital communications systems of the European digital hierarchy, to optimize, *i.e.*, modify, the dynamic range of an analog signal for digitizing. *Note 1:* The wide dynamic range of speech does not lend itself well to efficient linear digital encoding. A-law encoding effectively reduces the dynamic range of the signal, thereby increasing the coding efficiency and resulting in a signal-to-distortion ratio that is superior to that obtained by linear encoding for a given number of bits.

ALE: *Abbreviation for* **automatic link establishment.**

algorithmic language: An artificial language established for expressing a given class of algorithms.

aligned bundle: A bundle of optical fibers in which the relative spatial coordinates of each fiber are the same at the two ends of the bundle. (188) *Note:* Such a bundle may be used for the transmission of images. *Synonym* **coherent bundle.**

Allan variance: One half of the time average over the sum of the squares of the differences between successive readings of the frequency deviation sampled over the sampling period. *Note:* The Allan variance is conventionally expressed by $\sigma_y^2(\tau)$. The samples are taken with no dead-time between them. *Synonym* **two-sample variance.**

allcall: In adaptive high-frequency (HF) radio automatic link establishment (ALE), a general broadcast that does not request responses and does not designate any specific addresses. *Note:* This essential function is required for emergencies ("HELP"), sounding-type data exchanges, and propagation and connectivity tracking. [After FED-STD-1045A] (188)

all-glass fiber: *Synonym* **all-silica fiber.**

allocation (of a frequency band): **1.** Entry in the Table of Frequency Allocations of a given frequency band for the purpose of its use by one or more (terrestrial or space) radiocommunication services or the radio astronomy service under specified conditions. This term shall also be applied to the frequency band concerned. [NTIA] [RR] **2.** The process of designating radio-frequency bands for use by specific radio services. (188)

allotment (of a radio frequency or radio frequency channel): Entry of a designated frequency channel in an agreed plan, adopted by a component Conference, for use by one or more administrations for a (terrestrial or space) radiocommunication service in one or more identified countries or geographical areas and under specified conditions. [NTIA]

all-silica fiber: An optical fiber composed of a silica-based core and cladding. *Note:* The presence of a protective polymer overcoat does not disqualify a fiber as an all-silica fiber, nor does the presence of a tight buffer. [FAA] *Synonym* **all-glass fiber**.

all trunks busy (ATB): An equipment condition in which all trunks (paths) in a given trunk group are busy. *Note:* All-trunks-busy registers do not indicate subsequent attempts to reach trunk groups.

alphabet: **1.** An ordered set of all the letters used in a language, including letters with diacritical signs where appropriate, but not including punctuation marks. **2.** An ordered set of all the symbols used in a language, including punctuation marks, numeric digits, nonprinting control characters, and other symbols. *Note:* Examples of alphabets include the Roman alphabet, the Greek alphabet, the Morse Code, and the 128 characters of the American Standard Code for Information Interchange (ASCII) [IA No. 5]. (188)

alphabetic character set: A character set that contains letters and may contain control characters, special characters, and the space character, but not digits.

alphabetic code: A code according to which data are represented through the use of an alphabetic character set.

alphabetic string: **1.** A string consisting solely of letters from the same alphabet. **2.** A character string consisting solely of letters and associated special characters from the same alphabet.

alphabetic word: **1.** A word consisting solely of letters from the same alphabet. **2.** A word that consists of letters and associated special characters, but not digits.

alphabet translation: *Deprecated synonym for* **alphabet transliteration.** *See* **alphabet transliteration.**

alphabet transliteration: The substitution of the characters of one alphabet for the corresponding characters of a different alphabet, usually accomplished on a character-by-character basis. (188) *Note 1:* An example of alphabet transliteration is the substitution of the Roman letters a, b, and p for the Greek letters α, β, and π, respectively. *Note 2:* Alphabet transliteration is reversible. *Note 3:* Alphabet transliteration often becomes necessary in telecommunications systems because of the different alphabets and codes used worldwide. *Note 4:* In alphabet transliteration, no consideration is given to the meaning of the characters or their combinations.

alphanumeric: **1.** Pertaining to a character set that contains letters, digits, and sometimes other characters, such as punctuation marks. (188) **2.** Pertaining to a set of unique bit patterns that are used to represent letters of an alphabet, decimal digits, punctuation marks, and other special signs and symbols used in grammar, business, and science, such as those displayed on conventional typewriter keyboards.

alphanumeric character set: A character set that contains both letters and digits, special characters, and the space character.

alphanumeric code: **1.** A code derived from an alphanumeric character set. **2.** A code that, when used, results in a code set that consists of alphanumeric characters.

alphanumeric data: Data represented by letters, digits, and sometimes by special characters and the space character.

alpha profile: *See* **power-law index profile.**

altazimuth mount: A mounting, *e.g.*, for a directional antenna, in which slewing takes place in (a) the plane tangent to the surface of the Earth or other frame of reference and (b) elevation about, *i.e.*, above or below, that plane. *Synonym* **x-y mount.**

alternate mark inversion (AMI) signal: A pseudoternary signal, representing binary digits, in which (a) successive *"marks"* are of alternately positive and negative polarity and the absolute values of their amplitudes are normally equal and (b) *"spaces"* are of zero amplitude.

alternate party: In multilevel precedence and preemption, the call receiver, *i.e.*, the destination user, to which a precedence call will be diverted. *Note 1:* Diversion will occur when the response timer expires, when the call receiver is busy on a call of equal or higher precedence, or when the call receiver is busy with access resources that are non-preemptable. *Note 2:* Alternate party diversion is an optional terminating feature that is subscribed to by the call receiver. Thus, the alternate party is specified by the call receiver at the time of subscription. (188)

alternate routing: The routing of a call or message over a substitute route when a primary route is unavailable for immediate use. (188)

altitude of the apogee or of the perigee: The altitude of the apogee or perigee above a specified reference surface serving to represent the surface of the Earth. [NTIA] [RR] *Note:* In technical usage, the definite article is not used with the term *apogee* or *perigee* alone. A body orbiting the Earth is said simply to be *"at apogee"* or *"at perigee."* It may, however, properly be said to be *"at the point of apogee"* or *"at the point of perigee."*

ALU: *Abbreviation for* **arithmetic and logic unit.**

AM: *Abbreviation for* **amplitude modulation.**

AMA: *Abbreviation for* **automatic message accounting.**

amateur-satellite service: A radiocommunication service using space stations on Earth satellites for the same purposes as those of the amateur service. [NTIA] [RR]

amateur service: A radiocommunication service for the purpose of self-training, intercommunication and technical investigation carried out by amateurs, that is, by duly authorized persons interested in radio technique solely with a personal aim and without pecuniary interest. [NTIA] [RR]

amateur station: A station in the amateur service. [NTIA] [RR]

ambient noise level: The level of acoustic noise existing at a given location, such as in a room, in a compartment, or at a place out of doors. *Note 1:* Ambient noise level is measured with a sound level meter. *Note 2:* Ambient noise level is usually measured in dB above a reference pressure level of 0.00002 Pa, *i.e.*, 20 μPa (micropascals) in SI units. A pascal is a newton per square meter. *Note 3:* In the centimeter-gram-second system of units, the reference level for measuring ambient noise level is 0.0002 dyn/cm^2. (188) *Synonym* **room noise level.**

ambient temperature: The temperature of air or other media in a designated area, particularly the area surrounding equipment. (188)

AME: *Abbreviation for* **amplitude modulation equivalent, automatic message exchange.** *See* **compatible sideband transmission.**

American National Standards Institute (ANSI): The U.S. standards organization that establishes procedures for the development and coordination of voluntary American National Standards.

American Standard Code for Information Interchange (ASCII): *See* **ASCII.**

AMI: *Abbreviation for* **alternate mark inversion.** *See* **alternate mark inversion signal.**

AMI violation: A *"mark"* that has the same polarity as the previous *"mark"* in the transmission of alternate mark inversion (AMI) signals. *Note:* In some transmission protocols, AMI violations are deliberately introduced to facilitate synchronization or to signal a special event.

amplifier: *See* **fiber amplifier, optical repeater**.

amplitude distortion: Distortion occurring in a system, subsystem, or device when the output amplitude is not a linear function of the input amplitude under specified conditions. (188) *Note:*

Amplitude distortion is measured with the system operating under steady-state conditions with a sinusoidal input signal. When other frequencies are present, the term *"amplitude"* refers to that of the fundamental only.

amplitude equalizer: A corrective network that is designed to modify the amplitude characteristics of a circuit or system over a desired frequency range. (188) *Note:* Such devices may be fixed, manually adjustable, or automatic.

amplitude hit: *See* **hit (def. #1).**

amplitude intensity modulation (AIM): *Deprecated term. See* **intensity modulation.**

amplitude keying: Keying in which the amplitude of a signal is varied among the members of a set of discrete values. (188)

amplitude modulation (AM): Modulation in which the amplitude of a carrier wave is varied in accordance with some characteristic of the modulating signal. (188) *Note:* Amplitude modulation implies the modulation of a coherent carrier wave by mixing it in a nonlinear device with the modulating signal to produce discrete upper and lower sidebands, which are the sum and difference frequencies of the carrier and signal. The envelope of the resultant modulated wave is an analog of the modulating signal. The instantaneous value of the resultant modulated wave is the vector sum of the corresponding instantaneous values of the carrier wave, upper sideband, and lower sideband. Recovery of the modulating signal may be by direct detection or by heterodyning.

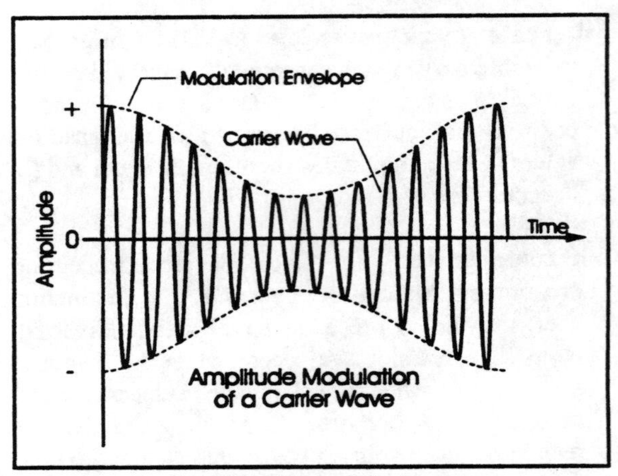

amplitude modulation

amplitude modulation equivalent (AME): *Synonym* **compatible sideband transmission.**

amplitude-vs.-frequency distortion: Distortion in a transmission system caused by nonuniform attenuation, or gain, in the system with respect to frequency under specified operating conditions. (188) *Synonym* **frequency distortion.**

AMPS: *Abbreviation for* **automatic message processing system.**

AMTS: *Abbreviation for* **automated maritime telecommunications system.**

analog computer: A device that performs operations on data that are represented within the device by continuous variables having a physical resemblance to the quantities being represented. *Note:* The earliest analog computers were constructed with purely mechanical components, such as levers, cogs, cams, discs, and gears. These components represented the quantities being manipulated or the operator-inserted values. Modern analog computers usually employ electrical parameters, such as voltages, resistances, or currents to represent the quantities being manipulated.

analog data: Data represented by a physical quantity that is considered to be continuously variable and has a magnitude directly proportional to the data or to a suitable function of the data. (188)

analog decoding: The portion of the digital-to-analog conversion process that generates an analog signal value from the digital signal that resulted from analog encoding. (188) *Note:* Further action is required to integrate these samples to obtain a continuous approximation of the original signal, because analog decoding does not smooth the signal.

analog encoding: The portion of the analog-to-digital conversion process that samples an analog signal and creates a digital signal that represents the value of the sample. (188) *Note:* Multiple samples are needed to digitize a waveform over a time interval.

analog facsimile equipment: Facsimile equipment in which (a) analog techniques are used to encode the image detected by the scanner and (b) the output is an analog signal. *Note:* Examples of analog facsimile equipment are CCITT Group 1 and CCITT Group 2 facsimile equipment.

analog signal: **1.** A signal that has a continuous nature rather than a pulsed or discrete nature. *Note:* Electrical or physical analogies, such as continuously varying voltages, frequencies, or phases, may be used as analog signals. (188) **2.** A nominally continuous electrical signal that varies in some direct correlation with another signal impressed on a transducer. (188) *Note:* For example, an analog signal may vary in frequency, phase, or amplitude in response to changes in physical phenomena, such as sound, light, heat, position, or pressure.

analog switch: Switching equipment designed, designated, or used to connect circuits between users for real-time transmission of analog signals. (188)

analog-to-digital (A-D) coder: *Synonym* **analog-to-digital converter (ADC).**

analog-to-digital converter (ADC): A device that converts an analog signal to a digital signal that represents equivalent information. (188) *Synonyms* **analog-to-digital (A-D) coder, analog-to-digital (A-D) encoder.**

analog-to-digital (A-D) encoder: *Synonym* **analog-to-digital converter (ADC).**

analog transmission: Transmission of a continuously varying signal as opposed to transmission of a discretely varying signal.

angle modulation: Modulation in which the phase or frequency of a sinusoidal carrier is varied. (188) *Note:* Phase and frequency modulation are particular forms of angle modulation.

angle of deviation: In optics, the net angular deflection experienced by a light ray after one or more refractions or reflections. (188)

angle of incidence: The angle between an incident ray and the normal to a reflecting or refracting surface. (188)

angstrom (Å): A unit of length equal to 10^{-10} m. *Note 1:* The angstrom is not an SI (International System) unit, and it is not accepted for government use (Fed. Std. 376B). *Note 2:* The angstrom is, and historically has been, used in the fields of optics, spectroscopy, and microscopy.

angular misalignment loss: Power loss caused by the deviation from optimum angular alignment of the axes of source to waveguide, waveguide to waveguide, or waveguide to detector. *Note 1:* The waveguide may be dielectric (an optical fiber) or metallic. *Note 2:* Angular misalignment loss does not include lateral offset loss and longitudinal offset loss.

ANI: *Abbreviation for* **automatic number identification.**

anisochronous: Pertaining to transmission in which the time interval separating any two significant instants in sequential signals is not necessarily related to the time interval separating any other two significant instants. *Note:* Isochronous and anisochronous are characteristics, while synchronous and asynchronous are relationships.

anisochronous transmission: *See* **asynchronous transmission.**

anisotropic: Pertaining to a material whose electrical or optical properties vary with (a) the direction of propagation of a traveling wave or with (b) different polarizations of a traveling wave. *Note 1:*

Anisotropy is exhibited by non-cubic crystals, which have different refractive indices for lightwaves propagating in different directions or with different polarizations. *Note 2:* Anisotropy may be induced in certain materials under mechanical strain.

anomalous propagation (AP): Abnormal propagation caused by fluctuations in the properties (such as density and refractive index) of the propagation medium. (188) *Note:* AP may result in the reception of signals well beyond the distances usually expected.

ANS: Abbreviation for **American National Standard.**

ANSI: *Abbreviation for* **American National Standards Institute.**

ANSI/EIA/TIA-568: A U.S. industry standard that specifies a generic telecommunications cabling system, which will support a multiproduct, multivendor environment, for commercial buildings. *Note 1:* The standard specifies performance characteristics for unshielded twisted pair telecommunications cabling, including categories allowing data communications up to 100 Mb/s. These categories are designated 3, 4, and 5. Categories 1 and 2 have not been defined. *Note 2:* The standard has been adopted as FIPS PUB 174.

answer back: A signal sent by receiving equipment to the sending station to indicate that the receiver is ready to accept transmission.

answer signal: A supervisory signal returned from the called telephone to the originating switch when the call receiver answers. *Note 1:* The answer signal stops the ringback signal from being returned to the caller. *Note 2:* The answer signal is returned by means of a closed loop.

antenna: Any structure or device used to collect or radiate electromagnetic waves. (188)

antenna aperture: *See* **aperture (def #1).**

antenna array: An assembly of antenna elements with dimensions, spacing, and illumination sequence such that the fields for the individual elements combine to produce a maximum intensity in a particular direction and minimum field intensities in other directions.

antenna blind cone: The volume of space, usually approximately conical with its vertex at the antenna, that cannot be scanned by an antenna because of limitations of the antenna radiation pattern and mount. *Note:* An example of an antenna blind cone is that of an air route surveillance radar (ARSR). The horizontal radiation pattern of an ARSR antenna is very narrow. The vertical radiation pattern is fan-shaped, reaching approximately 70° of elevation above the horizontal plane. As the antenna is rotated about a vertical axis, it can illuminate targets only if they are 70° or less from the horizontal plane. Above that elevation, they are in the antenna blind cone. *Synonym* **cone of silence.**

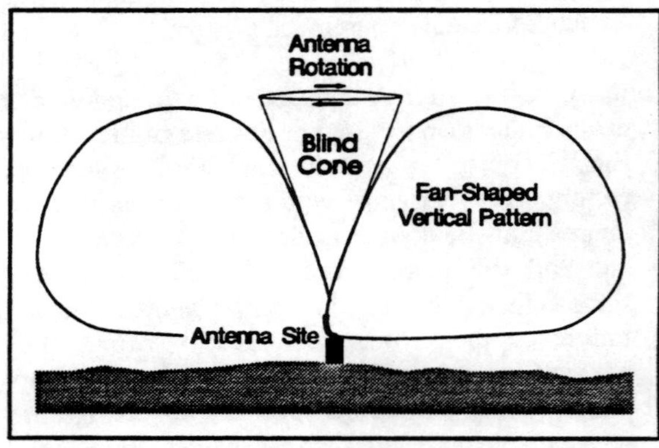

antenna blind cone

antenna coupler: A device used to match the impedance of a transmitter and/or receiver to an antenna to provide maximum power transfer.

antenna dissipative loss: A power loss resulting from changes in the measurable impedance of a practical antenna from a value theoretically calculated for a perfect antenna.

antenna effective area: The functionally equivalent area from which an antenna directed toward the source of the received signal gathers or absorbs the energy of an incident electromagnetic wave. *Note 1:* Antenna effective area is usually expressed in square meters. *Note 2:* In the case of parabolic and horn-parabolic antennas, the antenna effective area is about 0.35 to 0.55 of the geometric area of the antenna aperture.

antenna efficiency: The ratio of the total radiated power to the total input power. *Note:* The total radiated power is the total input power less antenna dissipative losses.

antenna electrical beam tilt: The shaping of the radiation pattern in the vertical plane of a transmitting antenna by electrical means so that maximum radiation occurs at an angle below the horizontal plane. [47CFR]

antenna gain: The ratio of the power required at the input of a loss-free reference antenna to the power supplied to the input of the given antenna to produce, in a given direction, the same field strength at the same distance. *Note 1:* Antenna gain is usually expressed in dB. *Note 2:* Unless otherwise specified, the gain refers to the direction of maximum radiation. The gain may be considered for a specified polarization. Depending on the choice of the reference antenna, a distinction is made between:

—absolute or isotropic gain (G_i), when the reference antenna is an isotropic antenna isolated in space;
—gain relative to a half-wave dipole (G_d) when the reference antenna is a half-wave dipole isolated in space and with an equatorial plane that contains the given direction;
—gain relative to a short vertical antenna (G_r), when the reference antenna is a linear conductor, much shorter than one quarter of the wavelength, normal to the surface of a perfectly conducting plane which contains the given direction. [RR] (188) *Synonyms* **gain of an antenna, power gain of an antenna.**

antenna gain-to-noise-temperature (G/T): In the characterization of antenna performance, a figure of merit, where G is the antenna gain in decibels at the receive frequency, and T is the equivalent noise temperature of the receiving system in kelvins.

antenna height above average terrain: The antenna height above the average terrain elevations from 3.2 to 16 kilometers (2 to 10 miles) from the antenna for the eight directions spaced evenly for each 45° of azimuth starting with true north. (188) *Note:* In general, a different antenna height above average terrain will be determined in each direction from the antenna. The average of these eight heights is the antenna height above average terrain. In some cases, such as seashore, fewer than eight directions may be used.

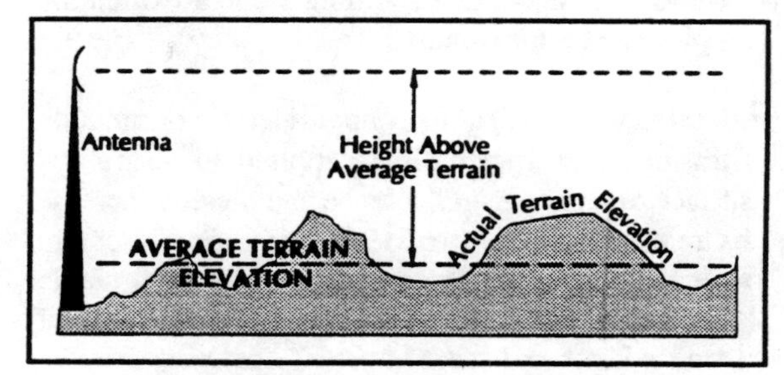

antenna height above average terrain

antenna lobe: A three-dimensional section of the radiation pattern of a directional antenna, bounded by one or more cones of nulls or by regions of diminished irradiance. (188)

antenna matching: The process of adjusting impedance so that the input impedance of an antenna equals or approximates the characteristic impedance of its transmission line over a specified range of frequencies. (188) *Note:* The impedance of either the transmission line, or the antenna, or both, may be adjusted to effect the match.

antenna noise temperature: The temperature of a hypothetical resistor at the input of an ideal noise-free receiver that would generate the same output noise power per unit bandwidth as that at the antenna output at a specified frequency. (188) *Note 1:* The antenna noise temperature depends on antenna coupling to all noise sources in its environment as well as on noise generated within the antenna. *Note 2:* The antenna noise temperature is a measure of noise whose value is equal to the actual temperature of a passive device.

anti-clockwise polarized wave: *Synonym* **left-hand polarized wave.**

anti-interference: Pertaining to equipment, processes, or techniques used to reduce the effect of natural and man-made noise on radio communications.

anti-jam: Measures to ensure that intended transmitted information can be received despite deliberate jamming attempts. [NIS] (188)

anti-node: A point in a standing wave at which the amplitude is a maximum.

antireflection coating: A thin, dielectric or metallic film, or several such films, applied to an optical surface to reduce its reflectance and thereby increase its transmittance. [After 2196] *Note:* For minimum reflection of a normal incident wave of a single wavelength, the antireflection coating may consist of a single layer and must have (a) a refractive index equal to the square root of the refractive indices of the materials bounding the coating, and (b) a thickness equal to one-quarter the wavelength in question (*i.e.*, the wavelength within the material of which the coating consists). For minimum reflection of multiple wavelengths, additional layers must be added.

anti-spoof: Measures to prevent an opponent's participation in a telecommunications network, or operation/control of a cryptographic or COMSEC system. [NIS]

anycall: In adaptive high-frequency (HF) radio automatic link establishment, a broadcast in which (a) the called stations are unspecified, (b) stations receiving the call stop scanning, and (c) each station automatically responds in pseudorandom time slots. (188)

AP: *Abbreviation for* **anomalous propagation.**

APC: *Abbreviation for* **adaptive predictive coding.**

APD: *Abbreviation for* **avalanche photodiode**. *Note:* **apd** and **a.p.d.** are also used.

aperiodic antenna: An antenna designed to have an approximately constant input impedance over a wide range of frequencies. *Note:* Examples of aperiodic antennas include terminated rhombic antennas and wave antennas. *Synonym* **nonresonant antenna.**

aperture: 1. In a directional antenna, the portion of a plane surface very near the antenna normal to the direction of maximum radiant intensity, through which the major part of the radiation passes. (188) **2.** In an acoustic device that launches a sound wave, the passageway, determined by the size of a hole in the inelastic material and the wavelength. [After 2196]

aperture distortion: In facsimile, the distortion of the recorded image caused by the shape and finite size of the scanning and recording apertures. *Note:* The distortion may occur in one or more attributes of the recorded image, such as in resolution, density, or shape.

aperture illumination: 1. The field distribution, in amplitude and phase, over the antenna physical aperture. **2.** The phase and amplitude of the element feed voltages or the distribution of the currents in an array of elements.

aperture-to-medium coupling loss: The difference between the theoretical gain of a very large antenna, such as the antennas in beyond-the-horizon microwave links, and the gain that can be realized in operation. *Note 1:* Aperture-to-medium coupling loss is related to the ratio of the scatter angle to the antenna beamwidth. *Note 2:* The "very large antennas" are referred to in wavelengths; thus, this loss can apply to line-of-sight systems also. (188)

API: *Abbreviation for* **application program interface.**

apogee: In an orbit of a satellite orbiting the Earth, the point that is farthest from the gravitational center of the Earth.

apogee altitude: *See* **altitude of the apogee or of the perigee**.

apparent power: In alternating-current power transmission and distribution, the product of the rms voltage and amperage. *Note 1:* When the applied voltage and the current are in phase with one another, the apparent power is equal to the effective power, *i.e.*, the real power delivered to or consumed by the load. If the current lags or leads the applied voltage, the apparent power is greater than the effective power. *Note 2:* Only effective power, *i.e.*, the real power delivered to or consumed by the load, is expressed in watts. Apparent power is properly expressed only in volt-amperes, never watts. *See diagram under* **effective power.**

Application Layer: *See* **Open Systems Interconnection—Reference Model.**

application program interface (API): A formalized set of software calls and routines that can be referenced by an application program in order to access supporting network services.

applique: Circuit components added to an existing system to provide additional or alternate functions. (188) *Note:* Applique may be used to modify carrier telephone equipment designed for ringdown manual operation to allow for use between points having dial equipment.

approved circuit: *Deprecated synonym for* **protected distribution system.**

aramid yarn: Generic name for a tough synthetic yarn that is often used in optical cable construction for the strength member, protective braid, and/or rip cord for jacket removal. [FAA]

Archie: Distributed-system-based software that searches indices of files available on public servers on the Internet. *Note 1: Archie* servers may provide access via telnet, E-mail, or a special Archie client. *Note 2:* Using Archie requires a user to be familiar with where the indices are located, *i.e.*, the user must provide an Archie server domain name or IP address.

architecture: *See* **computer architecture, network architecture.**

archiving: The storing of files, records, and other data for reference and alternative backup.

area broadcast shift: The changing from listening to transmissions intended for one broadcast area to listening to transmissions intended for another broadcast area. *Note 1:* An area broadcast shift may occur when a ship or aircraft crosses the boundary between listening areas. *Note 2:* Shift times, on the date a ship or aircraft is expected to pass into another area, must be strictly observed or the ship or aircraft will miss messages intended for it. *Synonym* **radio watch shift.**

area code: *See* **access code, code, country code, NXX code.**

area loss: When optical fibers are joined by a splice or a pair of mated connectors, a power loss that is caused by any mismatch in size or shape of the cross section of the cores of the mating fibers. *Note 1:* Any of the above conditions may allow light from the core of the "transmitting" fiber to enter the cladding of the "receiving" fiber, where it is quickly lost. *Note 2:* Area loss may be dependent on the direction of propagation. For example, in coupling a signal from an optical fiber having a smaller core to an otherwise identical one having a larger core, there will be no area loss, but in the opposite direction, there will be area loss. [After FAA]

argument: **1.** An independent variable. **2.** Any value of an independent variable. *Note:* Examples of arguments include search keys, numbers that identify the location of a data item in a table, and the θ in $\sin \theta$.

arithmetic and logic unit (ALU): A part of a computer that performs arithmetic, logic, and related operations.

arithmetic overflow: **1.** In a digital computer, the condition that occurs when a calculation produces a result that is greater than a given register or storage location can store or represent. (188) **2.** In a digital computer, the amount that a calculated value is greater than a given register or storage location can store or represent. *Note:* The overflow may be placed at another location. (188) *Synonym* **overflow.**

arithmetic register: A register that holds the operands or the results of operations such as arithmetic operations, logic operations, and shifts.

arithmetic shift: A shift, applied to the representation of a number in a fixed radix numeration system and in a fixed-point representation system, and in which only the characters representing the fixed-point part of the number are moved. An arithmetic shift is usually equivalent to multiplying the number by a positive or a negative integral power of the radix, except for the effect of any rounding; compare the logical shift with the arithmetic shift, especially in the case of floating-point representation.

arithmetic underflow: In a digital computer, the condition that occurs when a calculation produces a non-zero result that is less than the smallest non-zero quantity that a given register or storage location can store or represent.

arithmetic unit: In a processor, the part that performs arithmetic operations; sometimes the unit performs both arithmetic and logic operations.

Armed Forces Radio Service (AFRS): A radio broadcasting service that is operated by and for the personnel of the armed services in the area covered by the broadcast. *Note:* An example of an AFRS is the radio service operated by the U.S. Army for U.S. and allied military personnel on duty in overseas areas.

armor: Of a communications cable, a component intended to protect the critical internal components, *e.g.*, buffer tubes or fibers, or electrical conductors, from damage from external mechanical attack, *e.g.*, rodent attack or abrasion. [After FAA] *Note:* Armor usually takes the form of a steel or aluminum tape wrapped about an inner jacket that covers the critical internal components. An outer jacket usually covers the armor.

ARP: *Abbreviation for* **address resolution protocol.**

ARPANET: *Abbreviation for* **Advanced Research Projects Agency Network.** A packet-switching network used by the Department of Defense, later evolved into the Internet.

ARQ: *Abbreviation for* **automatic repeat-request.** Error control for data transmission in which the receiver detects transmission errors in a message and automatically requests a retransmission from the transmitter. *Note:* Usually, when the transmitter receives the ARQ, the transmitter retransmits the message until it is either correctly received or the error persists beyond a predetermined number of retransmissions. (188) *Synonyms* **error-detecting-and-feedback system, repeat-request system.**

array: **1.** An arrangement of elements in one or more dimensions. **2.** In a programming language, an aggregate that consists of data objects with identical attributes, each of which may be uniquely referenced by subscription.

array processor: A processor capable of executing instructions in which the operands may be arrays rather than data elements. *Synonym* **vector processor.**

arrester: A device that protects hardware, such as systems, subsystems, circuits, and equipment, from voltage or current surges produced by lightning or electromagnetic pulses. *Note:* If the hardware is adequately protected, associated software may also be adequately protected. (188) *Synonym* **surge suppressor.**

ARS: *Abbreviation for* **automatic route selection.**

articulation index: A measure of the intelligibility of voice signals, expressed as a percentage of speech units that are understood by the listener when heard out of context. (188) *Note:* The articulation index is affected by noise, interference, and distortion.

articulation score (AS): A subjective measure of the intelligibility of a voice system in terms of the percentage of words correctly understood over a channel perturbed by interference. *Note:* Articulation scores have been experimentally obtained as functions of varying word content, bandwidth, audio signal-to-noise ratio and the experience of the talkers and listeners involved.

artificial intelligence (AI): The capability of a device to perform functions that are normally associated with human intelligence, such as reasoning and optimization through experience. *Note:* AI is the branch of computer science that attempts to approximate the results of human reasoning by organizing and manipulating factual and heuristic knowledge. Areas of AI activity include expert systems, natural language understanding, speech recognition, vision, and robotics.

artificial transmission line: A four-terminal electrical network, *i.e.* an electrical circuit, that has the characteristic impedance, transmission time delay, phase shift, and/or other parameter(s) of a real transmission line and therefore can be used to simulate a real transmission line in one or more of these respects. *Colloquial synonym* **art line.**

art line: *Colloquial synonym for* **artificial transmission line.**

ARU: *Abbreviation for* **audio response unit.**

ASCII: *Acronym for* **American Standard Code for Information Interchange.** The standard code used for information interchange among data processing systems, data communications systems, and associated equipment in the United States. (188) *Note 1:* The ASCII character set contains 128 coded characters. *Note 2:* Each ASCII character is a 7-bit coded unique character; 8 bits when a parity check bit is included. *Note 3:* The ASCII character set consists of control characters and graphic characters. *Note 4:* When considered simply as a set of 128 unique bit patterns, or 256 with a parity bit, disassociated from the character equivalences in national implementations, the ASCII may be considered as an alphabet used in machine languages. *Note 5:* The ASCII is the U.S. implementation of International Alphabet No. 5 (IA No. 5) as specified in CCITT Recommendation V.3.

ASP: *Abbreviation for* **adjunct service point.**

aspect ratio: In facsimile or television, the ratio of the width to the height of a picture, document, or scanning field.

assemble: To translate a computer program expressed in an assembly language into a machine language.

assembler: A computer program that is used to assemble. *Synonym* **assembly program.**

assembly: In logistics, an item forming a portion of an equipment that can be provisioned and replaced as an entity and which normally incorporates replaceable parts or groups of parts. [JP1]

assembly language: A computer-oriented language (a) in which instructions are symbolic and usually in one-to-one correspondence with sets of machine language instructions and (b) that may provide other facilities, such as the use of macro instructions. (188) *Synonym* **computer-dependent language.**

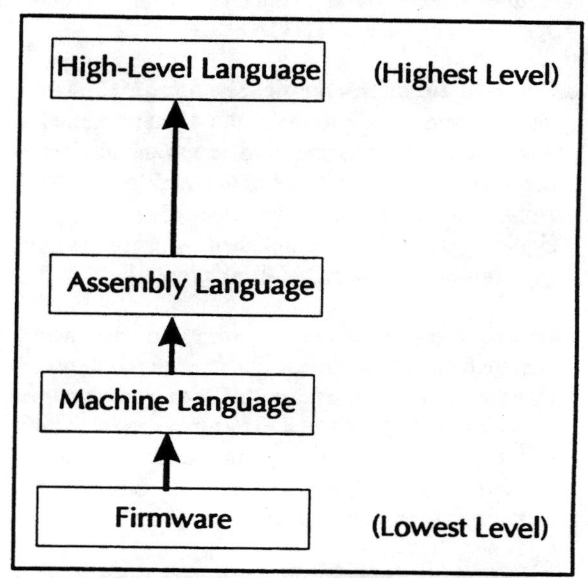

a hierarchy of levels of computer language

assembly program: *Synonym* **assembler.**

assembly time: The elapsed time taken for the execution of an assembler.

assigned frequency: **1.** The center of the assigned frequency band assigned to a station. [RR] **2.** The frequency of the center of the radiated bandwidth. (188) *Note:* The frequency of the rf carrier, whether suppressed or radiated, is usually given in parentheses following the assigned frequency, and is the frequency appearing in the dial settings of rf equipment intended for single-sideband or independent-sideband transmission.

assigned frequency band: The frequency band within which the emission of a station is authorized; the width of the band equals the necessary bandwidth plus twice the absolute value of the frequency tolerance. Where space stations are concerned, the assigned frequency band includes twice the maximum Doppler shift that may occur in relation to any point of the Earth's surface. [NTIA] [RR]

assignment: For NS/EP, the designation of priority level(s).

assignment (of a radio frequency or radio frequency channel): Authorization given by an administration for a radio station to use a radio frequency or radio frequency channel under specified conditions. [NTIA] [RR]

associated common-channel signaling: Common-channel signaling in which the signal channel is associated with a specific trunk group and terminates at the same pair of switches as the trunk group. (188) *Note:* In associated common-channel signaling, the signaling is usually accomplished by using the same facilities as the associated trunk group.

associative storage: **1.** A storage device whose storage locations are identified by their contents, or by a part of their contents, rather than by their names or positions. *Note:* Associative storage can also refer to this process as well as to the device. *Synonym* **content-addressable storage.** **2.** Storage that supplements another storage.

asymmetrical modulator: *Synonym* **unbalanced modulator.**

asynchronous communications system: A data communications system that uses asynchronous operation. *Note 1:* In an asynchronous communications system, extra signal elements are usually appended to the data for the purpose of synchronizing individual data characters or blocks. (188) *Note 2:* The time spacing between successive data characters or blocks may be of arbitrary duration. *Synonym* **start-stop system.**

asynchronous network: A network in which the clocks do not need to be synchronous or mesochronous. (188) *Synonym* **nonsynchronous network.**

asynchronous operation: **1.** A sequence of operations in which operations are executed out of time coincidence with any event. (188) **2.** An operation that occurs without a regular or predictable time relationship to a specified event; *e.g.,* the calling of an error diagnostic routine that may receive control at any time during the execution of a computer program. (188) *Synonym* **asynchronous working.**

asynchronous time-division multiplexing (ATDM): Time-division multiplexing in which asynchronous transmission is used.

asynchronous transfer mode (ATM): A high-speed multiplexing and switching method utilizing fixed-length cells of 53 octects to support multiple types of traffic. *Note:* ATM, specified in international standards, is asynchronous in the sense that cells carrying user data need not be periodic.

asynchronous transmission: Data transmission in which the instant that each character, or block of characters, starts is arbitrary; once started, the time of occurrence of each signal representing a bit within the character, or block, has the same relationship to significant instants of a fixed time frame. (188)

asynchronous working: *Synonym* **asynchronous operation.**

ATB: *Abbreviation for* **all trunks busy.**

AT Commands: A *de facto* standard for modem commands from an attached CPU, used in most 1,200 and 2,400 b/s modems.

ATDM: *Abbreviation for* **asynchronous time-division multiplexing.**

ATM: *Abbreviation for* **asynchronous transfer mode.**

atmospheric duct: A horizontal layer in the lower atmosphere in which the vertical refractive index gradients are such that radio signals (a) are guided or focused within the duct, (b) tend to follow the curvature of the Earth, and (c) experience less attenuation in the ducts than they would if the ducts were not present. (188) *Note:* The reduced refractive index at the higher altitudes bends the signals back toward the Earth. Signals in a higher refractive index layer, *i.e.,* duct, tend to remain in that layer because of the reflection and refraction encountered at the boundary with a lower refractive index material.

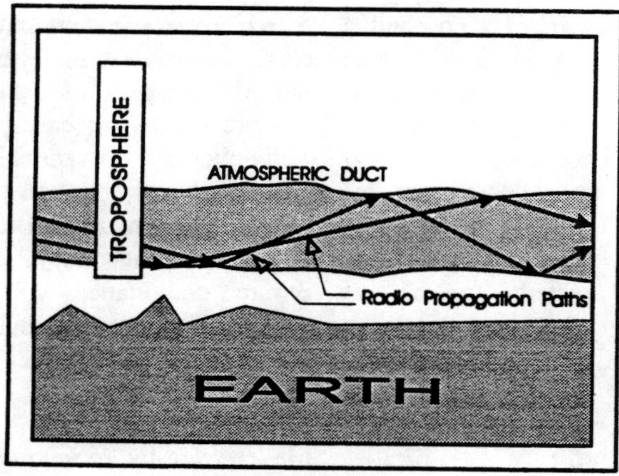

atmospheric ducting

atmospheric noise: Radio noise caused by natural atmospheric processes, primarily lightning discharges in thunderstorms. (188)

atomic time: *See* **International Atomic Time.**

ATOW: *Acronym for* **acquisition and tracking orderwire.** A downlink circuit that provides a terminal with information regarding uplink acquisition and synchronization status. (188)

attachment unit interface (AUI): In a local area network, the interface between the medium access unit (MAU) and the data terminal equipment within a data station.

attack time: The time between (a) the instant that a signal at the input of a device or circuit exceeds the activation threshold of the device or circuit and (b) the instant that the device or circuit reacts in a specified manner, or to a specified degree, to the input. *Note:* Attack time occurs in devices such as clippers, peak limiters, compressors, and voxes.

attendant access loop: A switched circuit that provides an attendant with a manual means for call completion and control. *Note:* An attendant access loop might be given a specific telephone number.

attendant conference: A network-provided service feature that allows an attendant to establish a conference connection of three or more users.

attendant position: The part of a switching system used by an attendant, *i.e.*, an operator, to assist users in call completion and use of special services. (188)

attention signal: The attention signal to be used by AM, FM, and TV broadcast stations to actuate muted receivers for inter-station receipt of emergency cuing announcements and broadcasts involving a range of emergency contingencies posing a threat to the safety of life or property. [47CFR]

attenuation: The decrease in intensity of a signal, beam, or wave as a result of absorption of energy and of scattering out of the path to the detector, but not including the reduction due to geometric spreading. [After JP1] *Note 1:* Attenuation is usually expressed in dB. *Note 2: "Attenuation"* is often used as a misnomer for **"attenuation coefficient,"** which is expressed in dB per kilometer. *Note 3:* A distinction must be made as to whether the attenuation is that of signal power or signal electric field strength.

attenuation coefficient: The rate of diminution of average power with respect to distance along a transmission path. *Synonym* **attenuation rate.**

attenuation constant: **1.** The real part of the propagation constant in any electromagnetic propagation medium. *Note 1:* The attenuation constant is usually expressed as a numerical value per unit length. *Note 2:* The attenuation constant may be calculated or experimentally determined for each medium. **2.** For a particular propagation mode in an optical fiber, the real part of the axial propagation constant.

attenuation-limited operation: The condition that prevails when attenuation, rather than bandwidth, limits the performance of a communications link. (188)

attenuation rate: *Synonym* **attenuation coefficient.**

attenuator: **1.** In electrical systems, a network that reduces the amplitude of a signal without appreciably distorting its waveform. *Note 1:* Electrical attenuators are usually passive devices. *Note 2:* The degree of attenuation may be fixed, continuously adjustable, or incrementally adjustable. Fixed attenuators are often called pads, especially in telephony. *Note 3:* The input and output impedances of an attenuator are usually matched to the impedances of the signal source and load, respectively. **2.** In optical systems, a device that reduces the amplitude of a signal without appreciably distorting its waveform. *Note 1:* Optical attenuators are usually passive devices. *Note 2:* The degree of attenuation may be fixed, continuously adjustable, or incrementally adjustable.

attribute: **1.** In database management, a property inherent in an entity or associated with that entity for database purposes. (188) **2.** In network management, a property of a managed object that has a value. (188) *Note 1:* Mandatory initial values for attributes can be specified as part of the managed object class definition. *Note 2:* Attributes may be either mandatory or conditional.

ATV: *Abbreviation for* **advanced television.**

audible: *Synonym for* **audible ringing tone.**

audible ringing tone: In telephony, a signal, usually consisting of an audio tone interrupted at a slow repetition rate, provided to a caller to indicate that the called-party instrument is being sent a ringing signal. (188) *Note:* The audible ringing tone may be generated by the called-party servicing switch or by the calling-party servicing switch, but it is not generated by the called telephone instrument. *Synonyms* **audible, ringback tone.**

audio frequency (AF): The band of frequencies (approximately 20 Hz to 20 kHz) that, when transmitted as acoustic waves, can be heard by the normal human ear. (188)

audio response unit (ARU): A device that provides synthesized voice responses to dual-tone multifrequency signaling input by processing calls based on (a) the call-originator input, (b) information received from a host data base, and (c) information in the incoming call, such as the time of day. *Note:* ARUs are used to increase the number of information calls handled and to provide consistent quality in information retrieval.

audit: **1.** To conduct an independent review and examination of system records and activities in order to test the adequacy and effectiveness of data security and data integrity procedures, to ensure compliance with established policy and operational procedures, and to recommend any necessary changes. **2.** Independent review and examination of records and activities to assess the adequacy of system controls, to ensure compliance with established policies and operational procedures, and to recommend necessary changes in controls, policies, or procedures. [NIS]

audit review file: A file created by executing statements included in a computer program for the explicit purpose of providing data for auditing.

audit trail: **1.** A record of both completed and attempted accesses and service. **2.** Data in the form of a logical path linking a sequence of events, used to trace the transactions that have affected the contents of a record. **3.** [In INFOSEC, a] chronological record of system activities to enable the reconstruction and examination of the sequence of events and/or changes in an event. *Note:* Audit trail may apply to information in an AIS, or to the transfer of COMSEC material. [NIS]

AUI: *Abbreviation for* **attachment unit interface.**

aurora: Sporadic radiant emission from the upper atmosphere that usually occurs about the North and South magnetic poles of the Earth. *Note 1:* Auroras are most intense at times of intense magnetic storms caused by sunspot activity. The distribution of auroral intensity with altitude shows a pronounced maximum near 100 km above the Earth. Auroras may occasionally be observed within 40° or less of the equator. *Note 2:* Auroras interfere with radio communications. *Note 3:* In the Northern hemisphere, the aurora is called the Aurora Borealis (Northern Lights). In the Southern hemisphere, the aurora is called the Aurora Australis (Southern Lights).

authenticate: **1.** To establish, usually by challenge and response, that a transmission attempt is authorized and valid. **2.** Verify the identity of a user, user device, or other entity, or the integrity of data stored, transmitted, or otherwise exposed to unauthorized modification in an automated information system, or establish the validity of a transmitted message. [NIS] **3.** A challenge given by voice or electrical means to attest to the authenticity of a message or transmission. [JP1]

authentication: [Any] Security measure designed to establish the validity of a transmission, message, or originator, or a means of verifying an individual's eligibility to receive specific categories of information. [NIS]

authenticator: **1.** A symbol or group of symbols, or a series of bits, selected or derived in a prearranged manner and usually inserted at a predetermined point within a message or transmission for the purpose of attesting to the validity of the message or transmission. [JP1] **2.** A letter, numeral, group of letters or numerals, or any combinations of these, attesting to the authenticity of a message or transmission. [After JP1] **3.** [In INFOSEC,] means used to confirm the identity or eligibility of a station, originator, or individual. [NIS]

authorization: **1.** The rights granted to a user to access, read, modify, insert, or delete certain data, or to execute certain programs. **2.** Access rights granted to a user, program, or process. [NIS]

authorized bandwidth: Authorized bandwidth is, for purposes of this Manual, the necessary bandwidth (bandwidth required for transmission and reception of intelligence) and does not include allowance for transmitter drift or Doppler shift. [NTIA]

authorized frequency: A frequency that is allocated and assigned by a competent authority to a specific user for a specific purpose. (188) *See* **assigned frequency.**

AUTODIN: *Acronym for* **automatic digital network.** *See* **Defense Data Network.**

automated data medium: *Synonym* **machine-readable medium.**

automated information system (AIS): **1.** An assembly of computer hardware, software, firmware, or any combination of these, configured to accomplish specific information-handling operations, such as communication, computation, dissemination, processing, and storage of information. **2.** [In INFOSEC,] any equipment or interconnected system or subsystems of equipment that is used in the automatic acquisition, storage, manipulation, management, movement, control, display, switching, interchange, transmission or reception of data and includes computer software, firmware, and hardware. *Note:* Included are computers, word processing systems, networks, or other electronic information handling systems, and associated equipment. [NIS]

automated information systems security: **1.** Measures and controls that ensure confidentiality, integrity, and availability of the information processed and stored by automated information systems. *Note 1:* The unauthorized disclosure, modification, or destruction may be accidental or intentional. *Note 2:* Automated information systems security includes consideration of all hardware and software functions, characteristics and features; operational procedures; accountability procedures; and access controls at the central computer facility, remote computer, and terminal facilities; management constraints; physical structures and devices, such as computers, transmission lines, and power sources; and personnel and communications controls needed to provide an acceptable level of risk for the automated information system and for the data and information contained in the system. Automated information systems security also includes the totality of security safeguards needed to provide an acceptable protection level for an automated information system and for the data handled by an automated information system. **2.** In INFOSEC, *synonym* **computer security.**

automated maritime telecommunications system (AMTS): An automatic, integrated and interconnected maritime communications system serving ship stations on specified inland and coastal waters of the United States. [47CFR]

automated radio: A radio that can be automatically controlled by electronic devices and that requires little or no human intervention.

automated tactical command and control system: A command and control system, or part thereof, that manipulates the movement of information from source to user without human intervention. *Note:* In an automated tactical command and control system, automated execution of a decision without human intervention is not mandatory.

automatic answering: A service feature in which the called terminal automatically responds to the calling signal and the call may be established whether or not the called terminal is attended by an operator.

automatic callback: A service feature that permits a user, when encountering a busy condition, to instruct the system to retain the called and calling numbers and to establish the call when there is an available line. *Note 1:* Automatic callback may be implemented in the terminal, in the switching system, or shared between them. *Note 2:* Automatic callback is not the same as camp-on.

automatic call distributor (ACD): A device that distributes incoming calls to a specific group of terminals. *Note:* If the number of active calls is less than the number of terminals, the next call will be routed to the terminal that has been in the idle state the longest. If all terminals are busy, the incoming calls are held in a first-in-first-out queue until a terminal becomes available. (188)

automatic calling: Calling in which the elements of the selection signal are entered into the data network contiguously at the full data signaling rate. The selection signal is generated by the data terminal equipment. *Note:* A limit may be imposed by the design criteria of the network to prevent more than a permitted number of unsuccessful call attempts to the same address within a specified period.

automatic calling unit (ACU): A device that enables equipment, such as computers and card dialers, to originate calls automatically over a telecommunications network.

automatic data handling (ADH): **1.** A generalization of automatic data processing to include the aspect of data transfer. [JP1] **2.** Combining data processing and data transfer.

automatic data processing (ADP): **1.** An interacting assembly of procedures, processes, methods, personnel, and equipment to perform automatically a series of data processing operations on data. *Note:* The data processing operations may result in a change in the semantic content of the data. (188) **2.** Data processing by means of one or more devices that use common storage for all or part of a computer program, and also for all or part of the data necessary for execution of the program; that execute user-written or user-designated programs; that perform user-designated symbol manipulation, such as arithmetic operations, logic operation, or character-string manipulations; and that can execute programs that modify themselves during their execution. *Note:* Automatic data processing may be performed by a stand-alone unit or by several connected units. **3.** Data processing largely performed by automatic means. [JP1] **4.** That branch of science and technology concerned with methods and techniques relating to data processing largely performed by automatic means. [JP1]

automatic data processing equipment (ADPE): Any equipment or interconnected system or subsystems of equipment that is used in the automatic acquisition, storage, manipulation, management, movement, control, display, switching, interchange, transmission, or reception, of data or information (i) by a Federal agency, or (ii) under a contract with a Federal agency which (i) requires the use of such equipment, or (ii) requires the performance of a service or the furnishing of a product which is performed or produced making significant use of such equipment. Such term includes (i) computer, (ii) ancillary equipment, (iii) software, firmware, and similar procedures, (iv) services, including support services, and (v) related resources as defined by regulations issued by the Administrator for General Services. . . . [Public Law 99-500, Title VII, Sec. 822 (a) Section 111(a) of the *Federal Property and Administrative Services Act of 1949 (40 U.S.C. 759(a)) revised.*] *See also paragraph #3 of the* **Foreword** *above.*

automatic dialing: *See* **automatic calling unit.**

Automatic Digital Network (AUTODIN): Formerly, a worldwide data communications network of the Defense Communications System, now replaced by the **Defense Switched Network (DSN).** (188)

automatic error correction: *See* **error-correcting code.**

automatic exchange: In a telephone system, an exchange in which communications among users are effected by means of switches set in operation by the originating user equipment without human intervention at the central office or branch exchange.

automatic frequency control (AFC): A device or circuit that maintains the frequency of an oscillator within the specified limits with respect to a reference frequency.

automatic function: A machine function or series of machine functions controlled by a program and carried out without assistance of an operator.

automatic gain control (AGC): A process or means by which gain is automatically adjusted in a specified manner as a function of a specified parameter, such as received signal level. (188)

automatic identified outward dialing (AIOD): A service feature of some switching or terminal devices that provides the user with an itemized statement of usage on directly dialed calls. *Note:* AIOD is facilitated by automatic number identification (ANI) equipment to provide automatic message accounting (AMA).

automatic link establishment (ALE): **1.** In high-frequency (HF) radio, the capability of a station to make contact, or initiate a circuit, between itself and another specified radio station, without human intervention and usually under processor control. *Note:* ALE techniques include automatic signaling, selective calling, and automatic handshaking. Other automatic techniques that are related to ALE are channel scanning and selection, link quality analysis (LQA), polling, sounding, message store-and-forward, address protection, and anti-spoofing. **2.** In HF radio, a link control system that includes automatic scanning, selective calling, sounding, and transmit channel selection using link quality analysis data. *Note:* Optional ALE functions include polling and the exchange of orderwire commands and messages. (188)

automatic message accounting (AMA): A service feature that automatically records data regarding user-dialed calls. (188)

automatic message exchange (AME): In an adaptive high-frequency (HF) radio network, an automated process allowing the transfer of a message from message injection to addressee reception, without human intervention. *Note:* Through the use of machine-addressable transport guidance information, *i.e.,* the message header, the message is automatically routed through an on-line direct connection through single or multiple transmission media.

automatic message processing system (AMPS): Any organized assembly of resources and methods used to collect, process, and distribute messages largely by automatic means. [JP1]

automatic number identification (ANI): A service feature in which the directory number or equipment number of a calling station is automatically obtained. *Note:* ANI is used in message accounting.

automatic operation: The functioning of systems, equipment, or processes in a desired manner at the proper time under control of mechanical or electronic devices that operate without human intervention.

automatic redial: A service feature that allows the user to dial, by depressing a single key or a few keys, the most recent telephone number dialed at that instrument. *Note:* Automatic redial is often associated with the telephone instrument, but may be provided by a PBX, or by the central office. *Synonym* **last number redial.** *Contrast with* **automatic calling unit.**

automatic reload: *See* **bootstrap (def.#3).**

automatic remote rekeying: [In INFOSEC, a] procedure to rekey a distant crypto-equipment electronically without specific actions by the receiving terminal operator. [NIS] *Note:* Automatic remote rekeying may also apply to non-crypto devices.

automatic remote reprogramming and rekeying: The procedure by which distant equipment is reprogrammed or rekeyed electronically without specific actions by the receiving terminal.

automatic repeat-request (ARQ): *See* **ARQ.**

automatic ringdown circuit: A circuit providing priority telephone service, typically for key personnel; the circuit is activated when the telephone handset is removed from the cradle causing a ringing signal to be sent to the distant unit(s). *See* **verified off-hook.**

automatic route selection (ARS): Electronic or mechanical selection and routing of outgoing calls without human intervention.

Automatic Secure Voice Communications Network (AUTOSEVOCOM): A worldwide, switched, secure voice network developed to fulfill DOD long-haul, secure voice requirements. [JP1] (188)

automatic sequential connection: A service feature in which the terminals at each of a set of specified addresses are automatically connected, in a predetermined sequence, to a single terminal at a specified address.

automatic signaling service: *Synonym* **hotline.**

automatic sounding: The testing of selected channels or paths by providing a very brief beacon-like identifying broadcast that may be used by other stations to evaluate connectivity, propagation, and availability, and to identify known working channels for possible later use for communications or calling. (188) *Note 1:* Automatic soundings are primarily intended to increase the efficiency of the automatic link establishment (ALE) function, thereby increasing system throughput. *Note 2:* Sounding information is used for identifying the specific channel to be used for a particular ALE connectivity attempt.

automatic switching system: **1.** In data communications, a switching system in which all the operations required to execute the three phases of information-transfer transactions are automatically executed in response to signals from a user end-instrument. *Note:* In an automatic switching system, the information-transfer transaction is performed without human intervention, except for initiation of the access phase and the disengagement phase by a user. **2.** In telephony, a system in which all the operations required to set up, supervise, and release connections required for calls are automatically performed in response to signals from a calling device. (188)

Automatic Voice Network (AUTOVON): Formerly, the principal long-haul, unsecure voice communications network within the Defense Communications System, now replaced by the Defense Switched Network (DSN). (188)

automation: **1.** The implementation of processes by automatic means. [JP1] **2.** The investigation, design, development, and application of methods of rendering processes automatic, self-moving, or self-controlling. **3.** The conversion of a procedure, a process, or equipment to automatic operation. [JP1]

AUTOSEVOCOM: *Acronym for* **Automatic Secure Voice Communications Network.**

AUTOVON: *Acronym for* **Automatic Voice Network.** *Superseded by* **Defense Switched Network.**

auxiliary operation: An offline operation performed by equipment not under control of the processing unit.

auxiliary power: Electric power that is provided by an alternate source and that serves as backup for the primary power source at the station main bus or prescribed sub-bus. (188) *Note 1:* An offline unit provides electrical isolation between the primary power source and the critical technical load whereas an online unit does not. *Note 2:* A Class A power source is a primary power source, *i.e.*, a source that assures an essentially continuous supply of power. *Note 3:* Types of auxiliary power services include Class B, a standby power plant to cover extended outages of the order of days; Class C, a 10-to-60-second quick-start unit to cover short-term outages of the order of hours; and Class D, an uninterruptible non-break unit using stored energy to provide continuous power within specified voltage and frequency tolerances.

auxiliary storage: **1.** Storage that is available to a processor only through its input/output channels. **2.** In a computer, any storage that is not internal memory, *i.e.*, is not random access memory (RAM). *Note:* Examples of auxiliary storage media are magnetic diskettes, optical disks including CD ROM, and magnetic tape cassettes.

availability: **1.** The degree to which a system, subsystem, or equipment is operable and in a committable state at the start of a mission, when the mission is called for at an unknown, *i.e.*, a random, time. *Note 1:* The conditions determining operability and committability must be specified. (188) *Note 2:* Expressed mathematically, availability is 1 minus the unavailability. **2.** The ratio of (a) the total time a functional unit is capable of being used during a given interval to (b) the length of the interval. *Note 1:* An example of availability is 100/168 if the unit is capable of being used for 100 hours in a week. *Note 2:* Typical availability objectives are specified in decimal fractions, such as 0.9998.

available line: **1.** In voice, video, or data communications, a circuit between two points that is ready for service, but is in the idle state. **2.** In facsimile transmission, the portion of the scanning line that can be specifically used for image signals. (188) *Synonym* **useful line.**

available time: From the point of view of a user, the time during which a functional unit can be used. *Note:* From the point of view of operating and maintenance personnel, the available time is the same as the uptime, *i.e.*, the time during which a functional unit is fully operational.

avalanche multiplication: A current-multiplying phenomenon that occurs in a semiconductor photodiode that is reverse-biased just below its breakdown voltage. *Note:* Under such a condition, photocurrent carriers, *i.e.*, electrons, are swept across the junction with sufficient energy to ionize additional bonds, creating additional electron-hole pairs in a regenerative action. [After FAA]

avalanche photodiode (APD): A photodiode that operates with a reverse-bias voltage that causes the primary photocurrent to undergo amplification by cumulative multiplication of charge carriers. *Note:* As the reverse-bias voltage increases toward the breakdown, hole-electron pairs are created by absorbed photons. An avalanche effect occurs when the hole-electron pairs acquire sufficient energy to create additional pairs when the incident photons collide with the ions, *i.e.*, the holes and electrons. Thus, a signal gain is achieved. [After 2196]

average picture level (APL): In video systems, the average level of the picture signal during active scanning time integrated over a frame period; defined as a percentage of the range between blanking and reference white level.

average rate of transmission: *Synonym* **effective transmission rate.**

avoidance routing: The assignment of a circuit path to avoid certain critical or trouble-prone circuit nodes.

AWGN: *Abbreviation for* **additive white gaussian noise.** *See* **white noise.**

axial propagation constant: In an optical fiber, the propagation constant evaluated along the optical axis of the fiber in the direction of transmission. *Note:* The real part of the axial propagation constant is the attenuation constant. The imaginary part is the phase constant. [After 2196]

axial ratio: Of an electromagnetic wave having elliptical polarization, the ratio of the magnitudes of the major axis and the minor axis of the ellipse described by the electric field vector.

axial ray: A light ray that travels along the optical axis. (188)

(this page intentionally left blank)

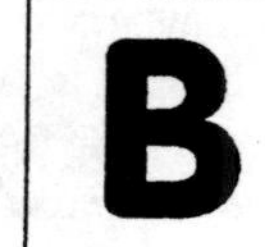

b: *Abbreviation for* **bit.**

B: *Abbreviation for* **bel, byte.**

babble: In transmission systems, the aggregate of crosstalk induced in a given line by all other lines.

backbone: **1.** The high-traffic-density connectivity portion of any communications network. (188) **2.** In packet-switched networks, a primary forward-direction path traced sequentially through two or more major relay or switching stations. *Note:* In packet-switched networks, a backbone consists primarily of switches and interswitch trunks.

background noise: The total system noise in the absence of information transmission. (188)

background processing: The execution of lower priority computer programs when higher priority programs are not using the system resources. *Note:* Priorities may be assigned by system software, application software, or the operator.

backscattering: **1.** Radio wave propagation in which the direction of the incident and scattered waves, resolved along a reference direction (usually horizontal) are oppositely directed. A signal received by backscattering is often referred to as "backscatter." [JP1] (188) **2.** In optics, the scattering of light into a direction generally opposite to the original one.

back-to-back connection: **1.** A direct connection between the output of a transmitting device and the input of an associated receiving device. (188) *Note:* When used for equipment measurements or testing purposes, such a back-to-back connection eliminates the effects of the transmission channel or medium. **2.** A direct connection between the output of a receiving device and the input to a transmitting device. *Note:* The term *"direct,"* as used in both definitions, may be construed as permitting a passive device such as a pad (attenuator) to accommodate power level constraints.

backup: *See* **backup file.**

backup file: A copy of a file made for purposes of later reconstruction of the file, if necessary. *Note:* A backup file may be used for preserving the integrity of the original file and may be recorded on any suitable medium. *Synonym* **job-recovery control file.**

backward channel: **1.** In data transmission, a secondary channel in which the direction of transmission is constrained to be opposite to that of the primary, *i.e.*, the forward (user-information) channel. *Note:* The direction of transmission in the backward channel is restricted by the control interchange circuit that controls the direction of transmission in the primary channel. **2.** In a data circuit, the channel that passes data in a direction opposite to that of its associated forward channel. (188) *Note 1:* The backward channel is usually used for transmission of supervisory, acknowledgement, or error-control signals. The direction of flow of these signals is opposite to that in which user information is being transferred. *Note 2:* The backward-channel bandwidth is usually less than that of the primary channel, *i.e.*, the forward (user information) channel.

backward recovery: The reconstruction of an earlier version of a file by using a newer version of data recorded in a journal.

backward signal: A signal sent from the called to the calling station, *i.e.*, from the original data sink to the original data source. (188) *Note:* Backward signals are usually sent via a backward channel and may consist of supervisory, acknowledgment, or control signals.

backward supervision: The use of supervisory signal sequences from a secondary to a primary station.

balance: In electrical circuits and networks, to adjust the impedance to achieve specific objectives, such as to reach specified return loss objectives at a hybrid junction of two-wire and four-wire circuits. (188)

balanced: Pertaining to electrical symmetry. (188)

balanced code: **1.** In PCM systems, a code constructed so that the frequency spectrum resulting from the transmission of any code word has no dc component. (188) **2.** In PCM, a code that has a finite digital sum variation.

balanced line: A transmission line consisting of two conductors in the presence of ground, capable of being operated in such a way that when the voltages of the two conductors at all transverse planes are equal in magnitude and opposite in polarity with respect to ground, the currents in the two conductors are equal in magnitude and opposite in direction. (188) *Note:* A balanced line may be operated in an unbalanced condition. *Synonym* **balanced signal pair.**

balanced modulator: A modulator constructed so that the carrier is suppressed and any associated carrier noise is balanced out. *Note 1:* The balanced modulator output contains only the sidebands. *Note 2:* Balanced modulators are used in AM transmission systems. (188)

balanced signal pair: *Synonym* **balanced line.**

balance return loss: **1.** A measure of the degree of balance between two impedances connected to two conjugate sides of a hybrid set, coil, network, or junction. **2.** A measure of the effectiveness with which a balancing network simulates the impedance of a two-wire circuit at a hybrid coil. (188)

balancing network: **1.** In a hybrid set, hybrid coil, or resistance hybrid, a circuit used to match, *i.e.*, to balance, the impedance of a uniform transmission line, *i.e.*, twisted metallic pair, over a selected range of frequencies. *Note:* A balancing network is required to ensure isolation between the two ports of the four-wire side of the hybrid. **2.** A device used between a balanced device or line and an unbalanced device or line for the purpose of transforming from balanced to unbalanced or from unbalanced to balanced. (188)

balun: *Abbreviation for* **balanced to unbalanced.** In radio frequency usage, a device used to couple a balanced device or line to an unbalanced device or line. (188)

band: **1.** In communications, the frequency spectrum between two defined limits. (188) **2.** A group of tracks on a magnetic drum or on one side of a magnetic disk. **3.** A set of frequencies authorized for use in a geographical area defined for common carriers for purposes of communications system management.

band-elimination filter: *Synonym* **band-stop filter.**

bandpass filter: A filter that ideally passes all frequencies between two non-zero finite limits and bars all frequencies not within the limits. *Note:* The cutoff frequencies are usually taken to be the 3-dB points. (188)

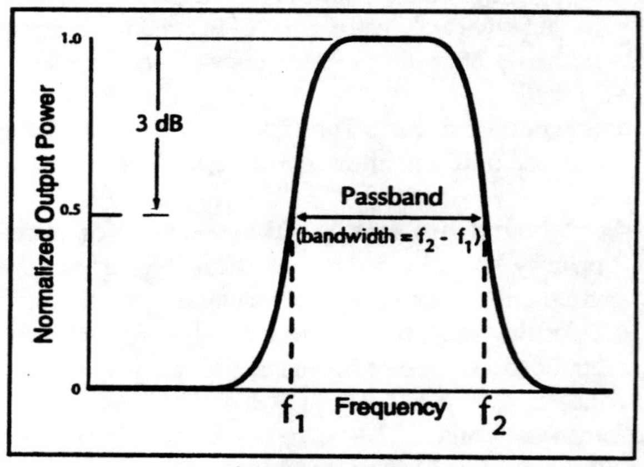

bandpass filter

bandpass limiter: A device that imposes hard limiting on a signal and contains a filter that suppresses the unwanted products (harmonics) of the limiting process.

band-rejection filter: *Synonym* **band-stop filter.**

band-stop filter: A filter that attenuates, usually to very low levels, all frequencies between two non-zero, finite limits and passes all frequencies not within the limits. (188) *Note:* A band-stop filter may be designed to stop the specified band of frequencies but usually only attenuates them below some specified level. *Synonyms* **band-elimination filter, band-rejection filter, band-suppression filter, notched filter.**

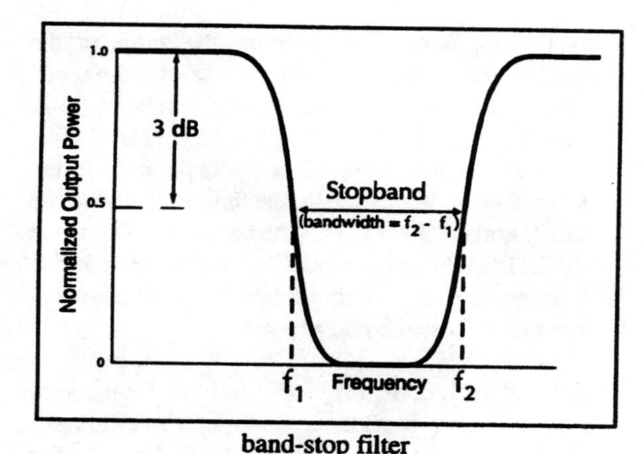

band-stop filter

band-suppression filter: *Synonym* **band-stop filter.**

bandwidth (BW): **1.** The difference between the limiting frequencies within which performance of a device, in respect to some characteristic, falls within specified limits. (188) **2.** The difference between the limiting frequencies of a continuous frequency band. (188)

bandwidth balancing mechanism: In a distributed-queue dual-bus network, a procedure in which a node occasionally skips the use of empty queued arbitrated slots, and which procedure effects sharing of the bandwidth mechanisms.

bandwidth compression: **1.** The reduction of the bandwidth needed to transmit a given amount of data in a given time. **2.** The reduction of the time needed to transmit a given amount of data in a given bandwidth. (188) *Note:* Bandwidth compression implies a reduction in normal bandwidth of an information-carrying signal without reducing the information content of the signal.

bandwidth•distance product: Of an optical fiber, under specified launching and cabling conditions, at a specified wavelength, a figure of merit equal to the product of the fiber's length and the 3-dB bandwidth of the optical signal. *Note 1:* The bandwidth•distance product is usually stated in megahertz•kilometer (MHz•km) or gigahertz•kilometer (GHz•km). *Note 2:* The bandwidth•distance product, which is normalized to 1 km, is a useful figure of merit for predicting the effective fiber bandwidth for other lengths, and for concatenated fibers. *Synonym* **bandwidth•length product.**

bandwidth•length product: *Synonym for* **bandwidth•distance product.**

bandwidth-limited operation: The condition prevailing when the system bandwidth limits performance. (188) *Note:* Bandwidth-limited operation occurs when the system distorts the signal waveform beyond specified limits. For linear systems, bandwidth-limited operation is equivalent to distortion-limited operation.

bandwidth (of an optical fiber): **1.** The lowest modulation frequency at which the RMS peak-to-valley amplitude (optical power) difference of an intensity-modulated monochromatic signal decreases, at the output of the fiber, to a specified fraction (usually one-half) of the RMS peak-to-valley amplitude (optical power) difference of a nearly-zero (arbitrarily low) modulation frequency, both modulation frequencies having the same RMS peak-to-valley amplitude (optical power) difference at the fiber input. *Note 1:* In multimode fibers, multimode distortion is usually the most significant parameter limiting fiber bandwidth, although material dispersion may also play a significant role, especially in the first (850-nm) window. *Note 2:* In multimode fibers, the bandwidth•distance product (colloquially, *"fiber bandwidth"*) is customarily specified by vendors for the bandwidth as limited by multimode distortion only. The spectral width of the optical source is assumed to be extremely narrow. In practice, the effective fiber bandwidth will also be limited by dispersion, especially in the first (850-nm) window, where material dispersion is relatively high, because optical sources have a finite spectral width. Laser diodes typically have a spectral width of several nanometers, FWHM. LEDs typically have a spectral width of 35 to 100 nm, FWHM. *Note 3:* The effective risetime of multimode fibers may be estimated fairly accurately as the square root of the sum of the squares of the material-dispersion-limited risetime and the multimode-distortion-limited risetime. *Note 4:* In single-mode fibers, the most important parameters affecting fiber bandwidth are material dispersion and waveguide dispersion. Practical fibers are designed so that material dispersion and waveguide dispersion cancel one another at the wavelength of interest. *Note 5:*

Regarding effective fiber bandwidth as it affects overall system performance, it should be recognized that optical detectors such as PIN diodes are square-law devices. Their photocurrent is proportional to the optical power of the detected signal. Because electrical power is a function of the square of the current, when the optical power decreases by one-half (a 3-dB decrease), the electrical power decreases by three-fourths (a 6-dB decrease). **2.** Loosely, *synonym* **bandwidth•distance product**.

bar code: A code representing characters by sets of parallel bars of varying thickness and separation that are read optically by transverse scanning. *Note:* Bar code uses include identifying merchandise, sorting mail, and inventorying supplies.

bar code

barrage jamming: Jamming accomplished by transmitting a band of frequencies that is large with respect to the bandwidth of a single emitter. *Note:* Barrage jamming may be accomplished by presetting multiple jammers on adjacent frequencies or by using a single wideband transmitter. Barrage jamming makes it possible to jam emitters on different frequencies simultaneously and reduces the need for operator assistance or complex control equipment. These advantages are gained at the expense of reduced jamming power at any given frequency.

base: 1. In the numeration system commonly used in scientific notation, the real number that is raised to a power denoted by the exponent and then multiplied by the coefficient to determine the value of the number represented without the use of exponents. *Note:* An example of a base is the number 6.25 in the expression $2.70 \times 6.25^{1.5} \approx 42.19$. The 2.70 is the coefficient and the 1.5 is the exponent. In the decimal numeration system, the base is 10 and in the binary numeration system, the base is 2. The value $e \approx 2.718$ is the natural base. **2.** A reference value. **3.** A number that is multiplied by itself as many times as indicated by an exponent.

baseband: 1. The original band of frequencies produced by a transducer, such as a microphone, telegraph key, or other signal-initiating device, prior to initial modulation. *Note 1:* In transmission systems, the baseband signal is usually used to modulate a carrier. *Note 2:* Demodulation re-creates the baseband signal. *Note 3: Baseband* describes the signal state prior to modulation, prior to multiplexing, following demultiplexing, and following demodulation. (188) *Note 4:* Baseband frequencies are usually characterized by being much lower in frequency than the frequencies that result when the baseband signal is used to modulate a carrier or subcarrier. **2.** In facsimile, the frequency of a signal equal in bandwidth to that between zero frequency and maximum keying frequency. (188)

baseband local area network: A local area network in which information is encoded, multiplexed, and transmitted without modulation of carriers.

baseband modulation: Intensity modulation of an optical source, *e.g.*, LED or ILD, directly, without first modulating the signal of interest onto an electrical carrier wave. [After FAA]

baseband signaling: Transmission of a digital or analog signal at its original frequencies; *i.e.*, a signal in its original form, not changed by modulation. (188)

basecom: *Abbreviation for* **base communications.**

base communications (basecom): Communications services, such as the installation, operation, maintenance, augmentation, modification, and rehabilitation of communications networks, systems, facilities, and equipment, including off-post extensions, provided for the operation of a military post, camp, installation, station, or activity. *Synonym* **communications base station.**

base Earth station: An Earth station in the fixed-satellite service or, in some cases, in the land mobile-satellite service, located at a specified fixed point or within a specified area on land to provide a feeder link for the land mobile-satellite service. [NTIA] [RR]

base station: **1.** A land station in the land mobile service. [NTIA] [RR] **2.** In personal communication service, the common name for all the radio equipment located at one fixed location, and that is used for serving one or several cells.

basic exchange telecommunications radio service (BETRS): A commercial service that can extend telephone service to rural areas by replacing the local loop with radio communications. *Note:* In the BETRS, non-government ultra-high frequency (UHF) and very high frequency (VHF) common carrier and the private radio service frequencies are shared.

basic group: *See* **group.**

basic mode link control: Control of data links by use of the control characters of the 7-bit character set for information processing interchange as given in ISO Standard 646-1983 and CCITT Recommendation V.3-1972.

basic rate interface (BRI): A CCITT Integrated Services Digital Network (ISDN) multipurpose user interface standard that denotes the capability of simultaneous voice and data services provided over two clear 64-kb/s channels and one clear 16-kb/s channel (2B+D) access arrangement to each user location.

basic service: **1.** A pure transmission capability over a communication path that is virtually transparent in terms of its interaction with customer-supplied information. **2.** The offering of transmission capacity between two or more points suitable for a user's transmission needs and subject only to the technical parameters of fidelity and distortion criteria, or other conditioning.

basic service element (BSE): **1.** An optional unbundled feature, generally associated with the basic serving arrangement (BSA), that an enhanced-service provider (ESP) may require or find useful in configuring an enhanced service. **2.** A fundamental (basic) communication network service; an optional network capability associated with a BSA. *Note:* BSEs constitute optional capabilities to which the customer may subscribe or decline to subscribe.

basic serving arrangement (BSA): **1.** The fundamental tariffed switching and transmission (and other) services that an operating company must provide to an enhanced service provider (ESP) to connect with its customers through the company network. **2.** In an open-network-architecture context, the fundamental underlying connection of an enhanced service provider (ESP) to and through the operating company's network including an ESP access link, the features and functions associated with that access link at the central office serving the ESP and/or other offices, and the transport (dedicated or switched) within the network that completes the connection from the ESP to the central office serving its customers or to capabilities associated with the customer's complementary network services. *Note:* Each component may have a number of categories of network characteristics. Within these categories of network characteristics are alternatives from among which the customer must choose. Examples of BSA components are ESP access link, transport and/or usage.

basic status: In data transmission, the status of the capability of a secondary station to send or receive a frame containing an information field.

batched communications: *Synonym* **batched transmission.**

batched transmission: The transmission of two or more messages from one station to another without intervening responses from the receiving station. *Synonym* **batched communications.**

batch processing: **1.** The processing of data or the accomplishment of jobs accumulated in advance in such a manner that the user cannot further influence the processing while it is in progress. **2.** The processing of data accumulated over a period of time. **3.** *Loosely,* the execution of computer programs serially. **4.** Pertaining to the technique of executing a set of computer programs such that each is completed before the next program of the set is started. (188) **5.** Pertaining to the sequential input of computer programs or data.

baud (Bd): 1. A unit of modulation rate. *Note:* One baud corresponds to a rate of one unit interval per second, where the modulation rate is expressed as the reciprocal of the duration in seconds of the shortest unit interval. **2.** A unit of signaling speed equal to the number of discrete signal conditions, variations, or events per second. (188) *Note 1:* If the duration of the unit interval is 20 milliseconds, the signaling speed is 50 bauds. If the signal transmitted during each unit interval can take on any one of *n* discrete states, the bit rate is equal to the rate in bauds times $\log_2 n$. The technique used to encode the allowable signal states may be any combination of amplitude, frequency, or phase modulation, but it cannot use a further time-division multiplexing technique to subdivide the unit intervals into multiple subintervals. In some signaling systems, non-information-carrying signals may be inserted to facilitate synchronization; *e.g.,* in certain forms of binary modulation coding, there is a forced inversion of the signal state at the center of the bit interval. In these cases, the synchronization signals are included in the calculation of the rate in bauds but not in the computation of bit rate. *Note 2: Baud* is sometimes used as a synonym for *bit-per-second.* This usage is deprecated.

Baudot code: A synchronous code in which five equal-length bits represent one character. *Note 1:* The Baudot code, which was developed circa 1880, has been replaced by the start-stop asynchronous International Alphabet No. 2 (IA No. 2). *Note 2:* IA No. 2 is not, and should not be identified as, the Baudot code. *Note 3:* The Baudot code has been widely used in teletypewriter systems.

BCC: *Abbreviation for* **block check character.**

BCD: *Abbreviation for* **binary coded decimal.**

B channel: 1. A communications channel used for the transmission of an aggregate signal generated by multichannel transmitting equipment. (188) **2.** The CCITT designation for a clear channel, 64-kb/s service capability provided to a subscriber under the Integrated Services Digital Network offering. *Note:* The B channel, also called the bearer channel, is intended for transport of user information, as opposed to signaling information.

BCH code: *Abbreviation for* **Bose-Chaudhuri-Hochquenghem code.** A multilevel, cyclic, error-correcting, variable-length digital code used to correct errors up to approximately 25% of the total number of digits. *Note:* BCH codes are not limited to binary codes, but may be used with multilevel phase-shift keying whenever the number of levels is a prime number or a power of a prime number, such as 2, 3, 4, 5, 7, 8, 11, and 13. A BCH code in 11 levels has been used to represent the 10 decimal digits plus a sign digit.

BCI: *Abbreviation for* **bit-count integrity.**

Bd: *Abbreviation for* **baud.**

beacon: *See* **radiobeacon station.**

beam: 1. The main lobe of an antenna radiation pattern. **2.** A column of light. *Note:* A beam may be parallel, divergent, or convergent. [After FAA]

beam diameter: Of an electromagnetic beam, along any specified line that (a) intersects the beam axis and (b) lies in any specified plane normal to the beam axis, the distance between the two diametrically opposite points at which the irradiance is a specified fraction, *e.g.,* ½ or $1/\epsilon$, of the beam's peak irradiance. (188) *Note 1:* Beam diameter is usually used to characterize electromagnetic beams in the optical regime, and occasionally in the microwave regime, *i.e.,* cases in which the aperture from which the beam emerges is very large with respect to the wavelength. *Note 2:* Beam diameter usually refers to a beam of circular cross section, but not necessarily so. A beam may, for example, have an elliptical cross section, in which case the orientation of the beam diameter must be specified, *e.g.,* with respect to the major or minor axis of the elliptical cross section.

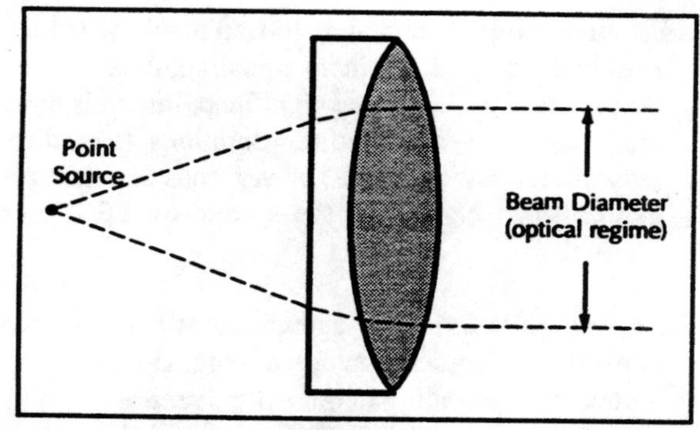

beam diameter

beam divergence: Of an electromagnetic beam, in any plane that intersects the beam axis, the increase in beam diameter with distance from the aperture from which the beam emerges. *Note 1:* Beam divergence is usually used to characterize electromagnetic beams in the optical regime, *i.e.*, cases in which the aperture from which the beam emerges is very large with respect to the wavelength. *Note 2:* Beam divergence usually refers to a beam of circular cross section, but not necessarily so. A beam may, for example, have an elliptical cross section, in which case the orientation of the beam divergence must be specified, *e.g.*, with respect to the major or minor axis of the elliptical cross section.

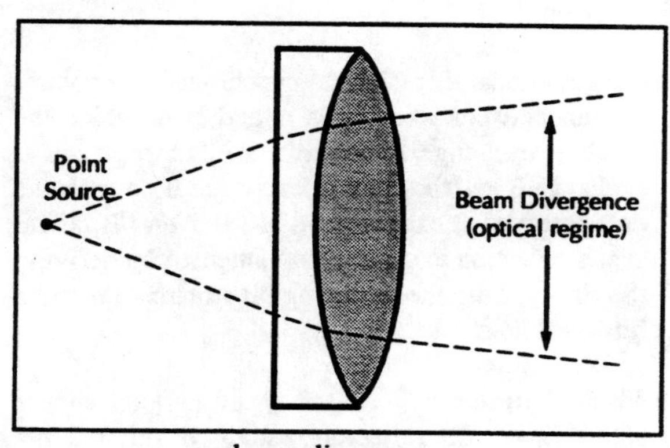

beam divergence

beamsplitter: A device for dividing an optical beam into two or more separate beams. *Note:* An example of a beamsplitter is a partially reflecting mirror.

beam steering: Changing the direction of the main lobe of a radiation pattern. *Note:* In radio systems, beam steering may be accomplished by switching antenna elements or by changing the relative phases of the rf signals driving the elements. In optical systems, beam steering may be accomplished by changing the refractive index of the medium through which the beam is transmitted or by the use of mirrors or lenses.

beamwidth: 1. In the radio regime, of an antenna pattern, the angle between the half-power (3-dB) points of the main lobe, when referenced to the peak effective radiated power of the main lobe. (188) *Note:* Beamwidth is usually expressed in degrees. It is usually expressed for the horizontal plane, but may also be expressed for the vertical plane. **2.** For the optical regime, *see* **beam divergence.**

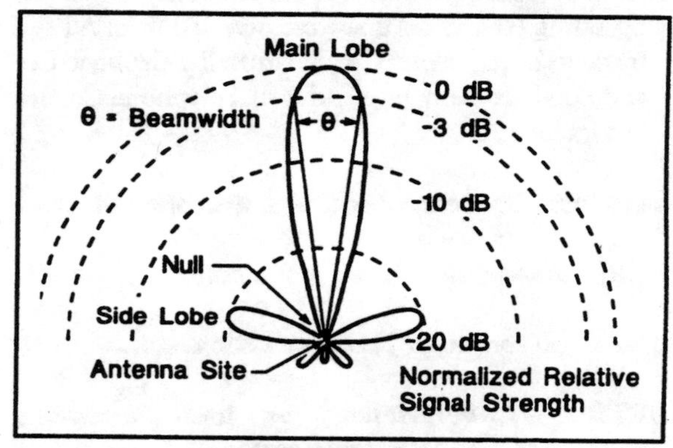

beamwidth

bearer channel: *See* **B channel.**

bearer service: A telecommunications service that allows transmission of user-information signals between user-network interfaces. *See* **B channel, service access.**

beating: *See* **heterodyne.**

beeping: *Synonym* **paging, radio paging.**

B8ZS: *Abbreviation for* **bipolar with eight-zero substitution.** A T-carrier line code in which bipolar violations are deliberately inserted if user data contains a string of 8 or more consecutive zeros. *Note 1:* B8ZS is used to ensure a sufficient number of transitions to maintain system synchronization when the user data stream contains an insufficient number of "ones" to do so. *Note 2:* B8ZS is used in the European hierarchy at the T1 rate.

bel (B): A unit of measure of ratios of power levels, *i.e.*, relative power levels. *Note 1:* The number of bels for a given ratio of power levels is calculated by taking the logarithm, to the base 10, of the ratio. Mathematically, the number of bels is calculated as $B = \log_{10}(P_1/P_2)$ where P_1 and P_2 are power levels. *Note 2:* The dB, equal to 0.1 B, is a more commonly used unit.

bell (BEL) character: A transmission control character that is used when there is a need to call for user or operator attention in a communications system, and that usually activates an audio or visual alarm or other attention-getting device.

Bell Operating Company (BOC): Any of the 22 operating companies that were divested from AT&T by court order. *Note:* Cincinnati Bell Telephone Co. and Southern New England Bell Telephone Co. are not included.

bend loss: *See* **macrobend loss, microbend loss.**

BER: *Abbreviation for* **bit error ratio.**

BERT: *Acronym for* **bit error ratio tester.**

BETRS: *Abbreviation for* **basic exchange telecommunications radio service.**

between-the-lines entry: Unauthorized access to a momentarily inactive terminal, of a legitimate user, assigned to a communications channel.

BEX: *Abbreviation for* **broadband exchange.**

bias: **1.** A systematic deviation of a value from a reference value. (188) **2.** The amount by which the average of a set of values departs from a reference value. (188) **3.** Electrical, mechanical, magnetic, or other force (field) applied to a device to establish a reference level to operate the device. (188) **4.** In telegraph signaling systems, the development of a positive or negative dc voltage at a point on a line that should remain at a specified reference level, such as zero. *Note:* A bias may be applied or produced by (i) the electrical characteristics of the line, (ii) the terminal equipment, and (iii) the signaling scheme. (188)

bias distortion: **1.** Signal distortion resulting from a shift in the bias. **2.** In binary signaling, distortion of the signal in which all the significant intervals have uniformly longer or shorter durations than their theoretical durations. (188) *Note:* Bias distortion is expressed in percent of the system-specified unit interval.

biconical antenna: An antenna consisting of two conical conductors, having a common axis and vertex, and extending in opposite directions. (188) *Note 1:* In a biconical antenna, excitation is applied at the common vertex. *Note 2:* If one of the cones is reduced to a plane, the antenna is called a discone.

bidirectional asymmetry: In data transmission, the condition that exists when information flow characteristics are different in each direction.

bidirectional coupler: *See* **directional coupler.**

bidirectional symmetry: The condition that exists when information flow characteristics are the same in each direction.

BIH: *French abbreviation for* **International Time Bureau.** *See* **International Atomic Time.**

bilateral control: *Synonym* **bilateral synchronization.**

bilateral synchronization: A synchronization control system between exchanges A and B in which the clock at exchange A controls the data received at exchange B and the clock at exchange B controls the data received at exchange A. (188) *Note:* Bilateral synchronization is usually implemented by deriving the timing from the incoming bit stream. *Synonym* **bilateral control.**

billboard antenna: An array of parallel dipole antennas with flat reflectors, usually positioned in a line or plane. *Note 1:* The spacing and dimensions of the dipoles depend on the wavelength. *Note 2:* The main lobe of a fixed billboard antenna may, within limits, be steered by appropriate phasing of the respective signals to individual elements of the array. *Synonym* **broadside antenna.**

binary: **1.** Pertaining to a selection, choice, or condition that has two possible different values or states. **2.** Pertaining to a fixed radix numeration system that has a radix of 2.

binary code: A code, the elements of which can assume either one of two possible states. (188)

binary-coded decimal (BCD): Pertaining to the representation of a decimal digit by a unique arrangement of no fewer than four binary digits. (188)

binary-coded decimal code: *Synonym* **binary-coded decimal notation.**

binary-coded decimal interchange code: *See* **binary-coded decimal notation.**

binary-coded decimal (BCD) notation: A binary notation in which each of the decimal digits is represented by a binary numeral. *Synonyms* **binary-coded decimal code, binary-coded decimal representation.**

binary-coded decimal representation: *Synonym* **binary-coded decimal notation.**

binary digit (bit): *See* **bit.**

binary element: A constituent element of data that takes either of two values or states. *Note: "Binary element"* should not be confused with *"binary digit."*.

binary exponential backoff: *See* **truncated binary exponential backoff.**

binary modulation: The process of varying a parameter of a carrier as a function of two finite, discrete states. (188)

binary notation: **1.** Any notation that uses two different characters, usually the digits 0 and 1. *Note:* Data encoded in binary notation need not be in the form of a pure binary numeration system; *e.g.*, they may be represented by a Gray code. **2.** A scheme for representing numbers, which scheme is characterized by the arrangements of digits in sequence, with the understanding that successive digits are interpreted as coefficients of successive powers of base 2. (188)

binary number: A number that is expressed in binary notation and is usually characterized by the arrangement of bits in sequence, with the understanding that successive bits are interpreted as coefficients of successive powers of the base 2. (188)

binary synchronous (bi-sync) communication: A character-oriented, data-link-layer protocol. *Note:* The bi-sync protocol is being phased out of most computer communication networks in favor of bit-oriented protocols such as SDLC, HDLC, and ADCCP.

binding: In computer, communications, and automatic data processing systems, assigning a value or referent to an identifier. *Note:* Examples of binding include assigning a value to a parameter, assigning an absolute address to a virtual or relative address, and assigning a device identifier to a symbolic address or label.

biphase modulation: *Synonym* **phase-shift keying.**

bipolar signal: A signal that may assume either of two polarities, neither of which is zero. (188) *Note 1:* A bipolar signal may have a two-state non-return-to-zero (NRZ) or a three-state return-to-zero (RZ) binary coding scheme. *Note 2:* A bipolar signal is usually symmetrical with respect to zero amplitude, *i.e.*, the absolute values of the positive and negative signal states are nominally equal.

birefringence: In a transparent material, anisotropism of the refractive index, which varies as a function of polarization as well as orientation with respect to the incident ray. (188) *Note 1:* The term *"birefringence"* means, literally, *"double refraction."* *Note 2:* All crystals except those of cubic lattice structure exhibit some degree of anisotropy with regard to their physical properties, including refractive index. Other materials, such as glasses or plastics, become birefringent when subjected to mechanical strain. *Note 3:* Birefringent materials, including crystals, have the ability to refract an unpolarized incident ray into two separate, orthogonally polarized rays, which in the general case take different paths, depending on orientation of the material with respect to the incident ray. The refracted rays are referred to as the "ordinary," or "O" ray, which obeys Snell's Law, and the "extraordinary," or "E" ray, which does not. [After FAA] *Synonym* **double refraction.**

birefringent medium: *See* **birefringence.**

B-ISDN: *Abbreviation for* **broadband ISDN.**

bistable: Pertaining to a device capable of assuming either one of two stable states.

bistable multivibrator: *Synonym* **flip-flop.**

bistable trigger circuit: *Synonym* **flip-flop.**

bi-sync: *Abbreviation for* **binary synchronous (communication).**

bit: *Abbreviation for* **binary digit. 1.** A character used to represent one of the two digits in the numeration system with a base of two, and only two, possible states of a physical entity or system. **2.** In binary notation either of the characters 0 or 1. (188) **3.** A unit of information equal to one binary decision or the designation of one of two possible and equally likely states of anything used to store or convey information. (188)

bit-by-bit asynchronous operation: In data transmission, an operation in which manual, semiautomatic, or automatic shifts in the data modulation rate are accomplished by gating or slewing the clock modulation rate. (188) *Note:* For example, bit-by-bit asynchronous operation may be at 50 b/s one moment and at 1200 b/s the next moment.

bit configuration: The sequence of bits used to encode a character.

bit-count integrity (BCI): 1. In message communications, the preservation of the exact number of bits that are in the original message. **2.** In connection-oriented services, preservation of the number of bits per unit time. (188) *Note:* Bit-count integrity is not the same as bit integrity, which requires that the delivered bits correspond exactly with the original bits.

bit density: The number of bits recorded per unit length, area, or volume. *Note:* Bit density is the reciprocal of bit pitch. *Synonym* **recording density.**

biternary transmission: Digital transmission in which two binary pulse trains are combined for transmission over a channel in which the available bandwidth is sufficient for transmission of only one of the two pulse trains at a time if they remain in binary form.

bit error rate: *Deprecated term. See* **bit error ratio.**

bit error ratio (BER): The number of erroneous bits divided by the total number of bits transmitted, received, or processed over some stipulated period. (188) *Note 1:* Examples of bit error ratio are (a) transmission BER, *i.e.,* the number of erroneous bits received divided by the total number of bits transmitted; and (b) information BER, *i.e.,* the number of erroneous decoded (corrected) bits divided by the total number of decoded (corrected) bits. *Note 2:* The BER is usually expressed as a coefficient and a power of 10; for example, 2.5 erroneous bits out of 100,000 bits transmitted would be 2.5 out of 10^5 or 2.5×10^{-5}.

bit error ratio tester (BERT): A testing device that compares a received data pattern with a known transmitted pattern to determine the level of transmission quality.

bit interval: *See* **bit, character interval, unit interval.**

bit inversion: 1. The changing of the state of a bit to the opposite state. (188) **2.** The changing of the state that represents a given bit, *i.e.,* a 0 or a 1, to the opposite state. (188) *Note:* For example, if a 1 is represented by a given polarity or phase at one stage in a circuit, the 1 is represented by the opposite polarity or phase at the next stage.

bit pairing: The practice of establishing, within a code set, a number of subsets that have an identical bit representation except for the state of a specified bit. (188) *Note:* An example of bit pairing occurs in the International Alphabet No. 5 and the American Standard Code for Information Interchange (ASCII), where the upper case letters are related to their respective lower case letters by the state of bit six.

bit position: A character position in a word in a binary notation.

bit rate (BR): In a bit stream, the number of bits occurring per unit time, usually expressed in bits per second. (188) *Note:* For n-ary operation, the bit rate is equal to $\log_2 n$ times the rate (in bauds), where n is the number of significant conditions in the signal.

bit-rate•distance product: *See* **bandwidth•distance product.**

bit robbing: In digital carrier systems, the practice or technique of preempting, at regular intervals and for the purpose of transmitting signaling information, one digit time slot that (a) is associated with the given user channel for which signaling is required, and (b) is used primarily for transporting encoded speech via that channel. *Note 1:* Bit robbing is an option in networks compatible with T-carrier, *e.g.,* an ISDN. *Note 2:* In conventional T-carrier systems, bit robbing uses, in every sixth frame, the time slot associated with the least significant bit. *Synonym* **speech digit signaling.**

bit-sequence independence: A characteristic of some digital data transmission systems that impose no restrictions on, or modification of, the transmitted bit sequence. *Note:* Bit-sequence-independent protocols are in contrast to protocols that reserve certain bit sequences for special meanings, such as the flag sequence, 01111110, for HDLC, SDLC, and ADCCP protocols.

bit slip: In digital transmission, the loss of a bit or bits, caused by variations in the respective clock rates of the transmitting and receiving devices. *Note:* One cause of bit slippage is overflow of a receive buffer that occurs when the transmitter's clock rate exceeds that of the receiver. This causes one or more bits to be dropped for lack of storage capacity.

bits per inch (b/in): A unit used to express the linear bit density of data in storage. *Note:* The abbreviation "*bpi*" is not in accordance with international standards, and is therefore deprecated.

bits per second (b/s): A unit used to express the number of bits passing a designated point per second. (188) *Note 1:* For example, for two-condition serial transmission in a single channel in which each significant condition represents a bit, *i.e.,* a 0 or a 1, the bit rate in bits per second and the baud have the same numerical value only if each bit occurs in a unit interval. In this case, the data signaling rate in bits per second is 1/T, where T is the unit interval. *Note 2:* The abbreviation *"bps"* is not in accordance with international standards, and is therefore deprecated.

bit-stepped: Pertaining to the control of digital equipment in which operations are performed one step at a time at the applicable bit rate. (188)

bit-stream transmission: **1.** In bit-oriented systems, the transmission of bit strings. **2.** In character-oriented systems, the transmission of bit streams that represent characters. *Note:* In bit-stream transmission, the bits usually occur at fixed time intervals, start and stop signals are not used, and the bit patterns follow each other in sequence without interruption.

bit string: A sequence of bits. *Note:* In a bit stream, individual bit strings may be separated by data delimiters.

bit stuffing: The insertion of noninformation bits into data. *Note 1:* Stuffed bits should not be confused with overhead bits. *Note 2:* In data transmission, bit stuffing is used for various purposes, such as for synchronizing bit streams that do not necessarily have the same or rationally related bit rates, or to fill buffers or frames. The location of the stuffing bits is communicated to the receiving end of the data link, where these extra bits are removed to return the bit streams to their original bit rates or form. Bit stuffing may be used to synchronize several channels before multiplexing or to rate-match two single channels to each other. (188) *Synonym* **positive justification.**

bit stuffing rate: *See* **nominal bit stuffing rate.**

bit synchronization: Synchronization in which the decision instant is brought into alignment with the received bit, *i.e.,* the basic signaling element. (188)

bit synchronous operation: Operation in which data circuit terminating equipment (DCE), data terminal equipment (DTE), and transmitting circuits are all operated in bit synchronism with a clock. *Note 1:* In bit synchronous operation, clock timing is usually delivered at twice the modulation rate, and one bit is transmitted or received during each clock cycle. *Note 2:* Bit synchronous operation is sometimes erroneously referred to as digital synchronization. (188)

BIU: *Abbreviation for* **bus interface unit.** *See* **network interface device.**

BLACK: [A] designation applied to telecommunications and automated information systems, and to associated areas, circuits, components, and equipment, in which only unclassified signals are processed. *Note:* Encrypted signals are unclassified. [NIS] (188)

blackbody: A totally absorbing body that does not reflect radiation. *Note:* In thermal equilibrium, a blackbody absorbs and radiates at the same rate; the radiation will just equal absorption when thermal equilibrium is maintained.

black facsimile transmission: **1.** In facsimile systems using amplitude modulation, that form of transmission in which the maximum transmitted power corresponds to the maximum density of the subject. (188) **2.** In facsimile systems using frequency modulation, that form of transmission in which the lowest transmitted frequency corresponds to the maximum density of the subject. (188)

black noise: Noise that has a frequency spectrum of predominately zero power level over all frequencies except for a few narrow bands or spikes. *Note:* An example of black noise in a facsimile transmission system is the spectrum that might be obtained when scanning a black area in which there are a few random white spots. Thus, in the time domain, a few random pulses occur while scanning.

black recording: **1.** In facsimile systems using amplitude modulation, recording in which the maximum received power corresponds to the maximum density of the record medium. (188) **2.** In a facsimile system using frequency modulation, recording in which the lowest received frequency corresponds to the maximum density of the record medium. (188)

black signal: In facsimile, the signal resulting from scanning a maximum-density area of the object. (188)

BLACK signal: A signal that represents only unclassified or encrypted information, usually in cryptographic systems. (188)

blanketing: The interference that is caused by the presence of an AM broadcast signal of one volt per meter (V/m) or greater strengths in the area adjacent to the antenna of the transmitting station. The 1 V/m contour is referred to as the blanket contour and the area within this contour is referred to as the *"blanket area."* [47CFR]

blanketing area: In the vicinity of a transmitting antenna, the area in which the signal from that antenna interferes with the reception of other signals. *Note:* The blanketing area around a given transmitting antenna depends on the selectivity and sensitivity of the receiver, and on the respective levels of the other signals in question.

blanking: In graphic display, the suppression of the display of one or more display elements or display segments.

blind transmission: Transmission without obtaining a receipt, *i.e.,* acknowledgement of reception, from the intended receiving station. *Note:* Blind transmission may occur or be necessary when security constraints, such as radio silence, are imposed, when technical difficulties with a sender's receiver or receiver's transmitter occur, or when lack of time precludes the delay caused by waiting for receipts.

blinking: In graphic display devices, an intentional periodic change in the intensity of one or more display elements or display segments.

block: **1.** A group of bits or digits that is transmitted as a unit and that may be encoded for error-control purposes. (188) **2.** A string of records, words, or characters, that for technical or logical purposes are treated as a unit. (188) *Note 1:* Blocks (a) are separated by interblock gaps, (b) are delimited by an end-of-block signal, and (c) may contain one or more records. *Note 2:* A block is usually subjected to some type of block processing, such as multidimensional parity checking, associated with it. **3.** In programming languages, a subdivision of a program that serves to group related statements, delimit routines, specify storage allocation, delineate the applicability of labels, or segment parts of the program for other purposes.

block character: *See* **end-of-transmission-block character.**

block check: In the processing or transmission of digital data, an error-control procedure that is used to determine whether a block of data is structured according to given rules. (188)

block check character (BCC): A character added to a transmission block to facilitate error detection. *Note:* In longitudinal redundancy checking and cyclic redundancy checking, block check characters are computed for, and added to, each message block transmitted. This block check character is compared with a second block check character computed by the receiver to determine whether the transmission is error free.

block code: An error detection and/or correction code in which the encoded block consists of N symbols, containing K information symbols ($K<N$) and $N-K$ redundant check symbols, such that most naturally occurring errors can be detected and/or corrected.

block diagram: A diagram of a system, a computer, or a device in which the principal parts are represented by suitably annotated geometrical figures to show both the basic functions of the parts and their functional relationships.

block distortion: In the received image in video systems, distortion characterized by the appearance of an underlying block encoding structure.

block efficiency: In a block, the ratio of the number of user information bits to the total number of bits. *Note:* For a given block scheme, block efficiency represents the maximum possible efficiency for a given block scheme transmitted over a perfect transmission link.

block-error probability: The expected block-error ratio. (188)

block-error ratio: The ratio of the number of incorrectly received blocks to the total number of blocks transferred. (188) *Note:* The block-error ratio is calculated using empirical measurements. Multiple block-error ratios may be used to predict block-error probability.

blocking: 1. The formatting of data into blocks for purposes of transmission, storage, checking, or other functions. **2.** Denying access to, or use of, a facility, system, or component. **3.** The failure of a telecommunications network to meet a user service demand, because of the lack of an available communications path.

blocking criterion: In telephone traffic engineering, a criterion that specifies the maximum number of calls or service demands that fail to receive immediate service. *Note:* The blocking criterion is usually expressed in probabilistic notation, such as P.001.

blocking factor: The number of records in a block. *Note:* The blocking factor is calculated by dividing the block length by the length of each record contained in the block. If the records are not of the same length, the average record length may be used to compute the blocking factor. *Synonym* **grouping factor.**

blocking formula: A specific probability distribution function intended to model calling patterns of users who fail to find available facilities. *Note:* There are several blocking formulas. The applicability of each to a given situation depends on its underlying assumptions regarding caller behavior.

blocking network: In telecommunications, a network that has fewer transmission paths than would be required if all users were to communicate simultaneously. *Note:* Blocking networks are used because not all users require service simultaneously. Certain statistical distributions apply to the patterns of user demand.

block length: The number of data units, such as bits, bytes, characters, or records, in a block.

block-loss probability: The ratio of the number of lost blocks to the total number of block transfer attempts during a specified period. (188)

block-misdelivery probability: The ratio of the number of misdelivered blocks to the total number of block transfer attempts during a specified period. (188)

block parity: The designation of one or more bits in a block as parity bits used to force the block into a selected parity, either odd or even. (188) *Note:* Block parity is used to assist in error detection or correction.

block transfer: The process, initiated by a single action, of transferring one or more blocks of data.

block transfer attempt: A coordinated sequence of user and telecommunication system activities undertaken to effect transfer of an individual block from a source user to a destination user. *Note:* A block transfer attempt begins when the first bit of the block crosses the functional interface between the source user and the telecommunication system. A block transfer attempt ends either in successful block transfer or in block transfer failure.

block transfer efficiency: The average ratio of user information bits to total bits in successfully transferred blocks.

block transfer failure: Failure to deliver a block successfully. *Note:* The principal block transfer failure outcomes are: lost block, misdelivered block, and added block.

block transfer rate: The number of successful block transfers during a performance measurement period divided by the duration of the period. (188)

block transfer time: The average value of the duration of a successful block transfer attempt. *Note:* A block transfer attempt is successful if (a) the transmitted block is delivered to the intended destination user within the maximum allowable performance period and (b) the contents of the delivered block are correct.

blue noise: In a spectrum of frequencies, a region in which the spectral density, *i.e.*, power per hertz, is proportional to the frequency.

blue-screening: *See* **chroma keying.**

blurring: In video systems, a global distortion, characterized by reduced sharpness of edges and limited spatial detail.

BOC: *Abbreviation for* **Bell Operating Company.**

Boltzmann's constant (k): The number that relates the average energy of a molecule to its absolute temperature. *Note:* Boltzmann's constant is approximately 1.38×10^{-23} J/K (joules/kelvin).

bond: An electrical connection that provides a low-resistance path between two conducting surfaces. (188)

bonding: **1.** In electrical engineering, the process of connecting together metal parts so that they make low resistance electrical contact for direct current and lower frequency alternating currents. [JP1] **2.** The process of establishing the required degree of electrical continuity between two or more conductive surfaces that are to be joined. (188)

Boolean function: **1.** A mathematical function that describes Boolean operations. **2.** A switching function in which the number of possible values of the function and each of its independent variables is two.

Boolean operation: **1.** Any operation in which each of the operands and the result take one of two values. *Note:* Typical states are "0 or 1," "on or off," "open or closed," or "present or absent." **2.** An operation that follows the rules of Boolean Algebra.

bootstrap: **1.** A technique or device designed to bring about a desired state by means of its own action. (188) **2.** That part of a computer program that may be used to establish another version of the computer program. **3.** The automatic procedure whereby the basic operating system of a processor is reloaded following a complete shutdown or loss of memory. **4.** A set of instructions that cause additional instructions to be loaded until the complete computer program is in storage. **5.** To initialize a system by means of a bootstrap.

boresight: **1.** The physical axis of a directional antenna. **2.** To align a directional antenna, using either an optical procedure or a fixed target at a known location.

Bose-Chaudhuri-Hochquenghem code: *See* **BCH code.**

bound mode: In an optical fiber, a mode that (a) has a field intensity that decays monotonically in the transverse direction everywhere external to the core and (b) does not lose power to radiation. *Note:* Except for single-mode fibers, the power in bound modes is predominantly contained in the core of the fiber. (188) [After 2196] *Synonyms* **guided mode, trapped mode.**

bound ray*:* *Synonym* **guided ray.**

bpi: *See* **bits per inch.**

braid: **1.** An essential part of many fiber-optic cable designs, consisting of a layer of woven yarn. *Note:* In the case of single-fiber loose-buffered or two-fiber "zip-cord" loose-buffered fiber-optic cables, the braid is situated between the buffer tube and jacket. In the case of cables having multiple buffer tubes, the braid is usually situated between the inner jacket and outer jacket. **2.** Loosely, an unwoven parallel bundle of yarn situated around the tight buffer of a single-fiber or two-fiber "zip-cord" fiber-optic cable. *Note 1:* The braid serves to add tensile strength to the cable. The braid may also be anchored to an optical connector or splice organizer assembly to secure the end of the cable. *Note 2:* The braid is often of an aramid yarn. [After FAA]

branch: **1.** In a computer program, a conditional jump or departure from the implicit or declared order in which instructions are being executed. (188) **2.** To select a branch, as in definition #1. **3.** A direct path joining two nodes of a network or graph. **4.** In a power distribution system, a circuit from a distribution device (power panel) of a lower power handling capability than that of the input circuits to the device. (188)

branching network: A network used for transmission or reception of signals over two or more channels. (188)

branching repeater: A repeater with two or more outputs for each input. (188) *Contrast with* **multiport repeater.**

breadboard: **1.** An assembly of circuits or parts used to prove the feasibility of a device, circuit, system, or principle with little or no regard to the final configuration or packaging of the parts. (188) **2.** To prepare a breadboard.

break out: To separate the individual fibers or buffer tubes of a fiber-optic cable for the purpose of splicing or installing optical connectors. [After FAA] *Synonyms* **fan out, furcate.**

break-out box: A testing device that permits a user to access individual leads of an interface cable, using jumper wires, in order to monitor, switch, or patch the electrical output of the cable.

breakout cable: A multifiber fiber-optic cable design in which individual fibers, usually tight-buffered, are surrounded by separate strength members and jackets, which are in turn enveloped by a common jacket. *Note 1:* The breakout cable facilitates easy installation of fiber-optic connectors. All that need be done to prepare the ends of the cable to receive connectors is to remove the outer jacket, exposing what are essentially individual single-fiber cables. *Note 2:* Because it tends to induce bends in the fibers, the breakout cable design usually results in slightly higher transmission losses, for a given fiber, than loose-buffer designs. [After FAA] *Synonym* **fanout cable.**

breakout kit: A kit of materials, composed of an outer jacket in which is contained a strength member consisting of a bundle of usually aramid yarn, which jacket and yarn may be slipped over a loose buffer tube containing a single fiber, to convert the buffer tube and fiber to a complete single-fiber cable to which a fiber-optic connector may be directly attached. *Note 1:* A heat-shrinkable plastic boot may also be used for cosmetic purposes, strain relief, and to seal the point where the individual cables so created, merge. *Note 2:* Use of a breakout kit

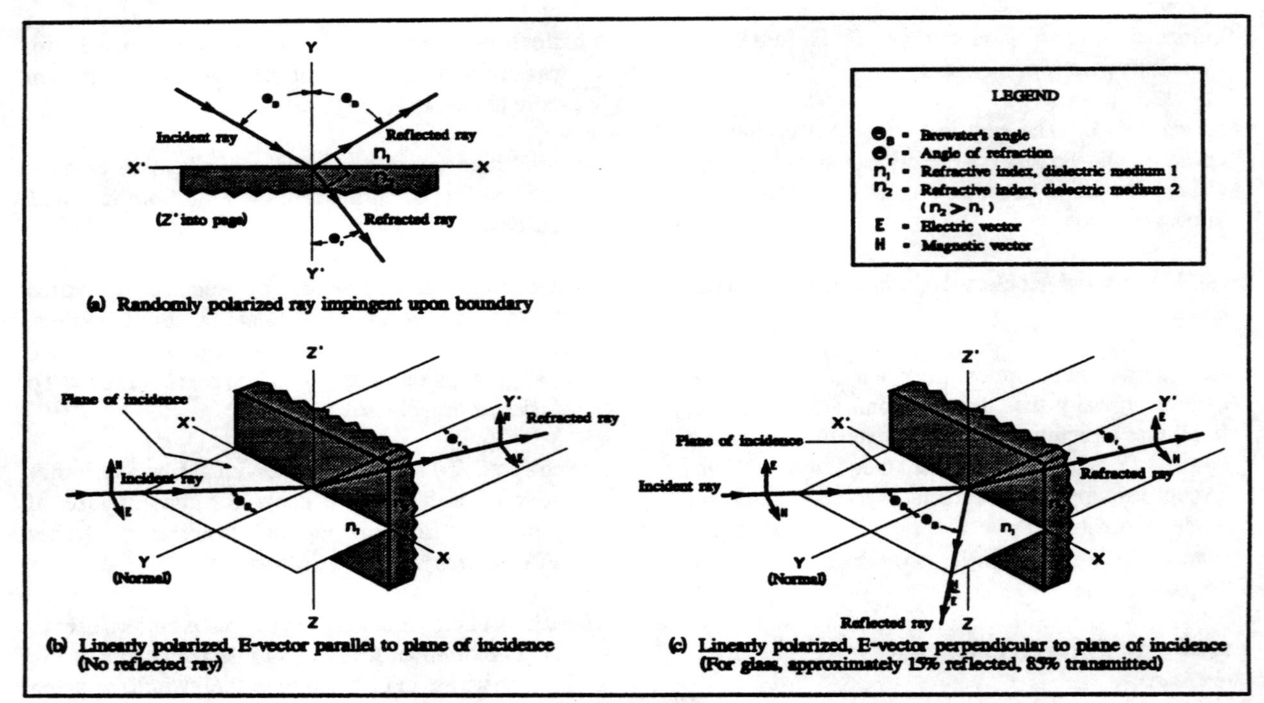

(a) Randomly polarized ray impingent upon boundary

(b) Linearly polarized, E-vector parallel to plane of incidence (No reflected ray)

(c) Linearly polarized, E-vector perpendicular to plane of incidence (For glass, approximately 15% reflected, 85% transmitted)

Brewster's angle

enables a fiber-optic cable containing multiple loose buffer tubes to receive connectors without the splicing of pigtails. [After FAA]

Brewster's angle: For a plane electromagnetic wavefront incident on a plane boundary between two dielectric media having different refractive indices, the angle of incidence at which transmittance from one medium to the other is unity when the wavefront is linearly polarized with its electric vector parallel to the plane of incidence. *Note 1:* Brewster's angle θ_B, is given by

$$\theta_B = \tan^{-1}\left(\frac{n_2}{n_1}\right) = \tan^{-1}\sqrt{\frac{\epsilon_2}{\epsilon_1}} ,$$

where n_1 and n_2 are the refractive indices of the respective media, and ϵ_1 and ϵ_2, their respective electric permittivities. *Note 2:* For a randomly polarized ray incident at Brewster's angle, the reflected and refracted rays are at 90° with respect to one another.

Brewster's law: *See* **Brewster's angle.**

BRI: *Abbreviation for* **basic rate interface.**

brick: A colloquial name for a hand-held radiotelephone unit.

bridge: **1.** In communications networks, a device that (a) links or routes signals from one ring or bus to another or from one network to another, (b) may extend the distance span and capacity of a single LAN system, (c) performs no modification to packets or messages, (d) operates at the data-link layer of the OSI—Reference Model (Layer 2), (e) reads packets, and (f) passes only those with addresses on the same segment of the network as the originating user. (188) **2.** A functional unit that interconnects two local area networks that use the same logical link control procedure, but may use different medium access control procedures. **3.** A balanced electrical network, *e.g.*, a Wheatstone bridge. *Note:* A bridge may be used for electrical measurements, especially resistances or impedances. **4.** *See* **hybrid coil.**

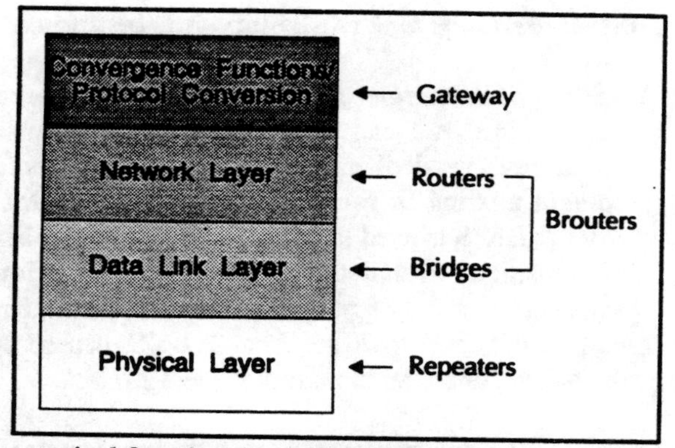

typical functional associations in an OSI network

bridged ringing: The part of a signaling system in which ringers associated with a particular line are connected across that line.

bridge lifter: A device that electrically or physically removes bridged telephone pairs. (188) *Note:* Relays, saturable inductors, and semiconductors are used as bridge lifters.

bridge-to-bridge station: A ship station operating in the port operations service in which messages are restricted to navigational communications and which is capable of operation from the ship's navigational bridge or, in the case of a dredge, from its main control station, operating on a frequency or frequencies in the 156-162 MHz band. [NTIA]

bridge transformer: *Synonym* **hybrid coil.**

bridging connection: A parallel connection used to extract some of the signal energy from a circuit, usually with negligible effect on the normal operation of the circuit. (188)

bridging loss: At a given frequency, the loss that results when an impedance is connected across a transmission line. (188) *Note:* Bridging loss is expressed as the ratio, in dB, of the signal power delivered, prior to bridging, to a given point in a system downstream from the bridging point, to the signal power delivered to the given point after bridging.

brightness: An attribute of visual perception in which a source appears to emit a given amount of light. *Note 1: "Brightness"* should be used only for nonquantitative references to physiological sensations and perceptions of light. *Note 2: "Brightness"* was formerly used as a synonym for the photometric term *"luminance"* and (incorrectly) for the radiometric term *"radiance."*

Brillouin diagram: *See* **Brillouin scattering.**

Brillouin scattering: In a physical medium, scattering of lightwaves, caused by thermally driven density fluctuations. *Note:* Brillouin scattering may cause frequency shifts of several gigahertz at room temperature. [From Weik '89]

broadband: *Synonym* **wideband.**

broadband exchange (BEX): A communications switch capable of interconnecting channels having bandwidths greater than voice bandwidth.

broadband ISDN (B-ISDN): An Integrated Services Digital Network (ISDN) offering broadband capabilities. *Note 1:* B-ISDN is a CCITT-proposed service that may (a) include interfaces operating at data rates from 150 to 600 Mb/s, (b) use asynchronous transfer mode (ATM) to carry all services over a single, integrated, high-speed packet-switched network, (c) have LAN interconnection capability, (d) provide access to a remote, shared disk server, (e) provide voice/video/data teleconferencing, (f) provide transport for programming services, such as cable TV, (g) provide single-user controlled access to remote video sources, (h) handle voice/video telephone calls, and (i) access shop-at-home and other information services. *Note 2:* Techniques used in the B-ISDN include code conversion, information compression, multipoint connections, and multiple-connection calls. Current proposals use a service-independent call structure that allows flexible arrangement and modular control of access and transport edges. The service components of a connection can provide each user with independent control of access features and can serve as the basis of a simplified control structure for multipoint and multiconnection calls. Such a network might be expected to offer a variety of ancillary information processing functions.

broadband system: *See* **wideband.**

broadcasting-satellite service: A radiocommunication service in which signals transmitted or retransmitted by space stations are intended for direct reception by the general public. In the broadcasting-satellite service, the term *"direct reception"* shall encompass both individual reception and community reception. [NTIA] [RR]

broadcasting satellite space station: A space station in the broadcasting-satellite service (sound broadcasting). [NTIA]

broadcasting service: A radiocommunication service in which the transmissions are intended for direct reception by the general public. This service may include sound transmissions, television transmissions or other types of transmissions. [NTIA] [RR]

broadcasting station: A station in the broadcasting service. [NTIA] [RR]

broadcast operation: The transmission of signals that may be simultaneously received by stations that usually make no acknowledgement. (188)

broadside antenna: *Synonym* **billboard antenna.**

brouter: A combined bridge and router that operates without protocol restrictions, routes data using a protocol it supports, and bridges data it cannot route.

browser: Any computer software program for reading hypertext. *Note 1:* Browsers are usually associated with the Internet and the World Wide Web (WWW). *Note 2:* A browser may be able to access information in many formats, and through different services including HTTP, FTP, Gopher, and Archie.

browsing: [The] act of searching through automated information system storage to locate or acquire information without necessarily knowing of the existence or the format of the information being sought. [NIS]

b/s: *Abbreviation for* **bits per second.**

BSA: *Abbreviation for* **basic serving arrangement.**

BSE: *Abbreviation for* **basic service element.**

BSI: *Abbreviation for* **British Standards Institution.**

B6ZS: *Abbreviation for* **bipolar with six-zero substitution.** A T-carrier line code in which bipolar violations are deliberately inserted if user data contain a string of 6 or more consecutive zeros. *Note 1:* B6ZS is used to ensure a sufficient number of transitions to maintain system synchronization when the user data stream contains an insufficient number of "ones" to do so. *Note 2:* B6ZS is used in the North American hierarchy at the T2 rate.

B3ZS: *Abbreviation for* **bipolar with three-zero substitution.** A T-carrier line code in which bipolar violations are deliberately inserted if user data contain a string of 3 or more consecutive zeros. *Note 1:* B3ZS is used to ensure a sufficient number of transitions to maintain system synchronization when the user data stream contains an insufficient number of "ones" to do so. *Note 2:* B3ZS is used in the North American hierarchy at the T3 rate.

budgeting: *Synonym* **proration (def. #1).**

buffer: **1.** A routine or storage medium used to compensate for a difference in rate of flow of data, or time of occurrence of events, when transferring data from one device to another. (188) *Note:* Buffers are used for many purposes, such as (a) interconnecting two digital circuits operating at different rates, (b) holding data for use at a later time, (c) allowing timing corrections to be made on a data stream, (d) collecting binary data bits into groups that can then be operated on as a unit, (e) delaying the transit time of a signal in order to allow other operations to occur. **2.** To use a buffer or buffers. (188) **3.** An isolating circuit, often an amplifier, used to minimize the influence of a driven circuit on the driving circuit. *Synonym* **buffer amplifier.** (188) **4.** In a fiber optic communication cable, one type of component used to encapsulate one or more optical fibers for the purpose of providing such functions as mechanical isolation, protection from physical damage and fiber identification. *Note:* The buffer may take the form of a miniature conduit, contained within the cable and called a loose buffer, or loose buffer tube, in which one or more fibers may be enclosed, often with a lubricating gel. A tight buffer consists of a polymer coating in intimate contact with the primary coating applied to the fiber during manufacture. (188)

buffer amplifier: *Synonym* **buffer (def. #3).**

bug: **1.** A concealed microphone or listening device or other audiosurveillance device. [JP1] (188) **2.** A mistake in a computer program. **3.** To install means for audiosurveillance. [JP1] **4.** A semiautomatic telegraph key. **5.** A mistake or malfunction. (188)

building out: The process of adding a combination of inductance, capacitance, and resistance to a cable pair so that its electrical length may be increased by a desired amount to control impedance and loss characteristics. (188) *Synonym* **line buildout.**

bulk encryption: Simultaneous encryption of all channels of a multichannel telecommunications trunk. [NIS] *Note:* A single encryption device can be used to encrypt the output signal from a multiplexer. (188)

bulletin board: A form of electronic messaging in which addressed messages or files are entered by users into a computer or network of computers. *Note:* Other users may obtain, at their convenience and request, messages or files available to them.

bunched frame-alignment signal: A frame-alignment signal in which the signal elements occupy consecutive digit positions.

bundle: A group of optical fibers or electrical conductors, such as wires and coaxial cables, usually in a single jacket. (188) *Note:* Multiple bundles of optical fibers or electrical conductors may be placed in the same cable. [After 2196]

buried cable: *See* **direct-buried cable.**

burst: **1.** In data communications, a sequence of signals, noise, or interference counted as a unit in accordance with some specific criterion or measure. (188) **2.** To separate continuous-form or multipart paper into discrete sheets.

burst isochronous: *Deprecated synonym for* **isochronous burst transmission.**

burst switching: In a packet-switched network, a switching capability in which each network switch extracts routing instructions from an incoming packet header to establish and maintain the appropriate switch connection for the duration of the packet, following which the connection is automatically released. *Note:* In concept, burst switching is similar to connectionless mode transmission, but it differs from the latter in that burst switching implies an intent to establish the switch connection in near real time so that only minimum buffering is required at the node switch.

burst transmission: **1.** Transmission that combines a very high data signaling rate with very short transmission times. (188) **2.** Operation of a data network in which data transmission is interrupted at intervals. *Note:* Burst transmission enables communications between data terminal equipment (DTEs) and a data network operating at dissimilar data signaling rates. *Synonym* **data burst.**

bus: One or more conductors or optical fibers that serve as a common connection for a group of related devices. (188)

bus interface unit (BIU): *See* **network interface device.**

bus network: *See* **network topology.**

bus topology: *See* **network topology.**

busy back: *Deprecated term. See* **busy signal.**

busy hour: In a communications system, the sliding 60-minute period during which occurs the maximum total traffic load in a given 24-hour period. (188) *Note 1:* The busy hour is determined by fitting a horizontal line segment equivalent to one hour under the traffic load curve about the peak load point. *Note 2:* If the service time interval is less than 60 minutes, the busy hour is the 60-minute interval that contains the service timer interval. *Note 3:* In cases where more than one busy hour occurs in a 24-hour period, *i.e.*, when saturation occurs, the busy hour or hours most applicable to the particular situation are used. *Synonym* **peak busy hour.**

busy season: During a 1-year cycle, the period of 3 consecutive months having the highest busy hour traffic.

busy signal: **1.** In telephony, an audible or visual signal that indicates that no transmission path to the called number is available. *Synonym* **busy tone.** **2.** In telephony, an audible or visual signal that indicates that the called number is occupied or otherwise unavailable. (188) *Synonym* **reorder tone.**

busy test: In telephony, a test made to determine whether certain facilities, such as a subscriber line or a central office trunk, are available for use.

busy tone: *Synonym* **busy signal.**

busy verification: In a public switched telephone network, a network-provided service feature that permits an attendant to verify the busy or idle state of station lines and to break into the conversation. *Note:* A 440-Hz tone is applied to the line for 2 seconds, followed by a 0.5-second burst every 10 seconds, to alert both parties that the attendant is connected to the circuit.

BW: *Abbreviation for* **bandwidth.**

bypass: **1.** The use of any telecommunications facilities or services that circumvents those of the local exchange common carrier. *Note:* Bypass facilities or services may be either customer-provided or vendor-supplied. **2.** An alternate circuit that is routed around equipment or system component. (188) *Note:* Bypasses are often used to allow system operation to continue when the bypassed equipment or a system component is inoperable or unavailable.

byte (B): A sequence of adjacent bits (usually 8) considered as a unit. (188) *Note:* In pre-1970 literature, *"byte"* referred to a variable-length bit string. Since that time the usage has changed so that now it almost always refers to an 8-bit string. This usage predominates in computer and data transmission literature; when so used, the term is synonymous with *"octet."*

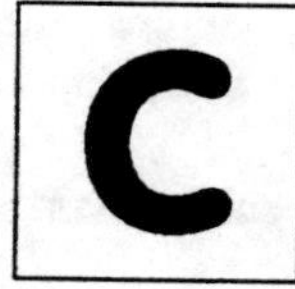

cable: **1.** An assembly of one or more insulated conductors, or optical fibers, or a combination of both, within an enveloping jacket. *Note:* A cable is constructed so that the conductors or fibers may be used singly or in groups. (188) **2.** A message sent by cable, or by any means of telegraphy.

cable assembly: A cable that is ready for installation in specific applications and usually terminated with connectors.

cable jacket: *See* **sheath.**

cable cutoff wavelength (λ_{cc}): For a cabled single-mode optical fiber under specified length, bend, and deployment conditions, the wavelength at which the fiber's second order mode is attenuated a measurable amount when compared to a multimode reference fiber or to a tightly bent single-mode fiber.

cable television relay service (CARS) station: A fixed or mobile station used for the transmission of television and related audio signals, signals of standard and FM broadcast stations, signals of instructional television fixed stations, and cablecasting from the point of reception to a terminal point from which the signals are distributed to the public. [47CFR]

cable TV (CATV): A television distribution method in which signals from distant stations are received, amplified, and then transmitted by (coaxial or fiber) cable or microwave links to users. *Note 1:* CATV originated in areas where good reception of direct broadcast TV was not possible. Now CATV also consists of a cable distribution system to large metropolitan areas in competition with direct broadcasting. *Note 2:* The abbreviation *CATV* originally meant *"community antenna television."* However, *CATV* is now usually understood to mean *cable TV.*

cache memory: A buffer, smaller and faster than main storage, used to hold a copy of instructions and data in main storage that are likely to be needed next by the processor and that have been obtained automatically from main storage.

call: **1.** In communications, any demand to set up a connection. **2.** A unit of traffic measurement. (188) **3.** The actions performed by a call originator. **4.** The operations required to establish, maintain, and release a connection. **5.** To use a connection between two stations. **6.** The action of bringing a computer program, a routine, or a subroutine into effect, usually by specifying the entry conditions and the entry point.

call abandoned: *See* **abandoned call.**

call accepted signal: A call control signal sent by the called terminal to indicate that it accepts the incoming call.

call associated signaling (CAS): Signaling required for supervision of a bearer service between two end points, including support for the functions of call origination, call delivery, and handover.

call attempt: In a telecommunications system, a demand by a user for a connection to another user. *Note:* In telephone traffic analysis, call attempts are counted during a specific time frame. The call-attempt count includes all completed, overflowed, abandoned, and lost calls.

call back: [The] procedure for identifying a remote AIS terminal, whereby the host system disconnects the caller and then dials the authorized telephone number of the remote terminal to reestablish the connection. [NIS]

call collision: **1.** The contention that occurs when a terminal and data circuit-terminating equipment (DCE) specify the same channel at the same time to transfer a call request and handle an incoming call. *Note:* When call collision occurs, the DCE proceeds with the call request and cancels the incoming call. **2.** The condition that occurs when a trunk or channel is seized at both ends simultaneously, thereby blocking a call. (188) *Deprecated synonym* **glare.**

call completion rate: The ratio of successfully completed calls to the total number of attempted calls. *Note:* This ratio is typically expressed as either a percentage or a decimal fraction.

call control character: One of a set of control characters used in call-control signaling. *Note:* The signals representing call control characters may be used under defined conditions on interchange circuits other than the originating circuit.

call control signal: A member of the set of network management signals used to establish, maintain, or release a call.

call delay: **1.** The delay that occurs when a call arrives at an automatic switching device and no channel or facility is immediately available to process the call. **2.** The time between the instant a system receives a call attempt and the instant of initiation of ringing at the call receiver end instrument.

call detail recording (CDR): A service feature in which call data on a specific telephone extension or group of subscribers are collected and recorded for cost-of-service accounting purposes.

call duration: **1.** The time between (a) the instant a connection, *i.e.,* off-hook condition at each end, is established between the call originator and the call receiver and (b) the instant the call originator or the call receiver terminates the call. **2.** In data transmission, the duration of the information transfer phase of an information transfer transaction.

called-line identification facility: A network-provided service feature in which the network notifies a calling terminal of the address to which the call has been connected.

called-line identification signal: A sequence of characters transmitted to the calling terminal to permit identification of the called line.

called party: *Synonym* **call receiver.**

called-party camp-on: A communication system service feature that enables the system to complete an access attempt in spite of issuance of a user blocking signal. *Note:* Systems that provide this feature monitor the busy user until the user blocking signal ends, and then proceed to complete the requested access. This feature permits holding an incoming call until the called party is free.

caller ID: A network service feature that permits the recipient of an incoming call to determine, even before answering, the number from which the incoming call is being placed.

caller identification: *See* **caller ID.**

call-failure signal: A signal sent in the backward direction indicating that a call cannot be completed because of a time-out, a fault, or a condition that does not correspond to any other particular signal.

call forwarding: A service feature, available in some switching systems, whereby calls can be rerouted automatically from one line, *i.e.,* station number, to another or to an attendant. *Note:* Call forwarding may be implemented in many forms.

call hold: A service feature in which a user may retain an existing call while accepting or originating another call using the same end instrument.

call identifier: A network utility that consists of a name assigned by the originating network for each established or partially established virtual call. *Note:* When a call identifier is used in conjunction with the calling data terminal equipment (DTE) address, the call identifier uniquely identifies the virtual call.

calling frequency: A radio frequency that a station uses to call another station.

calling-line identification facility: A network-provided service feature in which the network notifies a called terminal of the address from which the call has originated.

calling-line identification signal: A sequence of characters transmitted to the called terminal to permit identification of the calling line.

calling party: *Synonym* **call originator.**

calling-party camp-on: A service feature that enables the system to complete an access attempt in spite of temporary unavailability of system transmission or switching facilities required to establish the requested access. *Note:* Systems that provide calling party camp-on monitor the system facilities until the necessary facilities become available, and then proceed to complete the requested access. Such systems may or may not issue a system blocking signal to apprise the originating user of the access delay.

calling rate: The number of telephone calls originated during a specified time interval such as one hour.

calling sequence: A sequence of instructions together with any associated data necessary to perform a call.

calling signal: A call control signal transmitted over a circuit to indicate that a connection is desired.

call intensity: *Synonym* **traffic intensity.**

call management: **1.** In telegraphy, route selection, signaling, and circuit usage and availability for a call. **2.** In universal personal telecommunications, the ability of a user to inform the network how to handle incoming calls in accord with certain parameters, such as the call originator, the time of day, and the nature of the call. *Note:* Call management is accomplished by means of information in the user's service profile.

call-not-accepted signal: A call control signal sent by the called terminal to indicate that it does not accept the incoming call.

call originator: An entity, such as a person, equipment, or program that originates a call. (188) *Synonym* **calling party.**

call pickup: A service feature of some switching systems enabling a user, by dialing a predetermined code, to answer incoming calls that are directed to another user in a preselected call group.

call processing: **1.** The sequence of operations performed by a switching system from the acceptance of an incoming call through the final disposition of the call. **2.** The end-to-end sequence of operations performed by a network from the instant a call attempt is initiated until the instant the call release is completed. **3.** In data transmission, the operations required to complete all three phases of an information transfer transaction.

call progress signal: A call control signal transmitted by the called data circuit-terminating equipment (DCE) to the calling data terminal equipment (DTE) to report (a) the progress of a call by using a positive call progress signal or (b) the reason why a connection could not be established by using a negative call progress signal.

call progress tone: An audible signal returned by a network to a call originator to indicate the status of a call. *Note:* Examples of call progress tones include dial tones and busy signals.

call receiver: An entity, such as a person, equipment, or program to which a call is directed. *Synonym* **called party.**

call record: Recorded data pertaining to a single call. (188)

call release time: In communication systems, the time interval from initiation of a clearing signal by a terminal until the available-line condition appears on originating terminal equipment. (188)

call-request time: In the establishment of a connection or in the call setup, *i.e.*, placement of a call, the time from the initiation of a calling signal to the receipt of a proceed-to-select signal—such as a dial tone—by the call originator.

call restriction: A switching system service feature that prevents selected terminals from exercising one or more service features otherwise available from the switching system. (188)

calls-barred facility: A service feature that permits a terminal either to make outgoing calls or to receive incoming calls, but not both. (188)

call-second: A unit used to measure communications traffic. *Note 1:* A call-second is equivalent to 1 call 1 second long. (188) *Note 2:* One user making two 75-second calls is equivalent to two users each making one 75-second call. Each case produces 150 call-seconds of traffic. *Note 3:* The CCS, equivalent to 100 call-seconds, is often used. *Note 4:* 3600 call-seconds = 36 CCS = 1 call-hour. *Note 5:* 3600 call-seconds per hour = 36 CCS per hour = 1 call-hour per hour = 1 erlang = 1 traffic unit.

call-selection time: In the establishment of a connection or the placement of a call, the time from the receipt by the call originator of a proceed-to-select signal (dial tone), until all the selection signals have been transmitted (dialing has been completed).

call set-up time: **1.** The overall length of time required to establish a circuit-switched call between users. (188) **2.** For data communication, the overall length of time required to establish a circuit-switched call between terminals; *i.e.*, the time from the initiation of a call request to the beginning of the call message. *Note:* Call set-up time is the summation of: (a) call request time—the time from initiation of a calling signal to the delivery to the caller of a proceed-to-select signal; (b) selection time—the time from the delivery of the proceed-to-select signal until all the selection signals have been transmitted; and (c) post selection time—the time from the end of the transmission of the selection signals until the delivery of the call-connected signal to the originating terminal.

call sign: A station or address designator represented by a combination of characters or pronounceable words that is used to identify such entities as a communications facility, station, command, authority, activity, or unit.

call-sign allocation plan: The table of allocation of international call sign series contained in the current edition of the *International Telecommunication Union (ITU) Radio Regulations*. *Note:* In the table of allocation, the first two characters of each call sign (whether two letters or one number and one letter, in that order) identify the nationality of the station. In certain instances where the complete alphabetical block is allocated to a single nation, the first letter is sufficient for national identity. Individual assignments are made by appropriate national assignment authorities from the national allocation. [47CFR]

call spill-over: In common-channel signaling, the effect on a traffic circuit of the arrival at a switching center of an abnormally delayed call control signal relating to a previous call, while a subsequent call is being set up on the circuit. (188)

call splitting: A switching system service feature that allows a switch attendant to talk privately in either direction on an established call.

call tracing: A procedure that permits an entitled user to be informed about the routing of data for an established connection, identifying the entire route from the origin to the destination. *Note:* There are two types of call tracing. Permanent call tracing permits tracing of all calls. On-demand call tracing permits tracing, upon request, of a specific call, provided that the called party dials a designated code immediately after the call to be traced is disconnected, *i.e.*, before another call is received or placed.

call transfer: A switching system service feature that allows the calling or called user to instruct the local switching equipment or switch attendant to transfer an existing call to another terminal. *Note:* Call transfer may be available on a call-by-call basis or on a semipermanent basis.

call waiting: In telephony, a service feature that provides an indication to a terminal already engaged in an established call that one or more calls are awaiting connection.

CAMA: *Acronym for* **centralized automatic message accounting.**

camp-on: *See* **automatic callback, called-party camp-on, queue traffic.**

camp-on busy signal: **1.** A signal that informs a busy telephone user that another call originator is waiting for a connection. **2.** A teleprinter exchange facility signal that automatically causes a calling station to retry the call-receiver number after a given interval when the call-receiver teleprinter is occupied or the circuits are busy. *Synonym* **speed-up tone.**

camp-on-with-recall: A camp-on with the release of the call-originator terminal until the called-receiver terminal becomes free. *Note:* The call originator can thus establish other calls until the recall signal is obtained, rather than simply wait until the call-receiver line is available.

CAN: *Abbreviation for* **cancel character.**

Canadian Standards Association (CSA): An independent, nongovernment, not-for-profit association for the development, by consensus, of Canadian standards and product certification.

cancel character (CAN): **1.** A control character used by some conventions to indicate that the data with which it is associated are in error or are to be disregarded. **2.** An accuracy control character used to indicate that the data with which it is associated are in error, are to be disregarded, or cannot be represented on a particular device.

capacitive coupling: The transfer of energy from one circuit to another by means of the mutual capacitance between the circuits. (188) *Note 1:* The coupling may be deliberate or inadvertent. *Note 2:* Capacitive coupling favors transfer of the higher frequency components of a signal, whereas inductive coupling favors lower frequency components, and conductive coupling favors neither higher nor lower frequency components.

capacity: *See* **channel capacity, traffic capacity.**

capture effect: A phenomenon, associated with FM reception, in which only the stronger of two signals at or near the same frequency will be demodulated. (188) *Note 1:* The complete suppression of the weaker signal occurs at the receiver limiter, where it is treated as noise and rejected. *Note 2:* When both signals are nearly equal in strength, or are fading independently, the receiver may switch from one to the other. *Synonym* **FM capture effect.**

cardinal radial: Those eight radials at 0°, 45°, 90°, 135°, 180°, 225°, 270°, and 315° of azimuth with respect to true north. [47CFR] *Note:* The four radials at 0°, 90°, 180°, and 270° of azimuth with respect to true north are referred to as the cardinal points. The cardinal points are equivalent to true north, east, south, and west.

carrier: *Synonym* **common carrier.**

carrier (cxr): **1.** A wave suitable for modulation by an information-bearing signal. (188) **2.** An unmodulated emission. (188) *Note:* The carrier is usually a sinusoidal wave or a uniform or predictable series of pulses. *Synonym* **carrier wave.** **3.** Sometimes employed as a *synonym for* **carrier system.**

carrier dropout: A short-duration loss of carrier signal. (188)

carrier frequency: **1.** The nominal frequency of a carrier wave. (188) **2.** In frequency modulation, *synonym* **center frequency.**

carrier leak: The carrier remaining after carrier suppression in a suppressed carrier transmission system. (188) *Note:* Sometimes the residual carrier is used to provide the reference for an automatic frequency control system.

carrier level: The level of a carrier signal at a specified point in a communications system. *Note:* The carrier level is usually expressed in dB relative to a specified reference level. (188)

carrier multiplex: *See* **frequency-division multiplexing.**

carrier noise level: The noise level resulting from undesired variations of a carrier in the absence of any intended modulation. (188) *Synonym* **residual modulation.**

carrier power (of a radio transmitter): **1.** The average power supplied to the antenna transmission line by a transmitter during one radio frequency cycle taken under the condition of no modulation. [NTIA] [RR](188) *Note:* The concept does not apply to pulse modulation or frequency-shift keying. **2.** The average unmodulated power supplied to a transmission line.

carrier sense: In a local area network, an ongoing activity of a data station to detect whether another station is transmitting.

carrier sense multiple access (CSMA): A network control scheme in which a node verifies the absence of other traffic before transmitting.

carrier sense multiple access with collision avoidance (CSMA/CA): A network control protocol in which (a) a carrier sensing scheme is used, (b) a data station that intends to transmit sends a jam signal, (c) after waiting a sufficient time for all stations to receive the jam signal, the data station transmits a frame, and (d) while transmitting, if the data station detects a jam signal from another station, it stops transmitting for a random time and then tries again.

carrier sense multiple access with collision detection (CSMA/CD): A network control protocol in which (a) a carrier sensing scheme is used and (b) a transmitting data station that detects another signal while transmitting a frame, stops transmitting that frame, transmits a jam signal, and then waits for a random time interval before trying to send that frame again.

carrier shift: **1.** In the transmission of binary or teletypewriter signals, keying in which the carrier frequency is shifted in one direction for marking signals and in the opposite direction for spacing signals. (188) **2.** In amplitude modulation, a condition that results from imperfect modulation in which the positive and negative excursions of the modulating envelope are unequal in amplitude. *Note 1:* The carrier shift results in a change in carrier power. *Note 2:* The carrier shift may be a shift to a higher or to a lower frequency. (188)

carrier suppression: *See* **suppressed carrier transmission.**

carrier synchronization: In a radio receiver, the generation of a reference carrier with a phase closely matching that of a received signal.

carrier system: A multichannel telecommunications system in which a number of individual circuits (data, voice, or combination thereof) are multiplexed for transmission between nodes of a network. (188) *Note 1:* In carrier systems, many different forms of multiplexing may be used, such as time-division multiplexing and frequency-division multiplexing. *Note 2:* Multiple layers of multiplexing may ultimately be performed upon a given input signal; *i.e.*, the output resulting from one stage of modulation may in turn be modulated. *Note 3:* At a given node, specified channels, groups, supergroups, *etc.*, may be demultiplexed without demultiplexing the others. *Synonym (loosely)* **carrier.**

carrier-to-noise ratio (CNR): In radio receivers, the ratio of the level of the carrier to that of the noise in the intermediate frequency (IF) band before any nonlinear process, such as amplitude limitation and detection, takes place. (188) *Note:* The CNR is usually expressed in dB.

carrier-to-receiver noise density (*C/kT*): In satellite communications, the ratio of the received carrier power to the receiver noise power density. *Note 1:* The carrier-to-receiver noise density ratio is usually expressed in dB. *Note 2:* The carrier-to-receiver noise density is given by C/kT, where C is the received carrier power in watts, k is Boltzmann's constant in joules per kelvin, and T is the receiver system noise temperature in kelvins. *Note 3:* The receiver noise power density, kT, is the receiver noise power per hertz. (188)

carrier wave: *Synonym* **carrier (cxr) (def. #2).**

CARS: *Acronym for* **cable television relay service.**

Carson bandwidth rule: A rule defining the approximate bandwidth requirements of communications system components for a carrier signal that is frequency modulated by a continuous or broad spectrum of frequencies rather than a single frequency. *Note 1:* The Carson bandwidth rule is expressed by the relation $\text{CBR} = 2(\Delta f + f_m)$ where CBR is the bandwidth requirement, Δf is the carrier peak deviation frequency, and f_m is the highest modulating frequency. *Note 2:* The Carson bandwidth rule is often applied to transmitters, antennas, optical sources, receivers, photodetectors, and other communications system components.

CAS: *Abbreviation for* **centralized attendant services.**

CASE: *Acronym for* **computer-aided software engineering, computer-aided systems engineering.** Software used for the automated development of systems software, *i.e.*, computer code. *Note 1:* CASE functions include analysis, design, and programming. *Note 2:* CASE tools automate methods for designing, documenting, and producing structured computer code in the desired programming language.

case shift: **1.** In data equipment, the change from letters to other characters, or vice versa. (188) **2.** In typewriting or typesetting, the change from lower case letters to upper case letters, or vice versa.

CASE technology: Technology that makes use of computer assisted software engineering (CASE) to enhance the development of systems design and development.

Cassegrain antenna: An antenna in which the feed radiator is mounted at or near the surface of a concave main reflector and is aimed at a convex secondary reflector slightly inside the focus of the main reflector. *Note 1:* Energy from the feed unit illuminates the secondary reflector, which reflects it back to the main reflector, which then forms the desired forward beam. *Note 2:* The Cassegrain antenna design is adapted from optical telescope technology and allows the feed radiator to be more easily supported.

catastrophic degradation: The rapid reduction of the ability of a system, subsystem, component, equipment, or software to perform its intended function. *Note:* Catastrophic degradation usually results in total failure to perform any function.

Category 3: The ANSI/EIA/TIA-568 designation for 100-ohm unshielded twisted-pair cables and associated connecting hardware whose characteristics are specified for data transmission up to 16 Mb/s.

Category 4: The ANSI/EIA/TIA-568 designation for 100-ohm unshielded twisted-pair cables and associated connecting hardware whose characteristics are specified for data transmission up to 20 Mb/s.

Category 5: The ANSI/EIA/TIA-568 designation for 100-ohm unshielded twisted-pair cables and associated connecting hardware whose characteristics are specified for data transmission up to 100 Mb/s.

CATV: *Abbreviation for* **cable TV.**

cavity: A volume defined by conductor-dielectric or dielectric-dielectric reflective boundaries, or a combination of both, and having dimensions designed to produce specific interference effects (constructive or destructive) when excited by an electromagnetic wave.

C-band: *Colloquially,* a frequency band between 4 GHz and 6 GHz used in satellite communications. *Note 1:* For procurement purposes, the radio frequency band(s) must be specified using the upper and lower limits of the band, per 47 CFR 300. *Note 2:* Letter designators of radio frequency bands are imprecise, deprecated, and legally obsolete.

CCH: *Abbreviation for* **connections per circuit hour.**

CCIR: *Abbreviation for* International Radio Consultative Committee; a predecessor organization of the ITU-T.

CCIS: *Abbreviation for* **common-channel interoffice signaling.**

CCITT: *Abbreviation for* International Telegraph and Telephone Consultative Committee; a predecessor organization of the ITU-T.

CCS: *Abbreviation for* **hundred call-seconds.** *See* **call-second.**

CCSA: *Abbreviation for* **common control switching arrangement.**

CDF: *Abbreviation for* **combined distribution frame.**

CDMA: *Abbreviation for* **code-division multiple access.**

CDPSK: *Abbreviation for* **coherent differential phase-shift keying.**

CDR: *Abbreviation for* **call detail recording.**

CD ROM: *Abbreviation for* **compact disk read-only memory.** An optical digital storage device, of high capacity, capable of being read from but not written to.

C-E: *Abbreviation for* **communications-electronics.**

CEI: *Abbreviation for* **comparably efficient interconnection.**

cell: **1.** In cellular mobile, the geographical area covered by the smaller of: a base station, or a subsystem (sector antenna) of that base station corresponding to a specific logical identification on the radio path. *Note:* Mobile stations in a cell may be reached by the corresponding radio equipment of the base station. **2.** In communications, a string that contains a header and user information. *Note 1:* A cell is dedicated to one user for one session. Cells for a given system are usually of fixed length and smaller than a frame, such as 424 bits for a cell, compared to 1024 for a frame. *Note 2:* In asynchronous transfer mode (ATM) systems, a cell consists of 53 bytes, *i.e.*, a 5-byte header field and a 48-byte information field. *Note 3:* A cell does not have error-correction capability and is therefore suited for low-BER communications systems, such as digital fiber optic systems. **3.** In OSI, a fixed-length block labeled at the Physical Layer of the Open Systems Interconnection—Reference Model (OSI—RM). **4.** In computer systems, an addressable, internal hardware location. **5.** In computer applications, a single location on a spreadsheet.

cell relay: A statistically multiplexed interface protocol for packet switched data communications that uses fixed-length packets, *i.e.*, cells, to transport data. *Note 1:* Cell relay transmission rates usually are between 56 kb/s and 1.544 Mb/s, *i.e.*, the data rate of a DS1 signal. *Note 2:* Cell relay protocols (a) have neither flow control nor error correction capability, (b) are information-content independent, and (c) correspond only to layers one and two of the ISO Open Systems Interconnection—Reference Model. *Note 3:* Cell relay systems enclose variable-length user packets in fixed-length packets, *i.e.*, cells, that add addressing and verification information. Frame length is fixed in hardware, based on time delay and user packet-length considerations. One user data message may be segmented over many cells. *Note 4:* Cell relay is an implementation of fast packet technology that is used in (a) connection-oriented broadband integrated services digital networks (B-ISDN) and (b) connectionless IEEE 802.6, switched multi-megabit data service (SMDS). *Note 5:* Cell relay is used for time-sensitive traffic such as voice and video.

cellular mobile: A mobile communications system that uses a combination of radio transmission and conventional telephone switching to permit telephone communication to and from mobile users within a specified area. *Note:* In cellular mobile systems, large geographical areas are segmented into many smaller areas, *i.e.*, cells, each of which has its own radio transmitters and receivers and a single controller interconnected with the public switched telephone network. *Synonyms* **cellular phone, cellular radio, cellular telephone**.

cellular phone: *Synonym* **cellular mobile.**

cellular radio: *Synonym* **cellular mobile.**

cellular telephone: *Synonym* **cellular mobile.**

CELP: *Acronym for* **code-excited linear prediction.**

center frequency: **1.** In frequency modulation, the rest frequency, *i.e.*, the frequency of the unmodulated carrier. (188) *Synonym* **carrier frequency. 2.** In facsimile systems, the frequency midway between the picture-black and picture-white frequencies. (188)

centralized attendant services (CAS): A function of a usually centrally located attendant console that permits the control of multiple switches, some of which may be geographically remote.

centralized automatic message accounting (CAMA): An automatic message accounting system that serves more than one switch from a central location. *Note:* When using CAMA, human intervention may be required.

centralized operation: Operation of a communication network in which transmission may occur between the control station and any tributary station, but not between tributary stations.

centralized ordering group (COG): An organization provided by some communications service providers to coordinate services between the companies and vendors.

central office (C.O.): A common carrier switching center in which trunks and loops are terminated and switched. *Note:* In the DOD, "common carrier" is called *"commercial carrier." Synonyms* **exchange, local central office, local exchange, local office, switching center** (except in DOD DSN [formerly AUTOVON] usage), **switching exchange, telephone exchange.** *Deprecated synonym* **switch.**

central office connecting facility: *Loosely*, in the sense of a trunk between public and private switches, *a synonym for* **central office trunk.** *See* **trunk.**

central office trunk: *Loosely*, in the sense of a trunk between public and private switches, *a synonym for* **central office connecting facility.** *See* **trunk.**

central processing unit: *See* **CPU.**

central processor: *Synonym* **CPU.**

Centrex® (CTX) service: A service offered by Bell Operating Companies that provides functions and features comparable to those provided by a PBX or a PABX. *Note: "Centrex® C.O."* indicates that all equipment except the attendant's position and station equipment is located in the central office. *"Centrex® C.U."* indicates that all equipment, including the dial switching equipment, is located on the customer's premises.

certification: [The] comprehensive evaluation of the technical and nontechnical security features of an AIS [automated information system] and other safeguards, made in support of the accreditation process, to establish the extent to which a particular design and implementation meets a set of specified security requirements. [NIS]

certified network engineer: In computer networking, one who has met proprietary training and certification requirements pertinent to network design or maintenance. *Note:* The training requirements embrace both software and hardware configuration.

cesium clock: A clock containing a cesium standard as a frequency-determining element. (188)

cesium standard: A primary frequency standard in which electronic transitions between the two hyperfine ground states of cesium-133 atoms is used to control the output frequency. (188) *Note*: The energy level between the two hyperfine ground states corresponds, in the absence of external influences (*e.g.*, the magnetic field of the Earth), to a frequency of 9,192,631,770 Hz.

chad: The material separated from a punched tape or a punched card when forming a hole. (188)

chadless tape: **1.** Punched tape that has been punched in such a way that chad is not formed. **2.** A punched tape in which only partial perforation is performed so that the chad remains attached to the tape. (188) *Note:* The partial perforation is deliberate, and should not be confused with imperfect chadding.

chad tape: Punched tape used in telegraphy/teletypewriter operation. (188)

channel: **1.** A connection between initiating and terminating nodes of a circuit. (188) **2.** A single path provided by a transmission medium via either (a) physical separation, such as by multipair cable or (b) electrical separation, such as by frequency- or time-division multiplexing. (188) **3.** A path for conveying electrical or electromagnetic signals, usually distinguished from other parallel paths. (188) **4.** Used in conjunction with a predetermined letter, number, or codeword to reference a specific radio frequency. (188) **5.** The portion of a storage medium, such as a track or a band, that is accessible to a given reading or writing station or head. **6.** In a communications system, the part that connects a data source to a data sink.

channel-associated signaling: Signaling in which the signals necessary to switch a given circuit are transmitted via the circuit itself or via a signaling channel permanently associated with it. (188)

channel bank: The part of a carrier-multiplex terminal that performs the first step of modulation by multiplexing a group of channels into a higher bandwidth analog channel or higher bit-rate digital channel and, conversely, demultiplexes these aggregates back into individual channels. (188)

channel bits: Binary data transmitted over a communications link. *Note:* Channel bits are derived from user information by FEC (forward error correction) coding and interleaving.

channel capacity: The maximum possible information transfer rate through a channel, subject to specified constraints. (188)

channel gate: A device for connecting a channel to a highway, or a highway to a channel, at specified times.

channelization: The use of a single wideband, *i.e.*, high-capacity, facility to create many relatively narrowband, *i.e.*, lower capacity, channels by subdividing the wideband facility. (188)

channel noise level: **1.** The ratio of the channel noise at any point in a transmission system to an arbitrary level chosen as a reference. (188) *Note 1:* The channel noise level may be expressed in (a) dB above reference noise (dBrn), (b) dB above reference noise with C-message weighting (dBrnC), or (c) adjusted dB (dBa). *Note 2:* Each unit used to measure channel noise level reflects a circuit noise reading of a specialized instrument designed to account for different interference effects that occur under specified conditions. **2.** The noise power density spectrum in the frequency range of interest. (188) **3.** The average noise power in the frequency range of interest. (188)

channel offset: The constant frequency difference between a channel frequency and a reference frequency which may frequency hop. (188)

channel packing: Maximizing the use of voice frequency channels for data transmission by multiplexing a number of channels of lower data rate into a single voice frequency channel of higher data rate. (188)

channel reliability (ChR): The percentage of time a channel was available for use in a specified period of scheduled availability. *Note 1:* Channel reliability is given by

$$ChR = 100\left(1 - \frac{T_o}{T_s}\right) = 100\frac{T_a}{T_s} ,$$

where T_o is the channel total outage time,$_s$ T is the channel total scheduled time, and T_a is the channel total available time. *Note 2:* $T_s = T_a + T_o$. (188)

channel service unit (CSU): A line bridging device that (a) is used to perform loop-back testing, (b) may perform bit stuffing, (c) may also provide a framing and formatting pattern compatible with the network, and (d) is the last signal regeneration point, on the loop side, coming from the central office, before the regenerated signal reaches a multiplexer or data terminal equipment (DTE).

channel supergroup: *See* **group.**

channel time slot: A time slot that starts at a particular instant in a frame and is allocated to a channel for transmitting data, such as a character or an in-slot signal. (188)

character: **1.** A letter, digit, or other symbol that is used as part of the organization, control, or representation of data. (188) **2.** One of the units of an alphabet. (188)

character check: A method of error detection using the preset rules for the formulation of characters. (188)

character-count integrity: The preservation of the exact number of characters that are originated in a message in the case of message communications, or per unit time, in the case of a user-to-user connection. (188) *Note:* Character-count integrity is not the same as character integrity, which requires that the characters delivered are, in fact, exactly the same as they were originated.

character filter: Software that is capable of selectively removing characters from a data stream, *e.g.*, software that removes communications-control characters so that they are not printed.

character generator: A functional unit that converts the coded representation of a character into the graphic representation of the character for display.

character integrity: Preservation of a character during processing, storage, and transmission.

character interval: In a communications system, the total number of unit intervals required to transmit any given character, including synchronizing, information, error checking, or control characters, but not including signals that are not associated with individual characters. *Note:* An example of a time interval that is excluded when determining character interval is any time added between the end of a stop signal and the beginning of the next start signal to accommodate changing transmission conditions, such as a change in data signaling rate or buffering requirements. This added time is defined as a part of the intercharacter interval.

characteristic distortion: In telegraphy, the distortion caused by transients that, as a result of previous modulation, are present in the transmission channel. (188) *Note:* Characteristic distortion effects are not consistent. Their effects on a given signal transition are dependent upon transients remaining from previous signal transitions.

characteristic frequency: A frequency which can be easily identified and measured in a given emission. A carrier frequency may, for example, be designated as the characteristic frequency. [NTIA] [RR] (188)

characteristic impedance (Z_o): **1.** The impedance of a circuit that, when connected to the output terminals of a uniform transmission line of arbitrary length, causes the line to appear infinitely long. *Note 1:* A uniform line terminated in its characteristic impedance will have no standing waves, no reflections from the end, and a constant ratio of voltage to current at a given frequency at every point on the line. *Note 2:* If the line is not uniform, the iterative impedance must be used. **2.** For Maxwell's equations, the impedance of a linear, homogeneous, isotropic, dielectric propagation medium free of electric charge, given by the relation $Z=(\mu/\epsilon)^{1/2}$ where μ is the magnetic permeability and ϵ is the electric permittivity of the medium.

character recognition: The identification of characters by automatic means.

character set: **1.** A finite set of different characters that is complete for a given purpose. (188) *Note:* A character set may or may not include punctuation marks or other symbols. **2.** An ordered set of unique representations. (188) *Note:* Examples of character sets include the 26 letters of the English alphabet, Boolean characters 0 and 1, the 128 ASCII characters, and International Telegraph Alphabet 5 (ITA-5), published as CCITT Recommendation V.3 and ISO 646.

characters per inch (cpi): In a recording medium, a unit of linear packing density of characters. (188)

characters per second (cps): A unit of signaling speed used to express the number of characters passing a designated point per second.

character-stepped: Pertaining to control of start-stop teletypewriter equipment in which a device is stepped one character at a time. *Note:* The step interval is equal to or greater than the character interval at the applicable signaling rate. (188)

charging reference location: In Universal Personal Telecommunications Service, the geographical location that may be used by the UPT service providers to determine the distance-related charges applying to the call originator and/or to the destination UPT user.

check: A process for determining accuracy.

check bit: A bit, such as a parity bit, derived from and appended to a bit string for later use in error detection and possibly error correction. (188)

check character: A character, derived from and appended to a data item, for later use in error detection and possibly error correction. (188)

check digit: A digit, derived from and appended to a data item, for later use in error detection and possibly error correction. (188)

checksum: **1.** The sum of a group of data items, which sum is used for checking purposes. *Note 1:* A checksum is stored or transmitted with the group of data items. *Note 2:* The checksum is calculated by treating the data items as numeric values. *Note 3:* Checksums are used in error detecting and correcting. **2.** [The] value computed, via some parity or hashing algorithm, on information requiring protection against error or manipulation. *Note:* Checksums are stored or transmitted with data and are intended to detect data integrity problems. [NIS]

chip: **1.** *Synonym* **integrated circuit.** **2.** In satellite communications systems, the smallest element of data in an encoded signal. **3.** The most elemental component of a spread spectrum signal when it is decompressed in time; that is, the longest duration signal in which signal parameters are approximately constant.. **4.** In micrographic and display systems, a relatively small and separate piece of microform that contains microimages and coded information for search, identification, and retrieval purposes.

chip rate: **1.** The rate of encoding. [NTIA] **2.** In direct-sequence-modulation spread-spectrum systems, the rate at which the information signal bits are transmitted as a pseudorandom sequence of chips. *Note:* The chip rate is usually several times the information bit rate.

chip time: In spread-spectrum systems, the duration of a chip produced by a frequency-hopping signal generator.

chirping: **1.** The rapid changing, as opposed to long-term drifting, of the frequency of an electromagnetic wave. *Note:* Chirping is most often observed in pulsed operation of a source. **2.** A pulse compression technique that uses (usually linear) frequency modulation during the pulse.

chroma keying: In television, nearly instantaneous switching between multiple video signals, based on the state, *i.e.*, phase, of the color (chroma) signal of one, to form a single composite video signal. *Note 1:* Chroma keying is used to create an overlay effect in the final picture, *e.g.*, to insert a false background, such as a weather map or scenic view, behind the principal subject being photographed. *Note 2:* The principal subject is photographed against a background having a single color or a relatively narrow range of colors, usually in the blue. When the phase of the chroma signal corresponds to the preprogrammed state or states associated with the background color, or range of colors, behind the principal subject, the signal from the alternate, *i.e.*, false, background is inserted in the composite signal and presented at the output. When the phase of the chroma signal deviates from that associated with the background color(s) behind the principal subject, video associated with the principal subject is presented at the output. *Synonym* **color keying.** *Colloquial synonym* **blue-screening.** *Contrast with* **chrominance signal, composite video.**

chromatic dispersion: *A commonly used (but redundant) synonym for* **material dispersion.** *See* **dispersion.**

chrominance signal: In color television, that signal or portion of the composite signal that bears the color information.

cipher: **1.** [A] cryptographic system in which units of plain text are substituted according to a predetermined key. [NIS] *Note:* A cipher is any cryptographic system in which arbitrary symbols, or groups of symbols, represent units of plain text of regular length, usually single letters, or in which units of plain text are rearranged, or both, in accordance with certain predetermined rules. **2.** The result of using a cipher. *Note:* An example of a cipher is an enciphered message or text.

cipher system: Any cryptosystem that requires the use of a key to convert, unit by unit, plain text, encoded text, or signals into an unintelligible form for secure transmission. *Note:* The capability to decipher must be available at the receiving site.

cipher text: Enciphered information. [NIS] *Note:* Cipher text is the result obtained from enciphering plain or encoded text.

ciphony: The process of enciphering audio information. *Note: "Ciphony" is a contraction of "ciphered telephony."*

circuit: **1.** The complete path between two terminals over which one-way or two-way communications may be provided. (188) **2.** An electronic path between two or more points, capable of providing a number of channels. **3.** A number of conductors connected together for the purpose of carrying an electrical current. **4.** An electronic closed-loop path among two or more points used for signal transfer. (188) **5.** A number of electrical components, such as resistors, inductances, capacitors, transistors, and power sources connected together in one or more closed loops.

circuit noise level: At any point in a transmission system, the ratio of the circuit noise at that point to an arbitrary level chosen as a reference. (188) *Note:* The circuit noise level is usually expressed in dBrn0, signifying the reading of a circuit noise meter, or in dBa0, signifying circuit noise meter reading adjusted to represent an interfering effect under specified conditions.

circuit reliability (*CiR*): The percentage of time a circuit was available for use in a specified period of scheduled availability. *Note 1:* Circuit reliability is given by

$$CiR = 100\left(1 - \frac{T_o}{T_s}\right) = 100\frac{T_a}{T_s} \quad ,$$

where T_o is the circuit total outage time, T_s is the cir cuit total scheduled time, and T_a is the circuit total available time. *Note 2:* $T_s = T_a + T_o$. (188) *Synonym* **time availability.**

circuit restoration: The process by which a communications circuit is established between two users after disruption or loss of the original circuit. *Note:* Circuit restoration is usually performed in accordance with planned procedures and priorities. Restoration may be effected automatically, *e.g.*, by switching to a hot standby, or manually, *e.g.*, by manual patching.

circuit routing: In open systems architecture, the logical path of a message in a communications network based on a series of gates at the physical network layer in the Open Systems Interconnection—Reference Model and the GOSIP FIPS PUB 146-1.

circuit-switched data transmission service: A data transmission service requiring the establishment of a circuit-switched connection before data can be transferred from source data terminal equipment (DTE) to a sink DTE. *Note:* A circuit-switched data transmission service uses a connection-oriented network.

circuit switching: 1. A method of routing traffic through a switching center, from local users or from other switching centers, whereby a connection is established between the calling and called stations until the connection is released by the called or calling station. (188) **2.** A process that, on demand, connects two or more data terminal equipments (DTEs) and permits the exclusive use of a data circuit between them until the connection is released.

circuit switching center: *See* **circuit switching, switching center.**

circuit switching unit (CSU): Equipment used for routing messages over common-user circuits that interconnect a source data terminal equipment (DTE) to a sink DTE for information interchange.

circuit transfer mode: In ISDN applications, a transfer mode by means of permanent allocation of channels or bandwidth between connections.

circular polarization: In electromagnetic wave propagation, polarization such that the tip of the electric field vector describes a helix. *Note 1:* The magnitude of the electric field vector is constant. (188) *Note 2:* The projection of the tip of the electric field vector upon any fixed plane intersecting, and normal to, the direction of propagation, describes a circle. *Note 3:* A circularly polarized wave may be resolved into two linearly polarized waves in phase quadrature with their planes of polarization at right angles to each other. *Note 4:* Circular polarization may be referred to as *"right-hand"* or *"left-hand,"* depending on whether the helix describes the thread of a right-hand or left-hand screw, respectively.

circulator: 1. A passive junction of three or more ports in which the ports can be accessed in such an order that when power is fed into any port it is transferred to the next port, the first port being counted as following the last in order. (188) **2.** In radar, a device that switches the antenna alternately between the transmitter and receiver.

civision: 1. The application of cryptography to television signals. **2.** Television signals that have been enciphered to preserve the confidentiality of the transmitted information.

***C/kT*:** *Abbreviation for* **carrier-to-receiver noise density.**

cladding: 1. Of an optical fiber, one or more layers of material of lower refractive index, in intimate contact with a core material of higher refractive index. (188) **2.** A process of covering one metal with another (usually achieved by pressure rolling, extruding, drawing, or swaging) until a bond is achieved. (188)

cladding diameter: In the cross section of a realizable optical fiber, ideally circular, but in practice assumed to a first approximation to be elliptical, the average of the diameters of the smallest circle that can be circumscribed about the cladding, and the largest circle that can be inscribed within the cladding.

cladding eccentricity: *See* **ovality.**

cladding mode: An undesired mode that is confined to the cladding of an optical fiber by virtue of the fact that the cladding has a higher refractive index than the surrounding medium, *i.e.*, air or primary polymer overcoat. *Note:* Modern fibers have a primary polymer overcoat with a refractive index that is slightly higher, rather than lower, than that of the cladding, in order to strip off cladding modes after only a few centimeters of propagation.

cladding mode stripper: A device for converting optical fiber cladding modes to radiation modes; as a result, the cladding modes are removed from the fiber. *Note:* Often a material such as the fiber coating or jacket having a refractive index equal to or greater than that of the fiber cladding will perform this function.

cladding noncircularity: *See* **ovality.**

cladding ovality: *Synonym* **cladding noncircularity**. *See* **ovality.**

cladding ray: *See* **cladding mode**.

C-language: A general-purpose, high-level, structured computer programming language. *Note:* C-language was originally designed for and implemented on the UNIX™ operating system.

CLASS: *Acronym for* **custom local area signaling services.**

class d address: *Synonym (in Internet protocol)* **multicast address.**

classmark: A designator used to describe the service feature privileges, restrictions, and circuit characteristics for lines or trunks that access a switch. *Note:* Examples of classmarks include precedence level, conference privilege, security level, and zone restriction. (188) *Synonym* **class-of-service mark.**

class of emission: The set of characteristics of an emission, designated by standard symbols, *e.g.*, type of modulation of the main carrier, modulating signal, type of information to be transmitted, and also, if appropriate, any additional signal characteristics. [NTIA] [RR]

class of office: A ranking, assigned to each switching center in a communications network, determined by the center switching functions, interrelationships with other offices, and transmission requirements.

class of service: **1.** A designation assigned to describe the service treatment and privileges given to a particular terminal. (188) **2.** A subgrouping of telephone users for the purpose of rate distinction. *Note:* Examples of class of service subgrouping include distinguishing between (a) individual and party lines, (b) Government and non-Government lines, (c) those permitted to make unrestricted international dialed calls and those not so permitted, (d) business, residence, and coin-operated, (e) flat rate and message rate, and (f) restricted and extended area service. **3.** A category of data transmission provided by a public data network in which the data signaling rate, the terminal operating mode, and the code structure, are standardized. *Note:* Class of service is defined in CCITT Recommendation X.1. *Synonym* **user service class.**

class-of-service mark: *Synonym* **classmark.**

clear: To cause one or more storage locations to be in a prescribed state, usually that corresponding to a zero or that corresponding to the space character.

clear channel: **1.** In radio broadcasting, a frequency assigned for the exclusive use of one entity. **2.** In networking, a signal path that provides its full bandwidth for a user's service. *Note:* No control or signaling is performed on this path.

clear collision: Contention that occurs when a DTE and a DCE simultaneously transfer a clear request packet and a clear indication packet specifying the same logical channel. *Note:* The DCE will consider that the clearing is completed and will not transfer a DCE clear confirmation packet.

clear confirmation signal: A call control signal used to acknowledge reception of the data-terminal-equipment (DTE) clear request by the data circuit-terminating equipment (DCE) or to acknowledge the reception of the DCE clear indication by the DTE.

clearing: **1.** A sequence of events used to disconnect a call and return to the ready state. (188) **2.** Removal of data from an AIS, its storage devices, and other peripheral devices with storage capacity, in such a way that the data may not be reconstructed using normal system capabilities (*i.e.*, through the keyboard). *Note:* An AIS need not be disconnected from any external network before clearing takes place. Clearing enables a product to be reused within, but not outside of, a secure facility. It does not produce a declassified product by itself, but may be the first step in the declassification process. [NIS]

clear message: **1.** A message that (a) is sent in the forward direction and the backward direction, (b) contains a circuit-released signal or circuit-released acknowledgment signal, and (c) usually contains an indication of whether the message is in the forward or the backward direction. **2.** A message in plain language, *i.e.*, not enciphered.

clear text: *Synonym* **plain text.**

cleave: **1.** In an optical fiber, a deliberate, controlled break, intended to create a perfectly flat endface, perpendicular to the longitudinal axis of the fiber. *Note:* A cleave is made by first introducing a microscopic fracture ("nick") into the fiber with a special tool, called a *"cleaving tool,"* which has a sharp blade of hard material, such as diamond, sapphire, or tungsten carbide. If proper tension is applied to the fiber as the nick is made, or immediately afterward (this may be done by the cleaving tool in some designs, or manually in other designs), the fracture will propagate in a controlled fashion, creating the desired endface. **2.** To break a fiber in such a controlled fashion. *Note:* A good cleave is required for a successful splice of an optical fiber, whether by fusion or mechanical means. Also, some types of fiber-optic connectors do not employ abrasives and polishers. Instead, they use some type of cleaving technique to trim the fiber to its proper length, and produce a smooth, flat perpendicular endface.

client: In networking, a software application that allows the user to access a service from a server computer, *e.g.*, a server computer on the Internet.

client-server: Any hardware/software combination that generally adheres to a client-server architecture, regardless of the type of application.

client-server architecture: Any network-based software system that uses client software to request a specific service, and corresponding server software to provide the service from another computer on the network.

clipper: A circuit or device that limits the instantaneous output signal amplitude to a predetermined maximum value, regardless of the amplitude of the input signal. (188)

clipper chip: An IC designed for secure communications.

clipping: **1.** In telephony, the loss of the initial or final parts of a word, words, or syllable, usually caused by the nonideal operation of voice-actuated devices. **2.** The limiting of instantaneous signal amplitudes to a predetermined maximum value. (188) **3.** In a display device, the removal of those parts of display elements that lie outside of a given boundary.

clock: **1.** A reference source of timing information. (188) **2.** A device providing signals used in a transmission system to control the timing of certain functions such as the duration of signal elements or the sampling rate. (188) **3.** A device that generates periodic, accurately spaced signals used for such purposes as timing, regulation of the operations of a processor, or generation of interrupts.

clock error: The difference between local clock time or value and a designated reference clock time or value. *Note:* Subtracting the clock difference from the local clock brings the local clock into agreement with the reference clock.

clock phase slew: The rate of relative phase change between a given clock signal and a stable reference signal. (188) *Note:* The two signals are generally at or near the same frequency or have an integral multiple frequency relationship.

clock rate: The rate at which a clock issues timing pulses. *Note:* Clock rates are usually expressed in pulses per second, such as 4.96 Mpps (megapulses per second).

clock tolerance: The maximum permissible departure of a clock indication from a designated time reference such as Coordinated Universal Time (UTC).

clock track: A track on which a pattern of signals is recorded to provide a timing reference.

clockwise polarized wave: *Synonym* **right-hand (or clockwise) polarized wave.**

closed captioning: In broadcast and cable television, the insertion, into the blank lines between frames, of information that may be decoded and displayed on the screen as written words corresponding to those being spoken and transmitted via the conventional audio subcarrier. *Note:* Closed captioning, developed for the hearing-impaired, requires a special decoder, which may be external to, or built into, the television receiver. Closed captioning is mandated by the Americans with Disabilities Act of 1990.

closed circuit: **1.** In radio and television transmission, pertaining to an arrangement in which programs are directly transmitted to specific users and not broadcast to the general public. (188) **2.** In telecommunications, a circuit dedicated to specific users. (188) **3.** A completed electrical circuit.

closed-loop noise bandwidth: The integral, over all frequencies, of the absolute value of the closed-loop transfer function of a phase-locked loop. *Note:* The closed-loop noise bandwidth, when multiplied by the noise spectral density, gives the output noise power in a phase-locked loop.

closed-loop transfer function: A mathematical expression (algorithm) describing the net result of the effects of a closed (feedback) loop on the input signal to the circuits enclosed by the loop. *Note 1:* The closed-loop transfer function is measured at the output. *Note 2:* The output signal waveform can be calculated from the closed-loop transfer function and the input signal waveform.

closed user group: In a network, a group of users permitted to communicate with each other but not with users outside the group. *Note:* A user data terminal equipment (DTE) may belong to more than one closed user group.

closed user group with outgoing access: A closed user group in which at least one member of the group has a facility that permits communication with one or more users external to the closed user group.

closed waveguide: An electromagnetic waveguide (a) that is tubular, usually with a circular or rectangular cross section, (b) that has electrically conducting walls, (c) that may be hollow or filled with a dielectric material, (d) that can support a large number of discrete propagating modes, though only a few may be practical, (e) in which each discrete mode defines the propagation constant for that mode, (f) in which the field at any point is describable in terms of the supported modes, (g) in which there is no radiation field, and (h) in which discontinuities and bends cause mode conversion but not radiation.

closure: *Synonym* **splice closure.**

cloud attenuation: In the transmission of electromagnetic signals, attenuation caused by absorption and scattering by water or ice particles in clouds. *Note:* The amount of cloud attenuation depends on many factors, including (a) the density, particle size, and turbulence of the clouds and (b) the transmission path length in the clouds.

C-message weighting: A noise spectral weighting used in a noise power measuring set to measure noise power on a line that is terminated by a 500-type set or similar instrument. (188) *Note:* The instrument is calibrated in dBrnC.

CMRR: *Abbreviation for* **common-mode rejection ratio.**

CNR: *Abbreviation for* **carrier-to-noise ratio, combat net radio.**

CNS: *Abbreviation for* **complementary network services.**

C.O.: *Abbreviation for* **central office.**

COAM: *Acronym for* **customer owned and maintained equipment.** *Deprecated term. See* **customer premises equipment.**

coast Earth station: An Earth station in the fixed-satellite service or, in some cases, in the maritime mobile-satellite service, located at a specified fixed point on land to provide a feeder link for the maritime mobile-satellite service. [NTIA] [RR]

coasting mode: In timing-dependent systems, a free-running operational timing mode in which continuous or periodic measurement of clock error, *i.e.*, of timing error, is not made. *Note:* Operation in the coasting mode may be enhanced for a period of time by using clock-error data or clock-correction data (obtained during a prior period of operation in the tracking mode) to estimate clock corrections.

coast station: A land station in the maritime mobile service. [RR]

coating: *See* **primary coating.**

coax: *See* **coaxial cable.**

coaxial cable (coax): A cable consisting of a center conductor surrounded by an insulating material and a concentric outer conductor. (188) *Note:* Coaxial cable is used primarily for wideband, video, or rf applications.

coaxial patch bay: *See* **patch bay.**

COBOL: *Acronym for* **common business oriented language**. A programming language designed for business data processing.

co-channel interference: Interference resulting from two or more simultaneous transmissions on the same channel.

code: **1.** A set of unambiguous rules specifying the manner in which data may be represented in a discrete form. *Note 1:* Codes may be used for brevity or security. *Note 2:* Use of a code provides a means of converting information into a form suitable for communications, processing, or encryption. (188) **2.** [Any] system of communication in which arbitrary groups of letters, numbers, or symbols represent units of plain text of varying length. *Note:* Codes may or may not provide security. Common uses include: (a) converting information into a form suitable for communications or encryption, (b) reducing the length of time required to transmit information, (c) describing the instructions which control the operation of a computer, and (d) converting plain text to meaningless combinations of letters or numbers and vice versa. [NIS] **3.** A cryptosystem in which the cryptographic equivalents, (usually called "code groups") typically consisting of letters or digits (or both) in otherwise meaningless combinations, are substituted for plain text elements which are primarily words, phrases, or sentences. **4.** A set of rules that maps the elements of one set, the coded set, onto the elements of another set, the code element set. *Synonym* **coding scheme.** **5.** A set of items, such as abbreviations, that represents corresponding members of another set. **6.** To represent data or a computer program in a symbolic form that can be accepted by a processor. **7.** To write a routine.

codec: *Acronym for* **coder-decoder**. **1.** An assembly consisting of an encoder and a decoder in one piece of equipment. (188) **2.** A circuit that converts analog signals to digital code and vice versa. **3.** An electronic device that converts analog signals, such as video and voice signals, into digital form and compresses them to conserve bandwidth on a transmission path. (188) *Note:* Codecs in this sense are used in this sense for video conferencing systems.

code character: A character that (a) is used to represent a discrete value or symbol and (b) is derived in accordance with a code. (188)

code conversion: **1.** Conversion of signals, or groups of signals, in one code into corresponding signals, or groups of signals, in another code. (188) **2.** A process for converting a code of some predetermined bit structure, such as 5, 7, or 14 bits per character interval, to another code with the same or a different number of bits per character interval. *Note:* In code conversion, alphabetical order is not significant.

coded character set: A character set established in accordance with unambiguous rules that define the character set and the one-to-one relationships between the characters of the set and their coded representations. (188)

coded image: A representation of a display image in a form suitable for storage and processing.

code-division multiple access (CDMA): A coding scheme, used as a modulation technique, in which multiple channels are independently coded for transmission over a single wideband channel. (188) *Note 1:* In some communication systems, CDMA is used as an access method that permits carriers from different stations to use the same transmission equipment by using a wider bandwidth than the individual carriers. On reception, each carrier can be distinguished from the others by means of a specific modulation code, thereby allowing for the reception of signals that were originally overlapping in frequency and time. Thus, several transmissions can occur simultaneously within the same bandwidth, with the mutual interference reduced by the degree of orthogonality of the unique codes used in each transmission. *Note 2:* CDMA permits a more uniform distribution of energy in the emitted bandwidth.

coded set: A set of elements onto which another set of elements has been mapped according to a code. *Note:* Examples of coded sets include the list of names of airports that is mapped onto a set of corresponding three-letter representations of airport names, the list of classes of emission that is mapped onto a set of corresponding standard symbols, and the names of the months of the year mapped onto a set of two-digit decimal numbers.

code element: One of a set of parts of which the characters in a given code may be composed. (188)

code-excited linear prediction (CELP): An analog-to-digital voice coding scheme.

code group: [A] group of letters, numbers, or both in a code system used to represent a plain text word, phrase, or sentence. [NIS] (188) *Note:* Code groups may include symbols and other elements.

code-independent data communication: *Synonym* **code-transparent data communication.**

code restriction: A service feature by which certain terminals are prevented from accessing certain features of the network.

code set: The complete set of representations defined by a particular code and language. (188)

code-transparent data communication: A mode of data communication that uses protocols that do not depend for their correct functioning on the data character set or data code used. *Synonym* **code-independent data communication.**

code word: **1.** In a code, a word that consists of a sequence of symbols assembled in accordance with the specific rules of the code and assigned a unique meaning. *Note:* Examples of code words are error-detecting-or-correcting code words and communication code words, such as SOS, MAYDAY, ROGER, TEN-FOUR, and OUT. (188) **2.** A cryptonym used to identify sensitive intelligence data. [JP1] (188) **3.** A word that has been assigned a classification and a classified meaning to safeguard intentions and information regarding a classified plan or operation. [JP1]

coding: **1.** In communications systems, the altering of the characteristics of a signal to make the signal more suitable for an intended application, such as optimizing the signal for transmission, improving transmission quality and fidelity, modifying the signal spectrum, increasing the information content, providing error detection and/or correction, and providing data security. *Note:* A single coding scheme usually does not provide more than one or two specific capabilities. Different codes have different advantages and disadvantages. **2.** In communications and computer systems, implementing rules that are used to map the elements of one set onto the elements of another set, usually on a one-to-one basis. **3.** The digital encoding of an analog signal and, conversely, decoding to an analog signal.

coding scheme: *Synonym* **code (def. #4).**

codress message: In military communications systems, a message in which the entire address is encrypted with the message text.

COG: *Abbreviation for* **centralized ordering group.**

coherence area: Pertaining to an electromagnetic wave, the area of a surface perpendicular to the direction of propagation, over which the electromagnetic wave maintains a specified degree of coherence. *Note:* The specified degree of coherence is usually taken to be 0.88 or greater. (188)

coherence degree: *See* **degree of coherence.**

coherence length: The propagation distance from a coherent source to a point where an electromagnetic wave maintains a specified degree of coherence. (188) *Note 1:* In long-distance transmission systems, the coherence length may be reduced by propagation factors such as dispersion, scattering, and diffraction. *Note 2:* In optical communications, the coherence length, L, is given approximately by $L = \lambda^2/(n\Delta\lambda)$, where λ is the central wavelength of the source, n is the refractive index of the medium, and $\Delta\lambda$ is the spectral width of the source. *Note 3: Coherence length* is usually applied to the optical regime.

coherence time: For an electromagnetic wave, the time over which a propagating wave may be considered coherent. (188) *Note 1:* In long-distance transmission systems, the coherence time may be reduced by propagation factors such as dispersion, scattering, and diffraction. *Note 2:* In optical communications, coherence time, τ, is calculated by dividing the coherence length by the phase velocity of light in a medium; approximately given by $\tau = \lambda^2/(c\Delta\lambda)$ where λ is the central wavelength of the source, $\Delta\lambda$ is the spectral width of the source, and c is the velocity of light in vacuum. *Note 3:* "*Coherence time*" is usually applied to the optical regime.

coherent: Pertaining to a fixed phase relationship between corresponding points on an electromagnetic wave. (188) *Note:* A truly coherent wave would be perfectly coherent at all points in space. In practice, however, the region of high coherence may extend over only a finite distance.

coherent bundle: *Synonym* **aligned bundle.**

coherent differential phase-shift keying (CDPSK): Phase-shift keying (a) that is used for digital transmission, (b) in which the phase of the carrier is discretely modulated in relation to the phase of a reference signal and in accordance with data to be transmitted, and (c) in which the modulated carrier is of constant amplitude and frequency. (188) *Note:* A phase comparison is made of successive pulses, and information is recovered by examining the phase transitions between the carrier and successive pulses rather than by the absolute phases of the pulses.

coherent pulse operation: In pulsed carrier transmission, a method of operation in which a fixed phase relationship of the carrier wave is maintained from one pulse to the next. (188)

coherent radiation: *See* **coherent.**

cold standby: Pertaining to spare electronic equipment that is available for substitute use, but is not powered or warmed up and ready for use.

collective address: *Synonym* **group address.**

collective routing: Routing in which a switching center automatically delivers messages to a specified list of destinations. *Note 1:* Collective routing avoids the need to list each single address in the message heading. *Note 2:* Major relay stations usually transmit messages bearing collective-routing indicators to tributary, minor, and other major relay stations.

collimation: The process by which a divergent or convergent beam of electromagnetic radiation is converted into a beam with the minimum divergence or convergence possible for that system (ideally, a bundle of parallel rays). (188)

collimator: A device that renders divergent or convergent rays more nearly parallel. (188) *Note:* The degree of collimation (parallelism) should be stated.

collinear antenna array: An array of dipole antennas mounted in such a manner that every element of each antenna is in an extension, with respect to its long axis, of its counterparts in the other antennas in the array. *Note:* A collinear array is usually mounted vertically, in order to increase overall gain and directivity in the horizontal direction. When stacking dipole antennas in such a fashion, doubling their number will, with proper phasing, produce a 3-dB increase in directive gain.

collision: **1.** In a data transmission system, the situation that occurs when two or more demands are made simultaneously on equipment that can handle only one at any given instant. (188) **2.** In a computer, the situation that occurs when an attempt is made to store simultaneously two different data items at a given address that can hold only one of the items.

color errors: In video systems, distortion of hues in all or a portion of the received image.

color keying: *Synonym* **chroma keying.**

combat-net radio (CNR): A radio operating in a network that (a) provides a half-duplex circuit and (b) uses either a single radio frequency or a discrete set of radio frequencies when in a frequency hopping mode. (188) *Note:* CNRs are primarily used for push-to-talk-operated radio nets for command and control of combat, combat support, and combat service support operations among ground, sea, and air forces.

combinational logic element: A device having at least one output channel and one or more input channels, all characterized by discrete states, such that at any instant the state of each output channel is completely determined by the states of the input channels at the same instant.

combined communications: The common use of communications facilities by two or more military services, each belonging to a different nation. *Note:* Such use might be specified by a combined communications-electronics agency.

combined distribution frame (CDF): A distribution frame that combines the functions of main and intermediate distribution frames and contains both vertical and horizontal terminating blocks. (188) *Note 1:* The vertical blocks are used to terminate the permanent outside lines entering the station. Horizontal blocks are used to terminate inside plant equipment. This arrangement permits the association of any outside line with any desired terminal equipment. These connections are made either with twisted pair wire, normally referred to as jumper wire, or with optical fiber cables, normally referred to as jumper cables. *Note 2:* In technical control facilities, the vertical side may be used to terminate equipment as well as outside lines. The horizontal side is then used for jackfields and battery terminations.

combined station: In high-level data-link control (HDLC) operation, the station that is usually responsible for performing balanced link-level operations and that generates commands, interprets responses, interprets received commands, and generates responses.

combiner: *See* **maximal-ratio combiner.**

COMINT: *Acronym for* **communications intelligence.**

comma-free code: A code constructed so that any partial code word, beginning at the start of a code word but terminating prior to the end of that code word, is not a valid code word. *Note 1:* The comma-free property permits the proper framing of transmitted code words when (a) external synchronization is provided to identify the start of the first code word in a sequence of code words and (b) no uncorrected errors occur in the symbol stream. *Note 2:* Examples of comma-free are the variable-length Huffman codes. *Synonym* **prefix-free code.**

command: **1.** An order for an action to take place. **2.** In data transmission, an instruction sent by the primary station instructing a secondary station to perform some specific function. (188) **3.** In signaling systems, a control signal. **4.** In computer programming, that part of a computer instruction word that specifies the operation to be performed. **5.** *Loosely,* a mathematical or logic operator.

command and control (C^2): The exercise of authority and direction by a properly designated commander over assigned and attached forces in the accomplishment of the mission. Command and control functions are performed through an arrangement of personnel, equipment, communications, facilities, and procedures employed by a commander in planning, directing, coordinating, and controlling forces and operations in the accomplishment of the mission. [JP1-A]

command and control (C^2) system: The facilities, equipment, communications, procedures, and personnel essential to a commander for planning, directing, and controlling operations of assigned forces pursuant to the missions assigned. [JP1]

command and control warfare (C^2W): The integrated use of operations security (OPSEC), military deception, psychological operations (PSYOP), electronic warfare (EW), and physical destruction, mutually supported by intelligence, to deny information to, influence, degrade, or destroy adversary command and control capabilities, while protecting friendly command and control capabilities against such actions. *Note:* Command and control warfare applies across the operational continuum and all levels of conflict. C^2W is both offensive and defensive:
(a) **counter-C^2:** To prevent effective C^2 of adversary forces by denying information to, influencing, degrading, or destroying the adversary C^2 systems.
(b) **C^2-protection:** To maintain effective command and control of own forces by turning to friendly advantage or negating adversary efforts to deny information to, influence, degrade, or destroy the friendly C^2 system. [JP1]

command, control and communications (C^3): The capabilities required by commanders to exercise command and control of their forces. [JCS Pub 18, *Operations Security,* Dec. 1982.]

command, control, communications, computers, and intelligence (C^4I): The facilities, computer equipment, communications equipment, display devices, and intelligence systems necessary to support military operations.

command, control, communications, and computer systems (C^4S): Integrated systems of doctrine, procedures, organizational structures, personnel, equipment, facilities, and communications designed to support a commander's exercise of command and control, through all phases of the operational continuum. *Synonym* **C^4 systems.** [JP1]

command frame: In data transmission, a frame, containing a command, transmitted by a primary station.

command menu: A list of all the different commands that may be given to a computer or communications system by an operator. *Note:* Commands on a command menu may be selected by the operator by (a) using an electromechanical pointer, such as a light pen, (b) touching the display screen with a finger, (c) speaking to a voice-recognition system, or (d) positioning a cursor or reverse-video bar by using a keyboard or mouse, and depressing one or more keys on the keyboard or mouse.

command net: A communications network which connects an echelon of command with some or all of its subordinate echelons for the purpose of command and control. [JP1]

command protocol data unit: A protocol data unit (PDU) transmitted by a logical link control (LLC) sublayer in which the PDU command/response (C/R) bit is equal to "0".

comm center: *Synonym* **communications center.**

commercial carrier: *Synonym* **common carrier.**

commercial refile: In military communications systems, the processing of a message from (a) a given military network, such as a tape-relay network, a point-to-point telegraph network, a radio-telegraph network, or the DSN to (b) a commercial communications network. *Note:* Commercial refiling of a message will usually require a reformatting of the message, particularly the heading.

commit transaction: The application, *i.e.,* insertion, of information into any data repository of an integrated database management system in a distributed local communications network.

commonality: **1.** A quality that applies to materiel or systems: (a) possessing like and interchangeable characteristics enabling each to be utilized, or operated and maintained by personnel trained on the others without additional specialized training; (b) having interchangeable repair parts and/or components; (c) applying to consumable items interchangeably equivalent without adjustment. **2.** Pertaining to equipment or systems that have the quality of one entity possessing like and interchangeable parts with another equipment or system entity. (188) **3.** Pertaining to system design in which a given part can be used in more than one place in the system, *i.e.,* subsystems and components have parts in common. *Note:* Examples of commonality include the use of a firing pin that fits in many different weapons and the use of a light source that fits in many different types of fiber optic transmitters.

common battery: A single electrical power source used to energize more than one circuit, component, equipment, or system. (188) *Note 1:* A common battery is usually an electrolytic device and is usually centrally located to the equipment that it serves. *Note 2:* In many telecommunications applications, the common battery is at a nominal –48 vdc. *Note 3:* A central office common battery supplies power to operate all directly connected instruments. *Note 4: Common battery* may include one or more power conversion devices to transform commercial power to direct current, with an electrolytic battery floating across the output.

common-battery signaling: Signaling in which the signaling power of a telephone is supplied by the serving switch. (188) *Note:* In common-battery signaling, "talking power" may be supplied by common or local battery.

common carrier: In a telecommunications context, a telecommunications company that holds itself out to the public for hire to provide communications transmission services. *Note:* In the United States, such companies are usually subject to regulation by Federal and state regulatory commissions. *Synonyms* **carrier, commercial carrier, communications common carrier.**

common-channel interoffice signaling (CCIS): In multichannel switched networks, a method of transmitting all signaling information for a group of trunks by encoding it and transmitting it over a separate channel using time-division digital techniques.

common-channel signaling: In a multichannel communications system, signaling in which one channel in each link is used for signaling to control, account for, and manage traffic on all channels of the link. (188) *Note:* The channel used for common-channel signaling does not carry user information.

common control: An automatic switching arrangement in which the control equipment necessary for the establishment of connections is shared by being associated with a given call only during the period required to accomplish the control function for the given call. (188) *Note:* In common control, the channels that are used for signaling, whether frequency bands or time slots, are not used for message traffic.

common control switching arrangement (CCSA): An arrangement in which switching for a private network is provided by one or more common control switching systems. *Note:* The switching systems may be shared by several private networks and also may be shared with the public telephone networks.

common control system: An automatic switching system that makes use of common equipment to establish a connection. (188) *Note:* The common equipment then becomes available to establish other connections.

common equipment: Equipment used by more than one system, subsystem, component, or other equipment, such as a channel or switch. (188)

Common Management Information Protocol (CMIP): A protocol used by an application process to exchange information and commands for the purpose of managing remote computer and communications resources. (188)

common management information service (CMIS): A service that specifies the service interface to the Common Management Information Protocol (CMIP). (188) *Note:* To transfer management information between open systems using CMIS/CMIP, peer connections, *i.e.,* associations, must be established. This requires the establishment of an Application Layer association, a Session Layer connection, a Transport Layer connection, and, depending on supporting communications technology, Network Layer and Link Layer connections.

common-mode interference: **1.** Interference that appears between signal leads, or the terminals of a measuring circuit, and ground. (188) **2.** A form of coherent interference that affects two or more elements of a network in a similar manner (*i.e.,* highly coupled) as distinct from locally generated noise or interference that is statistically independent between pairs of network elements.

common-mode rejection ratio (CMRR): The ratio of the common-mode interference voltage at the input of a circuit, to the corresponding interference voltage at the output. (188)

common-mode voltage: **1.** The voltage common to both input terminals of a device. (188) **2.** In a differential amplifier, the unwanted part of the voltage between each input connection point and ground that is added to the voltage of each original signal. *Synonym* **longitudinal voltage.**

common return: A return path that is common to two or more circuits and that serves to return currents to their source or to ground. (188)

common return offset: In a line or circuit, the dc potential difference between ground and the common return. (188)

common user: In communications systems, pertaining to communications facilities and services provided to essentially all users in the area served by the system, rather than to one user or to a relatively small number of users, such as a closed user group with outgoing access.

common user circuit: A circuit designated to furnish a communication service to a number of users. (188)

common user network: A system of circuits or channels allocated to furnish communication paths between switching centers to provide communication service on a common basis to all connected stations or subscribers. It is sometimes described as a general purpose network. [JP1]

communications: **1.** Information transfer, among users or processes, according to agreed conventions. (188) **2.** The branch of technology concerned with the representation, transfer, interpretation, and processing of data among persons, places, and machines. *Note:* The meaning assigned to the data must be preserved during these operations.

communications blackout: **1.** A cessation of communications or communications capability caused by a lack of power to a communications facility or equipment. **2.** A total lack of communications capability caused by propagation anomalies, *e.g.,* those present during strong auroral activity or during the re-entry of a spacecraft into the Earth's atmosphere.

communications center: **1.** An agency charged with the responsibility for handling and controlling communications traffic. The center normally includes message center, transmitting, and receiving facilities. [JP1] **2.** A facility that (a) serves as a node for a communications network, (b) is equipped for technical control and maintenance of the circuits originating, transiting, or terminating at the node, (c) may contain message-center facilities, and (d) may serve as a gateway. (188) *Synonym* **comm center.**

communications channel: *See* **channel.**

communications common carrier: *Synonym* **common carrier.**

communications deception: **1.** Deliberate transmission, retransmission, or alteration of communications to mislead an adversary's interpretation of the communications. [NIS] **2.** Use of devices, operations, and techniques with the intent of confusing or misleading the user of a communications link or a navigation system. [JP1]

communications-electronics (C-E): The specialized field concerned with the use of electronic devices and systems for the acquisition or acceptance, processing, storage, display, analysis, protection, disposition, and transfer of information. (188) *Note:* C-E includes the wide range of responsibilities and actions relating to (a) electronic devices and systems used in the transfer of ideas and perceptions, (b) electronic sensors and sensory systems used in the acquisition of information devoid of semantic influence, and (c) electronic devices and systems intended to allow friendly forces to operate in hostile environments and to deny to hostile forces the effective use of electromagnetic resources.

communications intelligence (COMINT): Technical and intelligence information derived from foreign communications by other than the intended recipients. [JP1]

communications jamming (COMJAM): **1.** The portion of electronic jamming that is directed against communications circuits and systems. **2.** The prevention of successful radio communications by the use of electromagnetic signals, *i.e.,* the deliberate radiation, reradiation, or reflection of electromagnetic energy with the objective of impairing the effective use of electronic communications systems. *Note:* The aim of communications jamming is to prevent communications by electromagnetic means, or at least to degrade communications sufficiently to cause delays in transmission and reception. Jamming may be used in conjunction with deception to achieve an overall electronic-countermeasure (ECM) plan implementation.

communications net: An organization of stations capable of direct communication on a common channel or frequency. [JP1] *Synonym* **net.**

communications net operation: *See* **net operation.**

communications network: An organization of stations capable of intercommunications but not necessarily on the same channel. [JP1]

communications processor unit (CPU): A computer embedded in a communications system. *Note 1:* An example of a CPU is the message data processor of a DDN switching center. (188) *Note 2: CPU* is also an abbreviation for *central processing unit* of a computer.

communications protection: The application of communications security (COMSEC) measures to telecommunications systems in order (a) to deny unauthorized persons access to sensitive unclassified information of value, (b) to prevent disruption of telecommunications services, or (c) to ensure the authenticity of information handled by telecommunications systems.

communications protocol: *See* **protocol.**

communications reliability: The probability that information transmitted from a communications station will arrive at the intended destination in a timely manner without loss of content.

communications satellite: An orbiting vehicle that relays signals between (a) terrestrial communications stations, (b) a terrestrial communications station and another communications satellite, or (c) other communications satellites.

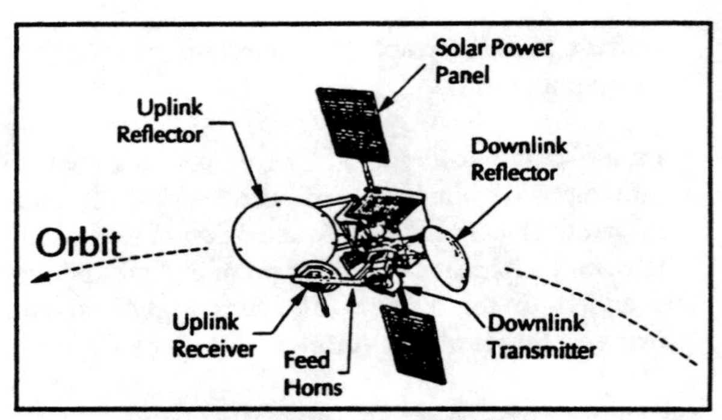

communications satellite

communications saturation: *See* **saturation.**

communications security (COMSEC): Measures and controls taken to deny unauthorized persons information derived from telecommunications and ensure the authenticity of such telecommunications. *Note:* Communications security includes cryptosecurity, transmission security, emission security, and physical security of COMSEC material. [NIS] (188)
(a) cryptosecurity: [The] component of communications security that results from the provision of technically sound cryptosystems and their proper use. [NIS]
(b) emission security: Protection resulting from all measures taken to deny unauthorized persons information of value which might be derived from intercept and analysis of compromising emanations from crypto-equipment, AIS, and telecommunications systems. [NIS]
(c) physical security: The component of communications security that results from all physical measures necessary to safeguard classified equipment, material, and documents from access thereto or observation thereof by unauthorized persons. [JP1]
(d) transmission security: [The] component of communications security that results from the application of measures designed to protect transmissions from interception and exploitation by means other than cryptanalysis. [NIS]

communications security equipment: *See* **COMSEC equipment.**

communications security material: *See* **COMSEC material.**

communications silence: The avoidance of any type of transmission, emission, or radiation by any means, including radiation from receiving equipment. *Note:* An example of communications silence is the maintaining of a listening watch only if the receivers do not radiate beyond a specified level.

communications sink: *See* **sink.**

communications source: *See* **source.**

communications subsystem: A functional unit or operational assembly that is smaller than the larger assembly under consideration. (188) *Note:* Examples of communications subsystems in the Defense Communications System (DCS) are (a) a satellite link with one Earth terminal in CONUS and one in Europe, (b) the interconnect facilities at each Earth terminal of the satellite link, and (c) an optical fiber cable with its driver and receiver in either of the interconnect facilities.

communications survivability: The ability of communications systems to continue to operate effectively under adverse conditions, though portions of the system may be damaged or destroyed. *Note:* Various methods may be used to maintain communications services, such as using alternate routing, different transmission media or methods, redundant equipment, and sites and equipment that are radiation hardened.

communications system: A collection of individual communications networks, transmission systems, relay stations, tributary stations, and data terminal equipment (DTE) usually capable of interconnection and interoperation to form an integrated whole. (188) *Note:* The components of a communications system serve a common purpose, are technically compatible, use common procedures, respond to controls, and operate in unison.

communications system engineering: The translation of user requirements for the exchange of information into cost-effective and low-risk technical solutions in terms of equipment and subsystems. (188) *Note:* Communications system engineering encompasses the integration of these parts into a complete entity resulting in a minimum investment for the entire system lifecycle required to satisfy the requirements of a majority of users of the communication system.

communications system survivability: *See* **survivability.**

communications theory: Theory that is devoted to the probabilistic characteristics of the transmission of data in the presence of noise, and that is used to advance the design, development, and operation of communications systems.

communications watch: The monitoring of one or more communications lines, frequencies, or channels to obtain information by listening to or receiving all transmissions on them and transmitting and receiving messages as required.

communications zone: [The] rear part of theater of operations (behind but contiguous to the combat zone), which contains the lines of communications, establishments for supply and evacuation, and other agencies required for the immediate support and maintenance of the field forces. [JP1]

community antenna television (CATV): *See* **cable TV.**

community of interest: A grouping of users who generate a majority of their traffic in calls to other members of the group. *Note:* The community of interest may be related to a geographic area or to an administrative organization. *Synonym* **special interest group.**

community reception (in the broadcasting-satellite service): The reception of emissions from a space station in the broadcasting-satellite service by receiving equipment, which in some cases may be complex and have antennae larger than those used for individual reception, and intended for use:

- by a group of the general public at one location;

  or
- through a distribution system covering a limited area. [NTIA] [RR]

compact: *See* **data compaction.**

compact disk read-only memory: *See* **CD ROM.**

compaction: *See* **data compaction.**

companding: An operation in which the dynamic range of signals is compressed before transmission and is expanded to the original value at the receiver. (188) *Note:* The use of companding allows signals with a large dynamic range to be transmitted over facilities that have a smaller dynamic range capability. Companding reduces the noise and crosstalk levels at the receiver.

compandor: A device that incorporates a compressor and an expander, each of which may be used independently. (188)

comparably efficient interconnection (CEI): An equal-access concept developed by the FCC stating that, ". . . if a carrier offers an enhanced service, it should be required to offer network interconnection (or collocation) opportunities to others that are comparably efficient to the interconnection that its enhanced service enjoys. Accordingly, a carrier would be required to implement CEI only as it introduces new enhanced services." [FCC *Report and Order* June 16, 1986]

comparator: **1.** In analog computing, a functional unit that compares two analog variables and indicates the result of that comparison. **2.** A device that compares two items of data and indicates the result of that comparison. **3.** A device for determining the dissimilarity of two items such as two pulse patterns or words.

compartmentation: The segregation of components, programs, and information to provide isolation. *Note:* Compartmentation provides some protection against overall compromise, contamination, or unauthorized access.

compatibility: **1.** Capability of two or more items or components of equipment or material to exist or function in the same system or environment without mutual interference. [JP1] (188) **2.** In computing, the ability to execute a given program on different types of computers without modification of the program or the computers. **3.** The capability that allows the substitution of one subsystem (storage facility), or of one functional unit (*e.g.*, hardware, software), for the originally designated system or functional unit in a relatively transparent manner, without loss of information and without the introduction of errors.

compatible sideband transmission: Independent sideband transmission in which the carrier is deliberately reinserted at a lower level after its normal suppression to permit reception by conventional AM receivers. (188) *Note:* Compatible sideband transmission is usually single-sideband (SSB) amplitude-modulation-equivalent (AME) transmission consisting of the emission of the carrier plus the upper sideband. *Synonym* **amplitude modulation equivalent.**

compelled signaling: Signaling in which the transmission of each signal in the forward direction from an originating terminal is inhibited until an acknowledgement of the satisfactory receipt of the previous signal is received by the originating terminal.

competitive access provider (CAP): A company that provides exchange access services in competition with an established U.S. telephone local exchange carrier.

competitive clip: In time-assignment speech interpolation (TASI) or digital speech interpolation (DSI), truncation of the initial part of a speech spurt, caused when all channels in a given direction of transmission are busy and the transmission of the spurt must wait for an available channel.

compile: **1.** To translate a computer program expressed in a high-level language into a program expressed in a lower level language, such as an intermediate language, assembly language, or a machine language. **2.** To prepare a machine language program from a computer program written in another programming language by making use of the overall logic structure of the program or by generating more than one computer instruction for each symbolic statement as well as performing the function of an assembler.

compiler: A computer program for compiling. *Synonym* **compiling program.**

compiling program: *Synonym* **compiler.**

complementary network service (CNS): A means for an enhanced-service provider customer to connect to a network and to the enhanced service provider.

Note: Complementary network services usually consist of the customer local service, such as a business or residence, and several associated service features, such as a call-forwarding service.

component: **1.** An assembly, or part thereof, that is essential to the operation of some larger assembly and is an immediate subdivision of the assembly to which it belongs. (188) *Note:* For example, a radio receiver may be a component of a complete radio set consisting of a combined transmitter-receiver, *i.e.*, a transceiver. The same radio receiver could also be a subsystem of the combined transmitter-receiver, in which case the IF amplifier section would be a component of the receiver but not of the radio set. Similarly, within the IF amplifier section, items, such as resistors, capacitors, vacuum tubes, and transistors, are components of that section. **2.** In logistics, a part, or combination of parts having a specified function, that can only be installed or replaced as an entity. [JP1]. **3.** In material, an assembly or any combination of parts, subassemblies, and assemblies mounted together in manufacture, assembly, maintenance, or rebuild. [JP1-A]

composite cable: A communications cable having both optical and metallic signal-carrying components. *Note 1:* A cable having optical fiber(s) and a metallic component, *e.g.*, a metallic twisted pair, used solely for conduction of electric power to repeaters, does qualify as a composite cable. *Note 2:* A cable having optical fiber(s) , plus a metallic strength member or armor, does not qualify as a composite cable. *Contrast with* **hybrid cable.**

composited circuit: A circuit that can be used simultaneously either for telephony and dc telegraphy or for telephony and signaling. *Note:* Separation of the two may be accomplished by frequency discrimination. *Synonym* **voice-plus circuit.**

composite signaling (CX): Signaling in which an arrangement is made to provide direct current signaling and dial pulsing beyond the range of conventional loop signaling. (188) *Note:* Composite signaling, like DX signaling, permits duplex operation, *i.e.*, permits simultaneous two-way signaling. *Synonym* **CX signaling.**

composite two-tone test signal: A test signal composed of two different frequencies and used for intermodulation distortion measurements. (188)

composite video: In television, a video signal in which synchronizing information (pulses) and picture information, including chroma, *i.e.*, color, information are combined.

compound signal: In ac signaling, a signal consisting of the simultaneous transmission of more than one frequency. *Note:* An example of compound signaling is dual-tone multifrequency (DTMF) signaling.

compress: *See* **data compaction, data compression, signal compression.**

compressed video: Video that has been encoded so as to reduce the number of bits required for storage or transmission.

compression: *See* **data compression, signal compression.**

compression ratio: **1.** In signal compression, the ratio of the dynamic range of compressor input signals to the dynamic range of the compressor output signals. *Note:* The compression ratio is usually expressed in dB. For example, a 40-dB input range compressed to a 30-dB output range would be equivalent to a 10-dB compression. **2.** In digital facsimile, the ratio of the total pels scanned for the object to the total encoded bits sent for picture information. **3.** The ratio of the gain of a device at a low power level to the gain at some higher level. *Note:* The compression ratio is usually expressed in dB.

compressor: A nonlinear analog device that has a lower gain at higher input levels than at lower input levels. *Note:* A compressor is used to allow signals with a large dynamic range to be sent through devices or circuits with a smaller dynamic range .

compromise: **1.** The known or suspected exposure of clandestine personnel, installations, or other assets or of classified information or material, to an unauthorized person. [JP1] **2.** The disclosure of cryptographic information to unauthorized persons. **3.** The recovery of plain text of encrypted messages by unauthorized persons through cryptanalysis methods. **4.** The disclosure of information or data to unauthorized persons, or a violation of the security policy of a system in which unauthorized intentional or unintentional disclosure, modification, destruction, or loss of an object may have occurred. [NIS]

compromising emanations: Unintentional signals that, if intercepted and analyzed would disclose the information transmitted, received, handled, or otherwise processed by telecommunications or automated systems equipment. [NIS] (188)

computer: **1.** A device that accepts data, processes the data in accordance with a stored program, generates results, and usually consists of input, output, storage, arithmetic, logic, and control units. (188) **2.** A functional unit that can perform substantial computation, including numerous arithmetic operations or logic operations, without human intervention during a run. *Note 1:* This definition, approved by the Customs Council, distinguishes a computer from similar devices, such as hand-held calculators and certain types of control devices. *Note 2:* Computers have been loosely classified into microcomputers, minicomputers, and main-frame computers, based on their size. These distinctions are rapidly disappearing as the capabilities of even the smaller units have increased. Microcomputers now are usually more powerful and versatile than the minicomputers and the main-frame computers were a few years ago.

computer-aided software engineering: *See* **CASE.**

computer-aided systems engineering: *See* **CASE.**

computer architecture: Of a computer, the physical configuration, logical structure, formats, protocols, and operational sequences for processing data, controlling the configuration, and controlling the operations. *Note:* Computer architecture may also include word lengths, instruction codes, and the interrelationships among the main parts of a computer or group of computers.

computer conferencing: **1.** Teleconferencing supported by one or more computers. **2.** An arrangement in which access, by multiple users, to a common database is mediated by a controlling computer. **3.** The interconnection of two or more computers working in a distributed manner on a common application process. (188)

computer-dependent language: *Synonym* **assembly language.**

computer graphics: **1.** Graphics implemented through the use of computers. **2.** Methods and techniques for converting data to or from graphic displays via computers. **3.** The branch of science and technology concerned with methods and techniques for converting data to or from visual presentation using computers.

computer language: *Synonym* **programming language.**

computer network: **1.** A network of data processing nodes that are interconnected for the purpose of data communication. **2.** A communications network in which the end instruments are computers. (188)

computer network engineering: *See* **network engineering (def. # 2).**

computer network operating system (NOS): A specialized operating system designed for computer networking on minicopmputers and microcomputers in a local networking area / campus area network. *Note:* A NOS is usually designed to run on existing software designed for that computer and may require interface hardware for the workstation and server.

computer-oriented language: A programming language in which words and syntax are designed for use on a specific computer or class of computers. (188) *Synonyms* **low-level language, machine-oriented language.**

computer peripheral: *See* **peripheral equipment.**

computer program: *See* **program.**

computer program origin: The address assigned to the initial storage location of a computer program in main storage.

computer routine: *See* **routine.**

computer science: The discipline that is concerned with methods and techniques relating to data processing performed by automatic means.

computer security (COMPUSEC): 1. Measures and controls that ensure confidentiality, integrity, and availability of the information processed and stored by a computer. [NIS] *Synonym* **automated information systems security. 2.** The application of hardware, firmware and software security features to a computer system in order to protect against, or prevent, the unauthorized disclosure, manipulation, deletion of information or denial of service. (AC/35(WG/1)WP(88)4-7) [NATO]

computer system: A functional unit, consisting of one or more computers and associated software, that (a) uses common storage for all or part of a program and also for all or part of the data necessary for the execution of the program, (b) executes user-written or user-designated programs, and (c) performs user-designated data manipulation, including arithmetic and logic operations. *Note:* A computer system may be a stand-alone system or may consist of several interconnected systems. *Synonyms* **ADP system, computing system.**

computer system fault tolerance: The ability of a computer system to continue to operate correctly even though one or more of its components are malfunctioning. *Note:* System performance, such as speed and throughput, may be diminished until the faults are corrected. *Synonym* **computer system resilience.**

computer system resilience: *Synonym* **computer system fault tolerance.**

computer systems engineering: *See* **systems design.**

computer word: In computing, a group of bits or characters that occupies one or more storage locations and is treated by computers as a unit. (188) *Synonym* **machine word.**

computing system: *Synonym* **computer system.**

COMSEC: *Acronym for* **communications security.**

COMSEC equipment: Equipment designed to provide security to telecommunications by converting information to a form unintelligible to an unauthorized interceptor and, subsequently, by reconverting such information to its original form for authorized recipients; also, equipment designed specifically to aid in, or as an essential element of, the conversion process. *Note:* COMSEC equipment includes crypto-equipment, crypto-ancillary equipment, cryptoproduction equipment, and authentication equipment. [NIS]

COMSEC material: [An] item designed to secure or authenticate telecommunications. *Note:* COMSEC material includes, but is not limited to, key, equipment, devices, documents, firmware or software that embodies or describes cryptographic logic and other items that perform COMSEC functions. [NIS]

concentrator: 1. In data transmission, a functional unit that permits a common path to handle more data sources than there are channels currently available within the path. *Note:* A concentrator usually provides communication capability between many low-speed, usually asynchronous channels and one or more high-speed, usually synchronous channels. Usually different speeds, codes, and protocols can be accommodated on the low-speed side. The low-speed channels usually operate in contention and require buffering. **2.** A device that connects a number of circuits, which are not all used at once, to a smaller group of circuits for economy. (188)

concentricity error: In an optical fiber, the distance between the center of the two concentric circles that specify the cladding diameter and the center of the two concentric circles that specify the core diameter. *Note:* The concentricity error is used in conjunction with tolerance fields to specify or characterize optical fiber core and cladding geometry. *Synonyms* **core eccentricity, core-to-cladding concentricity, core-to-cladding eccentricity, core-to-cladding offset.**

concurrent operation: 1. *Synonym* **multitasking. 2.** In data link operations, the operation in which two or more data links are used during the same, usually short, time interval, while adhering to the protocols of each link without providing data forwarding among the links.

conditioned baseband representation: *Synonym* **non-return-to-zero mark.**

conditioned circuit: A communications circuit optimized to obtain desired characteristics for voice or data transmission. (188)

conditioned diphase modulation: A form of diphase modulation, combined with signal conditioning, that (a) eliminates the dc component of the signal, (b) enhances timing recovery, and (c) facilitates transmission over voice frequency (VF) circuits or coaxial cables. (188)

conditioned loop: A loop that has conditioning equipment to obtain the desired line characteristics for voice or data transmission. (188) *Note:* The conditioning equipment is used to improve the amplitude-vs.-frequency characteristics of the circuit and to match impedance.

conditioning equipment: **1.** At junctions of circuits, equipment used to obtain desired circuit characteristics, such as matched transmission levels, matched impedances, and equalization between facilities. (188) **2.** Corrective networks used to improve data transmission, such as equalization of the insertion-loss-vs.-frequency characteristic and the envelope delay distortion over a desired frequency range.

conducted interference: **1.** Interference resulting from noise or unwanted signals entering a device by conductive coupling, *i.e.*, by direct coupling. (188) **2.** An undesired voltage or current generated within, or conducted into, a receiver, transmitter, or associated equipment, and appearing at the antenna terminals. (188)

conduction band: **1.** In a semiconductor, the range of electron energy, higher than that of the valence band, sufficient to make the electrons free to move from atom to atom under the influence of an applied electric field and thus constitute an electric current. **2.** In the atomic structure of a material, a partially filled or empty energy level in which electrons are free to move, thus allowing the material to conduct an electrical current upon application of an electric field by means of an applied voltage.

conductive coupling: Energy transfer achieved by means of physical contact, *i.e.*, coupling other than inductive or capacitive coupling. (188) *Note 1:* Conductive coupling may be achieved by wire, resistor, or common terminal, such as a binding post or metallic bonding. *Note 2:* Conductive coupling passes the full spectrum of frequencies, including dc. *Synonym* **direct coupling.**

cone of silence: *Synonym* **antenna blind cone.**

CONEX: *Acronym for* **connectivity exchange.**

conference call: A service feature that allows a call to be established among three or more stations in such a manner that each of the stations is able to communicate with all the other stations. (188) *Synonym* **multiple call.**

conference operation: In a communications network, operation that allows a call to be established among three or more stations in such a manner that each of the stations is able to communicate directly with all the other stations. (188) *Note:* In radio systems, the stations may receive simultaneously, but must transmit one at a time. The common operational modes are "push-to-talk" for telephone operation and "push-to-type" for telegraph and data transmission.

configuration: In a communications or computer system, an arrangement of functional units according to their nature, number, and chief characteristics. *Note 1:* Configuration pertains to hardware, software, firmware, and documentation. *Note 2:* The configuration will affect system performance.

configuration control: **1.** After establishing a configuration, such as that of a telecommunications or computer system, the evaluating and approving changes to the configuration and to the interrelationships among system components. **2.** In distributed-queue dual-bus (DQDB) networks, the function that ensures the resources of all nodes of a DQDB network are configured into a correct dual-bus topology. *Note:* The functions that are managed include the head of bus, external timing source, and default slot generator functions.

configuration management: **1.** [The] management of security features and assurances through control of changes made to hardware, software, firmware, documentation, test, test fixtures and test documentation of an automated information system, throughout the development and operational life of a system. [NIS] **2.** The control of changes—including the recording thereof—that are made to the hardware, software, firmware, and documentation throughout the system lifecycle.

confirmation signaling: Signaling that ensures error-free transmission of dialed information by returning a unique digit-dependent signal from the far end as each digit is sent over a trunk. (188)

confirmation to receive: In facsimile, a signal from a CCITT Group 1, 2, or 3 facsimile receiver, indicating it is ready to receive picture signals.

conformance test: A test performed by an independent body to determine if a particular piece of equipment satisfies the criteria in a specified controlling document, such as a Federal standard, an American National Standard, a Military Standard, or a Military Specification. *Contrast with* **acceptance test.**

congestion: **1.** In a communications switch, a state or condition that occurs when more subscribers attempt simultaneously to access the switch than it is able to handle, even if unsaturated. (188) **2.** In a saturated communications system, the condition that occurs when an additional demand for service occurs. (188)

connecting arrangement: In the public switched telephone networks, the equipment provided by a common carrier to accomplish electrical interconnection between customer-provided equipment and the facilities of the common carrier.

connection: **1.** A provision for a signal to propagate from one point to another, such as from one circuit, line, subassembly, or component to another. **2.** An association established between functional units for conveying information.

connection-in-progress signal: A call control signal at the data circuit-terminating-equipment/data-terminal-equipment (DCE/DTE) interface that indicates to the DTE that the establishment of the data connection is in progress and that the ready-for-data signal will follow.

connectionless data transfer: *See* **connectionless mode transmission.**

connectionless mode transmission: In a packet-switched network, transmission in which each packet is encoded with a header containing a destination address sufficient to permit the independent delivery of the packet without the aid of additional instructions. *Note 1:* A packet transmitted in a connectionless mode is frequently called a datagram. *Note 2:* In connectionless mode transmission of a packet, the service provider usually cannot guarantee there will be no loss, error insertion, misdelivery, duplication, or out-of-sequence delivery of the packet. However, the risk of these hazards' occurring may be reduced by providing a reliable transmission service at a higher protocol layer, such as the Transport Layer of the Open Systems Interconnection—Reference Model.

connectionless transmission: *See* **connectionless-mode transmission.**

connection-mode transmission: *See* **connection-oriented mode transmission.**

connection-oriented data transfer protocol: A data-transfer protocol in which a logical connection is established between end user terminals.

connection-oriented mode transmission: In a packet-switched network, a mode of transmission in which there is a complete information transfer transaction for each packet or group of packets, *i.e.,* the information transfer phase is preceded by an access phase and followed by a disengagement phase. *Note 1:* During the information transfer phase of connection-oriented mode transmission, more than one packet may be transmitted. The header of each information packet contains a sequence number and an identifier field that associates the packet with the connection that was established during the access phase before the information transfer phase begins. *Note 2:* Connection-oriented mode transmission usually enables detection of lost, erroneous, duplicated, or out-of-sequence packets because a connection is established from end to end before transmission begins. *Note 3:* The CCITT X.25 protocols are widely used to implement connection-oriented mode transmission on packet-switched public data networks. The protocols are implemented at

Layers 1, 2, and 3 of the Open Systems Interconnection—Reference Model.

connections per circuit hour (CCH): **1.** A unit of traffic measurement expressed as the number of connections established at a switching point per hour. (188) **2.** A unit of traffic measurement used to express the rate at which circuits are established at a switch. *Note:* The magnitude of the CCH is an instantaneous value subject to change as a function of time, *i.e.*, from moment to moment.

connectivity exchange (CONEX): In an adaptive or manually operated high-frequency (HF) radio network, the automatic or manual exchange of information concerning routes to stations that are not directly reachable by the exchange originator. *Note:* The purpose of the exchange is to identify indirect paths and/or possible relay stations to those stations that are not directly reachable.

connector: A device for mating and demating electrical power connections or communications media. (188) *Note:* A connector is distinguished from a splice, which is a permanent joint.

conservation of radiance: A basic principle of optics, that no passive optical system can increase the quantity L/n^2, where L is the radiance of a beam and n is the local refractive index. (188) *Note:* "*Conservation of radiance*" was formerly called "*conservation of brightness*" or the "*brightness theorem.*"

consolidated local telecommunications service: Local communications service provided by GSA to all Federal agencies located in a building, complex, or geographical area.

consultation hold: A service feature that allows a speaker on an extension instrument to place one call on hold and to speak with another caller on a separate line.

content-addressable storage: *Synonym* **associative storage (def. #1).**

contention: **1.** A condition that arises when two or more data stations attempt to transmit at the same time over a shared channel, or when two data stations attempt to transmit at the same time in two-way alternate communication. *Note:* A contention can occur in data communications when no station is designated a master station. In contention, each station must monitor the signals and wait for a quiescent condition before initiating a bid for master status. **2.** Competition by users of a system for use of the same facility at the same time. (188)

continuity check: A check made of a circuit to verify that a communication or power path exists. (188)

continuously variable slope delta (CVSD) modulation: A type of delta modulation in which the size of the steps of the approximated signal is progressively increased or decreased as required to make the approximated signal closely match the input analog wave. (188)

continuous operation: **1.** Operation in which certain components, such as nodes, facilities, circuits, or equipment, are in an operational state at all times. (188) *Note:* Continuous operation usually requires that there be fully redundant configuration, or at least a sufficient X out of Y degree of redundancy for compatible equipment, where X is the number of spare components and Y is the number of operational components. **2.** In data transmission, operation in which the master station need not stop for a reply from a slave station after transmitting each message or transmission block.

continuous presence: In teleconferencing, the concurrent presence of two or more video images, such as two images that may appear on a single monitor on a split screen or on two separate monitors.

continuous tone copy: In facsimile, an object, *i.e.*, an original, or a recorded copy, that contains shades of gray, *i.e.*, contains densities between black and white, such as in a photographic print. (188)

continuous wave (cw): A wave of constant amplitude and constant frequency. (188)

contouring: In digital facsimile, density step lines in the recorded copy resulting from analog-to-digital conversion when the object, *i.e.*, the original, has observable shades of gray between the smallest density steps of the digital system.

contrast: **1.** In display systems, the relation between (a) the intensity of color, brightness, or shading of an area occupied by a display element, display group, or display image on the display surface of a display device and (b) the intensity of an area not occupied by a display elements, a display group, or a display image. *Deprecated synonym* **brightness ratio.** **2.** In optical character recognition, the difference between color or shading of the printed material on a document and the background on which it is printed.

contribution: In B-ISDN applications, the use of broadband transmission of audio or video information to the user for post-production processing and distribution.

control ball: *Synonym* **trackball.**

control character: A character that initiates, modifies, or stops a function, event, operation, or control operation. (188) *Note:* Control characters may be recorded for use in subsequent actions. They are not graphic characters but may have a graphic representation in some circumstances.

control communications: The branch of technology devoted to the design, development, and application of communications facilities used specifically for control purposes, such as for controlling (a) industrial processes, (b) movement of resources, (c) electric power generation, distribution, and utilization, (d) communications networks, and (e) transportation systems.

control field: In a protocol data unit (PDU), the field that (a) contains data interpreted by the receiving destination logical-link controller (LLC) and (b) may be the field immediately following the destination service access point (DSAP) and source service access point (SSAP) address fields of the PDU.

control function: *Synonym* **control operation.**

controlled access: Access in which the resources of an area or system is limited to authorized personnel, users, programs, processes, or other systems, and denied to all others. (188)

controlled area: **1.** An area (a) in which uncontrolled movement will not result in compromise of classified information, (b) that is designed to provide administrative control and safety, or (c) that serves as a buffer for controlling access to limited-access areas. **2.** An area to which security controls have been applied to protect an information-processing system's equipment and wirelines, equivalent to that required for the information transmitted through the system. (188)

controlled not-ready signal: A signal, sent in the backward direction, to indicate that a call cannot be completed because the called line is not in a ready condition, but is under control, as opposed to not being in a ready condition and not under control.

controlled security mode: *See* **controlled security operation.**

controlled security operation: In a communications system, operation in which (a) internal security controls prevent inadvertent disclosures, (b) personnel, physical, and administrative controls are used to prevent unauthorized access, (c) both cleared and uncleared users are serviced, and (d) if required, both secured and unsecured remote terminal areas are serviced. (188)

controlled space: Three-dimensional space surrounding telecommunications and automated information systems equipment, within which unauthorized persons are denied unrestricted access and are either escorted by authorized persons or are under continuous physical or electronic surveillance. *Synonym* **restricted area.** [NIS]

controller: In an automated radio, the device that commands the radio transmitter and receiver, and that performs processes, such as automatic link establishment, channel scanning and selection, link quality analysis, polling, sounding, message store and forward, address protection, and anti-spoofing.

control of electromagnetic radiation: **1.** Measures taken to minimize unintended electromagnetic radiation emanating from a system or component and to minimize electromagnetic interference. *Note:* Control of electromagnetic radiation is exercised for purposes of security and the reduction of interference, especially on ships and aircraft. **2.** A national operational plan to minimize the use of electromagnetic radiation in the United States and its possessions and the Panama Canal Zone in the event of attack or imminent threat thereof, as an aid to the navigation of hostile aircraft, guided missiles, or other devices.

control operation: An operation that affects the recording, processing, transmission, or interpretation of data. *Note:* Examples of control operations include starting and stopping a process; executing a carriage return, a font change, or a rewind; and transmitting an end-of-transmission (EOT) control character. *Synonym* **control function.**

control station: In a communications network, the station that selects the master station and supervises operational procedures, such as polling, selecting, and recovery. *Note:* The control station has the overall responsibility for the orderly operation of the entire network.

conversational mode: A mode of communication analogous to a conversation between two persons.

conversational service: In telecommunications, a service that provides two-way, interactive, real-time, end-to-end information transfer.

convolutional code: A type of error-correction code in which (a) each *m*-bit information symbol (each *m*-bit string) to be encoded is transformed into an *n*-bit symbol, where $n>m$ and (b) the transformation is a function of the last *k* information symbols, where *k* is the constraint length of the code. *Note:* Convolutional codes are often used to improve the performance of radio and satellite links.

convoy internal communications: In a land or maritime convoy, communications (a) that is among the elements of the convoy, (b) includes radio, visual, and sound transmissions, (c) in which radio intervehicle or intership communications are usually by radiotelephone using the receipt method of operation, (d) in which the convoy commander vehicle or ship usually has the net-control station aboard and (e) that usually use a single assigned frequency.

cooperation factor: In facsimile systems, the product of the total scanning length and the scanning density, given by $CF = L\sigma$, where L is the scanning line length and σ is the scanning line density, both in compatible units. *Note:* For example, a 20-cm line and a line density of 6 scanning pitches per centimeter would yield a cooperation factor of 120.

coordinated clock: One of a set of clocks distributed over a spatial region, producing time scales that are synchronized to the time scale of a reference clock at a specified location. (188)

coordinated time scale: A time scale synchronized within given tolerances to a reference time scale. (188)

Coordinated Universal Time (UTC): Time scale, based on the second (SI), as defined and recommended by the CCIR, and maintained by the Bureau International des Poids et Mesures (BIPM). For most practical purposes associated with the *Radio Regulations,* UTC is equivalent to mean solar time at the prime meridian (0° longitude), formerly expressed in GMT. [NTIA] [RR] *Note 1:* The maintenance by BIPM includes cooperation among various national laboratories around the world. *Note 2:* The full definition of UTC is contained in CCIR Recommendation 460-4. (188) *Note 3*: The second was formerly defined in terms of astronomical phenomena. When this practice was abandoned in order to take advantage of atomic resonance phenomena ("atomic time") to define the second more precisely, it became necessary to make occasional adjustments in the "atomic" time scale to coordinate it with the workaday mean solar time scale, UT-1, which is based on the somewhat irregular rotation of the Earth. Rotational irregularities usually result in a net decrease in the Earth's average rotational velocity, and ensuing lags of UT-1 with respect to UTC. *Note 4:* Adjustments to the atomic, *i.e.,* UTC, time scale consist of an occasional addition or deletion of one full second, which is called a *leap second.* Twice yearly, during the last minute of the day of June 30 and December 31, Universal Time, adjustments may be made to ensure that the accumulated difference between UTC and UT-1 will not exceed 0.9 s before the next scheduled adjustment. Historically, adjustments, when necessary, have usually consisted of adding an extra second to the UTC time scale in order to allow the rotation of the Earth to "catch up." Therefore, the last minute of the UTC time scale, on the day when an adjustment is made, will have 59 or

61 seconds. *Synonyms* **World Time, Z Time, Zulu Time.**

coordination area: The area associated with an Earth station outside of which a terrestrial station sharing the same frequency band neither causes nor is subject to interfering emissions greater than a permissible level. [NTIA] [RR]

coordination contour: **1.** The line enclosing the coordination area. [NTIA] [RR] **2.** The perimeter of the coordination area.

coordination distance: Distance on a given azimuth from an Earth station beyond which a terrestrial station sharing the same frequency band neither causes nor is subject to interfering emissions greater than a permissible level. [NTIA] [RR]

copy: **1.** To receive a message. **2.** A recorded message or a duplicate of it. **3.** To read data from a source, leaving the source data unchanged at the source, and to write the same data elsewhere, though they may be in a physical form that differs from that of the source. **4.** To understand a transmitted message.

copy-protected: Of a data medium such as a diskette, pertaining to the use of one or more schemes designed to thwart copying in violation of copyright law or security considerations.

copy watch: A radio- or video-communications watch in which an operator is required to maintain a continuous receiver watch and to keep a complete log.

cord circuit: A switchboard circuit in which a plug-terminated cord is used to establish connections manually between user lines or between trunks and user lines. (188) *Note:* A number of cord circuits are furnished as part of the switchboard position equipment. The cords may be referred to as front cord and rear cord or trunk cord and station cord. In modern cordless switchboards, the cord-circuit function is switch operated and may be programmable.

cord lamp: The lamp associated with a cord circuit that indicates supervisory conditions for the respective part of the connection.

cordless switchboard: A telephone switchboard in which manually operated keys are used to make connections. (188)

core: **1.** The central region about the longitudinal axis of an optical fiber, which region supports guiding of the optical signal. (188) *Note 1:* For the fiber to guide the optical signal, the refractive index of the core must be slightly higher than that of the cladding. *Note 2:* In different types of fibers, the core and core-cladding boundary function slightly differently in guiding the signal. Especially in single-mode fibers, a significant fraction of the energy in the bound mode travels in the cladding. **2.** A piece of ferromagnetic material, usually toroidal in shape, used as a component in a computer memory device. *Note:* The type of memory referred to has very limited application in today's computer environment. It has been largely replaced by semiconductor and other technologies. **3.** The material at the center of an electromechanical relay or solenoid, about which the coil is wound. (188)

core area: The part of the cross-sectional area of an optical fiber within which the refractive index is everywhere greater than that of the innermost homogeneous cladding, by a specified fraction of the difference between the maximum refractive index of the core and the refractive index of the innermost cladding. *Note 1:* Artifacts of the manufacturing process, such as refractive index dip, are ignored in computing the points (refractive indices) of demarcation. *Note 2:* The core area is the cross-sectional area within which the refractive index is given by

$$n_3 > [n_2 + m(n_1 - n_2)] \ ,$$

where n_1 is the maximum refractive index of the core, n_2 is the refractive index of the homogeneous cladding adjacent to the core, n_3 is the defining refractive index, and m is a fraction, usually not greater than 0.05. [After 2196]

core-cladding offset: *See* **concentricity error**.

core diameter: In the cross section of a realizable optical fiber, ideally circular, but assumed to a first approximation to be elliptical, the average of the diameters of the smallest circle that can be circumscribed about the core-cladding boundary, and the largest circle that can be inscribed within the core-cladding boundary.

core dump: A printout, usually in hexadecimal characters, of the contents of core memory. *Note:* A core dump is useful for analyzing an abnormally terminated computer run or finding bugs in a computer program.

core eccentricity: *Synonym* **concentricity error.** *See* **ovality.**

core noncircularity: *See* **ovality.**

core storage: *See* **magnetic core storage.**

core-to-cladding concentricity: *Synonym* **concentricity error.**

core-to-cladding eccentricity: *Synonym* **concentricity error.**

core-to-cladding offset: *Synonym* **concentricity error.**

corner reflector: **1.** A reflector consisting of three mutually perpendicular intersecting conducting flat surfaces, which returns a reflected electromagnetic wave to its point of origin. (188) *Note:* Such reflectors are often used as radar targets. **2.** A directional antenna using two mutually intersecting conducting flat surfaces. **3.** A device, normally consisting of three metallic surfaces or screens perpendicular to one another, designed to act as a radar target or marker. [JP1] **4.** In radar interpretation, an object that, by means of multiple reflections from smooth surfaces, produces a radar return of greater magnitude than might be expected from the physical size of the object. [JP1] **5.** A passive optical mirror, that consists of three mutually perpendicular flat, intersecting reflecting surfaces, which returns an incident light beam in the opposite direction.

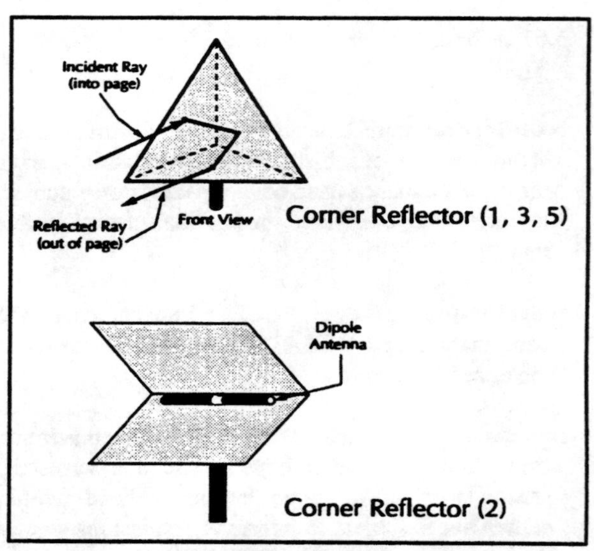

corner reflector

corner reflector antenna: (188) *See* **corner reflector (def #2).**

corrective maintenance: **1.** Maintenance actions carried out to restore a defective item to a specified condition. **2.** Tests, measurements, and adjustments made to remove or correct a fault. (188)

cosine emission law: *Synonym* **Lambert's cosine law.**

cosite: Collocation of electronic equipment on the same vehicle, station, or base. *Note:* Equipment so located is often subject to interference because of its proximity to other equipment.

cosmic noise: Random noise that originates outside the Earth's atmosphere. (188) *Note:* Cosmic noise characteristics are similar to those of thermal noise. Cosmic noise is experienced at frequencies above about 15 MHz when highly directional antennas are pointed toward the Sun or to certain other regions of the sky such as the center of the Milky Way Galaxy. *Synonym* **galactic radio noise.**

Costas loop: A phase-locked loop used for carrier phase recovery from suppressed-carrier modulation signals, such as from double-sideband suppressed carrier signals. *Note:* In the usual implementation of a Costas loop, a local voltage-controlled oscillator provides quadrature outputs, one to each of two phase detectors, *i.e.,* product detectors. The same phase of the input signal is also applied to both phase detectors and the output of each phase detector is passed through a low-pass filter. The outputs of these low-pass filters are inputs to another phase detector, the output of which passes through a loop filter before being used to control the voltage-controlled oscillator.

Coulomb's law: The universal law of attraction and repulsion of electric charges.

counterpoise: A conductor or system of conductors used as a substitute for earth or ground in an antenna system. [From Weik '89]

country code: **1.** In international direct telephone dialing, a code that consists of 1-, 2-, or 3-digit numbers in which the first digit designates the region and succeeding digits, if any, designate the country. **2.** In international record carrier transmissions, a code consisting of 2- or 3-letter abbreviations of the country names, or 2- or 3-digit numbers that represent the country names, that follow the geographical place names.

coupled modes: **1.** In fiber optics, a mode that shares energy among one or more other modes, all of which propagate together. (188) [After 2196] *Note:* The distribution of energy among the coupled modes changes with propagation distance. **2.** In microwave transmission, a condition where energy is transferred from the fundamental mode to higher order modes. *Note:* Energy transferred to coupled modes is undesirable in usual microwave transmission in a waveguide.

coupler: *See* **directional coupler.**

coupling: The desirable or undesirable transfer of energy from one medium, such as a metallic wire or an optical fiber, to another like medium, including fortuitous transfer. (188) *Note:* Examples of coupling include capacitive (electrostatic) coupling, inductive (magnetic) coupling, conducted (resistive or hard-wire) coupling, and fiber-optic coupling.

coupling coefficient: A number that expresses the degree of electrical coupling that exists between two circuits. *Note:* The coupling coefficient is calculated as the ratio of the mutual impedance to the square root of the product of the self-impedances of the coupled circuits, all impedances being expressed in the same units.

coupling efficiency: In fiber optics, the efficiency of optical power transfer between two optical components. (188) [After 2196] *Note 1:* The transfer may take place (a) between an active component, such as an LED, and a passive component, such as an optical fiber, or (b) between two passive components such as two optical fibers. *Note 2:* Coupling efficiency is usually expressed as the ratio, converted to percent, of the input power, *i.e.,* the available power from one component, to the power transferred to the other component.

coupling loss: **1.** The loss that occurs when energy is transferred from one circuit, circuit element, or medium to another. *Note:* Coupling loss is usually expressed in the same units—such as watts or dB—as in the originating circuit element or medium. (188) **2.** In fiber optics, the power loss that occurs when coupling light from one optical device or medium to another. [After 2196]

cover: The technique of concealing or altering the characteristics of communications patterns for the purpose of denying an unauthorized receiver information that would be of value. *Note: Cover* is a process of modulo two addition of a pseudorandom bit stream generated by a cryptographic device with bits from the control message. (188)

coverage: In radio communications, the geographical area within which service from a radio communications facility can be received.

CPE: *Abbreviation for* **customer premises equipment.**

cpi: *Abbreviation for* **characters per inch.**

cps: *Abbreviation for* **characters per second.** *Note: Formerly, abbreviation for* **cycles per second,** the unit used to express frequency. However, ***hertz***, an SI unit, is the proper unit for frequency.

CPU: *Abbreviation for* **central processing unit. 1.** The portion of a computer that includes circuits controlling the interpretation and execution of instructions. (188) **2.** The portion of a computer that executes programmed instructions, performs arithmetic and logical operations on data, and controls input/output functions. *Synonym* **central processor.** **3.** *Abbreviation for* **communications processor unit.** The portion of a digital communications switch that executes programmed instructions, performs arithmetic and logical operations on signals, and controls input/output functions.

CR: *Abbreviation for* **channel reliability, circuit reliability.**

CRC: *Abbreviation for* **cyclic redundancy check.**

critical angle: In geometric optics, at a refractive boundary, the smallest angle of incidence at which total internal reflection occurs. (188) *Note 1:* The angle of incidence is measured with respect to the normal at the refractive boundary. *Note 2:* The critical angle is given by

$$\theta_c = \sin^{-1}\left(\frac{n_1}{n_2}\right) ,$$

where θ_c is the critical angle, n_1 is the refractive index of the less dense medium, and n_2 is the refractive index of the denser medium. *Note 3:* The incident ray is in the denser medium. *Note 4:* If the incident ray is precisely at the critical angle, the refracted ray is tangent to the boundary at the point of incidence.

critical area: An operational area that requires specific environmental control because of the equipment or information contained therein. (188)

critical frequency: **1.** In radio propagation by way of the ionosphere, the limiting frequency at or below which a wave component is reflected by, and above which it penetrates through, an ionospheric layer. (188) **2.** At vertical incidence, the limiting frequency at or below which incidence, the wave component is reflected by, and above which it penetrates through, an ionospheric layer. (188) *Note:* The existence of the critical frequency is the result of electron limitation, *i.e.*, the inadequacy of the existing number of free electrons to support reflection at higher frequencies.

critical service: *Synonym* **essential service.**

critical technical load: That part of the total technical power load required for synchronous communications and automatic switching equipment. (188)

critical wavelength: The free-space wavelength that corresponds to the critical frequency. *Note:* The critical wavelength is equal, in meters, to the speed of light (3×10^8 m/s) divided by the critical frequency in hertz.

cross assembler: An assembler that can run symbolic-language input on one type of computer and produce machine-language output for another type of computer.

cross-band radiotelegraph procedure: A radiotelegraph network operational procedure in which calling stations, such as ship stations, call other stations, such as shore stations, using one frequency, and then shift to another frequency to transmit their messages; the called stations answer using a third frequency.

crossbar switch: A switch that has a plurality of vertical paths, a plurality of horizontal paths, and electromagnetical means, *i.e.*, relays, for interconnecting any one of the vertical paths to any one of the horizontal paths. (188)

cross-connect: *Synonym* **cross-connection.**

cross-connection: Connections between terminal blocks on the two sides of a distribution frame, or between terminals on a terminal block. (188) *Note:* Connections between terminals on the same block are also called *"straps."* *Synonyms* **cross-connect, jumper.**

cross coupling: The coupling of a signal from one channel, circuit, or conductor to another, where it is usually considered to be an undesired signal. (188)

crosslink: A data link between two satellites. (188)

cross modulation: Intermodulation caused by the modulation of the carrier of a desired signal by an undesired signal. (188)

cross-office trunk: A trunk that has its terminations within a single facility. (188)

crosspoint: A single element that (a) is in the array of elements that compose a switch and (b) consists of a set of physical or logical contacts that operate together to extend the speech and signaling channels in a switched network.

cross-polarized operation: The operation of two transmitters on the same frequency, but with polarizations in the opposite sense, *e.g.*, plane polarization with one transmitter-receiver pair being vertically polarized and the other pair horizontally polarized. (188)

cross-site link: In a satellite communications system, the signal power and control connections between the components of an Earth station. *Note:* Examples of cross-site links are (a) links between transmitters and antennas and (b) links between control consoles and transmitters.

crosstalk (XT): **1.** Undesired capacitive, inductive, or conductive coupling from one circuit, part of a circuit, or channel, to another. **2.** Any phenomenon by which a signal transmitted on one circuit or channel of a transmission system creates an undesired effect in another circuit or channel. (188) *Note:* In telephony, crosstalk is usually distinguishable as speech or signaling tones.

crosstalk coupling: The ratio of the power in a disturbing circuit to the induced power in the disturbed circuit observed at specified points of the circuits under specified terminal conditions. *Note:* Crosstalk coupling is usually expressed in dB. (188) *Synonym* **crosstalk coupling loss.**

crosstalk coupling loss: *Synonym* **crosstalk coupling.**

cryptanalysis: **1.** Operations performed in converting encrypted messages to plain text without initial knowledge of the crypto-algorithm and/or key employed in the encryption. [NIS] **2.** The study of encrypted texts.

CRYPTO: [The] marking or designator identifying COMSEC keying material used to secure or authenticate telecommunications carrying classified or sensitive U.S. Government or U.S. Government-derived information. *Note:* When written in all upper case letters, CRYPTO has the meaning stated above. When written in lower case as a prefix, crypto and crypt are abbreviations for cryptographic. [NIS]

cryptochannel: A complete system of crypto-communications between two or more holders. The basic unit for naval cryptographic communication. It includes: (a) the cryptographic aids prescribed; (b) the holders thereof; (c) the indicators or other means of identification; (d) the area or areas in which effective; (e) the special purpose, if any, for which provided; and (f) pertinent notes as to distribution, usage, *etc.* A cryptochannel is analogous to a radio circuit. [JP1]

cryptographic information: All information significantly descriptive of cryptographic techniques and processes or of cryptographic systems and equipment, or their functions and capabilities, and all cryptomaterial. [JP1]

cryptography: **1.** [The] principles, means, and methods for rendering plain information unintelligible and for restoring encrypted information to intelligible form. [NIS] **2.** The branch of cryptology that treats of the principles, means, and methods of designing and using cryptosystems.

crypto key: *Deprecated term. See* **key.**

cryptologic: Of or pertaining to cryptology. [JP1]

cryptology: The science that deals with hidden, disguised, or encrypted communications. It embraces communications security and communications intelligence. [JP1] (188)

cryptomaterial: All material including documents, devices, equipment, and apparatus essential to the encryption, decryption, or authentication of telecommunications. When classified, it is designated CRYPTO and subject to special safeguards. [JP1]

cryptonet: Stations that hold a specific key for use. *Note:* Activities that hold key for other than use, such as cryptologistic depots, are not cryptonet members for that key. Controlling authorities are *de facto* members of the cryptonets they control. [NIS]

cryptosecurity: *See* **communications security.**

cryptosystem: Associated COMSEC items interacting to provide a single means of encryption or decryption. [NIS] (188)

crystal oscillator (XO): An oscillator in which the frequency is controlled by a piezoelectric crystal. *Note 1:* A crystal oscillator may require controlled temperature because its operating frequency is a function of temperature. *Note 2:* Types of crystal oscillators include voltage-controlled crystal oscillators (VCXO), temperature-compensated crystal oscillators (TCXO), oven-controlled crystal oscillators (OCXO), temperature-compensated-voltage controlled crystal oscillators (TCVCXO), oven-controlled voltage-controlled crystal oscillators (OCVCXO), microcomputer-compensated crystal oscillators (MCXO), and rubidium crystal oscillators (RbXO).

CSMA: *Abbreviation for* **carrier sense multiple access.**

CSMA/CA: *Abbreviation for* **carrier sense multiple access with collision avoidance.**

CSMA/CD: *Abbreviation for* **carrier sense multiple access with collision detection.**

CSU: *Abbreviation for* **channel service unit, circuit switching unit, customer service unit.**

C2W: *Abbreviation for* **command and control warfare.**

CTX: *Abbreviation for* **Centrex® service.**

cursor: A movable, visible mark used to indicate a position of interest on a display surface. *Note:* A cursor may have a controllable shape, such as an underline, a rectangle, or a pointer and usually indicates where the next character or graphic will be entered or revised.

curvature loss: *Synonym* **macrobend loss.**

curve-fitting compaction: Data compaction accomplished by substituting an analytical expression for the data to be stored or transmitted. *Note:* Examples of curve-fitting compaction are (a) the breaking of a continuous curve into a series of straight line segments and specifying the slope, intercept, and range for each segment and (b) using a mathematical expression, such as a polynomial or a trigonometric function, and a single point on the corresponding curve instead of storing or transmitting the entire graphic curve or a series of points on it.

customer access: In an Integrated Services Digital Network (ISDN), the portion of the ISDN access that a network provider supplies to connect the customer, *i.e.,* subscriber, installation to the network. *Note:* Customer access includes those network elements or portions of elements that extend from the access switch to the network interface.

customer management complex: In network management, a complex that (a) is controlled by a customer and (b) is responsible for, and performs, maintenance for the customer installation.

customer office terminal: **1.** Termination equipment that (a) is located on the customer premises and (b) performs a function that may be integrated into the common carrier equipment. *Note:* An example of a customer office terminal is a stand-alone multiplexer located on the customer premises. **2.** The digital loop carrier (DLC) multiplexing function that is near the exchange termination (ET) when provided by a stand-alone multiplexer. *Note:* This function may be integrated into the ET.

customer owned and maintained equipment (COAM): *Deprecated term. See* **customer premises equipment.**

customer premises equipment (CPE): Terminal and associated equipment and inside wiring located at a subscriber's premises and connected with a carrier's communication channel(s) at the demarcation point (*"demarc"*). *Note 1:* The demarc is a point established in a building or complex to separate customer equipment from telephone company equipment. *Note 2:* Excluded from CPE are over-voltage protection equipment and pay telephones.

customer-provided equipment: *Deprecated term. See* **customer premises equipment.**

customer service unit (CSU): A device that provides an accessing arrangement at a user location to either switched or point-to-point, data-conditioned circuits at a specifically established data signaling rate. *Note:* A CSU provides local loop equalization, transient protection, isolation, and central office loop-back testing capability.

custom local area signaling service (CLASS): One of an identified group of network-provided enhanced services. *Note:* A CLASS group for a given network usually includes several enhanced service offerings, such as incoming-call identification, call trace, call blocking, automatic return of the most recent incoming call, call redial, and selective forwarding and programming to permit distinctive ringing for incoming calls.

cutback technique: A destructive technique for determining certain optical fiber transmission characteristics, such as attenuation and bandwidth, by (a) performing the desired measurements on a long length of the fiber under test, (b) cutting the fiber under test at a point near the launching end, (c) repeating the measurements on the short length of fiber, and (d) subtracting the results obtained on the short length to determine the results for the residual long length. *Note 1:* The cut should not be made less than 1 meter from the launch end. However, cutting the fiber so close to the launch end (in a multimode fiber) will introduce errors in the measurements because at that point, modal equilibrium conditions have not been established. The errors so introduced will result in conservative results (*i.e.*, higher transmission losses and lower bandwidths) than would be realized under equilibrium conditions. *Note 2:* Several characteristics may be determined using the same test fiber. *Note 3:* A variation of the cutback technique is the substitution method, in which measurements are made on a full length of fiber, and then on a short length of fiber having the same characteristics (core size, numerical aperture), with the results from the short length being subtracted to give the results for the full length.

cutoff attenuator: A waveguide, of adjustable length, which varies the attenuation of signals passing through it.

cutoff frequency: **1.** The frequency either above which or below which the output of a circuit, such as a line, amplifier, or filter, is reduced to a specified level. (188) **2.** The frequency below which a radio wave fails to penetrate a layer of the ionosphere at the incidence angle required for transmission between two specified points by reflection from the layer. (188)

cutoff mode: The highest order mode that will propagate in a given waveguide at a given frequency. (188)

cutoff wavelength: **1.** The wavelength corresponding to the cutoff frequency. (188) **2.** In an uncabled single-mode optical fiber, the wavelength greater than which a particular waveguide mode ceases to be a bound mode. (188) *Note 1:* The cutoff wavelength is usually taken to be the wavelength at which the normalized frequency is equal to 2.405. *Note 2:* The *cabled* cutoff wavelength is usually considered to be a more functional parameter because it takes into consideration the effects of cabling the fiber.

cutover: The physical changing of circuits or lines from one configuration to another. (188)

CVSD: *Abbreviation for* **continuously variable slope delta modulation.**

cw: *Abbreviation for* **carrier wave, continuous wave.**

CX: *Abbreviation for* **composite signaling.**

cxr: *Abbreviation for* **carrier.**

CX signaling: *Synonym* **composite signaling.**

cyclic distortion: In telegraphy, distortion that (a) is periodic and (b) is not characteristic, not biased, and not fortuitous. (188) *Note:* Causes of cyclic distortion include irregularities in the duration of contact time of the brushes of a transmitter distributor and interference by disturbing alternating currents.

cyclic redundancy check (CRC): An error-detection scheme that (a) uses parity bits generated by polynomial encoding of digital signals, (b) appends those parity bits to the digital signal, and (c) uses decoding algorithms that detect errors in the received digital signal. *Note:* Error correction, if required, may be accomplished through the use of an automatic repeat-request (ARQ) system.

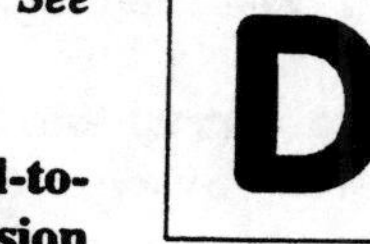

D*: *(Pronounced "D-Star") See* **specific detectivity.**

D-A: *Abbreviation for* **digital-to-analog.** *See* **digital transmission system.**

DACS: *Acronym for* **digital access and cross-connect system.**

DAMA: *Abbreviation for* **demand assignment multiple access.**

damping: **1.** The progressive diminution with time of certain quantities characteristic of a phenomenon. **2.** The progressive decay with time in the amplitude of the free oscillations in a circuit. (188)

dark current: The external current that, under specified biasing conditions, flows in a photoconductive detector when there is no incident radiation. (188)

data: Representation of facts, concepts, or instructions in a formalized manner suitable for communication, interpretation, or processing by humans or by automatic means. Any representations such as characters or analog quantities to which meaning is or might be assigned. [JP1]

data access arrangement: **1.** In public switched telephone networks, a single item or group of items at the customer side of the network interface for data transmission purposes, including all equipment that may affect the characteristics of the interface. **2.** A data circuit-terminating equipment (DCE) supplied or approved by a common carrier that permits a DCE or data terminal equipment (DTE) to be attached to the common carrier network. *Note:* Data access arrangements are an integral part of all modems built for the public telephone network.

data attribute: A characteristic of a data element such as length, value, or method of representation.

data bank: **1.** A set of data related to a given subject and organized in such a way that it can be consulted by users. **2.** A data repository accessible by local and remote users. (188) *Note:* A data bank may contain information on single or multiple subjects, may be organized in any rational manner, may contain more than one database, and may be geographically distributed. More than one data bank may be required to build a comprehensive database.

database: **1.** A set of data that is required for a specific purpose or is fundamental to a system, project, enterprise, or business. (188) *Note:* A database may consist of one or more data banks and be geographically distributed among several repositories. **2.** A formally structured collection of data. *Note:* In automated information systems, the database is manipulated using a database management system.

database engineering: The discipline involving (a) the conception, modeling, and creation, *i.e.*, programming, of a database, (b) data analysis and administration of the database, and (c) database documentation.

database management system (DBMS): A software system that facilitates (a) the creation and maintenance of a database or databases, and (b) the execution of computer programs using the database or databases. (188)

data burst: *Synonym* **burst transmission (def. #2).**

data bus: A bus used to transfer data within or to and from a processing unit or storage device.

data circuit connection: The interconnection of any combination of links and trunks, on a tandem basis, by means of switching equipment to facilitate information interchange. (188)

data circuit-terminating equipment: *See* **DCE.**

data collection facility: A facility for gathering and organizing data from a group of sources.

data communication: The transfer of information between functional units by means of data transmission according to a protocol. (188) *Note:* Data are transferred from one or more sources to one or more sinks over one or more data links.

data communication control procedure: A means used to control the orderly communication of information among stations in a data communication network.

data communications control character: *See* **control character.**

data communications equipment: *Deprecated term. See* **DCE.**

data compaction: The reduction of the number of data elements, bandwidth, cost, and time for the generation, transmission, and storage of data without loss of information by eliminating unnecessary redundancy, removing irrelevancy, or using special coding. *Note 1:* Examples of data compaction methods are the use of fixed-tolerance bands, variable-tolerance bands, slope-keypoints, sample changes, curve patterns, curve fitting, variable-precision coding, frequency analysis, and probability analysis. *Note 2:* Simply squeezing noncompacted data into a smaller space, for example by increasing packing density or by transferring data on punched cards onto magnetic tape, is not data compaction. *Note 3:* Whereas data compaction reduces the amount of data used to represent a given amount of information, data compression does not.

data compression: **1.** Increasing the amount of data that can be stored in a given domain, such as space, time, or frequency, or contained in a given message length. **2.** Reducing the amount of storage space required to store a given amount of data, or reducing the length of message required to transfer a given amount of information. *Note 1:* Data compression may be accomplished by simply squeezing a given amount of data into a smaller space, for example, by increasing packing density or by transferring data on punched cards onto magnetic tape. *Note 2:* Data compression does not reduce the amount of data used to represent a given amount of information, whereas data compaction does. Both data compression and data compaction result in the use of fewer data elements for a given amount of information.

data concentrator: A functional unit that permits a common transmission medium to serve more data sources than there are channels currently available within the transmission medium.

data conferencing repeater: A device that enables any one user of a group of users to transmit a message to all other users in that group. (188) *Synonym* **technical control hubbing repeater.**

data contamination: *Synonym* **data corruption.**

data corruption: The violation of data integrity. (188) *Synonym* **data contamination.**

data country code: A 3-digit numerical country identifier that is part of the 14-digit network terminal numbering plan. *Note:* The data country code prescribed numerical designation further constitutes a segment of the overall 14-digit X.121 numbering plan for a CCITT X.25 network.

data-dependent protection: The application of protective data elements to a data stream in such a manner that the composition of the data stream determines the amount or type of protective elements to be added.

data dictionary: **1.** A part of a database management system that provides a centralized repository of information about data in a database, such as meaning, relationship to other data. **2.** An inventory that describes, defines, and lists all of the data elements that are stored in a database.

data directory: An inventory that specifies the source, location, ownership, usage, and destination of all of the data elements that are stored in a database.

data element: **1.** A named unit of data that, in some contexts, is considered indivisible and in other contexts may consist of data items. **2.** A named identifier of each of the entities and their attributes that are represented in a database. **3.** A basic unit of information built on standard structures having a unique meaning and distinct units or values. [JP1] **4.** In electronic recordkeeping, a combination of characters or bytes referring to one separate item of information, such as name, address, or age. [JP1]

Data Encryption Standard (DES): [A] cryptographic algorithm for the protection of unclassified computer data and published by the National Institute of Standards and Technology in Federal Information Processing Standard Publication 46-1. [NIS] *Note:* DES is **not** approved for protection of national security classified information.

data forwarder: A device that (a) receives data from one data link and retransmits data representing the same information, using proper format and link protocols, to another data link and (b) may forward data between (a) links that are identical, *i.e.*, TADIL B to TADIL B, (b) links that are similar, *i.e.*, TADIL A to TADIL B, or (c) links that are dissimilar, *i.e.*, TADIL A to TADIL J.

datagram: In packet switching, a self-contained packet, independent of other packets, that contains information sufficient for routing from the originating data terminal equipment (DTE) to the destination DTE without relying on prior exchanges between the equipment and the network. *Note:* Unlike virtual call service, when datagrams are sent there are no call establishment or clearing procedures. Thus, the network may not be able to provide protection against loss, duplication, or misdelivery.

data integrity: **1.** [The] condition that exists when data is unchanged from its source and has not been accidentally or maliciously modified, altered, or destroyed. [NIS] **2.** The condition in which data are identically maintained during any operation, such as transfer, storage, and retrieval. (188) **3.** The preservation of data for their intended use. **4.** Relative to specified operations, the *a priori* expectation of data quality.

data item: **1.** A named component of a data element; usually the smallest component. **2.** A subunit of descriptive information or value classified under a data element. For example the data element "military personnel grade" contains data items such as sergeant, captain, and colonel. [JP1]

data link: **1.** The means of connecting one location to another for the purpose of transmitting and receiving data. [JP1] **2.** An assembly, consisting of parts of two data terminal equipments (DTEs) and the interconnecting data circuit, that is controlled by a link protocol enabling data to be transferred from a data source to a data sink. (188)

data link escape character (DLE): A transmission control character that changes the meaning of a limited number of contiguously following characters or coded representations.

Data Link Layer: *See* **Open Systems Interconnection—Reference Model.**

data logging: The dating, time-labeling, and recording of data. (188)

data management: The control of data handling operations—such as acquisition, analysis, translation, coding, storage, retrieval, and distribution of data—but not necessarily the generation and use of data. [From Weik '89]

data management system: *See* **database management system.**

data medium: The material in or on which one or more characteristics of the material may be used to represent information statically or dynamically. *Note:* Examples of data media are films, compact optical disks, cards, magnetic disks, magnetic drums, and paper.

data mode: In a communications network, the state of data circuit-terminating equipment (DCE) when connected to a communications channel and ready to transmit data, usually digital data. *Note:* When in the data mode, the DCE is not in a talk or dial mode.

data network identification code (DNIC): In the CCITT International X.121 format, the first four digits of the international data number, the three digits that may represent the data country code, and the 1-digit network code, *i.e.*, the network digit.

data numbering plan area (DNPA): In the U.S. implementation of a CCITT X.25 network, the first three digits of a network terminal number (NTN). *Note:* The 10-digit NTN is the specific addressing information for an end-point terminal in an X.25 network.

data phase: The phase of a data call during which data may be transferred between data terminal equipments (DTEs) that are interconnected via the network. *Note:* The data phase of a data call corresponds to the information transfer phase of an information transfer transaction.

data processing: The systematic performance of operations upon data such as handling, merging, sorting, and computing. (188) *Note:* The semantic content of the original data should not be changed. The semantic content of the processed data may be changed. *Synonym* **information processing**.

data rate: *See* **data signaling rate.**

data register: *See* **register (def. #2).**

data scrambler: A device used in digital transmission systems to convert digital signals into a pseudorandom sequence that is free from long strings of simple patterns, such as marks and spaces. *Note:* The data scrambler facilitates timing extraction, reduces the accumulation of jitter, and prevents baseline drift.

data security: [The] protection of data from unauthorized (accidental or intentional) modification, destruction, or disclosure. [NIS]

data service unit (DSU): **1.** A device used for interfacing data terminal equipment (DTE) to the public telephone network. (188) **2.** A type of short-haul, synchronous-data line driver, usually installed at a user location, that connects user synchronous equipment over a 4-wire circuit at a preset transmission rate to a servicing central-office. *Note:* This service can be for a point-to-point or multipoint operation in a digital data network.

data set: *Deprecated term. See* **DCE.**

data signaling rate (DSR): The aggregate rate at which data pass a point in the transmission path of a data transmission system. *Note 1:* The DSR is usually expressed in bits per second. *Note 2:* The data signaling rate is given by

$$\sum_{i=1}^{m} \frac{\log_2 n_i}{T_i} \quad ,$$

where m is the number of parallel channels, n_i is the number of significant conditions of the modulation in the I-th channel, and T_i is the unit interval, expressed in seconds, for the I-th channel. *Note 3:* For serial transmission in a single channel, the DSR reduces to $(1/T)\log_2 n$; with a two-condition modulation, *i.e.*, n=2, the DSR is $1/T$. *Note 4:* For parallel transmission with equal unit intervals and equal numbers of significant conditions on each channel, the DSR is $(m/T)\log_2 n$; in the case of a two-condition modulation, this reduces to m/T. *Note 5:* The DSR may be expressed in bauds, in which case, the factor $\log_2 n_i$ in the above summation formula should be deleted when calculating bauds. *Note 6:* In synchronous binary signaling, the DSR in bits per second may be numerically the same as the modulation rate expressed in bauds. Signal processors, such as four-phase modems, cannot change the DSR, but the modulation rate depends on the line modulation scheme, in accordance with Note 4. For example, in a 2400 b/s 4-phase sending modem, the signaling rate is 2400 b/s on the serial input side, but the modulation rate is only 1200 bauds on the 4-phase output side.

data signaling rate transparency: *See* **transparency.**

data sink: *See* **sink.**

data source: *See* **source.**

data station: Data terminal equipment (DTE), data circuit-terminating equipment (DCE), and any intermediate equipment connected at one location. *Note:* The DCE may be connected directly to a data processing system or it may be a part of the data processing system.

data stream: A sequence of digitally encoded signals used to represent information in transmission.

data subscriber terminal equipment: In the DDN, a general purpose terminal device that consists of (a) all the equipment necessary to provide interface functions, perform code conversion, and transform messages on various data media, such as punched cards, magnetic tapes, and paper tapes, to electrical signals for transmission and (b) all the equipment necessary to convert received electrical signals into data stored or recorded on various data media. [From Weik '89]

data switching exchange (DSE): The equipment installed at a single location to perform switching functions such as circuit switching, message switching, and packet switching.

data terminal equipment: *See* **DTE**.

data transfer rate: The average number of bits, characters, or blocks per unit time passing between corresponding equipment in a data transmission system. (188)

data transfer request signal: A call control signal transmitted by the data circuit-terminating equipment (DCE) to the data terminal equipment (DTE) indicating that a request signal, originated by a distant DTE, has been received from a distant DCE to exchange data with the station.

data transfer time: The time between (a) the instant at which a user data unit, such as a character, word, block, or message, is made available to a network for transfer by a transmitting data terminal equipment (DTE) and (b) the receipt of that complete data unit by a receiving DTE. (188)

data transmission: The sending of data from one place to another by means of signals over a channel. (188)

data transmission circuit: The transmission media and the intervening equipment used for the transfer of data between data terminal equipments (DTEs). (188) *Note 1:* A data transmission circuit includes any required signal conversion equipment. *Note 2:* A data transmission circuit may transfer information in (a) one direction only, (b) either direction but one way at a time, or (c) both directions simultaneously.

data volatility: Pertaining to the rate of change in the values of stored data over a period of time.

date: An instant in the passage of time, identified with desired precision by a clock and a calendar. *Note:* An example of a date is 23 seconds after 3:14 PM on February 9, 1926. This date might be represented as 1926FEB091514.23.

date-time group (DTG): In a message, a set of characters, usually in a prescribed format, used to express the day of the month, the hour of the day, the minute of the hour, the time zone, and the year. *Note 1:* The DTG is usually placed in the header of the message. *Note 2:* The DTG may be used as a message identifier if it is unique for each message. *Note 3:* The DTG may indicate either the date and time a message was dispatched by a transmitting station or the date and time it was handed into a transmission facility by a user or originator for dispatch.

dating format: The format used to express the time of an event. (188) *Note:* The time of an event on the UTC time scale is given in the following sequence: hour, day, month, year; *e.g.,* 0917 UT, 30 August 1997. The hour is designated by the 24-hour system.

dB: *Abbreviation for* **decibel(s).** One tenth of the common logarithm of the ratio of relative powers, equal to 0.1 B (bel). *Note 1:* The decibel is the conventional relative power ratio, rather than the bel, for expressing relative powers because the decibel is smaller and therefore more convenient than the bel. The ratio in dB is given by

$$dB = 10 \log_{10}\left(\frac{P_1}{P_2}\right),$$

where P_1 and P_2 are the actual powers. Power ratios may be expressed in terms of voltage and impedance, E and Z, or current and impedance, I and Z, since

$$P = I^2 Z = \frac{E^2}{Z}.$$

Thus dB is also given by

$$dB = 10 \log_{10}\left(\frac{E_1^2/Z_1}{E_2^2/Z_2}\right) = 10 \log_{10}\left(\frac{I_1^2 Z_1}{I_2^2 Z_2}\right).$$

If $Z_1 = Z_2$, these become

$$dB = 20 \log_{10}\left(\frac{E_1}{E_2}\right) = 20 \log_{10}\left(\frac{I_1}{I_2}\right).$$

Note 2: The dB is used rather than arithmetic ratios or percentages because when circuits are connected in tandem, expressions of power level, in dB, may be arithmetically added and subtracted. For example, in an optical link, if a known amount of optical power, in dBm, is launched into a fiber, and the losses, in dB, of each component (*e.g.*, connectors, splices, and lengths of fiber) are known, the overall link loss may be quickly calculated with simple addition and subtraction.

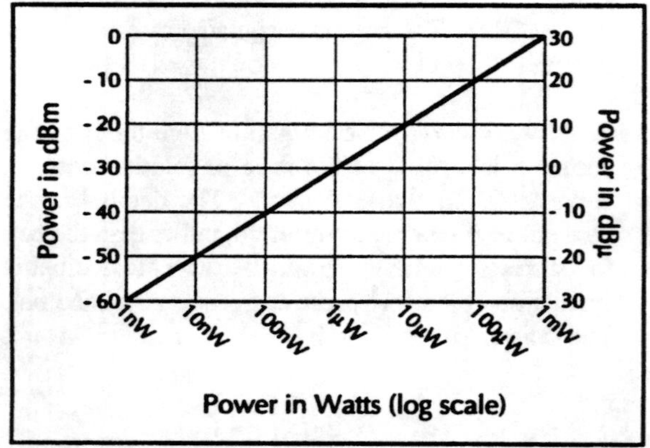

relationship between absolute power and dBm, dBμ

dBa: *Abbreviation for* **decibels adjusted.** Weighted absolute noise power, calculated in dB referenced to 3.16 picowatts (–85 dBm), which is 0 dBa. (188) *Note:* The use of F1A-line or HA1-receiver weighting must be indicated in parentheses as required. A one-milliwatt, 1000-Hz tone will read +85 dBa, but the same power as white noise, randomly distributed over a 3-kHz band (nominally 300 to 3300 Hz), will read +82 dBa, due to the frequency weighting. *Synonym* **dBrn adjusted.**

dBa(F1A): Weighted absolute noise power in dBa, measured by a noise measuring set with F1A-line weighting. (188) *Note:* F1A weighting is no longer used for DOD applications.

dBa(HA1): Weighted noise power in dBa, measured across the receiver of a 302-type or similar subscriber set, by a noise measuring set with HA1-receiver weighting. (188) *Note:* HA1 weighting is no longer used in DOD applications.

dBa0: Noise power in dBa referenced to or measured at zero transmission level point (0TLP), also called a point of zero relative transmission level (0 dBr). (188) *Note:* It is preferred to convert noise readings from dBa to dBa0, as this makes it unnecessary to know or state the relative transmission level at the point of actual measurement.

dBc: *Abbreviation.* dB relative to the carrier power.

dBi: *Abbreviation.* In the expression of antenna gain, the number of decibels of gain of an antenna referenced to the zero dB gain of a free-space isotropic radiator.

dBm: *Abbreviation.* dB referenced to one milliwatt. (188) *Note 1:* dBm is used in communication work as a measure of absolute power values. Zero dBm equals one milliwatt. *Note 2:* In DOD practice, unweighted measurement is normally understood, applicable to a certain bandwidth, which must be stated or implied. *Note 3:* In European practice, psophometric weighting may be implied, as indicated by context; equivalent to dBm0p, which is preferred.

dBm(psoph): Noise power in dBm, measured with psophometric weighting where

$$dBm\,(psoph) = 10 \log_{10} (pWp) - 90$$
$$= dBa - 84 ,$$

where *pWp* is power in picowatts psophometrically weighted and *dBa* is the weighted noise power in dB referenced to 3.16 picowatts.

DBMS: *Abbreviation for* **database management system.**

dBmV: *Abbreviation.* dB referenced to one millivolt across 75 ohms. (188) *Note:* This reference is not equivalent to dBm; it is, in fact, 1.33×10^{-5} milliwatts.

dBm0: Power in dBm referred to or measured at a zero transmission level point (0TLP). *Note 1:* A 0TLP is also called a point of zero relative transmission level (0 dBr0). (188) *Note 2:* Some international documents use dBm0 to mean noise power in dBm0p (psophometrically weighted dBm0). In the United States, dBm0 is not so used.

dBm0p: Noise power in dBm0, measured by a psophometer or noise measuring set having psophometric weighting. (188)

dBr: *Abbreviation.* The power ratio, expressed in dB, between any point and a reference point selected as the zero relative transmission level point. (188) *Note:* Any power expressed in dBr does not specify the absolute power. It is a relative measurement only.

dBrn: *Abbreviation.* dB above reference noise. *Note 1:* Weighted noise power in dB is referred to 1.0 picowatt. Thus, 0dBrn = –90 dBm. Use of 144-line, 144-receiver, or C-message weighting, or flat weighting, must be indicated in parentheses as required. (188) *Note 2:* With C-message weighting, a one-milliwatt, 1000-Hz tone will read +90 dBrn, but the same power as white noise, randomly distributed over a 3-kHz band will read approximately +88.5 dBrn (rounded off to +88 dBrn), because of the frequency weighting. *Note 3:* With 144 weightings, a one-milliwatt, 1000-Hz white noise tone will also read +90 dBrn, but the same 3-kHz power will only read +82 dBrn, because of the different frequency weighting.

dBrn adjusted: *Synonym* **dBa.**

dBrnC: Weighted noise power in dBrn, measured by a noise measuring set with C-message weighting. (188)

dBrnC0: Noise power in dBrnC referred to or measured at a zero transmission level point (0TLP). (188)

dBrn(f_1-f_2): Flat noise power in dBrn, measured over the frequency band between frequencies f_1 and f_2. (188)

dBrn(144-line): Weighted noise power in dBrn, measured by a noise measuring set with 144-line weighting. (188)

dBv: *Abbreviation.* dB relative to 1 volt peak-to-peak. *Note:* The dBv is usually used for television video signal level measurements. [From Weik '89]

dBW: *Abbreviation.* dB referenced to one watt. (188)

dBx: *Abbreviation.* dB above reference coupling. *Note:* dBx is used to express the amount of crosstalk coupling in telephone circuits. dBx is measured with a noise measuring set. [From Weik '89]

dc: *Abbreviation for* **direct current.**

DCA: *Abbreviation for* **Defense Communications Agency.** *Now* **DISA** (Defense Information Systems Agency).

DCE: *Abbreviation for* **data circuit-terminating equipment.** **1.** In a data station, the equipment that performs functions, such as signal conversion and coding, at the network end of the line between the data terminal equipment (DTE) and the line, and that may be a separate or an integral part of the DTE or of intermediate equipment. **2.** The interfacing equipment that may be required to couple the data terminal equipment (DTE) into a transmission circuit or channel and from a transmission circuit or channel into the DTE. (188) *Synonyms* **data communications equipment** *(deprecated)*, **data set** *(deprecated).*

DCE clear signal: A call control signal transmitted by data circuit-terminating equipment (DCE) to indicate that it is clearing the associated circuit after a call is finished.

DCE waiting signal: A call control signal at the data-circuit-terminating-equipment/data-terminal-equipment (DCE/DTE) interface that indicates that the DCE is ready for another event in the call establishment procedure.

D channel: In ISDN, the 16-kb/s segment of a 144-kb/s, full-duplex subscriber service channel that is subdivided into 2B+D channels, *i.e.*, into two 64-kb/s clear channels and one 16-kb/s channel for the ISDN basic rate. *Note 1:* The D channel is usually used for out-of-band signaling. The two 64-kb/s clear channels are used for subscriber voice and data services. *Note 2:* The D-channel specifications are addressed in the CCITT Recommendation for the Integrated Services Digital Network (ISDN). *Note 3:* The D-channel may be 64 kb/s for the primary rate ISDN service.

dc patch bay: A patch bay in which dc circuits are grouped. (188)

DCS: *Abbreviation for* **Defense Communications System.**

DDD: *Abbreviation for* **direct distance dialing.**

DDN: *Abbreviation for* **Defense Data Network.**

deadlock: **1.** Unresolved contention for the use of a system or component. [From Weik '89] **2.** In computer and data processing systems, an error condition such that processing cannot continue because each of two components or processes is waiting for an action or response from the other. [From Weik '89] **3.** A permanent condition in which a system cannot continue to function unless some corrective action is taken. [From Weik '89]

dead sector: In facsimile, the interval between (a) the end of scanning of one object line and (b) the start of scanning of the following line. (188)

dead space: The area, zone, or volume of space that is within the expected range of a radio, radar, or other transmitted signal but in which the signal is not detectable and therefore cannot be received. [From Weik '89]

debug: To detect, trace, and eliminate mistakes. (188)

deception: *See* **electronic deception.**

deception repeater: A device that can (a) receive a signal, (b) amplify, delay, or otherwise manipulate the signal, and (c) retransmit it solely for creating deception. [From Weik '89]

decibel: *See* **dB.**

decipher: [To] convert enciphered text to the equivalent plain text by means of a cipher system. [NIS] *Note:* This does not include solution by cryptanalysis.

decision circuit: A circuit that measures the probable value of a signal element and makes an output signal decision based on the value of the input signal and a predetermined criterion or criteria. (188)

decision instant: In the reception of a digital signal, the instant at which a decision is made by a receiving device as to the probable value of a signal condition. (188) *Synonym* **selection position.**

decode: **1.** To convert data by reversing the effect of previous encoding. (188) **2.** To interpret a code. **3.** [To] convert encoded text into equivalent plain text by means of a code. [NIS] (188) *Note:* Decoding does not include deriving plain text by cryptanalysis.

decollimation: In a beam with the minimum possible ray divergence or convergence, any mechanism by which rays are caused to diverge or converge from parallelism. *Note 1:* Decollimation may be deliberate for systems reasons, or may be caused by many factors, such as refractive index inhomogeneities, occlusions, scattering, deflection, diffraction, reflection, and refraction. *Note 2:* Decollimation occurs in applications such as radio, radar, sonar, and optical communications.

decrypt: **1.** [A] generic term encompassing decode and decypher. [NIS] **2.** To convert encrypted text into its equivalent plain text by means of a cryptosystem. (This does not include solution by cryptanalysis.) *Note:* The term *"decrypt"* covers the meanings of *"decipher"* and *"decode."* [JP1]

dedicated circuit: A circuit designated for exclusive use by specified users. *Note:* DOD normally considers a dedicated circuit to be between two users only. (188)

dedicated service: In a communications system, a specified set of functions provided to designated users. (188) *Note:* Dedicated service is usually specified in a communications format, such as voice, digital data, facsimile, or video.

deemphasis: In FM transmission, the process of restoring (after detection) the amplitude-vs.-frequency characteristics of the signal. (188)

deeply depressed cladding fiber: An optical fiber construction, usually a single-mode fiber, that has an outer cladding of approximately the same refractive index as the core, and an inner cladding of very low (depressed) refractive index material between them.

deep space: Space at distances from the Earth equal to or greater than 2×10^6 kilometers. [NTIA] [RR]

***de facto* standard:** A standard that is widely accepted and used, but lacks formal approval by a recognized standards organization.

default: Pertaining to the pre-defined initial, original, or specific setting, condition, value, or action a system will assume, use, or take in the absence of instructions from the user. [From Weik '89]

Defense Communications System (DCS): Department of Defense long-haul voice, data, and record traffic system which includes the Defense Data Network, Defense Satellite Communications Systems, and Defense Switched Network.

Defense Data Network (DDN): A component of the Defense Communications System used for switching Department of Defense automated data processing systems. [JP1-A]

Defense Switched Network (DSN): A component of the Defense Communications System that handles Department of Defense voice, data, and video communications.

definition: A figure of merit for image quality. (188) *Note:* In an image, definition is usually expressed in terms of the smallest resolvable element, such as lines per inch, or pels per square inch.

deflection: **1.** A change in the direction of a traveling particle, usually without loss of particle kinetic energy, representing a change in velocity without a change in the scalar speed of the particle. **2.** A change in the direction of a wave, beam, electron, or other entity, such as might be accomplished by an electric or magnetic field. *Note:* If the deflection is caused by a prism (refraction), a mirror (reflection), or optical grating (diffraction), the specific applicable term should be used. [From Weik '89]

degradation: **1.** The deterioration in quality, level, or standard of performance of a functional unit. **2.** In communications, a condition in which one or more of the required performance parameters fall outside predetermined limits, resulting in a lower quality of service. *Note:* Degradation is usually categorized as either *"graceful"* or *"catastrophic."*

degraded service state: The condition that exists when one or more of the required service performance parameters fall outside predetermined limits, resulting in a lower quality of service. (188) *Note:* A degraded service state is considered to exist when a specified level of degradation persists for a specified period of time.

degree of coherence: A dimensionless unit, expressed as a ratio, used to indicate the extent of coherence of an electromagnetic wave such as a lightwave. (188) [After 2196] *Note 1:* For lightwaves, the magnitude of the degree of coherence is equal to the visibility, *V*, of the fringes of a two-beam interference test, as given by

$$V = \frac{I_{max} - I_{min}}{I_{max} + I_{min}} ,$$

where I_{max} is the intensity at a maximum of the interference pattern, and I_{min} is the intensity at a minimum. *Note 2:* Light is considered to be highly coherent when the degree of coherence exceeds 0.88, partially coherent for values less than 0.88 but more than nearly zero values, and incoherent for nearly zero and zero values.

degree of isochronous distortion: In data transmission, the ratio of (a) the absolute value of the maximum measured difference between the actual and the theoretical intervals separating any two significant instants of modulation (or demodulation) to (b) the unit interval. *Note 1:* These instants are not necessarily consecutive. (188) *Note 2:* The degree of isochronous distortion is usually expressed as a percentage. *Note 3:* The result of the measurement should be qualified by an indication of the period, usually limited, of the observation. For a prolonged modulation (or demodulation), it will be appropriate to consider the probability that an assigned value of the degree of distortion will be exceeded.

degree of start-stop distortion: **1.** In asynchronous data transmission, the ratio of (a) the absolute value of the maximum measured difference between the actual and theoretical intervals separating any significant instant of modulation (or demodulation) from the significant instant of the start element immediately preceding it to (b) the unit interval. **2.** The highest absolute value of individual distortion affecting the significant instants of a start-stop modulation. *Note:* The degree of distortion of a start-stop modulation (or demodulation) is usually expressed as a percentage. Distinction can be made between the degree of late (positive) distortion and the degree of early (negative) distortion.

dehop: To modify a frequency-hopping signal so that it has a constant center frequency. (188)

dejitterizer: A device that reduces jitter in a digital signal. *Note 1:* A dejitterizer usually consists of an elastic buffer in which the signal is temporarily stored and then retransmitted at a rate based on the average rate of the incoming signal. *Note 2:* A dejitterizer is usually ineffective in dealing with low-frequency jitter, such as waiting-time jitter.

delay: **1.** The amount of time by which an event is retarded. **2.** The time between the instant at which a given event occurs and the instant at which a related aspect of that event occurs. (188) *Note 1:* The events, relationships, and aspects of the entity being delayed must be precisely specified. *Note 2:* Total delay may be demonstrated by the impulse response of a device or system. *Note 3:* In analog systems, total delay is described in terms of the transfer functions in the frequency domain. *Synonym* **delay time.** **3.** In radar, the electronic delay of the start of the time base used to select a particular segment of the total.

delay distortion: In a waveform consisting of two or more wave components at different frequencies, distortion caused by the difference in arrival times of the frequency components at the output of a transmission system. (188) *Synonyms* **time-delay distortion.**

delayed-delivery facility: In a communications network, a facility that stores data, destined for delivery to one or more addresses, for delivery at a later time.

delay encoding: The encoding of binary data to form a two-level signal such that (a) a "0" causes no change of signal level unless it is followed by another "0" in which case a transition to the other level takes place at the end of the first bit period; and (b) a "1" causes a transition from one level to the other in the middle of the bit period. *Note:* Delay encoding is used primarily for encoding radio signals because the frequency spectrum of the encoded signal contains less low-frequency energy than a conventional non-return-to-zero (NRZ) signal and less high-frequency energy than a biphase signal.

delay equalizer: A corrective network designed to make the phase delay or envelope delay of a circuit or system substantially constant over a desired frequency range. (188)

delay line: **1.** A transmission line, or equivalent device, used to delay a signal. (188) **2.** A single-input-channel device, such as a single-input sequential logic element, in which the output channel state at a given instant, *t*, is the same as the input channel state at the instant *t–n*, where n is a number of time units, *i.e.*, the input sequence undergoes a delay of *n* time units, such as *n* femtoseconds, nanoseconds, or microseconds. *Note:* The delay line may have additional taps yielding output channels with values less than *n*. (188)

delay modulation: *See* **delay encoding.**

delay time: *Synonym* **delay (def. #1).**

delay working: In telephone switchboard operations, operation intended to ensure fair distribution of the time of one or more lines among groups of call originators. *Note:* An example of delay working is the withdrawing of one or more lines from general use and placing them under the control of a delay operator so that when other operators book call demands on tickets, the tickets can be passed to the delay operator for connection in the order in which they are booked. [From Weik '89]

deleted bit: A bit not delivered to the intended destination. (188)

deleted block: A block not delivered to the intended destination. (188)

delimiter: **1.** A character used to indicate the beginning and end of a character string, *i.e.*, a symbol stream, such as words, groups of words, or frames. **2.** A flag that separates and organizes items of data.

delivered block: A successfully transferred block.

delivered overhead bit: A bit that (a) is successfully transferred to a destination user, (b) performs its primary function within the telecommunications system, and (c) does not represent user information.

delivered overhead block: A block that (a) is successfully transferred to a destination user, (b) performs its primary function within the telecommunications system, and (c) does not contain user information bits.

delivery confirmation: Information returned to the originator indicating that a given unit of information has been delivered to the intended addressee(s).

Dellinger effect: An effect—lasting from several minutes to several hours—that causes electromagnetic sky wave signals to disappear rapidly as a result of greatly increased ionization in the ionosphere caused by increased noise from solar storms. *Synonyms:* **Dellinger fadeout, Dellinger fading.** [From Weik '89]

Dellinger fadeout: *Synonym* **Dellinger effect.**

Dellinger fading: *Synonym* **Dellinger effect.**

delta modulation (DM): Analog-to-digital signal conversion in which (a) the analog signal is approximated with a series of segments, (b) each segment of the approximated signal is compared to the original analog wave to determine the increase or decrease in relative amplitude, (c) the decision process for establishing the state of successive bits is determined by this comparison, and (d) only the change of information is sent, *i.e.*, only an increase or decrease of the signal amplitude from the previous sample is sent whereas a no-change condition causes the modulated signal to remain at the same 0 or 1 state of the previous sample. (188) *Note:* Examples of delta modulation are continuously variable slope delta modulation, delta-sigma modulation, and differential modulation.

delta-sigma modulation: Delta modulation in which the integral of the input signal is encoded rather than the signal itself. (188) *Note:* Delta-sigma modulation may be achieved by preceding a conventional delta-modulation encoder with an integrating network.

demand assignment: An operation in which several users share access to a communications channel on a real-time basis, *i.e.*, a user needing to communicate with another user on the same network requests the required circuit, uses it, and when the call is finished, the circuit is released, making the circuit available to other users. *Note:* Demand assignment is similar to conventional telephone switching, in which common trunks are provided for many users, on a demand basis, through a limited-size trunk group. (188)

demand assignment multiple access (DAMA): In a communications system, a technique for allocating use of bandwidth among multiple users, based on demand. *Note:* DAMA can be implemented in many ways including TDM and FDM.

demand factor: **1.** The ratio of (a) the maximum real power consumed by a system to (b) the maximum real power that would be consumed if the entire load connected to the system were to be activated at the same time. (188) *Note:* The maximum real power is usually integrated over a specified time interval, such as 15 or 30 minutes, and is usually expressed in kilowatts. The real power that would be consumed if the entire load connected to the system were to be activated at the same time is obtained by summing the power required by all the connected equipment. This load is expressed in kilowatts if the consumed real power is expressed in kilowatts. **2.** The ratio of (a) the maximum power, integrated over a specified time interval, such as 15 or 30 minutes, and usually expressed in kilowatts, consumed by a system, to (b) the maximum volt-amperes, expressed in kilovolt-amperes if the power is expressed in kilowatts, integrated over a time interval of the same duration, though not necessarily during the same interval. *Note:* Charges for electrical power may be based on the demand factor as well as the kilowatt-hours of electrical energy consumed.

demand load: **1.** In general, the total power required by a facility. (188) *Note:* The demand load is the sum of the operational load (including any tactical load) and nonoperational demand loads. It is determined by applying the proper demand factor to each of the connected loads and a diversity factor to the sum total. **2.** At a communications center, the power required by all automatic switching, synchronous, and terminal equipment (operated simultaneously on-line or in standby), control and keying equipment, plus lighting, ventilation, and air-conditioning equipment required to maintain full continuity of communications. (188) **3.** The power required for ventilating equipment, shop lighting, and other support items that may be operated simultaneously with the technical load. (188) **4.** The sum of the technical demand and nontechnical demand loads of an operating facility. (188) **5.** At a receiver facility, the power required for all receivers and auxiliary equipment that may be operated on prime or spare antennas simultaneously, those in standby condition, multicouplers, control and keying equipment, plus lighting, ventilation, and air conditioning equipment required for full continuity of communications. (188) **6.** At a transmitter facility, the power required for all transmitters and auxiliary equipment that may be operated on prime or spare antennas or dummy loads simultaneously, those in standby condition, control and keying equipment, plus lighting, ventilation, and air conditioning equipment required for full continuity of communications. (188)

demand service: In ISDN applications, a telecommunications service that establishes an immediate communication path in response to a user request made through user-network signaling.

demarc: *Acronym for* **demarcation point.**

demarcation point (demarc): That point at which operational control or ownership of communications facilities changes from one organizational entity to another. *Note:* The demarcation point is usually the interface point between customer-premises equipment and external network service provider equipment. *Synonym* **network terminating interface.**

democratically synchronized network: A mutually synchronized network in which all clocks in the network are of equal status and exert equal amounts of control on the others, the network operating clock pulse repetition rate being the mean of the natural (uncontrolled) clock pulse repetition rates of the population of clocks.

demodulation: The recovery, from a modulated carrier, of a signal having substantially the same characteristics as the original modulating signal. (188)

demultiplex (DEMUX): *See* **demultiplexing.**

demultiplexing: The separation of two or more channels previously multiplexed; *i.e.*, the reverse of multiplexing. (188)

deMUX: *Acronym for* **demultiplex** and **demultiplexer.**

dense binary code: A binary code in which all possible bit patterns that can be made from a fixed number of bits are used to encode user information but no overhead information. *Note:* Examples of dense binary codes are (a) a pure binary representation for sexadecimal digits using all sixteen possible patterns and (b) an octal representation using all eight patterns. A binary representation of decimal numbers using four binary digits of which only 10 of the possible 16 patterns are used is not a dense binary code. If a binary code is not dense, the unused patterns can be used to detect errors inasmuch as they should only occur if there is an error. [From Weik '89]

density: In a facsimile system, a measure of the light transmission or reflection properties of an area of an object. (188) *Note 1:* Density is usually expressed as the logarithm to the base 10 of the ratio of incident to transmitted or reflected irradiance. *Note 2:* There are many types of density, such as diffuse, double diffuse, and specular density, each of which will usually have different numerical values for different materials. The relevant type of density depends on the type of optical system, the component materials of the object, and the surface characteristics of the object.

Department of Defense (DOD) master clock: *See* **DOD master clock.**

departure angle: The angle between the axis of the main lobe of an antenna pattern and the horizontal plane at the transmitting antenna. (188) *Synonym* **takeoff angle.**

depolarization: 1. Reducing or randomizing the polarization of an electromagnetic wave. *Note:* Depolarization may be caused by transmission through a nonhomogeneous medium or a depolarizer. (188) **2.** Prevention of polarization in an electric cell or battery. (188)

depressed-cladding fiber: *Synonym* **doubly clad fiber.**

depressed-inner-cladding fiber: *Synonym* **doubly clad fiber.**

dequeue: *Abbreviation for* **double-ended queue.**

deregulation: A reduction in regulation of (a) tariffs, (b) market entry and exit, and/or (c) facilities in public telecommunication services.

DES: *Abbreviation for* **Data Encryption Standard.**

descrambler: The inverse of a scrambler. *Note:* The descrambler output is a signal restored to the state that it had when it entered the associated scrambler, provided that no errors have occurred.

desensitation: The reduction of desired signal gain as a result of receiver reaction to an undesired signal. *Note:* The gain reduction is generally due to overload of some portion of the receiver (*e.g.*, the AGC circuitry) resulting in desired signal suppression because the receiver will no longer respond linearly to incremental changes in input voltage.

design margin: The additional performance capability above required standard basic system parameters that may be specified by a system designer to compensate for uncertainties. (188)

design objective (DO): In communications systems, a desired performance characteristic for communications circuits and equipment that is based on engineering analyses, but (a) is not considered feasible to mandate in a standard, or (b) has not been tested. (188) *Note 1:* DOs are used because applicable systems standards are not in existence. *Note 2:* Examples of reasons for designating a performance characteristic as a DO rather than as a standard are (a) it may be bordering on an advancement in the state of the art, (b) the requirement may not have been fully confirmed by measurement or experience with operating circuits, and (c) it may not have been demonstrated that the requirement can be met considering other constraints, such as cost and size. *Note 3:* A DO is sometimes established in a standard for developmental consideration. A DO may also specify a performance characteristic that may be used in the preparation of specifications for development or procurement of new equipment or systems.

despotically synchronized network: A synchronized network in which a unique master clock controls all other clocks in the network. [From Weik '89]

despun antenna: In a rotating communications satellite, an antenna with a main beam that is continuously redirected with respect to the satellite so that the antenna illuminates a given area on the surface of the Earth, *i.e.*, the footprint does not move with respect to the Earth. *Note:* An antenna may be despun mechanically or electrically.

destination routing: In communications system operations, the routing of messages based on the name of the destination office, the destination user, or the address on the message, *i.e.*, the addressee. [From Weik '89]

destination user: In an information transfer transaction, the user that receives information from the source, *i.e.*, from the originating user.

de-stuffing: The controlled deletion of stuffing bits from a stuffed digital signal, to recover the original signal. *Synonyms* **negative justification, negative pulse stuffing.**

detection: 1. The recovery of information from an electrical or electromagnetic signal. (188) *Note:* Conventional radio waves are usually detected by heterodyning, *i.e.*, coherent reception/detection. In this method of reception/detection, the received signal is mixed, in some type of nonlinear device, with a signal from a local oscillator, to produce an intermediate frequency, *i.e.*, beat frequency, from which the modulating signal is recovered, *i.e.*, detected. The inherent instabilities of available optical sources have, until relatively recently, prevented practical use of coherent reception/detection in optical communication receivers. At present, coherent optical receivers, using sophisticated technology, are just beginning to emerge from the laboratory into the field. Virtually all existing optical receivers employ direct detection; that is, the received optical signal impinges directly onto a detector. Direct detection is less sensitive than coherent detection. [After FAA] **2.** In tactical operations, the perception of an object of possible military interest but unconfirmed by recognition. [JP1] **3.** In surveillance, the determination and transmission by a surveillance system that an event has occurred. [JP1]

detectivity: The reciprocal of noise equivalent power.

detector: **1.** A device that is responsive to the presence or absence of a stimulus. **2.** In an AM radio receiver, a circuit or device that recovers the signal of interest from the modulated wave. *Note:* In FM reception, a circuit called a discriminator is used to convert frequency variations to amplitude variations. **3.** In an optical communications receiver, a device that converts the received optical signal to another form. *Note:* Currently, this conversion is from optical to electrical power; however, optical-to-optical techniques are under development.

deterministic routing: **1.** In a switched network, switching in which the routes between given pairs of nodes are pre-programmed, *i.e.*, are determined, in advance of transmission. *Note:* The routes used to complete a given call through a network are identified, in advance of transmission, in routing tables maintained in each switch database. The tables assign the trunks that are to be used to reach each switch code, area code, and International Access Prefix (IAP), usually with one or two alternate routes. **2.** In a non-switched network, routing in which the routes between given pairs of nodes are determined in advance. *Note:* The routes used to send a given message through a network are identified in advance in routing tables maintained in a database.

deterministic transfer mode: An asynchronous transfer mode in which the maximum information transfer capacity of a telecommunication service is provided throughout a call.

Deutsches Institut für Normung (DIN): Germany's standards-setting organization, equivalent to the American National Standards Institute (ANSI).

deviation ratio: In a frequency modulation system, the ratio of the maximum frequency deviation of the carrier to the maximum modulating frequency of the system under specified conditions. (188)

D4: *See* **channel bank.**

D-4: A framing standard for traditional time-division multiplexing, which standard describes user channels multiplexed onto a trunk that has been segmented (framed) into 24 bytes of 8 bits each. *Note:* The multiplexing function is performed in the D-4 framing structure by interleaving bits of consecutive bytes as they are presented from individual circuits into each D-4 frame.

DFSK: *Abbreviation for* **double-frequency shift keying.**

diad: *Synonym* **dibit.**

diagnostic program: **1.** A program used to investigate the cause or the nature of conditions or problems within specified elements of a system. (188) **2.** A computer program that detects, locates, or identifies a fault in equipment, an error in input data, or an error in a computer program. (188)

dialing: In a communications system, using a device that generates signals for selecting and establishing connections. *Note:* The term *"dialing"* is often used to designate or refer to all calling devices used for inserting data to establish connections. [From Weik '89]

dial mode: Operation of data circuit-terminating equipment (DCE) so that circuitry, such as data terminal equipment (DTE), associated with call origination is directly connected to a communications channel.

dial pulse: A dc pulse produced by an end instrument that interrupts a steady current at a sequence and rate determined by the selected digit and the operating characteristics of the instrument. (188)

dial pulsing: Pulsing in which a direct-current pulse train is produced by interrupting a steady signal according to a fixed or formatted code for each digit and at a standard pulse repetition rate. *Note:* Dial pulsing originated with rotary mechanical devices integrated into telephone instruments, for the purpose of signaling. Subsequent applications use electronic circuits to generate dial pulses. *Synonym* **pulsing.**

dial service assistance (DSA): A network-provided service feature, associated with the switching center equipment, in which services, such as directory assistance, call interception, random conferencing, and precedence calling assistance, are rendered by an attendant. (188)

dial signaling: Signaling in which dual tone multifrequency (DTMF) signals or pulse trains are transmitted to a switching center. *Note 1:* Rotary dials produce pulse trains. Keypads may produce either DTMF signals or pulse trains. *Note 2:* Dial signaling traditionally refers to pulse trains only. (188)

dial switching equipment: Switching equipment actuated by electrical impulses generated by a dial or key pulsing arrangement. [47CFR]

dial through: A technique, applicable to access circuits, that permits an outgoing routine call to be dialed by the PBX user after the PBX attendant has established the initial connection. (188)

dial tone: A tone employed in a dial telephone system to indicate to the calling party that the equipment is ready to receive dial or tone pulses.

dial-tone delay: The time between the instant of going off-hook and the instant of receiving a dial tone.

dial-up: **1.** A service feature in which a user initiates service on a previously arranged trunk or transfers, without human intervention, from an active trunk to a standby trunk. (188) **2.** A service feature that allows a computer terminal to use telephone systems to initiate and effect communications with other computers.

diametral index of cooperation: *Synonym* **index of cooperation.**

dibit: A group of two bits. *Note:* The four possible states for a dibit are 00, 01, 10, and 11. (188) *Synonym* **diad.**

dichroic filter: An optical filter that reflects one or more optical bands or wavelengths and transmits others, while maintaining a nearly zero coefficient of absorption for all wavelengths of interest. *Note:* A dichroic filter may be high-pass, low-pass, band-pass, or band rejection.

dichroic mirror: A mirror used to reflect light selectively according to its wavelength. (188)

DID: *Abbreviation for* **direct inward dialing.**

dielectric: **1.** A substance in which an electric field may be maintained with zero or near-zero power dissipation, *i.e.*, the electrical conductivity is zero or near zero. *Note 1:* A dielectric material is an electrical insulator. *Note 2:* In a dielectric, electrons are bound to atoms and molecules, hence there are few free electrons. **2.** Pertaining to a substance that has a zero or near zero electrical conductivity.

dielectric filter: *See* **interference filter.**

dielectric lens: In the radio regime, a lens made of dielectric material that refracts radio waves in the same manner that an optical lens refracts light waves. (188)

dielectric strength: **1.** Of an insulating material, the maximum electric field strength that it can withstand intrinsically without breaking down, *i.e.*, without experiencing failure of its insulating properties. *Note:* The theoretical dielectric strength of a material is an intrinsic property of the bulk material and is not dependent on the configuration of the material or the electrodes with which the field is applied. **2.** For a given configuration of dielectric material and electrodes, the minimum electric field that produces breakdown. *Note 1:* At breakdown, the electric field frees bound electrons, turning the material into a conductor. *Note 2:* The field strength at which breakdown occurs in a given case is dependent on the respective geometries of the dielectric (insulator) and the electrodes with which the electric field is applied, as well as the rate of increase at which the electric field is applied. *Note 3:* The electric field strength is usually expressed in volts per meter. (188)

dielectric waveguide: A waveguide that consists of a dielectric material surrounded by another dielectric material, such as air, glass, or plastic, with a lower refractive index. *Note 1:* An example of a dielectric waveguide is an optical fiber. *Note 2:* A metallic waveguide filled with a dielectric material is not a dielectric waveguide.

differential encoding: Encoding in which signal significant conditions represent binary data, such as "0" and "1", and are represented as changes to succeeding values rather than with respect to a given reference. *Note:* An example of differential encoding is phase-shift keying (PSK) in which the information is not conveyed by the absolute phase of the signal with respect to a reference, but by the difference between phases of successive symbols, thus eliminating the requirement for a phase reference at the receiver.

differentially coherent phase-shift keying: *See* **coherent differential phase-shift keying.**

differentially encoded baseband: *Synonym* **non-return-to-mark.**

differential Manchester encoding: Encoding in which (a) data and clock signals are combined to form a single self-synchronizing data stream, (b) one of the two bits, *i.e.*, "0" or "1", is represented by no transition at the beginning of a pulse period and a transition in either direction at the midpoint of a pulse period, and (c) the other is represented by a transition at the beginning of a pulse period and a transition at the midpoint of the pulse period. *Note:* In differential Manchester encoding, if a "1" is represented by one transition, a "0" is represented by two transitions, and vice versa.

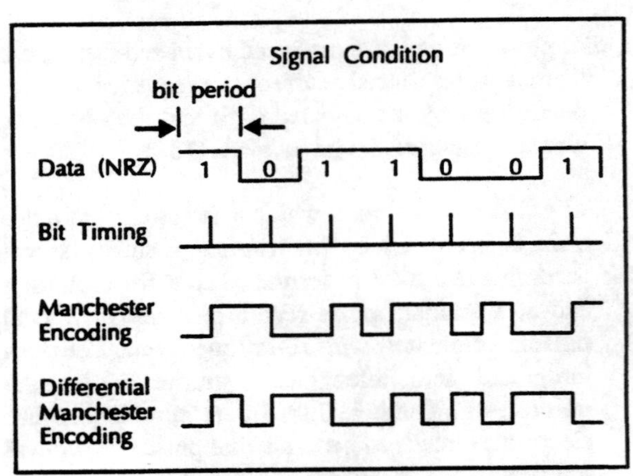

differential Manchester encoding

differential mode attenuation: In an optical fiber, the variation in attenuation among the propagating modes. (188)

differential mode delay: In an optical fiber, the variation in propagation delay that occurs because of the different group velocities of different modes. (188) *Synonym* **multimode group delay.**

differential mode interference: **1.** Interference that causes a change in potential of one side of a signal transmission path relative to the other side. (188) **2.** Interference resulting from an interference current path coinciding with the signal path. (188)

differential modulation: Modulation in which the choice of the significant condition for any signal element is dependent on the significant condition for the previous signal element. (188) *Note:* An example of differential modulation is delta modulation.

differential phase-shift keying (DPSK): Phase-shift keying that is used for digital transmission in which the phase of the carrier is discretely varied (a) in relation to the phase of the immediately preceding signal element and (b) in accordance with the data being transmitted. (188)

differential pulse-code modulation (DPCM): Pulse-code modulation in which an analog signal is sampled and the difference between the actual value of each sample and its predicted value, derived from the previous sample or samples, is quantized and converted, by encoding, to a digital signal. *Note:* There are several variations of differential pulse-code modulation.

differential quantum efficiency: In an optical source or detector, the slope of the curve relating output quanta to input quanta. (188)

differentiating network: A network, or circuit, that produces an output waveform that is the time derivative of the input waveform. *Note:* Differentiating networks are used in signal processing, such as for producing short timing pulses from square waves.

diffraction: The deviation of an electromagnetic wavefront from the path predicted by geometric optics when the wavefront interacts with, *i.e.*, is restricted by, a physical object such as an opening (aperture) or an edge. (188) *Note:* Diffraction is usually most noticeable for openings of the order of a wavelength. However, diffraction may still be important for apertures many orders of magnitude larger than the wavelength.

diffraction grating: An array of fine, parallel, equally spaced grooves ("rulings") on a reflecting or transparent substrate, which grooves result in diffractive and mutual interference effects that concentrate reflected or transmitted electromagnetic energy in discrete directions, called *"orders,"* or *"spectral orders." Note 1:* The groove dimensions and spacings are on the order of the wavelength in question. In the optical regime, in which the use of diffraction gratings is most common, there are many hundreds, or thousands, of grooves per millimeter. *Note 2:* Order zero corresponds to direct transmission or specular reflection. Higher orders result in deviation of the incident beam from the direction predicted by geometric (ray) optics. With a normal angle of incidence, the angle θ, the deviation of the diffracted ray from the direction predicted by geometric optics, is given by

$$\theta = \pm \sin^{-1}\left(\frac{n\lambda}{d}\right) ,$$

where n is the spectral order, λ is the wavelength, and d is the spacing between corresponding parts of adjacent grooves. *Note 3:* Because the angle of deviation of the diffracted beam is wavelength-dependent, a diffraction grating is dispersive, *i.e.*, it separates the incident beam spatially into its constituent wavelength components, producing a spectrum. *Note 4:* The spectral orders produced by diffraction gratings may overlap, depending on the spectral content of the incident beam and the number of grooves per unit distance on the grating. The higher the spectral order, the greater the overlap into the next-lower order. *Note 5:* By controlling the cross-sectional shape of the grooves, it is possible to concentrate most of the diffracted energy in the order of interest. This technique is called *"blazing."*

diffraction limited: 1. In optics, pertaining to a light beam in which the far-field beam divergence is equal to that predicted by diffraction theory. **2.** In focusing optics, pertaining to a light beam in which the impulse response or resolution limit is equal to that predicted by diffraction theory. (188) [After 2196]

diffraction region: In radio propagation, a region outside the line-of-sight region in which radio reception is made possible by the diffraction of the radiated waves.

diffuse reflection: Reflection from a rough or irregular surface which does not maintain the integrity of the incident wavefront. *Contrast with* **specular reflection.**

digit: A symbol, numeral, or graphic character that represents an integer. *Note 1:* Examples of digits include any one of the decimal characters "0" through "9" and either of the binary characters "0" or "1." (188) *Note 2:* In a given numeration system, the number of allowable different digits, including zero, is always equal to the base (radix).

digital: Characterized by discrete states.

digital access and cross-connect system (DACS): In communications systems, a digital system in which (a) access is performed by T-1 hardware architecture in private and public networks with centralized switching and (b) cross-connection is performed by D3/D4 framing for switching digital-signal-0 (DS-0) channels to other DS-0 channels. *Note:* Modern digital access and cross-connect systems are not limited to the T-carrier system, and may accommodate high data rates such as those of SONET.

digital alphabet: A coded character set in which the characters of an alphabet have a one-to-one relationship with their digitally coded representations. (188)

digital circuit patch bay: A patch bay in which low-level digital data circuits can be patched, monitored, and tested. (188) *Note:* A digital circuit patch bay can be either "D" type (unbalanced) or "K" type (balanced).

digital combining: A method of interfacing digital data signals, in either synchronous or asynchronous mode, without converting the data into a quasi-analog signal. (188)

digital computer: A computer that consists of one or more central processing units (CPUs), that is controlled by internally stored programs, and that stores and processes data in digital form.

digital data: **1.** Data represented by discrete values or conditions, as opposed to analog data. (188) **2.** Discrete representations of quantized values of variables, *e.g.*, the representation of numbers by digits, perhaps with special characters and the "space" character.

digital error: An inconsistency between the digital signal actually received and the digital signal that should have been received. (188)

digital facsimile equipment: Facsimile equipment that digitally encodes the picture signal, *i.e.*, encodes the baseband signal resulting from scanning the object. *Note:* The facsimile equipment output may be either (a) analog, as defined by CCITT Group 3 protocol, or (b) digital, as defined by CCITT Group 4, STANAG 5000 Type I, and STANAG 5000 Type II protocols. (188)

digital filter: A filter (usually linear), in discrete time, that is normally implemented through digital electronic computation. (188) *Note:* Digital filters differ from continuous time filters only in application. The parameters of digital filters are generally more stable than the parameters of commonly used analog (continuous) filters. Digital filters can be applied as optimal estimators. Commonly used forms are finite impulse response (FIR) and infinite impulse response (IIR).

digital frequency modulation: The transmission of digital data by frequency modulation of a carrier, as in binary frequency-shift keying. (188)

digital group: *See* **digroup.**

digital loop carrier (DLC): The equipment, including lines, that is used for digital multiplexing of telephone circuits, and that is provided by the network as part of the subscriber access. (188)

digital milliwatt: **1.** In digital telephony, a test signal consisting of eight 8-bit words corresponding to one cycle of a sinusoidal signal approximately 1 kHz in frequency and one milliwatt, rms, in power. *Note 1:* The digital milliwatt is stored in ROM. A continuous signal of arbitrary length, *i.e.*, an indefinite number of cycles, may be realized by continually reading out and concatenating the stored information into a data stream to be converted into analog form. *Note 2:* The digital milliwatt is used in lieu of separate test equipment. It has the advantage of being tied in frequency and amplitude to the relatively stable digital clock signal and power (voltage) supply, respectively, that are used by the digital channel bank. **2.** A digital signal that is the coded representation of a 0-dBm, 1000-Hertz sine wave. [47CFR]

digital modulation: The process of varying one or more parameters of a carrier wave as a function of two or more finite and discrete states of a signal. (188)

digital multiplexer: A device for combining several digital signals into an aggregate bit stream. (188) *Note:* Digital multiplexing may be implemented by interleaving bits, in rotation, from several digital bit streams either with or without the addition of extra framing, control, or error detection bits.

digital multiplex hierarchy: A hierarchy consisting of an ordered repetition of tandem digital multiplexers that produce signals of successively higher data rates at each level of the hierarchy. (188) *Note 1:* Digital multiplexing hierarchies may be implemented in many different configurations depending on (a) the number of channels desired, (b) the signaling system to be used, and (c) the bit rate allowed by the communications media. *Note 2:* Some currently available digital multiplexers have been designated as Dl-, DS-, or M-series, all of which operate at T-carrier rates. *Note 3:* In the design of digital multiplex hierarchies, care must be exercised to ensure interoperability of the multiplexers used in the hierarchy.

digital network: *See* **integrated digital network.**

digital phase-locked loop: A phase-locked loop in which the reference signal, the controlled signal, or the controlling signal, or any combination of these, is in digital form.

digital phase modulation: Modulation in which the instantaneous phase of the modulated wave is shifted between a set of predetermined discrete values in accordance with the significant conditions of the modulating signal.

digital primary patch bay: A patch bay that provides (a) the first access of most local user digital circuits in a technical control facility and (b) patching, monitoring, and testing capabilities for both high-level and low-level digital circuits.

digital selective calling (DSC): A synchronous system developed by the International Radio Consultative Committee (CCIR), used to establish contact with a station or group of stations automatically by means of radio. The operational and technical characteristics of this system are contained in CCIR Recommendation 493. [47CFR]

digital signal (DS): A signal in which discrete steps are used to represent information. (188) *Note 1:* In a digital signal, the discrete steps may be further characterized by signal elements, such as significant conditions, significant instants, and transitions. *Note 2:* Digital signals contain m-ary significant conditions.

digital signal 0 (DS0): In T-carrier, a basic digital signaling rate of 64 kb/s, corresponding to the capacity of one voice-frequency-equivalent channel. *Note 1:* The DS0 rate forms the basis for the North American digital multiplex transmission hierarchy. *Note 2:* The DS0 rate may support twenty 2.4-kb/s channels, or ten 4.8-kb/s channels, or five 9.67-kb/s channels, or one 56-kb/s channel, or one 64-kb/s clear channel.

digital signal 1 (DS1): A digital signaling rate of 1.544 Mb/s, corresponding to the North American and Japanese T1 designator.

digital signal 1C (DS1C): A digital signaling rate of 3.152 Mb/s, corresponding to the North American T1C designator.

digital signal 2 (DS2): A digital signaling rate of 6.312 Mb/s, corresponding to the North American and Japanese T2 designator.

digital signal 3 (DS3): **1.** A digital signal rate of 44.736 Mb/s, corresponding to the North American T3 designator. **2.** A digital signaling rate of 32.064 Mb/s, corresponding to the Japanese T3 designator.

digital signal 4 (DS4): **1.** A digital signal rate of 274.176 Mb/s, corresponding to the North American T4 designator. **2.** A digital signaling rate of 97.728 Mb/s, corresponding to the Japanese T4 designator.

digital slip: In the reception of a digital data stream, the loss of a bit, or the insertion by the receiver of a bit that was not transmitted, because of a difference in the bit rates of the incoming data stream and the local clock.

digital speech interpolation (DSI): In digital speech transmission, the use of periods of inactivity or constant signal level to increase the transmission efficiency by insertion of additional signals.

digital subscriber line (DSL): In Integrated Services Digital Networks (ISDN), equipment that provides full-duplex service on a single twisted metallic pair at a rate sufficient to support ISDN basic access and additional framing, timing recovery, and operational functions. (188) *Note:* The physical termination of the DSL at the network end is the line termination; the physical termination at the customer end is the network termination.

digital switch: A switch that performs time-division-multiplexed switching of digitized signals. (188) *Note 1:* When used with analog inputs, analog-to-digital and digital-to-analog conversion are required. These functions may be performed by the digital switch. *Note 2:* Implementation is accomplished by the interchange of time slots between input and output ports on a sequential basis under the direction of control systems. The control systems may be automatic, semiautomatic, or manual.

digital switching: Switching in which digitized signals are switched without converting them to or from analog signals. (188)

digital synchronization: Synchronization based on the start of a transmitted digital data unit, such as a bit, character, block, or frame. (188)

digital-to-analog (D-A) converter: A device that converts a digital input signal to an analog output signal carrying equivalent information. (188)

digital transmission group: A group of digitized voice or data channels or both with bit streams that are combined into a single digital bit stream for transmission over communications media. (188) *Note:* Digital transmission groups usually are categorized by their maximum capacity, not by a specific number of channels. However, the maximum digital transmission group capacity must be equal to or greater than the sum of the individual multiplexer input channel capacities.

digital transmission system: A transmission system in which (a) all circuits carry digital signals and (b) the signals are combined into one or more serial bit streams that include all framing and supervisory signals. *Note:* A-D/D-A conversion, if required, is accomplished external to the system. (188)

digital transport: *See* **digital transmission system.**

digital voice transmission: Transmission of analog voice signals that have been converted into digital signals. *Note:* An example of digital voice transmission is transmission of pulse-code modulated (PCM) analog voice signals. (188)

digitize: To convert an analog signal into a digital signal carrying equivalent information. (188)

digitizer: **1.** A device that converts an analog signal into a digital representation of the analog signal. (188) *Note:* A digitizer usually samples the analog signal at a constant sampling rate and encodes each sample into a numeric representation of the amplitude value of the sample. **2.** A device that converts the position of a point on a surface into digital coordinate data. (188)

digit time slot: In a digital data stream, a time interval that is allocated to a single digit and that can be uniquely recognized and defined. (188)

digroup: *Abbreviation for* **digital group.** In telephony, a basic digital multiplexing group. *Note 1:* In the North American and Japanese T-carrier digital hierarchies, each digroup supports 12 PCM voice channels or their equivalent in other services. The DS1 line rate (2 digroups plus overhead bits) is 1.544 Mb/s, supporting 24 voice channels or their equivalent in other services. *Note 2:* In the European hierarchy, each digroup supports 15 PCM channels or their equivalent in other services. The DS1 line rate (2 digroups plus overhead bits) is 2.048 Mb/s, supporting 30 voice channels or their equivalent in other services.

DIN: *Abbreviation for* **Deutsches Institut für Normung.**

diode laser: *Synonym* **injection laser diode.**

DIP: *Abbreviation for* **dual in-line package.**

diplexer: A three-port frequency-dependent device that may be used as a separator or a combiner of signals. (188) *Note:* Duplex transmission through a diplexer is not possible.

diplex operation: The sharing of one common element, such as a single antenna or channel, for transmission or reception of two simultaneous, independent signals on two different frequencies. *Note:* An example of diplex operation is the use of one antenna for two radio transmitters on different frequencies. (188)

dipole antenna: Usually a straight, center-fed, one-half wavelength antenna. (188)

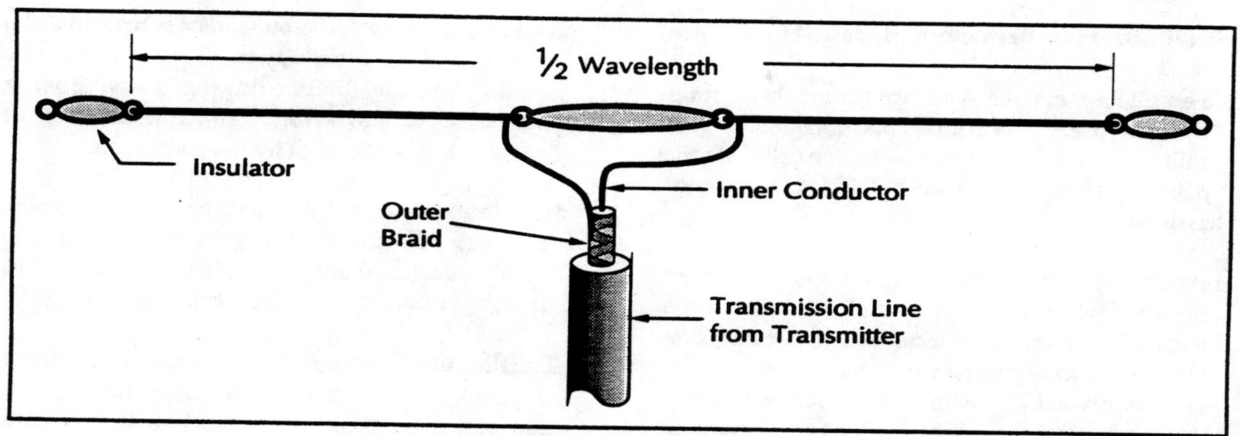

dipole antenna

DIP switch: A group of subminiature switches mounted in a package compatible with standard integrated-circuit sockets. *Note:* DIP switches are usually composed of rocker or slide-type switches.

dipulse coding: The coding of "1's" and "0's" in a message in which one full cycle of a square wave, *i.e.,* a positive pulse followed by a negative pulse in the same bit period, is transmitted when the message bit is a "1" and nothing is transmitted when the bit is a "0", or vice versa. *Note:* A dipulse signal can be generated by encoding the data into 50% return-to-zero (RZ) unipolar data and sending the bits through an AND gate with the system clock pulse. This RZ bit stream is then delayed one half-bit period and then added to the undelayed RZ stream. This produces the final dipulse waveform. The dipulse power spectrum is similar to that of the biphase coding power spectrum except dipulse coding produces a pulse-repetition rate equal to the bit rate. [From Weik '89]

Dirac delta function: *Synonym* **unit impulse.**

direct access: **1.** The capability to obtain data from a storage device, or to enter data into a storage device, in a sequence independent of their relative positions by means of addresses that indicate the physical location of the data. **2.** Pertaining to the organization and access method that must be used for a storage structure in which locations of records are determined by their keys, without reference to an index or to other records that may have been previously accessed.

direct address: In computing, an address that designates the storage location of an item of data to be treated as an operand.

direct bond: An electrical connection using continuous metal-to-metal contact between the members being joined. (188)

direct-buried cable: A communication cable manufactured or produced for the purpose of burial in direct contact with the earth. (188)

direct call: A facility-handled call in which the network interprets the call request signal as an instruction to establish a connection based on previously designated user information.

direct connect: **1.** In computer systems, a permanent communications link that connects directly to a mainframe computer through a terminal controller, usually at binary synchronous (bi-sync) transmission rates. **2.** In computer systems, a temporary connection between a microcomputer, *i.e.,* a desktop workstation, and a host bulletin board system or server.

direct coupling: *Synonym* **conductive coupling.**

direct current signaling (DX signaling): In telephony, a method whereby the signaling circuit E & M leads use the same cable pair(s) as the voice circuit and no filter is required to separate the control signals from the voice transmission. (188)

direct detection: *See* **detection (def. #1).**

direct dialing service: A service feature that permits a user to place information concerning credit card calls, collect calls, and special billing calls into the public telephone network without operator assistance.

direct distance dialing (DDD): A network-provided service feature in which a call originator may, without operator assistance, call any other user outside the local calling area. (188) *Note 1:* DDD extends beyond the boundaries of national public telephone networks. *Note 2:* DDD requires more digits in the number dialed than are required for calling within the local area.

directed broadcast address: An Internet Protocol address that specifies "all hosts" on a specified network. *Note:* A single copy of a directed broadcast is routed to the specified network, where it is broadcast to all terminals on that network.

directed net: A radio net in which no station other than the net control station may communicate with any other station without first obtaining permission from the net control station. *Note:* A directed net is established by the net control station. The net control station may restore the net to a free net. [From Weik '89]

direct inward dialing (DID): A service feature that allows inward-directed calls to a PBX to reach a specific PBX extension without human intervention. *Synonym* **network inward dialing.** (188)

directional antenna: An antenna in which the radiation pattern is not omnidirectional, *i.e.*, a nonisotropic antenna. (188)

directional coupler: A transmission coupling device for separately sampling (through a known coupling loss) either the forward (incident) or the backward (reflected) wave in a transmission line. (188) *Note:* A directional coupler may be used to sample either a forward or backward wave in a transmission line. A unidirectional coupler has available terminals or connections for sampling only one direction of transmission; a bidirectional coupler has available terminals for sampling both directions.

directionalization: The temporary conversion of a portion or all of a two-way trunk group to one-way trunks favoring traffic flowing away from a congested switch. *Note:* Adjacent nodes must cooperate to accomplish directionalization.

direction finding: A procedure for obtaining bearings of radio frequency emitters by using a highly directional antenna and a display unit on an intercept receiver or ancillary equipment. [JP1]

direction of scanning: In a facsimile transmitting apparatus, the scanning of the plane (developed in the case of a drum transmitter) of the message surface along lines running from right to left commencing at the top so that scanning commences at the top right-hand corner of the surface and finishes at the bottom left-hand corner; this is equivalent to scanning over a right-hand helix on a drum. *Note 1:* The orientation of the message on the scanning plane will depend upon its dimensions and is of no consequence. At the receiving apparatus, scanning takes place from right to left and top to bottom (in the above sense) for "positive" reception and from left to right and top to bottom (in the above sense) for "negative" reception. (188) *Note 2:* This is the CCITT Recommendation for phototelegraphic equipment.

directive gain: **1.** Of an antenna, the ratio of (a) 4π times the radiance, *i.e.*, power radiated per unit solid angle (watts per steradian), in a given direction to (b) the total power, *i.e.*, the power radiated to 4π steradians. (188) *Note 1:* The directive gain is usually expressed in dB. *Note 2:* The directive gain is relative to an isotropic antenna. *Note 3:* The power radiated to 4π steradians is the total power radiated by the antenna because 4π steradians constitute an entire sphere. **2.** Of an antenna, for a given direction, the ratio of the radiance, *i.e.*, the radiation intensity, produced in the given direction to the average value of the radiance in all directions. *Note 1:* If the direction is not specified, the direction of maximum radiance is assumed. *Note 2:* The directive gain is usually expressed in dB. (188)

directivity pattern: *Synonym* **radiation pattern.**

direct orbit: For a satellite orbiting the Earth, an orbit in which the projection of the satellite on the equatorial plane revolves about the Earth in the same direction as the rotation of the Earth. (188)

direct outward dialing (DOD): An automated PBX service feature that provides for outgoing calls to be dialed directly from the user terminal. (188) *Synonym* **network outward dialing.**

direct ray: A ray of electromagnetic radiation that follows the path of least possible propagation time between transmitting and receiving antennas. (188) *Note:* The path of least propagation time is not always the shortest distance path.

direct recording: In facsimile systems, recording in which a visible record is produced, without subsequent processing, in response to received signals. (188)

direct-sequence modulation: In spread-spectrum systems, modulation in which a sequence of binary pulses is used directly to modulate a carrier, usually by phase-shift keying. *Synonym* **direct-spread modulation.** [From Weik '89]

direct-sequence spread spectrum: **1.** A system (a) for generating spread-spectrum transmissions by phase-modulating a sine wave pseudorandomly with a continuous string of pseudonoise code symbols, each of duration much smaller than a bit and (b) that may be time-gated, where the transmitter is keyed periodically or randomly within a specified time interval. **2.** A signal structuring technique utilizing a digital code sequence having a chip rate much higher than the information signal bit rate. Each information bit of a digital signal is transmitted as a pseudorandom sequence of chips. [NTIA]

direct-spread modulation: *Synonym* **direct-sequence modulation.**

DISA: *Abbreviation for* **Defense Information Systems Agency,** *formerly* **DCA** (Defense Communications Agency).

disabling tone: A tone, transmitted over a communications path, used to control equipment. (188) *Note:* An example of a disabling tone is a tone that places an echo suppressor in a nonoperative condition during data transmission over a telephone circuit.

disc: *See* **diskette.**

DISC: *Abbreviation for* **disconnect command.**

discone antenna: *See* **biconical antenna.**

disconnect: In telephony, the disassociation or release of a switched circuit between two stations.

disconnect command (DISC): In Link-Layer protocols, such as high-level data link control (HDLC), synchronous data link control (SDLC), and advanced data communication control procedure (ADCCP), an unnumbered command used to terminate the operational mode previously set.

disconnect signal: In a switched telephone network, a supervisory signal transmitted from one end of a user line or trunk to indicate at the other end that the established connection should be disconnected. (188)

disconnect switch: In a power system, a switch used for closing, opening, or changing the connections in a circuit or system or for purposes of isolation. *Note:* It has no interrupting rating and is intended to be operated only after the circuit has been opened by some other means, such as by a circuit breaker or variable transformer. (188)

discriminator: The part of an FM receiver that extracts the desired signal from an incoming FM wave by changing frequency variations into amplitude variations. (188)

disengagement attempt: An attempt to terminate a telecommunications system access. *Note:* Disengagement attempts may be initiated by a user or the telecommunications system.

disengagement denial: After a disengagement attempt, a failure to terminate the telecommunications system access. *Note:* Disengagement denial is usually caused by excessive delay in the telecommunications system.

disengagement-denial probability: The ratio of disengagement attempts that result in disengagement denial to the total disengagement attempts counted during a measurement period.

disengagement failure: Failure of a disengagement attempt to return a communication system to the idle state, for a given user, within a specified maximum disengagement time.

disengagement originator: The user or functional unit that initiates a disengagement attempt. *Note 1:* A disengagement originator may be the originating user, the destination user, or the communications system. *Note 2:* The communications system may deliberately originate the disengagement because of preemption or inadvertently because of system malfunction.

disengagement phase: In an information transfer transaction, the phase during which successful disengagement occurs. *Note:* The disengagement phase is the third phase of an information transfer transaction.

disengagement request: A control or overhead signal issued by a disengagement originator for the purpose of initiating a disengagement attempt.

disengagement time: **1.** The average value of elapsed time between the start of a disengagement attempt for a particular source or destination user and the successful disengagement of that user. **2.** Elapsed time between the start of a disengagement attempt and successful disengagement.

diskette: In computer technology, a small disk of flexible plastic, coated with a magnetizable material and enclosed in a protective jacket, used to store digital data. *Note:* A diskette is distinguished from a hard disk by virtue of the fact that it is flexible, and unlike most hard disks, is removable from its drive. *Synonyms* **flexible disk, floppy disk.**

disk pack: An assembly of magnetic disks that can be removed as a whole from a disk drive together with a container from which the assembly must be separated when operating.

DISNET: *Abbreviation for* **Defense Integrated Secure Network.**

disparity: In pulse-code modulation (PCM), the digital sum, *i.e.*, the algebraic sum, of a set of signal elements. *Note:* The disparity will be zero and there will be no cumulative or drifting polarization if there are as many positive elements (those that represent 1) as there are negative elements (those that represent 0).

dispersion: Any phenomenon in which the velocity of propagation of an electromagnetic wave is wavelength dependent. *Note 1:* In communication technology, *"dispersion"* is used to describe any process by which an electromagnetic signal propagating in a physical medium is degraded because the various wave components (*i.e.*, frequencies) of the signal have different propagation velocities within the physical medium. *Note 2:* In an optical fiber, there are several significant dispersion effects, such as material dispersion, profile dispersion, and waveguide dispersion, that degrade the signal. *Note 3:* In optical fiber communications, the incorrect terms *"multimode dispersion"* and *"intermodal dispersion"* should not be used as synonyms for the correct term *"multimode distortion." Note 4:* In classical optics, *"dispersion"* is used to denote the wavelength dependence of refractive index in matter, ($dn/d\lambda$, where n is the refractive index and λ is the wavelength) caused by interaction between the matter and light. *"Dispersion,"* as used in fiber optic communications, should not be confused with *"dispersion"* as used by optical lens designers. *Note 5:* Three types of dispersion, relating to optical fibers, are defined as follows:

➢ **material dispersion:** In optical fiber communication, the wavelength dependence of the velocity of propagation (of the optical signal) on the bulk material of which the fiber is made. *Note 1:* Because every optical signal has a finite spectral width, material dispersion results in spreading of the signal. *Note 2:* Use of the redundant term *"chromatic dispersion"* is discouraged. *Note 3:* In pure silica, the basic material from which the most common telecommunication-grade fibers are made, material dispersion is minimum at wavelengths in the vicinity of 1.27 μm (slightly longer in practical fibers).

➢ **profile dispersion**: In an optical fiber, that dispersion attributable to the variation of refractive index contrast with wavelength. Profile dispersion is a function of the profile dispersion parameter.

➢ **waveguide dispersion:** Dispersion, of importance only in single-mode fibers, caused by the dependence of the phase and group velocities on core radius, numerical aperture, and wavelength. (188) *Note 1:* For circular waveguides, the dependence is on the ratio, *a*/λ, where *a* is the core radius and λ is the wavelength. *Note 2:* Practical single-mode fibers are designed so that material dispersion and waveguide dispersion cancel one another at the wavelength of interest.

dispersion coefficient: *See* **material dispersion coefficient.**

dispersion-limited operation: Operation of a communications link in which signal waveform degradation attributable to the dispersive effects of the communications medium is the dominant mechanism that limits link performance. (188) *Note 1:* The amount of allowable degradation is dependent on the quality of the receiver. *Note 2:* In fiber optic communications, *"dispersion-limited operation"* is often confused with *"distortion-limited operation."*

dispersion-shifted fiber: A single-mode optical fiber that has its minimum-dispersion wavelength shifted, by the addition of dopants, toward its minimum-loss wavelength. *Synonym* **EIA Class IVb fiber.**

dispersion-unmodified fiber: *Synonym* **dispersion-unshifted fiber.**

dispersion-unshifted fiber: A single-mode optical fiber that has a nominal zero-dispersion wavelength in the 1.3-μm transmission window. *Synonyms* **dispersion-unmodified fiber, EIA Class IVa fiber, nonshifted fiber.**

display device: An output unit that gives a visual representation of data.

distance learning: *Synonym* **teletraining.**

distance measuring equipment (DME): In radio location systems, equipment that ascertains the distance between an interrogator and a transponder. [From Weik '89]

distance training: *Synonym* **teletraining.**

distortion: **1.** In a system or device, any departure of the output signal waveform from that which should result from the input signal waveform's being operated on by the system's specified, *i.e.*, ideal, transfer function. (188) *Note:* Distortion may result from many mechanisms. Examples include nonlinearities in the transfer function of an active device, such as a vacuum tube, transistor, or operational amplifier. Distortion may also be caused by a passive component such as a coaxial cable or optical fiber, or by inhomogeneities, reflections, *etc.*, in the propagation path. **2.** In start-stop teletypewriter signaling, the shifting of the significant instants of the signal pulses from their proper positions relative to the beginning of the start pulse. *Note:* The magnitude of the distortion is expressed in percent of an ideal unit pulse length.

distortion-limited operation: The condition prevailing when distortion of a received signal, rather than its attenuated amplitude (or power), limits performance under stated operational conditions and limits. (188) *Note:* Distortion-limited operation is reached when the system distorts the shape of the waveform beyond specified limits. For linear systems, distortion-limited operation is equivalent to bandwidth-limited operation.

distributed control: Control of a network from multiple points. *Note:* Each point controls a portion of the network, using local information or information transmitted over the network from distant points.

distributed database: **1.** A database that is not entirely stored at a single physical location, but rather is dispersed over a network of interconnected computers. **2.** A database that is under the control of a central database management system in which storage devices are not all attached to a common processor.

distributed frame-alignment signal: A frame-alignment signal in which the signal elements occupy digit positions that are not consecutive.

distributed network: A network structure in which the network resources, such as switching equipment and processors, are distributed throughout the geographical area being served. (188) *Note:* Network control may be centralized or distributed.

distributed processing: Data processing in which an integrated set of functions is performed within multiple, physically separated devices. (188)

distributed-queue dual-bus (DQDB) [network]: A distributed multi-access network that (a) supports integrated communications using a dual bus and distributed queuing, (b) provides access to local or metropolitan area networks, and (c) supports connectionless data transfer, connection-oriented data transfer, and isochronous communications, such as voice communications.

distributed switching: Switching in which many processor-controlled switching units are distributed, usually close to concentrations of users, and operated in conjunction with a host switch. (188) *Note:* Distributed switching provides improved communications services for concentrations of users remote from the host switch, and reduces the transmission requirements, *i.e.*, the traffic, between such concentrations and the host switch.

distribution: In ISDN applications, the use of broadband transmission of audio or video information to the user without applying any post-production processing to the information. *Note:* *"Distribution"* is in contrast to *"contribution."*

distribution frame: In communications, a structure with terminations for connecting the permanent wiring of a facility in such a manner that interconnection by cross-connections may readily be made. (188)

distribution-quality television: Television conforming to the NTSC standard, the SECAM standard, the PAL standard, or the PAL-M standard. *Synonym [in CCITT usage]* **existing-quality television.**

distribution service: In ISDN applications, a telecommunications service that allows one-way flow of information from one point in the network to other points in the network with or without user individual presentation control.

distribution voltage drop: The voltage drop between any two defined points of interest in a power distribution system. (188)

distributor: In data transmission, a device that accepts a data stream from one line and places a sequence of signals, one or more at a time, on several lines, thus performing a spatial multiplexing of the original stream. *Note:* Examples of a distributor are (a) a mechanical unit with input to a rotor and output through many contacts wiped by the rotor and (b) a set of combinational logic circuits, such as a series of AND gates, that are sequentially enabled by a set of pulses and that are all connected to a common bus carrying the input signals. [From Weik '89]

disturbance voltage: An unwanted voltage induced in a system by natural or man-made sources. *Note:* In telecommunications systems, the disturbance voltage creates currents that limit or interfere with the interchange of information. An example of a disturbance voltage is a voltage that produces (a) false signals in a telephone, (b) noise in a radio receiver, or (c) distortion in a received signal.

diurnal phase shift: The phase shift of electromagnetic signals associated with daily changes in the ionosphere. (188) *Note 1:* The major changes usually occur during the period of time when sunrise or sunset is present at critical points along the path. *Note 2:* Significant phase shifts may occur on paths wherein a reflection area of the path is subject to a large tidal range. *Note 3:* In cable systems, significant phase shifts can be occasioned by diurnal temperature variance.

divergence: *See* **beam divergence.**

diversity: The property of being made up of two or more different elements, media, or methods. *Note:* In communications, diversity is usually used to provide robustness, reliability, or security.

diversity combiner: A circuit or device for combining two or more signals carrying the same information received via separate paths or channels with the objective of providing a single resultant signal that is superior in quality to any of the contributing signals. (188)

diversity factor: The ratio of the sum of the individual maximum demands of the various parts of a power distribution system to the maximum demand of the whole system. *Note:* The diversity factor is always greater than unity. (188)

diversity gain: In radio communications, the ratio of the signal field strength obtained by diversity combining to the signal strength obtained by a single path. *Note:* Diversity gain is usually expressed in dB. [From Weik '89]

diversity reception: Radio reception in which a resultant signal is obtained by combining or selecting signals, from two or more independent sources, that have been modulated with identical information-bearing signals, but which may vary in their fading characteristics at any given instant. *Note 1:* Diversity reception is used to minimize the effects of fading. (188) *Note 2:* The amount of received signal improvement when using diversity reception is directly dependent on the independence of the fading characteristics.

diversity transmission: Radio communication using a reception technique in which a resultant signal is obtained by combining signals (a) that originate from two or more independent sources that have been modulated with identical information-bearing signals and (b) that may vary in their transmission characteristics at any given instant. (188) *Note 1:* Diversity transmission and reception are used to obtain reliability and signal improvement by overcoming the effects of fading, outages, and circuit failures. *Note 2:* When using diversity transmission and reception, the amount of received signal improvement depends on the independence of the fading characteristics of the signal as well as circuit outages and failures.

divestiture: The court-ordered separation of the Bell Operating Telephone Companies from AT&T.

D layer: *See* **D region, ionosphere.**

DLE: *Abbreviation for* **data link escape character.**

DM: *Abbreviation for* **delta modulation.**

DO: *Abbreviation for* **design objective.**

DOD: *Abbreviation for* **Department of Defense, direct outward dialing.**

DOD master clock: The master clock to which time and frequency measurements for the U.S. Department of Defense are referenced, *i.e.*, are traceable. *Note 1:* The U.S. Naval Observatory master clock is designated as the DOD Master Clock. (188) *Note 2:* The U.S. Naval Observatory master clock is one of the two standard time and frequency references for the U.S. Government in accordance with Federal Standard 1002-A. The other standard time and frequency reference for the U.S. Government is the National Institute of Standards and Technology (NIST) master clock.

domain: **1.** The independent variable used to express a function. *Note:* Examples of domains are time, frequency, and space. **2.** In distributed networks, all the hardware and software under the control of a specified set of one or more host processors. [From Weik '89]

domain name server: A server that retains the addresses and routing information for TCP/IP LAN users.

Domain Name System (DNS): The online distributed database system that (a) is used to map human-readable addresses into Internet Protocol (IP) addresses, (b) has servers throughout the Internet to implement hierarchical addressing that allows a site administrator to assign machine names and addresses, (c) supports separate mappings between mail destinations and IP addresses, and (d) uses domain names that (i) consist of a sequence of names, *i.e.*, labels, separated by periods, *i.e.*, dots, (ii) usually are used to name Internet host computers uniquely, (iii) are hierarchical, and (iv) are processed from right to left, such as the host nic.ddn.mil has a name (nic — the Network Information Center), a subdomain (ddn — the Defense Data Network), and a primary domain (mil — the MILNET).

domestic fixed public service: A fixed service, the stations of which are open to public correspondence, for radiocommunications originating and terminating solely at points all of which lie within [. . . the entire United States and certain other geographic areas as specified in 47CFR].

domestic public radio services: The land mobile and domestic fixed public services, the stations of which are open to public correspondence. [47CFR]

dominant mode: In a waveguide that can support more than one propagation mode, the mode that propagates with the minimum degradation, *i.e.*, the mode with the lowest cutoff frequency. (188) *Note:* Designations for the dominant mode are TE_{10} for rectangular waveguides and TE_{11} for circular waveguides.

D1: *See* **channel bank.**

dopant: An impurity added to an optical medium to change its optical properties. *Note:* Dopants are used in optical fibers to control the refractive index profile and other refractive properties of the fiber. [FAA]

doppler effect: The change in the observed frequency (or wavelength) of a wave, caused by a time rate of change in the effective path length between the source and the point of observation.

doppler shift: The degree of observed change in frequency (or wavelength) of a wave due to the doppler effect.

double-current transmission: *Synonym* **polar direct-current telegraph transmission.**

double-ended control: *Synonym* **double-ended synchronization.**

double-ended queue (dequeue): A queue in which the contents may be changed by adding or removing items at either end.

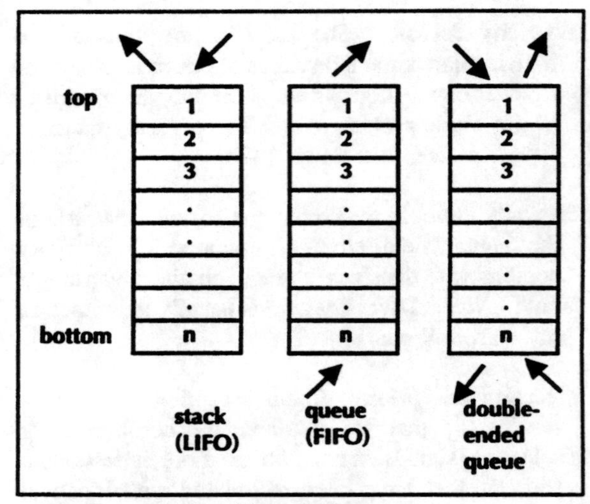

double-ended queue

double-ended synchronization: For two connected exchanges in a communications network, a synchronization control scheme in which the phase error signals used to control the clock at one exchange are derived by comparison with the phase of the incoming digital signal and the phase of the internal clocks at both exchanges. *Synonym* **double-ended control.**

double-frequency shift keying (DFSK): Frequency-shift keying in which two telegraph signals are multiplexed and transmitted simultaneously by frequency shifting among four frequencies. (188)

double modulation: Modulation in which (a) a subcarrier is modulated with an information-bearing signal and (b) the resulting modulated subcarrier is then used to modulate another carrier that has a higher frequency. (188)

double refraction: *Synonym* **birefringence.**

double-sideband reduced carrier (DSB-RC) transmission: Transmission in which (a) the frequencies produced by amplitude modulation are symmetrically spaced above and below the carrier and (b) the carrier level is reduced for transmission at a fixed level below that which is provided to the modulator. *Note:* In DSB-RC transmission, the carrier is usually transmitted at a level suitable for use as a reference by the receiver, except for the case in which it is reduced to the minimum practical level, *i.e.*, the carrier is suppressed.

double-sideband suppressed-carrier (DSB-SC) transmission: Transmission in which (a) frequencies produced by amplitude modulation are symmetrically spaced above and below the carrier frequency and (b) the carrier level is reduced to the lowest practical level, ideally completely suppressed. (188) *Note:* DSB-SC transmission is a special case of double-sideband reduced carrier transmission.

double-sideband (DSB) transmission: AM transmission in which both sidebands and the carrier are transmitted. (188)

doubly clad fiber: A single-mode fiber that has two claddings. *Note 1:* Each cladding has a refractive index that is lower than that of the core. Of the two claddings, inner and outer, the inner cladding has the lower refractive index. *Note 2*: A doubly clad fiber has the advantage of very low macrobending losses. It also has two zero-dispersion points, and low dispersion over a much wider wavelength range than a singly clad fiber. *Synonyms* **depressed-cladding fiber, depressed-inner-cladding fiber, W-profile fiber** (from the fact that a symmetrical plot of its refractive index profile superficially resembles the letter W).

down-converter: A device for performing frequency translation in such a manner that the output frequencies are lower in the spectrum than the input frequencies. (188)

downlink: 1. A data link from a satellite or other spacecraft to a terrestrial terminal. (188) **2.** A data link from an airborne platform to a ground-based terminal.

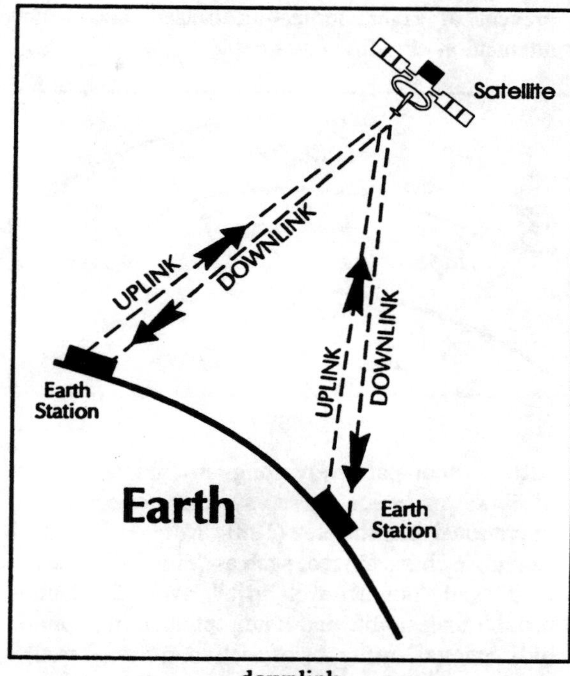

downlink

downstream: 1. In communications, the direction of transmission flow from the source toward the sink. **2.** With respect to the flow of data in a communications path: at a specified point, the direction toward which data are received later than at the specified point.

downtime: The interval during which a functional unit is inoperable. (188)

DPCM: *Abbreviation for* **differential pulse-code modulation.**

DPSK: *Abbreviation for* **differential phase-shift keying.**

DQDB: *Abbreviation for* **distributed-queue dual-bus.**

D region: That portion of the ionosphere existing approximately 50 to 95 km above the surface of the Earth. (188) *Note:* Attenuation of radio waves, caused by ionospheric free-electron density generated by solar radiation, is pronounced during daylight hours. Because solar radiation is not present at night, ionization ceases, and hence attenuation of radio waves ceases.

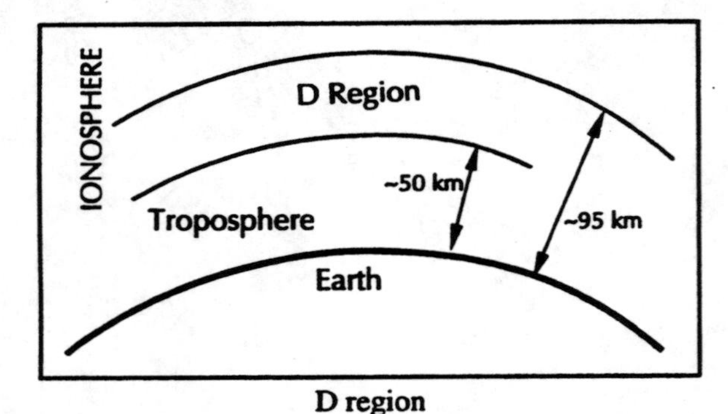

D region

drift: A comparatively long-term change in an attribute or value of a system or equipment operational parameter. (188) *Note 1:* The drift should be characterized, such as "diurnal frequency drift" and "output level drift." *Note 2:* Drift is usually undesirable and unidirectional, but may be bidirectional, cyclic, or of such long-term duration and low excursion rate as to be negligible.

drop: **1.** In a communications network, the portion of a device directly connected to the internal station facilities, such as toward a switchboard or toward a switching center. (188) **2.** The central office side of test jacks. (188) **3.** A wire or cable from a pole or cable terminus to a building. (188) **4.** To delete, intentionally or unintentionally, or to lose part of a signal, such as dropping bits from a bit stream. (188)

drop and insert: In a multichannel transmission system, a process that diverts (drops) a portion of the multiplexed aggregate signal at an intermediate point, and introduces (inserts) a different signal for subsequent transmission in the same position, *e.g.*, time slot or frequency band, previously occupied by the diverted signal. (188) *Note 1:* Signals not of interest at the drop-and-insert point are not diverted. *Note 2:* The diverted signal may be demodulated or reinserted into another transmission system in the same or another time slot or frequency band. *Note 3:* The time slot or frequency band vacated by the diverted signal need not necessarily be reoccupied by another signal. Likewise, a previously unoccupied time slot or frequency band may be occupied by a signal inserted at the drop-and-insert point.

drop channel operation: Operation in which one or more channels of a multichannel system are terminated, *i.e.*, dropped, at an intermediate point between the end terminals of the system.

dropout: **1.** In a communications system, a momentary loss of signal. *Note:* Dropouts are usually caused by noise, propagation anomalies, or system malfunctions. (188) **2.** A failure to read properly a binary character from data storage. *Note:* A dropout is usually caused by a defect in the storage medium or by a malfunction of the read mechanism. **3.** In magnetic tape, disk, card, or drum systems, a recorded signal with an amplitude less than a predetermined percentage of a reference signal. (188)

drop repeater: In a multichannel communications system, a repeater that has the necessary equipment for the local termination, *i.e.*, the dropping, of one or more channels. (188)

drum factor: **1.** In facsimile systems in which drums are used, the ratio of drum length to drum diameter. **2.** In facsimile systems in which drums are not used, the ratio of (a) the page width to (b) the page length divided by π. (188)

drum speed: In facsimile systems, the rotation rate of the facsimile transmitter or recorder drum. *Note:* Drum speed is usually expressed in revolutions per minute. (188)

DS: *Abbreviation for* **digital signal.**

DSA: *Abbreviation for* **dial service assistance.**

DSA board: A local dial office switchboard at which are handled assistance calls, intercepted calls, and calls from miscellaneous lines and trunks. It may also be employed for handling certain toll calls. [47CFR]

DSB: *Abbreviation for* **double sideband.** *See* **double-sideband transmission.**

DSB board: A switchboard of a dial system for completing incoming calls received from manual offices. [47CFR]

DSB-SC: *Abbreviation for* **double-sideband suppressed carrier.** *See* **double-sideband suppressed-carrier transmission.**

DSC: *Abbreviation for* **digital selective calling.**

DSE: *Abbreviation for* **data switching exchange.**

DSI: *Abbreviation for* **digital speech interpolation.**

DS0: *Abbreviation for* **digital signal 0.**

DS1 . . . DS4: *Abbreviations for* **digital signal 1. . . digital signal 4.**

DS1C: *Abbreviation for* **digital signal 1C.**

DSN: *Abbreviation for* **Defense Switched Network.**

DSR: *Abbreviation for* **data signaling rate.**

D-Star(D*): *Synonym* **specific detectivity.**

DSU: *Abbreviation for* **data service unit.**

DTE: *Abbreviation for* **data terminal equipment.** **1.** An end instrument that converts user information into signals for transmission or reconverts the received signals into user information. (188) **2.** The functional unit of a data station that serves as a data source or a data sink and provides for the data communication control function to be performed in accordance with link protocol. *Note 1:* The data terminal equipment (DTE) may be a single piece of equipment or an interconnected subsystem of multiple pieces of equipment that perform all the required functions necessary to permit users to communicate. *Note 2:* A user interacts with the DTE, or the DTE may be the user. The DTE interacts with the data circuit-terminating equipment (DCE).

DTE clear signal: A call control signal sent by data terminal equipment (DTE) to initiate clearing.

DTE waiting signal: A call control signal, sent by the data-circuit-terminating-equipment/data-terminal-equipment (DCE/DTE) interface, that indicates that the DTE is waiting for a call control signal from the DCE.

DTMF: *Abbreviation for* **dual-tone multifrequency (signaling).**

D2: *See* **channel bank.**

D3: *See* **channel bank.**

D-type patch bay: A patch bay designed for patching and monitoring of unbalanced data circuits at rates up to 1 Mb/s. (188)

dual access: **1.** The connection of a user to two switching centers by separate access lines using a single message routing indicator or telephone number. **2.** In satellite communications, the transmission of two carriers simultaneously through a single communication satellite repeater. (188)

dual bus: A pair of parallel buses arranged such that the direction of data flow in one bus is opposite to the direction of data flow in the other bus.

dual diversity: The simultaneous combining of (or selection from) two independently fading signals, so that the resultant signal can be detected through the use of space, frequency, angle, time, or polarization characteristics. (188)

dual homing: The connection of a terminal so that it is served by either of two switching centers. *Note:* In dual homing, a single directory number or a single routing indicator is used. (188)

dual in-line package (DIP): An electronic package with a rectangular housing and a row of pins along each of two opposite sides. *Note:* DIP packages may be used for integrated circuits, or for discrete components, such as resistors or toggle switches. An example of a DIP is a microcircuit package with two rows of seven vertical leads that is specially designed for mounting on a printed circuit board.

dual in-line package switch: *See* **DIP switch.**

dual-precedence message: A message that contains two precedence designations. *Note:* Usually the higher precedence message is for all action addressees and the lower for all information addressees. [From Weik '89]

dual-tone multifrequency (DTMF) signaling: In telephone systems, multifrequency signaling in which standard set combinations of two specific voice band frequencies, one from a group of four low frequencies and the other from a group of four higher frequencies, are used. (188) *Synonyms* **multifrequency pulsing, multifrequency signaling.** *Note 1:* DTMF signals, unlike dial pulses, can pass through the entire connection to the destination user, and therefore lend themselves to various schemes for remote control after access, *i.e.*, after the connection is established. *Note 2:* Telephones using DTMF usually have 12 keys. Each key corresponds to a different pair of frequencies. Each pair of frequencies corresponds to one of the ten decimal digits, or to the symbol "#" or "*", the "*" being reserved for special purposes. *Note 3:* The standard signal frequency pairs transmitted by DTMF equipment used by the public exchange carriers are as follows:

Button or Digit	Frequencies (Hz)
1	697/1209
2	697/1336
3	697/1477
4	770/1209
5	770/1336
6	770/1477
7	852/1209
8	852/1336
9	852/1477
0	941/1336
*	941/1209
#	941/1477

Note 4: Tactical telephones have 16 keys, the extra 4 being used for precedence. For DSN (Defense Switched Network) the signal frequency pairs transmitted for the ten decimal digits and the * and # are the same as those used by the public exchange carriers. The additional four keys, corresponding to four different frequency pairs and the precedence, are as follows:

Button or Key	Frequencies (Hz)
FO (Flash Override)	697/1633
F (Flash)	770/1633
I (Immediate)	852/1633
P (Priority)	941/1633

dual-use access line: A user access line normally used for analog voice communication, but which has special conditioning for use as a digital transmission circuit. (188)

duct: 1. In interfacility cabling, a conduit, which may be direct-earth buried or encased in concrete, used to enclose communications or power cables. *Note:* For maximum resistance to rodent attack, direct-earth-buried conduit should have an outside diameter equal to or greater than 6 cm (2.25 in.). **2.** *See* **atmospheric duct.**

ducting: The propagation of radio waves within an atmospheric duct. (188)

dumb terminal: An asynchronous terminal that (a) does not use a transmission control protocol and (b) sends or receives data sequentially one character at a time. *Note:* Dumb terminals usually handle ASCII characters.

dummy load: A dissipative impedance-matched network, usually used at the end of a transmission line to absorb all incident energy. *Note:* The dummy load usually converts the incident energy to thermal energy. (188)

duobinary signal: A pseudobinary-coded signal in which a "0" *("zero")* bit is represented by a zero-level electric current or voltage; a "1" *("one")* bit is represented by a positive-level current or voltage if the quantity of "0" bits since the last "1" bit is even, and by a negative-level current or voltage if the quantity of "0" bits since the last "1" bit is odd. (188) *Note 1:* Duobinary signals require less bandwidth than NRZ. *Note 2:* Duobinary signaling also permits the detection of some errors without the addition of error-checking bits.

duplex cable: A fiber-optic cable that contains two optical fibers. [After FAA]

duplex circuit: A circuit that permits transmission in both directions. *Note:* For simultaneous two-way transmission, *see* **full-duplex circuit.**

duplexer: In radar systems, a device that isolates the receiver from the transmitter while permitting them to share a common antenna. *Note 1:* A duplexer must be designed for operation in the frequency band used by the receiver and transmitter, and must be capable of handling the output power of the transmitter. *Note 2:* A duplexer must provide adequate rejection of transmitter noise occurring at the receive frequency, and must be designed to operate at, or less than, the frequency separation between the transmitter and receiver. *Note 3:* A duplexer must provide sufficient isolation to prevent receiver desensitization.

duplex operation: Operating method in which transmission is possible simultaneously, in both directions of a telecommunication channel. [NTIA] [RR] (188) *Note 1:* This definition is not limited to radio transmission. *Note 2:* In general, duplex operation and semi-duplex operation require two frequencies in radiocommunication; simplex operation may use either one or two. *Synonyms* **full-duplex operation, two-way simultaneous operation.**

duty cycle: **1.** In an ideal pulse train, *i.e.*, one having rectangular pulses, the ratio of the pulse duration to the pulse period. (188) *Note:* For example, the duty cycle is 0.25 for a pulse train in which the pulse duration is 1 µs and the pulse period is 4 µs. **2.** The ratio of (a) the sum of all pulse durations during a specified period of continuous operation to (b) the total specified period of operation. **3.** In a continuously variable slope delta (CVSD) modulation converter, the mean proportion of binary "1" digits at the converter output in which each "1" indicates a run of a specified number of consecutive bits of the same polarity in the digital output signal. (188) **4.** In a periodic phenomenon, the ratio of the duration of the phenomenon in a given period to the period. *Note:* In a piece of electrical equipment, *e.g.*, an electric motor, the period for which it may be operated without deleterious effects, *e.g.*, from overheating.

dwell time: The period during which a dynamic process remains halted in order that another process may occur.

DX signaling: *Abbreviation for* **direct current signaling.**

DX signaling unit: A duplex signaling unit that repeats "E" and "M" lead signals into a cable pair(s) via "A" and "B" leads. These signals are transmitted on the same cable pair(s) that transmit(s) the message. (188)

dynamically adaptive routing: In route determination for packet-switched networks, adaptive routing in which an algorithm is used that (a) automatically routes traffic around congested, damaged, or destroyed switches and trunks and (b) allows the system to continue to function over the remaining portions of the network.

dynamicizer: *See* **parallel-to-serial conversion.**

dynamic range: **1.** In a system or device, the ratio of (a) a specified maximum level of a parameter, such as power, current, voltage, or frequency to (b) the minimum detectable value of that parameter. *Note:* The dynamic range is usually expressed in dB. **2.** In a transmission system, the ratio of (a) the overload level, *i.e.*, the maximum signal power that the system can tolerate without distortion of the signal, to (b) the noise level of the system. *Note:* The dynamic range of transmission systems is usually expressed in dB. (188) **3.** In digital systems or devices, the ratio of maximum and minimum signal levels required to maintain a specified bit error ratio.

dynamic variation: A short-term variation (as opposed to long-term drift) in the characteristics of power delivered to electrical equipment. *Note:* Dynamic variations indicate a short-term departure from steady-state conditions. (188)

(this page intentionally left blank)

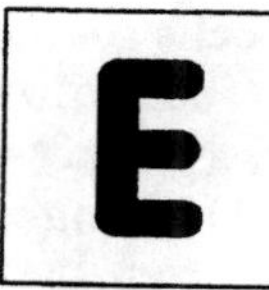

E & M signaling: In telephony, an arrangement that uses separate leads, called the "E" lead and "M" lead, for signaling and supervisory purposes. *Note 1:* The near end signals the far end by applying –48 vdc to the "M" lead, which results in a ground being applied to the far end's "E" lead. When –48 vdc is applied to the far end "M" lead, the near-end "E" lead is grounded. *Note 2:* The "E" originally stood for "ear," *i.e.*, when the near-end "E" lead was grounded, the far end was calling and "wanted your ear." The "M" originally stood for "mouth," because when the near-end wanted to call (*i.e.*, speak to) the far end, –48 vdc was applied to that lead.

earth: *See* **ground.**

Earth coverage (EC): In satellite communications, the coverage that occurs when the satellite-to-Earth beam is sufficiently wide to cover all of the surface of the Earth exposed to the satellite, *i.e.*, the footprint is as large as it can possibly be from a geographic standpoint. (188)

earth electrode subsystem: A network of electrically interconnected rods, plates, mats, or grids, installed and connected, for the purpose of establishing a low-resistance contact with earth. (188)

Earth exploration-satellite service: A radiocommunication service between Earth stations and one or more space stations, which may include links between space stations, in which:

- information relating to the characteristics of the Earth and its natural phenomena is obtained from active sensors or passive sensors on Earth satellites;
- similar information is collected from airborne or Earth-based platforms;
- such information may be distributed to Earth stations within the system concerned;
- platform interrogation may be included.

This service may also include feeder links necessary for its operation. [NTIA] [RR]

earth ground: *See* **ground.**

Earth station: A station located either on the Earth's surface or within the major portion of the Earth's atmosphere and intended for communication:

- with one or more space stations; or
- with one or more stations of the same kind by means of one or more reflecting satellites or other objects in space. [NTIA] [RR]

Earth terminal: In a satellite link, one of the non-orbiting communications stations that receives, processes, and transmits signals between itself and a satellite. (188) *Note:* Earth terminals may be at mobile, fixed, airborne, and waterborne Earth terminal complexes. *Synonym* **satellite Earth terminal.**

Earth terminal complex: In satellite communications systems, the assembly of equipment and facilities necessary to integrate an Earth terminal into a communications network. (188) *Note:* The Earth terminal complex includes the Earth terminal and its support equipment and any required interconnect facilities and their support equipment. It does not include facilities at the site that are not necessary to establish and integrate the satellite links with the network.

EAS: *Abbreviation for* **extended area service.**

EBCDIC: *Acronym for* **extended binary coded decimal interchange code.** An 8-bit alphanumeric coded character set.

E-bend: A smooth change in the direction of the axis of a waveguide, throughout which the axis remains in a plane parallel to the direction of electric E-field (transverse) polarization. (188) *Synonym* **E-plane bend.**

E_b/N_0: *See* **signal-to-noise ratio per bit.**

EBS: *Abbreviation for* **Emergency Broadcast System.**

EC: *Abbreviation for* **Earth coverage.**

ECCM: *Abbreviation for* **electronic counter-countermeasures.**

echo: **1.** A wave that has been reflected by a discontinuity in the propagation medium. **2.** A wave that has been reflected or otherwise returned with sufficient magnitude and delay to be perceived. (188) *Note 1:* Echoes are frequently measured in dB relative to the directly transmitted wave. *Note 2:*

Echoes may be desirable (as in radar) or undesirable (as in telephone systems). **3.** In computing, to print or display characters (a) as they are entered from an input device, (b) as instructions are executed, or (c) as retransmitted characters received from a remote terminal. **4.** In computer graphics, the immediate notification of the current values provided by an input device to the operator at the display console.

echo area: *See* **scattering cross section.**

echo attenuation: In a communication circuit (4- or 2-wire) in which the two directions of transmission can be separated from each other, the attenuation of echo signals that return to the input of the circuit under consideration. *Note:* Echo attenuation is expressed as the ratio of the transmitted power to the received echo power in dB. (188)

echo cancellation: In a system, the reduction of the power level of an echo or the elimination of an echo. (188) *Note:* Echo cancellation is usually an active process in which echo signals are measured and canceled or eliminated by combining an inverted signal with the echo signal.

echo canceler: *See* **echo suppressor.**

echo check: A check to determine the integrity of transmission of data, whereby the received data are returned to the source for comparison with the originally transmitted data. *Synonym* **loop check.**

echo command: A programming language command that causes an echo response from a computer to be displayed on a monitor or printer for analysis or monitoring of the progress of processing.

echo effect: *See* **ghost.**

echo line: In computer systems, on a hard-output or display device, a line of information that verifies (reflects) data entered into the computer.

echoplex: An echo check used in public switched networks operating in the full-duplex transmission mode, *i.e.*, the two-way-simultaneous mode.

echo return loss: *See* **return loss.**

echo sounding: The measurement of the depth of a body of water or the distance to an object in a body of water by measuring the time it takes sound or electromagnetic waves of known velocity to reflect from the bottom of the water body or from the distant object. *Note:* In echo sounding, damped cw transmission is usually used.

echo suppressor: A device for connection to a two-way telephone circuit to attenuate echo signals in one direction caused by signals in the other direction. (188)

ECM: *Abbreviation for* **electronic countermeasures.**

edge busyness: In a video display, distortion that is concentrated at or near the edge of objects, and that is categorized further by its temporal and spatial characteristics.

edge-emitting LED: An LED that has a physical structure superficially resembling that of an injection laser diode, operated below the lasing threshold and emitting incoherent light. *Note:* Edge-emitting LEDs have a relatively small beam divergence, and thus are capable of launching more optical power into a given fiber than are the conventional surface-emitting LEDs. [After FAA]

EDI: *Abbreviation for* **electronic document interchange.**

EDTV: *Abbreviation for* **extended-definition television.**

effective antenna gain contour (of a steerable satellite beam): An envelope of antenna gain contours resulting from moving the boresight of a steerable satellite beam along the limits of the effective boresight area. [NTIA] [RR]

effective boresight area (of a steerable satellite beam): An area on the surface of the Earth within which the boresight of a steerable satellite beam is pointed. There may be more than one unconnected effective boresight area to which a single steerable satellite beam can be pointed. [NTIA] [RR]

effective data transfer rate: The average number of units of data, such as bits, characters, blocks, or frames, transferred per unit time from a source and accepted as valid by a sink. (188) *Note:* The effective data transfer rate is usually expressed in bits, characters, blocks, or frames per second. The effective data transfer rate may be averaged over a period of seconds, minutes, or hours.

effective Earth radius: The radius of a hypothetical Earth for which the distance to the radio horizon, assuming rectilinear propagation, is the same as that for the actual Earth with an assumed uniform vertical gradient of atmospheric refractive index. (188) *Note:* For the standard atmosphere, the effective Earth radius is 4/3 that of the actual Earth radius.

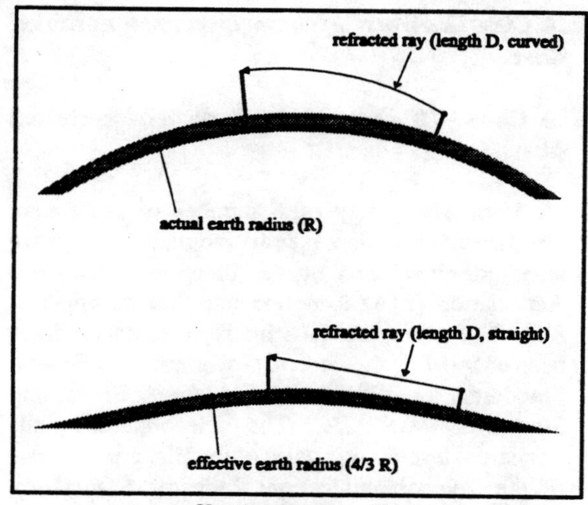

effective Earth radius

effective height: **1.** The height of the center of radiation of an antenna above the effective ground level. (188) **2.** In low-frequency applications involving loaded or nonloaded vertical antennas, the moment of the current distribution in the vertical section divided by the input current. (188) *Note:* For an antenna with symmetrical current distribution, the center of radiation is the center of distribution. For an antenna with asymmetrical current distribution, the center of radiation is the center of current moments when viewed from points near the direction of maximum radiation.

effective input noise temperature: The source noise temperature in a two-port network or amplifier that will result in the same output noise power, when connected to a noise-free network or amplifier, as that of the actual network or amplifier connected to a noise-free source. (188) *Note:* If F is the noise figure numeric and 290 K the standard noise temperature, then the effective noise temperature is given by $T_n = 290(F-1)$.

effective isotropically radiated power (e.i.r.p.): The arithmetic product of (a) the power supplied to an antenna and (b) its gain.

effective mode volume: For an optical fiber, the square of the product of the diameter of the near-field pattern and the sine of the radiation angle of the far-field pattern. The diameter of the near-field radiation pattern is defined here as the full width at half maximum and the radiation angle at half maximum intensity. *Note:* Effective mode volume is proportional to the breadth of the relative distribution of power amongst the modes in a multimode fiber. It is not truly a spatial volume but rather an *"optical volume"* equal to the product of area and solid angle.

effective monopole radiated power (e.m.r.p.): The product of the power supplied to the antenna and its gain relative to a short vertical antenna in a given direction. [RR]

effective power: In alternating-current power transmission and distribution, the product of the rms voltage and amperage, *i.e.,* the apparent power, multiplied by the power factor, *i.e.,* the cosine of the phase angle between the voltage and the current. *Note:* Only effective power, *i.e.,* the actual power delivered to or consumed by the load, is expressed in watts. Apparent power is properly expressed only in volt-amperes, never watts. *Synonym* **true power.** *See figure on following page.*

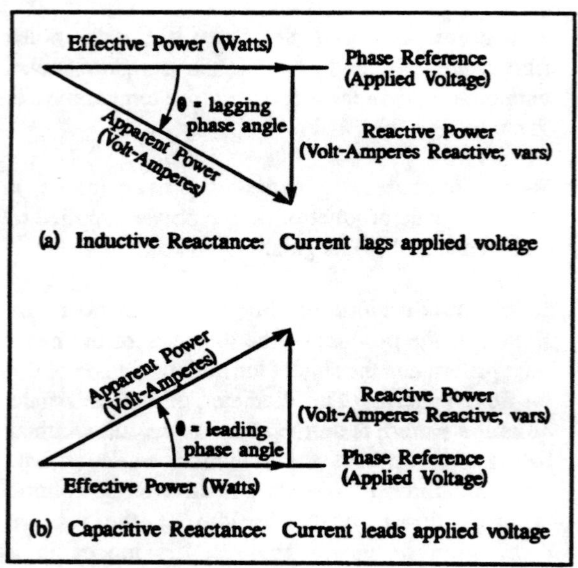

effective power

effective radiated power (e.r.p.) (in a given direction): **1.** The power supplied to an antenna multiplied by the antenna gain in a given direction. (188) *Note 1:* If the direction is not specified, the direction of maximum gain is assumed. *Note 2:* The type of reference antenna must be specified. **2.** The product of the power supplied to the antenna and its gain relative to a half-wave dipole in a given direction. [NTIA] [RR] *Note:* If the direction is not specified, the direction of maximum gain is assumed.

effective speed of transmission: *Synonym* **effective transmission rate.**

effective transmission rate: The rate at which information is processed by a transmission facility. *Note 1:* The effective transmission rate is calculated as (a) the measured number of units of data, such as bits, characters, blocks, or frames, transmitted during a significant measurement time interval divided by (b) the measurement time interval. *Note 2:* The effective transmission rate is usually expressed as a number of units of data per unit time, such as bits per second or characters per second. (188) *Synonyms* **average rate of transmission, effective speed of transmission.**

efficiency factor: In data communications, the ratio of (a) the time to transmit a text automatically at a specified modulation rate to (b) the time actually required to receive the same text at a specified maximum error rate. (188) *Note 1:* All of the communication facilities are assumed to be in the normal condition of adjustment and operation. *Note 2:* Telegraph communications may have different temporal efficiency factors for the two directions of transmission. *Note 3:* The practical conditions of measurement should be specified, especially the duration of the measurement.

EHF: *Abbreviation for* **extremely high frequency.**

EIA: *Abbreviation for* **Electronic Industries Association.**

EIA Class IVa fiber: *Synonym* **dispersion-unshifted fiber.**

EIA Class IVb fiber: *Synonym* **dispersion-shifted fiber.**

EIA interface: Any of a number of equipment interfaces compliant with voluntary industry standards developed by the Electronic Industries Association (EIA) to define interface parameters. *Note 1:* Some of the EIA interface standards have been adopted by the Federal Government as Federal Standards or Federal Information Processing Standards. *Note 2:* The telecommunication-standards-developing bodies of the EIA are now part of the Telecommunications Industry Association (TIA), and the standards are designated TIA/EIA-XXX.

eight-hundred (800) service: A service that allows call originators to place toll telephone calls to 800-service subscribers, from within specified rate areas, without a charge to the call originator.

elastic buffer: **1.** A buffer that has an adjustable capacity for data. **2.** A buffer that introduces an adjustable delay of signals.

E layer: *See* **E region, ionosphere.**

electrical length: **1.** Of a transmission medium, its length expressed as a multiple or submultiple of the wavelength of a periodic electromagnetic or electrical signal propagating within the medium. *Note 1:* The wavelength may be expressed in radians or in artificial units of angular measure, such as degrees. *Note 2*: In both coaxial cables and optical fibers, the velocity of propagation is approximately two-thirds that of free space. Consequently, the wavelength will be approximately two-thirds that in free space, and the electrical length, approximately 1.5 times the physical length. **2.** Of a transmission medium, its physical length multiplied by the ratio of (a) the propagation time of an electrical or electromagnetic signal through the medium to (b) the propagation time of an electromagnetic wave in free space over a distance equal to the physical length of the medium in question. *Note:* The electrical length of a physical medium will always be greater than its physical length. For example, in coaxial cables, distributed resistances, capacitances and inductances impede the propagation of the signal. In an optical fiber, interaction of the lightwave with the materials of which the fiber is made, and fiber geometry, affect the velocity of propagation of the signal. **3.** Of an antenna, the effective length of an element, usually expressed in wavelengths. *Note 1:* The electrical length is in general different from the physical length. *Note 2:* By the addition of an appropriate reactive element (capacitive or inductive), the electrical length may be made significantly shorter or longer than the physical length.

electrically powered telephone: A telephone in which the operating power is obtained either from a battery located at the telephone, *i.e.*, a local battery, or from a telephone central office, *i.e.*, a common battery.

electric field: The effect produced by the existence of an electric charge, such as an electron, ion, or proton, in the volume of space or medium that surrounds it. *Note:* Each of a distribution of charges contributes to the whole field at a point on the basis of superposition. A charge placed in the volume of space or in the surrounding medium has a force exerted on it.

electrochemical recording: Facsimile recording by means of a chemical reaction brought about by the passage of a signal-controlled current through the sensitized portion of the record sheet. (188)

electrographic recording: *See* **electrostatic recording.**

electroluminescence: Nonthermal conversion of electrical energy into light. *Note 1:* Electroluminescence is distinguished from incandescence, which is a thermal process. *Note 2:* One example of electroluminescence is the photon emission resulting from electron-hole recombination in a pn junction, as in a light-emitting diode (LED).

electrolytic recording: Electrochemical facsimile recording in which the recorded copy is made by the passage of a signal-controlled current through an electrolyte which causes metallic ions to be deposited, thus forming an image of the object. (188)

electromagnetic compatibility (EMC): **1.** Electromagnetic compatibility is the condition which prevails when telecommunications equipment is performing its individually designed function in a common electromagnetic environment without causing or suffering unacceptable degradation due to unintentional electromagnetic interference to or from other equipment in the same environment. [NTIA] **2.** The ability of systems, equipment, and devices that utilize the electromagnetic spectrum to operate in their intended operational environments without suffering unacceptable degradation or causing unintentional degradation because of electromagnetic radiation or response. It involves the application of sound electromagnetic spectrum management; system, equipment, and device design configuration that ensures interference-free operation; and clear concepts and doctrines that maximize operational effectiveness. [JP1]

electromagnetic emission control: The control of friendly electromagnetic emissions, such as radio and radar transmissions, for the purpose of preventing or minimizing their use by unintended recipients. (188)

electromagnetic environment (EME): **1.** For a telecommunications system, the spatial distribution of electromagnetic fields surrounding a given site. (188) *Note:* The electromagnetic environment may be expressed in terms of the spatial and temporal distribution of electric field strength (volts/meter), irradiance (watts/meter^2), or energy density (joules/meter^3). **2.** The resulting product of the power and time distribution, in various frequency ranges, of the radiated or conducted electromagnetic emission levels that may be encountered by a military force, system, or platform when performing its assigned mission in its intended operational environment. It is the sum of electromagnetic interference; electromagnetic pulse; hazards of electromagnetic radiation to personnel, ordnance, and volatile materials; and natural phenomena effects of lightning and p-static. [JP1]

electromagnetic interference (EMI): Any electromagnetic disturbance that interrupts, obstructs, or otherwise degrades or limits the effective performance of electronics/electrical equipment. It can be induced intentionally, as in some forms of electronic warfare, or unintentionally, as a result of spurious emissions and responses, intermodulation products, and the like. [JP1] *Synonym* **radio frequency interference.**

electromagnetic interference (EMI) control: The control of radiated and conducted energy such that emissions that are unnecessary for system, subsystem, or equipment operation are reduced, minimized, or eliminated. *Note:* Electromagnetic radiated and conducted emissions are controlled regardless of their origin within the system, subsystem, or equipment. Successful EMI control with effective susceptibility control leads to electromagnetic compatibility. (188)

electromagnetic intrusion: The intentional insertion of electromagnetic energy into transmission paths in any manner, with the objective of deceiving operators or of causing confusion. [JP1]

electromagnetic pulse (EMP): **1.** The electromagnetic radiation from a nuclear explosion caused by Compton-recoil electrons and photoelectrons from photons scattered in the materials of the nuclear device or in a surrounding medium. The resulting electric and magnetic fields may couple with electrical/electronic systems to produce damaging current and voltage surges. May also be caused by nonnuclear means. [JP1] **2.** A broadband, high-intensity, short-duration burst of electromagnetic energy. (188) *Note:* In the case of a nuclear detonation, the electromagnetic pulse consists of a continuous frequency spectrum. Most of the energy is distributed throughout the lower frequencies between 3 Hz and 30 kHz.

electromagnetic radiation (EMR): Radiation made up of oscillating electric and magnetic fields and propagated with the speed of light. Includes gamma radiation, X-rays, ultraviolet, visible, and infrared radiation, and radar and radio waves. [JP1]

electromagnetic radiation hazards (RADHAZ *or* EMR hazards): Hazards caused by a transmitter/antenna installation that generates electromagnetic radiation in the vicinity of ordnance, personnel, or fueling operations in excess of established safe levels or increases the existing levels to a hazardous level; or a personnel, fueling, or ordnance installation located in an area that is illuminated by electromagnetic radiation at a level that is hazardous to the planned operations or occupancy. These hazards will exist when an electromagnetic field of sufficient intensity is generated to: (a) induce or otherwise couple currents and/or voltages of magnitudes large enough to initiate electroexplosive devices or other sensitive explosive components of weapon systems, ordnance, or explosive devices; (b) cause harmful or injurious effects to humans and wildlife; (c) create sparks having sufficient magnitude to ignite flammable mixtures of materials that must be handled in the affected area. [JP1]

electromagnetic spectrum: The range of frequencies of electromagnetic radiation from zero to infinity. *Note:* The electromagnetic spectrum was, by custom and practice, formerly divided into 26 alphabetically designated bands. This usage still prevails to some degree. However, the ITU formally recognizes 12 bands, from 30 Hz to 3000 GHz. New bands, from 3 THz to 3000 THz, are under active consideration for recognition. *Refer to the figure on page E-20.*

electromagnetic survivability: The ability of a system, subsystem, or equipment to resume functioning without evidence of degradation following temporary exposure to an adverse electromagnetic environment. *Note:* The system, subsystem, or equipment performance may be degraded during exposure to the adverse electromagnetic environment, but the system will not experience permanent damage, such as component burnout, that will prevent proper operation when the adverse electromagnetic environment is removed. (188)

electromagnetic vulnerability (EMV): The characteristics of a system that cause it to suffer a definite degradation (incapability to perform the designated mission) as a result of having been subjected to a certain level of electromagnetic environmental effects. [JP1] (188)

electromagnetic wave (EMW): A wave produced by the interaction of time-varying electric and magnetic fields.

electromechanical recording: Recording by means of a signal-actuated mechanical device. (188)

electronically controlled coupling (ECC): The coupling of a lightwave from one dielectric waveguide into another dielectric waveguide upon the application of an electric field or electrical signal. *Note:* Devices that perform ECC can be used as switches.

electronic classroom: *Synonym* **teletraining.**

electronic commerce: Business transactions conducted by electronic means other than conventional telephone service, *e.g.,* facsimile or electronic mail (E-mail).

electronic counter-countermeasures (ECCM): That division of electronic warfare involving actions taken to ensure friendly effective use of the electromagnetic spectrum despite the enemy's use of electronic warfare.

electronic countermeasures (ECM): That division of electronic warfare involving actions taken to prevent or reduce an enemy's effective use of the electromagnetic spectrum.

electronic deception: **1.** The deliberate radiation, reradiation, alteration, suppression, absorption, denial, enhancement, or reflection of electromagnetic energy in a manner intended to convey misleading information and to deny valid information to an enemy or to enemy electronics-dependent weapons. *Note:* Among the types of electronic deception are: (a) manipulative electronic deception—Actions to eliminate revealing or convey misleading, telltale indicators that may be used by hostile forces; (b) simulative electronic deception—Actions to represent friendly notional or actual capabilities to mislead hostile forces; (c) imitative electronic deception—The introduction of electromagnetic energy into enemy systems that imitates enemy emissions. **2.** Deliberate activity designed to mislead an enemy in the interpretation or use of information received by his electronic systems.

electronic emission security: Those measures taken to protect all transmissions from interception and electronic analysis. (188)

electronic jamming: The deliberate radiation, reradiation, or reflection of electromagnetic energy for the purpose of disrupting enemy use of electronic devices, equipment, or systems.

electronic line of sight: The path traversed by electromagnetic waves that is not subject to reflection or refraction by the atmosphere. [JP1]

electronic line scanning: In facsimile, a method of scanning that provides motion of the scanning spot along the scanning line by electronic means. (188)

electronic mail (E-mail): An electronic means for communication in which (a) usually text is transmitted, (b) operations include sending, storing, processing, and receiving information, (c) users are allowed to communicate under specified conditions, and (d) messages are held in storage until called for by the addressee. (188)

electronic message system (EMS): A message system incorporating electronic mail to a central facility which then assumes responsibility for delivering the message in hard copy form. *Note:* In DOD, these messages have a specific format known as message text format (MTF).

electronic reconnaissance: The detection, identification, evaluation, and location of foreign electromagnetic radiations emanating from other than nuclear detonations or radioactive sources. [JP1]

electronics intelligence (ELINT): Technical and geolocation intelligence information derived from foreign noncommunications electromagnetic radiations emanating from other than nuclear detonations or radioactive sources. [JP1]

electronics security (ELSEC): The protection resulting from all measures designed to deny unauthorized persons information of value that might be derived from their interception, and study of noncommunications electromagnetic radiations, *e.g.*, radar. [JP1]

electronic switching system (ESS): 1. A telephone switching system based on the principles of time-division multiplexing of digitized analog signals. *Note:* An electronic switching system digitizes analog signals from subscribers' loops, and interconnects them by assigning the digitized signals to the appropriate time slots. It may also interconnect digital data or voice circuits. **2.** A switching system with major devices constructed of semiconductor components. *Note:* A semi-electronic switching system that has reed relays or crossbar matrices, as well as semiconductor components, is also considered to be an ESS. (188)

electronic warfare (EW): Any military action involving the use of electromagnetic and directed energy to control the electromagnetic spectrum or to attack the enemy. The three major subdivisions within electronic warfare are: electronic attack, electronic protection, and electronic warfare support. [After JP1]

electronic warfare support measures (ESM): 1. That division of electronic warfare involving actions taken under direct control of an operational commander to search for, intercept, identify, and locate sources of radiated electromagnetic energy for the purpose of immediate threat recognition. Thus, electronic warfare support measures (ESM) provide a source of information required for immediate decisions involving electronic countermeasures (ECM), electronic counter-countermeasures (ECCM), avoidance, targeting, and other tactical employment of forces. Electronic warfare support measures data can be used to produce signals intelligence (SIGINT), both communications intelligence (COMINT) and electronics intelligence (ELINT). **2.** That division of electronic warfare involving action taken to search for, intercept, identify, and locate radiated electromagnetic energy for the purpose of immediate threat recognition. It provides a source of information required for immediate decisions involving electronic countermeasures, electronic counter-countermeasures, and other tactical actions such as avoidance, targeting and homing.

electro-optical intelligence (ELECTRO-OPTINT): Intelligence information other than signals intelligence derived from the optical monitoring of the electromagnetic spectrum from ultraviolet (0.01 μm) through the far infrared (1000 μm). [JP1]

electro-optic detector: *Deprecated term. See* **optoelectronic.**

electro-optic effect: Any one of a number of phenomena that occur when an electromagnetic wave in the optical spectrum interacts with an electric field, or with matter under the influence of an electric field. (188) *Note 1:* Two of the most important electro-optic effects having application as modulation mechanisms in optical communication are the Kerr effect and the Pockels effect, in which birefringence is induced or modified in a liquid (Kerr effect) or solid (Pockels effect). *Note 2:* The term *"electro-optic"* is often erroneously used as a synonym for *"optoelectronic."*

electro-optic modulator: An optical device in which a signal-controlled element is used to modulate a beam of light. (188) *Note 1:* The modulation may be imposed on the phase, frequency, amplitude, or direction of the modulated beam. *Note 2:* Modulation bandwidths into the gigahertz range are possible using laser-controlled modulators.

electro-optics: The technology associated with those components, devices and systems which are designed to interact between the electromagnetic (optical) and the electric (electronic) state. [JP1] *Note 1:* The operation of electro-optic devices depends on modification of the refractive index of a material by electric fields. (188) *Note 2:* In a Kerr cell, the refractive index change is proportional to the square of the electric field, and the material is usually a liquid. *Note 3:* In a Pockels cell, the refractive index change varies linearly with the electric field, and the material is a crystal. *Note 4:* *"Electro-optic"* is often erroneously used as a synonym for *"optoelectronic"*.

ELECTRO-OPTINT: *Acronym for* **electro-optical intelligence.**

electrophotographic recording: Recording in which light is used to produce a change in electrostatic charge distribution to form a photographic image. (188) *Note:* Subsequent processing is usually required to make the image visible.

electrosensitive recording: Recording in which an electrical signal is directly impressed on the record medium.

electrostatic recording: Recording by means of a signal-controlled electrostatic field. (188) *Note:* Subsequent processing is usually required to make the image visible.

electrothermal recording: That type of recording produced principally by signal-controlled thermal action. (188)

elemental area: In facsimile transmission systems, any segment of a scanning line, the dimension of which along the line is exactly equal to the nominal line width. (188) *Note:* An elemental area is not necessarily the same as the scanning spot.

elementary signaling element: *See* **unit interval.**

elevated duct: An atmospheric duct consisting of a high-density air layer that starts at high altitudes and continues upward or remains at high altitudes, thus affecting primarily very-high-frequency (VHF) transmission. [From Weik '89]

ELF: *Abbreviation for* **extremely low frequency.**

ELINT: *Acronym for* **electronics intelligence.**

elliptical polarization: In electromagnetic wave propagation, polarization such that the tip of the electric field vector describes an ellipse in any fixed plane intersecting, and normal to, the direction of propagation. (188) *Note 1:* An elliptically polarized wave may be resolved into two linearly polarized waves in phase quadrature with their polarization planes at right angles to each other. [2196] *Note 2:* Circular and linear polarization are special cases of elliptical polarization.

ELSEC: *Acronym for* **electronics security.**

E-mail: *Abbreviation for* **electronic mail.**

emanations security (EMSEC): The protection resulting from all measures designed to deny unauthorized persons information of value that might be derived from intercept and analysis of compromising emanations from other than crypto-equipment and telecommunications systems.

embedded base equipment: Customer-premises equipment that had been provided by the Bell Operating Companies prior to January 1, 1984, that was ordered transferred from the BOCs to AT&T by court order.

embedded customer-premises equipment: Telephone-company-provided premises equipment in use or in inventory of a regulated telephone utility as of divestiture (December 31, 1981).

embedded processor: In non-ADP equipment, a CPU and firmware that are critical to the operation of the equipment. *Note:* An embedded processor is not subject to FIRMR regulation when used for control of devices such as weapons systems, communications devices, home appliances, automobile diagnostics, *etc.*

EMC: *Abbreviation for* **electromagnetic compatibility.**

EMC analysis: Analysis of a system, subsystem, facility, or equipment to determine its electromagnetic compatibility (EMC) status. (188) *Note:* The EMC analysis may be theoretical analysis before construction or an empirical analysis after construction.

EMCON: *Abbreviation for* **emission control.**

EMD: *Abbreviation for* **equilibrium mode distribution.**

EME: *Abbreviation for* **electromagnetic environment.**

Emergency Broadcast System (EBS): The EBS is composed of AM, FM, and TV broadcast stations; low-power TV stations; and non-Government industry entities operating on a voluntary, organized basis during emergencies at national, state, or operational (local) area levels. [47CFR]

emergency locator transmitter (ELT): A transmitter of an aircraft or survival craft actuated manually or automatically that is used as an alerting and locating aid for survival purposes. [NTIA] [RR]

emergency position-indicating radiobeacon station: A station in the mobile service the emissions of which are intended to facilitate search and rescue operations. [NTIA] [RR]

EMI: *Abbreviation for* **electromagnetic interference.**

emission: **1.** Electromagnetic energy propagated from a source by radiation or conduction. (188) *Note:* The emission may be either desired or undesired and may occur anywhere in the electromagnetic spectrum. **2.** Radiation produced, or the production of radiation, by a radio transmitting station. For example, the energy radiated by the local oscillator of a radio receiver would not be an emission but a radiation. [NTIA] [RR]

emission control (EMCON): The selective and controlled use of electromagnetic, acoustic, or other emitters to optimize command and control capabilities while minimizing, for operations security (OPSEC): (a) detection by enemy sensors; (b) to minimize mutual interference among friendly systems; and/or (c) to execute a military deception plan. [After JP1]

emission security: Protection resulting from all measures taken to deny unauthorized persons information of value which might be derived from intercept and analysis of compromising emanations from crypto-equipment, AIS, and telecommunications systems. [NIS]

emission spectrum: Of a radio emission, the distribution of power or energy as a function of frequency.

emissivity: The ratio of power radiated by a substance to the power radiated by a blackbody at the same temperature. (188)

EMP: *Abbreviation for* **electromagnetic pulse.**

emphasis: In FM transmission, the intentional alteration of the amplitude-vs.-frequency characteristics of the signal to reduce adverse effects of noise in a communication system. (188) *Note:* The high-frequency signal components are emphasized to produce a more equal modulation index for the transmitted frequency spectrum, and therefore a better signal-to-noise ratio for the entire frequency range.

EMR: *Abbreviation for* **electromagnetic radiation.**

EMR hazards: *Abbreviation for* **electromagnetic radiation hazards**

e.m.r.p.: *Abbreviation for* **effective monopole radiated power.**

EMS: *Abbreviation for* **electronic message system.**

EMSEC: *Acronym for* **emanations security.**

emulate: To duplicate the functions of one system with a different system, so that the second system appears to behave like the first system. *Note:* For example, a computer emulates another, different computer by accepting the same data, executing the same programs, and achieving the same results. *Contrast with* **simulate.**

EMV: *Abbreviation for* **electromagnetic vulnerability.**

enabling signal: A signal that permits the occurrence of an event.

en-bloc signaling: Signaling in which address digits are transmitted in one or more blocks, each block containing sufficient address information to enable switching centers to carry out progressive onward routing.

encapsulation: In open systems, the technique used by layered protocols in which a lower layer protocol accepts a message from a higher layer protocol and places it in the data portion of a frame in the lower layer.

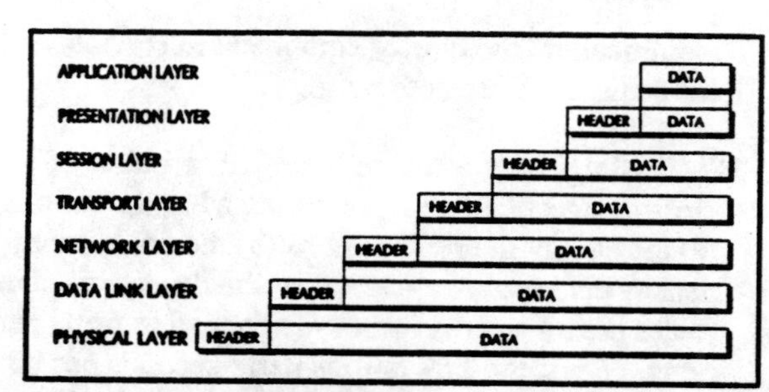

OSI—Reference Model example of data encapsulation

encipher: [To] Convert plain text into an unintelligible form by means of a cipher. [NIS]

encode: **1.** To convert data by the use of a code, frequently one consisting of binary numbers, in such a manner that reconversion to the original form is possible. (188) **2.** [To] Convert plain text to equivalent cipher text by means of a code. [NIS] **3.** To append redundant check symbols to a message for the purpose of generating an error detection and correction code.

encoder: *See* **analog-to-digital converter.**

encoding: *See* **analog encoding.**

encoding law: A law defining the relative values of the quantum steps used in quantizing and encoding signals.

encrypt: **1.** [A] generic term encompassing encipher and encode. [NIS] **2.** To convert plain text into unintelligible forms by means of a cryptosystem. *Note:* The term *"encrypt"* covers the meanings of *"encipher"* and *"encode."* [JP1]

end distortion: In start-stop teletypewriter operation, the shifting of the end of all marking pulses, except the stop pulse, from their proper positions in relation to the beginning of the next start pulse. (188) *Note 1:* Shifting of the end of the stop pulse is a deviation in character time and rate rather than an end distortion. *Note 2:* Spacing end distortion is the termination of marking pulses before the proper time. *Note 3:* Marking end distortion is the continuation of marking pulses past the proper time. *Note 4:* The magnitude of the distortion is expressed as a percentage of an ideal pulse length.

end exchange: *Synonym* **end office.**

end finish: For an optical fiber, the optical quality of the surface at the end of the fiber.

end instrument: A communication device that is connected to the terminals of a circuit. (188)

end office (EO): A central office at which user lines and trunks are interconnected. *Synonym* **end exchange.**

end-of-medium character: A control character that may be used to identify either the physical end of a data medium or the end of the usable or used portion of a data medium. [From Weik '89]

end-of-message function: In tape relay procedure, the letter and key functions, including the end-of-message indicator, that constitute the last format line. (188)

end-of-selection character: The character that indicates the end of the selection signal.

end-of-text character (ETX): A transmission control character used to terminate text.

end-of-transmission-block character (ETB): A transmission control character used to indicate the end of a transmission block of data when data are divided into such blocks for transmission purposes.

end-of-transmission character (EOT): A transmission control character used to indicate the conclusion of a transmission that may have included one or more texts and any associated message headings. *Note:* An EOT is often used to initiate other functions, such as releasing circuits, disconnecting terminals, or placing receive terminals in a standby condition.

endpoint node: In network topology, a node connected to one and only one branch. *Synonym* **peripheral node.**

end system (ES): A system containing the application processes that are the ultimate source and sink of user traffic. *Note:* The functions of an end system can be distributed among two or more processors or computers.

end-to-end encryption: The encryption of information at its origin and decryption at its intended destination without any intermediate decryption. (188)

end-to-end security: Safeguarding information in a secure telecommunication system by cryptographic or protected distribution system means from point of origin to point of destination. [NIS]

endurability: The property of a system, subsystem, equipment, or process that enables it to continue to function within specified performance limits for an extended period of time, usually months, despite a severe natural or man-made disturbance, such as a nuclear attack, or a loss of external logistic or utility support. (188) *Note:* Endurability is not compromised by temporary failures when the local capability exists to restore and maintain the system, subsystem, equipment, or process to an acceptable performance level.

endurable operation: *See* **endurability.**

end user: The ultimate user of a telecommunications service.

engineering channel: *Synonym* **orderwire circuit.**

engineering orderwire (EOW): *Synonym* **orderwire circuit.**

enhanced-quality television: *Synonym [in CCITT usage]* **improved-definition television.**

enhanced service: Service, offered over commercial carrier transmission facilities used in interstate communications, that employs computer processing applications that act on the format, content, code, protocol, or similar aspects of the subscriber's transmitted information; provides the subscriber with additional, different, or restructured information; or involves subscriber interaction with stored information. (188)

ENQ: *Abbreviation for* **enquiry character.**

enquiry character (ENQ): A transmission control character used as a request for a response from the station with which a connection has been set up. *Note:* The response may include station identification, the type of equipment in service, and the status of the remote station.

***E/N* ratio:** In the transmission of a pulse of an electromagnetic wave representing a bit, the ratio of (a) the energy in each bit, E, to (b) the noise energy density per hertz, N. *Note:* E is usually expressed in joules per bit and N is usually expressed in watts per hertz. Thus, the *E/N* ratio is hertz-seconds per bit. A joule is a watt-second and a hertz is a cycle per second. Thus, the *E/N* ratio is actually cycles per bit. However, if a cycle is a bit, then the *E/N* ratio is dimensionless. [From Weik '89]

entrance facility: The entrance to a building for both public and private network service cables (including antenna transmission lines, where applicable), including the entrance point at the building wall or floor, and continuing to the entrance room or entrance space. [After ANSI/TIA/EIA-568A]

entrance point: In a building, the point of emergence of telecommunications service cables through an exterior wall, floor slab, or from a rigid metal conduit or intermediate metal conduit. [After ANSI/TIA/EIA-568A]

entrance room: In a building, a space in which the joining of inter- and/or intrabuilding telecommunications backbone facilities takes place. *Note:* An entrance room may serve also as an equipment room. [After ANSI/TIA/EIA-568A]

envelope: The boundary of the family of curves obtained by varying a parameter of a wave. (188) *See figure under* **amplitude modulation.**

envelope delay distortion: Signal distortion that results when the rate of change of phase shift with frequency over the necessary bandwidth of the signal is not constant. (188) *Note:* Envelope delay distortion is usually expressed as one-half the difference between the delays of the two extremes of the necessary bandwidth.

environmental control: *See* **air-conditioning.**

environmental security: **1.** The security that is inherent in the physical surroundings in which a facility or functional unit is located, such as on ships, on aircraft, and in underground vaults, where locations by their nature provide a certain amount of protection against exploitation of compromising emanation even before other protective measures are implemented. 2. The application of electrical, acoustic, physical, and other safeguards to an area to minimize the risk of unauthorized interception of information from the area. [From Weik '89]

EO: *Abbreviation for* **end office.**

EOT: *Abbreviation for* **end-of-transmission character.**

EOW: *Abbreviation for* **engineering orderwire.**

E-plane bend: *Synonym* **E-bend.**

epoch date: A date in history, chosen as the reference date from which time is measured. *Note 1:* An example of an epoch date is the beginning instant of January 1, 1900, Universal Time, for Transmission Control Protocol/Internet Protocol (TCP/IP). *Note 2:* TCP/IP programs exchange date or time-of-day information with time expressed as the number of seconds past the epoch date.

equal gain combiner: A diversity combiner in which the signals on each channel are added. *Note:* The channel gains can be made to remain always and everywhere equal so that the resultant signal remains approximately constant. (188)

equalization: The maintenance of system transfer function characteristics within specified limits by modifying circuit parameters. (188) *Note:* Equalization includes modification of circuit parameters, such as resistance, inductance, or capacitance.

equal-length code: A telegraph or a data code in which (a) all the words or code groups are composed of the same number of unit elements, (b) each element has the same duration or spatial length, (c) each word or code group has the same duration or spatial length, and (d) usually each word or code group has the same number of characters. [From Weik '89]

equal-level patch bay: An analog patching facility at which all nominal input and output voice frequency levels are uniform. (188) *Note:* The use of an equal-level patch bay permits patching without making transmission level adjustments.

equatorial orbit: For a satellite orbiting the Earth, an orbit in the equatorial plane. *Note:* An equatorial orbit has an inclination angle of 0°. (188)

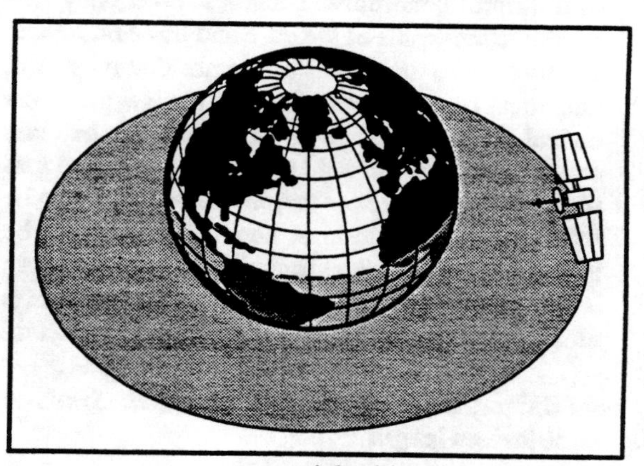

equatorial orbit

equilibrium coupling length: *Synonym* **equilibrium length.**

equilibrium length: For a specific excitation condition, the length of multimode optical fiber necessary to attain equilibrium mode distribution. (188) *Note:* Equilibrium length is sometimes used to refer to the longest such length, as would result from a worst-case, but undefined, excitation. *Synonyms* **equilibrium coupling length, equilibrium mode distribution length.**

equilibrium mode distribution (EMD): That condition in a multimode fiber wherein after propagation has taken place for a certain distance, called the *"equilibrium length,"* the relative power distribution among modes becomes statistically constant and remains so for the course of further propagation down the fiber. *Note 1:* In practice, the equilibrium length may vary from a fraction of a kilometer to more than a kilometer. *Note 2:* After the equilibrium length has been traversed, the numerical aperture of the fiber's output is independent of the numerical aperture of the optical source, *i.e.*, beam, that drives the fiber. This is because of mode coupling and stripping, primarily by small perturbations in the fiber's geometry which result from the manufacturing and cabling processes.

Note 3: In the ray-optics analogy, the equilibrium mode distribution may be loosely thought of as a condition in which the "outermost rays" in the fiber core are stripped off by such phenomena as microbends, and only the "innermost rays" continue to propagate. In a typical 50-µm core multimode graded-index fiber, light propagating under equilibrium conditions occupies essentially the middle seven-tenths of the core and has a numerical aperture approximately seven-tenths that of the full numerical aperture of the fiber. This is why in-line optical attenuators based on the principle of gap loss may be ineffective or induce a lower-than-rated loss if they are inserted near the optical receiver. To be fully effective, gap-loss attenuators should be inserted near the optical transmitter, where the core is fully filled. [After FAA] *Synonyms* **equilibrium mode power distribution, steady-state condition.**

equilibrium mode distribution length: *Synonym* **equilibrium length.**

equilibrium mode power distribution: *Synonym* **equilibrium mode distribution.**

equilibrium mode simulator: For an optical fiber, a device or optical system used to create an approximation of the equilibrium mode distribution.

equipment clock: A clock that satisfies the particular needs of equipment and, in some cases, may control the flow of data at the equipment interface. (188)

equipment intermodulation noise: Intermodulation noise introduced into a system by a specific piece of equipment. (188)

equipment room: In a building, a centralized space for telecommunications equipment that serves the occupants of the building. *Note:* An equipment room is considered distinct from a telecommunications closet because of the nature or complexity of the equipment housed by the equipment room. [After ANSI/TIA/EIA-568A]

equipment side: The portion of a device that is directly connected to facilities internal to a station, such as the data terminal equipment (DTE) side of the DTE/data-circuit-terminating (DCE) interface, switches, and user end instruments.

equipotential ground plane: A mass, or bonded masses, of conducting material that offer a negligible impedance to current flow. (188) *Note:* Equipotential ground planes may be in direct contact with the earth or may be physically isolated from the earth and suitably connected to it.

equivalent network: **1.** In a system, a network that may replace another network without altering the performance of the system. **2.** A network with external characteristics that are identical to those of another network. **3.** A theoretical representation of an actual network. (188)

equivalent noise resistance: A quantitative representation in resistance units of the spectral density of a noise-voltage generator, given by $R_n = (\pi W_n)/(kT_0)$, where W_n is the spectral density, k is Boltzmann's constant, T_0 is the standard noise temperature (290 K), and $kT_0 = 4.00 \times 10^{-21}$ watt-seconds. *Note:* The equivalent noise resistance in terms of the mean-square noise-generator voltage, e^2, within a frequency increment, Δf, is given by $R_n = e^2/(4kT_0\Delta f)$.

equivalent noise temperature: The temperature, usually expressed in kelvins, of a hypothetical matched resistance at the input of an assumed noiseless device, such as a noiseless amplifier, that would account for the measured output noise. [From Weik '89]

equivalent pulse code modulation (PCM) noise: The amount of thermal noise power on a frequency-division multiplexed (FDM) or wire channel necessary to approximate the same judgment of speech quality created by quantizing noise in a PCM channel. *Note 1:* The speech quality judgment is based on comparative tests. (188) *Note 2:* Generally, 33.5 dBrnC ± 2.5 dB is considered the approximate equivalent PCM noise of a 7-bit PCM system.

equivalent satellite link noise temperature: The noise temperature referred to the output of the receiving antenna of the Earth station corresponding to the radio-frequency noise power which produces the total observed noise at the output of the satellite link excluding noise due to interference coming from satellite links using other satellites and from terrestrial systems. [NTIA] [RR]

erase: **1.** To obliterate information from a storage medium, such as to clear or to overwrite. (188) **2.** In a magnetic storage medium, to remove all stored data by (a) changing the medium to an unmagnetized state or (b) changing the medium to a predetermined magnetized state. **3.** In paper tape and punched card storage, to punch a hole at every punch position.

erect position: In frequency-division multiplexing, a position of a translated channel in which an increase signal frequency in the untranslated channel causes an increase signal frequency in the translated channel. (188) *Synonym* **upright position.**

E region: That portion of the ionosphere existing between approximately 95 and 130 km above the surface of the Earth. *Note:* The E Region lies between the D and F regions. (188) *Synonyms* **Heaviside layer, Kennelly-Heaviside layer.**

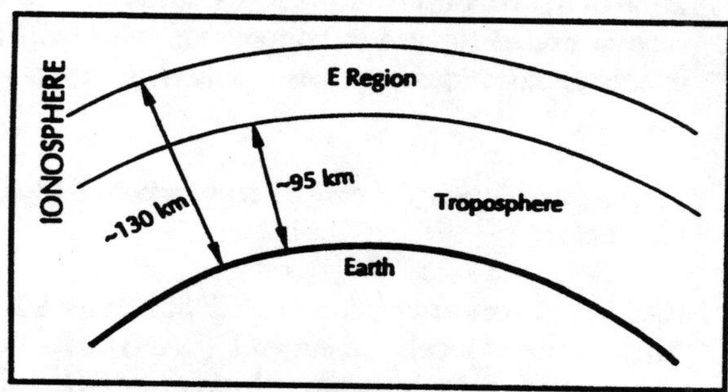

E region

erlang: A dimensionless unit of the average traffic intensity (occupancy) of a facility during a period of time, usually a busy hour. *Note 1:* Erlangs, a number between 0 and 1, inclusive, is expressed as the ratio of (a) the time during which a facility is continuously or cumulatively occupied to (b) the time that the facility is available for occupancy. (188) *Note 2:* Communications traffic, measured in erlangs for a period of time, and offered to a group of shared facilities, such as a trunk group, is equal to the average of the traffic intensity, in erlangs for the same period of time, of all individual sources, such as telephones, that share and are served exclusively by this group of facilities. *Synonym* **traffic unit.**

erroneous block: A block in which there are one or more erroneous bits.

error: **1.** The difference between a computed, estimated, or measured value and the true, specified, or theoretically correct value. (188) **2.** A deviation from a correct value caused by a malfunction in a system or a functional unit. *Note:* An example of an error is the occurrence of a wrong bit caused by an equipment malfunction. (188)

error blocks: In video systems, a form of block distortion in which a block or blocks in the received image bear no resemblance to the current or previous scene and may contrast greatly with adjacent blocks.

error budget: The bit-error-ratio requirement allocated to the respective segments of a communications system, such as trunking, switching, access, and terminal devices, in a manner that satisfies the specified system end-to-end bit-error-ratio requirement for transmitted traffic. (188)

error burst: A contiguous sequence of symbols, received over a data transmission channel, such that the first and last symbols are in error and there exists no contiguous subsequence of *m* correctly received symbols within the error burst. (188) *Note:* The integer parameter *m* is referred to as the guard band of the error burst. The last symbol in a burst and the first symbol in the following burst are accordingly separated by *m* correct bits or more. The parameter *m* should be specified when describing an error burst.

error control: Any technique that will detect or correct errors.

error-correcting code: A code in which each telegraph or data signal conforms to specific rules of construction so that departures from this construction in the received signal can generally be automatically detected and corrected. *Note 1:* If the number of errors is less than or equal to the maximum correctable threshold of the code, all errors will be corrected. (188) *Note 2:* Error-correcting codes require more signal elements than are necessary to convey the basic information. *Note 3:* The two main classes of error-correcting codes are block codes and convolutional codes.

error-correcting system: In digital data transmission, a system employing either forward error correction (FEC) or automatic repeat-request

(ARQ) techniques such that most transmission errors are automatically removed from the data unit prior to delivery to the destination facility. (188)

error-detecting-and-feedback system: *Synonym* **ARQ.**

error-detecting code: A code in which each telegraph or data signal conforms to specific rules of construction, so that departures from this construction in the received signal can generally be detected automatically. (188) *Note:* Error-detecting codes require more signal elements than are necessary to convey the basic information.

error-detecting system: A system employing an error-detecting code and so arranged that any signal detected as being in error is either deleted from the data delivered to the data sink, in some cases with an indication that such deletion has taken place, or delivered to the data sink together with an indication that the signal is in error.

error message: In a computer or communications system, a message that indicates that an error has been made and, sometimes, the nature or type of error. [From Weik '89]

error rate: *Deprecated term. See* **error ratio.**

error ratio: The ratio of the number of bits, elements, characters, or blocks incorrectly received to the total number of bits, elements, characters, or blocks sent during a specified time interval. (188) *Note:* For a given communication system, the bit error ratio will be affected by both the data transmission rate and the signal power margin.

error signal: In computer and communications systems, an audio or visual signal that indicates that an error has been made by the system or its operator. *Note:* In most systems, the error signal accompanies an error message and is used to draw operator attention to the error message. [From Weik '89]

e.r.p. [*or* ERP]: *Abbreviation for* **effective radiated power.**

ES: *Abbreviation for* **end system, expert system.**

ESC: *Abbreviation for* **escape character.**

escape character (ESC): **1.** In alphabet coding schemes, a specially designated character, the occurrence of which in the data signifies that one or more of the characters to follow are from a different character code, *i.e.*, have meanings other than normal. **2.** In a text-control sequence of characters, a control character that indicates the beginning of the sequence and the end of any preceding text.

ESF: *Abbreviation for* **extended superframe.**

ESM: *Abbreviation for* **electronic warfare support measure.**

ESS: *Abbreviation for* **electronic switching system.**

essential service: A network-provided service feature in which a priority dial tone is furnished. *Note 1:* Essential service is typically provided to fewer than 10 % of network users. *Note 2:* Essential service is recommended for use in conjunction with NS/EP telecommunications services. *Synonym* **critical service.**

ETB: *Abbreviation for* **end-of-transmission-block character.**

Ethernet: A standard protocol (IEEE 802.3) for a 10-Mb/s baseband local area network (LAN) bus using carrier-sense multiple access with collision detection (CSMA/CD) as the access method, implemented at the Physical Layer in the ISO Open Systems Interconnection—Reference Model, establishing the physical characteristics of a CSMA/CD network. *Note 1:* Ethernet is a standard for using various transmission media, such as coaxial cables, unshielded twisted pairs, and optical fibers. *Note 2:* The IEEE-802.3 standard is based on a proprietary product with a similar name.

ETX: *Abbreviation for* **end-of-text character.**

evanescent field: In a waveguide, a time-varying field having an amplitude that decreases monotonically as a function of transverse radial distance from the waveguide, but without an accompanying phase shift. (188) *Note 1:* The evanescent field is coupled, *i.e.*, bound, to an electromagnetic wave or mode propagating inside the waveguide. *Note 2:* The evanescent field is a surface wave. *Note 3:* In fiber optics, the evanescent field may be used to provide coupling to another fiber. [After 2196]

evanescent mode: A mode of the evanescent field. (188)

even parity: *See* **parity, parity check.**

event: **1.** An occurrence or happening, usually significant to the performance of a function, operation, or task. (188) **2.** In Integrated Services Digital Networks (ISDN), an instantaneous occurrence that changes at least one of the attributes of the global status of a managed object. *Note:* An event (a) may be persistent or temporary, thus allowing for functions, such as surveillance, monitoring, and performance measurement, (b) may generate reports, (c) may be spontaneous or planned, (d) may trigger other events, and (e) may be triggered by one or more other events.

EW: *Abbreviation for* **electronic warfare.**

exalted-carrier reception: A method of receiving either amplitude- or phase-modulated signals in which method the carrier is separated from the sidebands, filtered and amplified, and then combined with the sidebands again at a higher level prior to demodulation. *Synonym* **reconditioned carrier reception.** (188)

exception condition: In data transmission, the condition assumed by a device when it receives a command that it cannot execute.

excess insertion loss: *Deprecated term. See* **insertion loss.** *Note: Excess insertion loss* was used to indicate that, in an optical-fiber coupler, the loss occasioned by dividing the input power among the ports is not the total insertion loss.

exchange: **1.** A room or building equipped so that telephone lines terminating there may be interconnected as required. *Note:* The equipment may include manual or automatic switching equipment. (188) **2.** In the telephone industry, a geographic area (such as a city and its environs) established by a regulated telephone company for the provision of local telephone services. **3.** In the Modification of Final Judgment (MFJ), a local access and transport area.

exchange access: In telephone networks, access in which exchange services are provided for originating or terminating interexchange telecommunications within the exchange area.

exchange area: A geographic area served by one or more central offices within which local telephone service is furnished under regulation.

exchange facilities: The facilities included within a local access and transport area.

executive program: *Synonym* **supervisory program.**

exempted addressee: An organization, activity, or person included in the collective address group of a message and deemed by the message originator as having no need for the information in the message. *Note:* Exempted addressees may be explicitly excluded from the collective address group for the particular message to which the exemption applies.

existing quality television: *Synonym [in CCITT usage]* **distribution-quality television.**

expander: A device that restores the dynamic range of a compressed signal to its original dynamic range. (188)

expansion: The restoration of the dynamic range of a compressed signal to its original dynamic range.

expansion capability: The inherent limit for increasing the capacity of a system beyond its installed capacity. [NATO]

expedited data unit: In layered systems, a service data unit that is delivered to a peer entity in the destination open system before the delivery of any subsequent service data unit sent on that connection.

experimental station: A station utilizing radio waves in experiments with a view to the development of science or technique. This definition does not include amateur stations. [NTIA] [RR]

expert system (ES): A computer system that facilitates solving problems in a given field or application by drawing inference from a knowledge base developed from human expertise. *Note 1:* The term "*expert system*" is sometimes used synonymously with "*knowledge-based system,*" although it is usually taken to emphasize expert knowledge. *Note 2:* Some expert systems are able to improve their knowledge base and develop new inference rules based on their experience with previous problems.

express orderwire: A permanently connected voice circuit between selected stations for technical control purposes. (188)

extended area service (EAS): A network-provided service feature in which a user pays a higher flat rate to obtain wider geographical coverage without paying per-call charges for calls within the wider area.

extended binary coded decimal interchange code: *See* **EBCDIC.**

extended-definition television (EDTV): Television in which (a) improvements are made to the standard National Television System Committee (NTSC) television system, (b) the improvements are receiver-compatible with the standard NTSC television system, and (c) the improvements modify the standard NTSC television system emission standards. *Note 1:* EDTV improvements may include (a) a wider aspect ratio, (b) a higher picture definition than NTSC definition, and (c) any of the improvements used in improved-definition television (IDTV). *Note 2:* When EDTV signals are transmitted in the 4:3 aspect ratio, it is referred to as "*EDTV.*" When transmitted in a wider aspect ratio, it is referred to as "*EDTV-Wide.*"

extended superframe: A T-carrier framing technique in which framing requiring less frequent synchronization than the original T-carrier superframe format is provided for D-4 formatting and for on-line, real-time testing of circuit capability and operating condition. *Note:* Less-frequent synchronization frees overhead bits for use in testing and monitoring.

extension bell: In telephony, a user end device, separate from a subscriber telephone, which device produces an audible signal indicating that there is an incoming call from a switchboard or exchange. [From Weik '89]

extension facility: A facility that provides access to communications for a user or group of users isolated from a central communications node. (188)

extension terminal: A terminal that is added to an existing terminal and that uses the same circuit and address, *i.e.,* port and number, as the terminal to which it is added.

external timing reference: In a given communications system, a timing reference obtained from a source, such as a navigation system, external to the given system. *Note:* External timing references are usually referenced to Coordinated Universal Time .

extinction coefficient: The sum of the absorption coefficient and the scattering coefficient. [From Weik '89]

extinction ratio (r_e): The ratio of two optical power levels,

$$r_e = \frac{P_1}{P_2} ,$$

of a digital signal generated by an optical source, *e.g.,* a laser diode, where P_1 is the optical power level generated when the light source is "on," and P_2 is the power level generated when the light source is "off." *Note:* The extinction ratio may be expressed as a fraction or in dB. [2196].

extra bit: *Synonym* **added bit.**

extra block: *Synonym* **added block.**

extremely high frequency (EHF): Frequencies from 30 GHz to 300 GHz. (188)

extremely low frequency (ELF): Frequencies from 30 Hz to 300 Hz. (188)

extrinsic joint loss: For an optical fiber, that portion of a joint loss that is not intrinsic to the fibers, *e.g.*, loss caused by end separation, angular misalignment, or lateral misalignment.

eye pattern: An oscilloscope display in which a pseudorandom digital data signal from a receiver is repetitively sampled and applied to the vertical input, while the data rate is used to trigger the horizontal sweep. (188) *Note:* System performance information can be derived by analyzing the display. An open eye pattern corresponds to minimal signal distortion. Distortion of the signal waveform due to intersymbol interference and noise appears as closure of the eye pattern.

eyes only: A message marker for a special-category message that is intended for delivery only to a specific person, or authorized representative of that person, and therefore no one else. [From Weik '89]

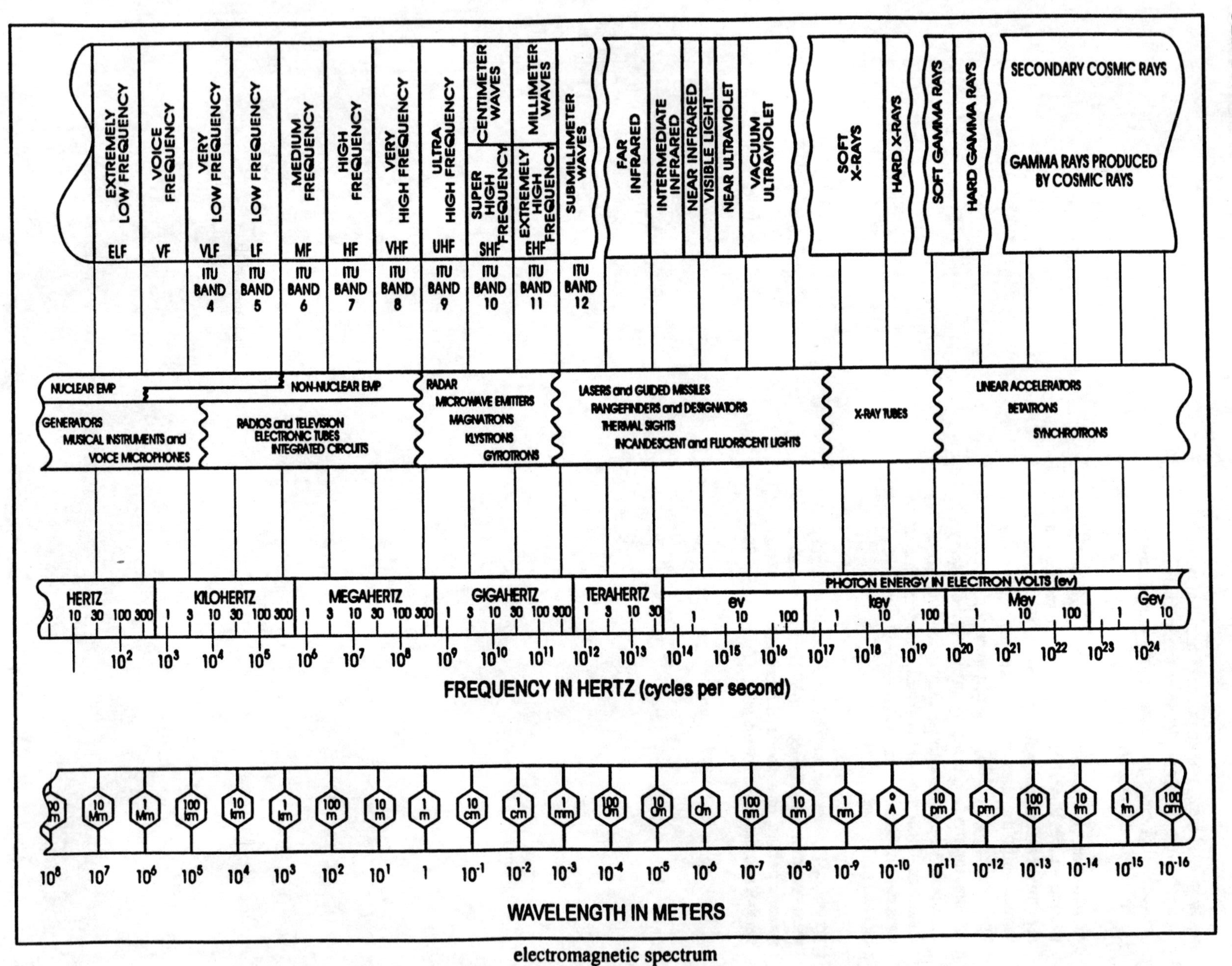

electromagnetic spectrum

facet erosion: In laser diodes, a phenomenon in which a high field intensity of stimulated optical radiation causes degradation of the facets, *i.e.*, those forming the cavity mirrors, decreasing reflectivity and resulting in a decrease of the internal quantum efficiency and an increase in the threshold current.

facility: 1. A fixed, mobile, or transportable structure, including (a) all installed electrical and electronic wiring, cabling, and equipment and (b) all supporting structures, such as utility, ground network, and electrical supporting structures. (188) 2. A network-provided service to users or the network operating administration. 3. A transmission pathway and associated equipment. 4. In a protocol applicable to a data unit, such as a block or frame, an additional item of information or a constraint encoded within the protocol to provide the required control. 5. A real property entity consisting of one or more of the following: a building, a structure, a utility system, pavement, and underlying land. [JP1]

facility grounding system: The electrically interconnected system of conductors and conductive elements that (a) provides multiple current paths to the earth electrode subsystem, and (b) consists of the earth electrode subsystem, the lightning protection subsystem, and the fault protection subsystem. (188)

facsimile (FAX): 1. A form of telegraphy for the transmission of fixed images, with or without half-tones, with a view to their reproduction in a permanent form. In this definition the term telegraphy has the same general meaning as defined in the Convention. [NTIA] [RR] 2. The process by which fixed graphic images, such as printed text and pictures, are scanned, and the information converted into electrical signals that may be transmitted over a telecommunications system and used to create a copy of the original, or an image so produced. (188) *Note 1:* Wirephoto and telephoto are facsimile via wire circuits. Radiophoto is facsimile via radio. *Note 2:* Technology now exists that permits the transmission and reception of facsimile data to or from a computer without requiring hard copy at either end. *Note 3:* Current facsimile systems are designated and defined as follows:

➢ **Group 1 Facsimile:** The mode of black and white facsimile operation, defined in CCITT Recommendation T.2, that uses double sideband modulation without any special measures to compress the bandwidth. *Note 1:* A 216 × 279-mm document, *i.e.*, an 8½ × 11-inch document, may be transmitted in approximately 6 minutes via a telephone-type circuit. Additional modes in this group may be designed to operate at a lower resolution suitable for the transmission of 216 × 279-mm documents in 3 to 6 minutes. *Note 2:* The CCITT frequencies used are 1300 Hz for white and 2300 Hz for black. The North American standard is 1500 Hz for white and either 2300 or 2400 Hz for black.

➢ **Group 2 Facsimile:** The mode of black and white facsimile operation, defined in CCITT Recommendation T.3, that accomplishes bandwidth compression by using encoding and vestigial sideband, but excludes processing of the document signal to reduce redundancy. *Note:* A 216 × 279-mm document, *i.e.*, an 8½ × 11-inch document, may be transmitted in approximately 3 minutes using a 2100-Hz AM/PM/VSB, over a telephone-type circuit.

➢ **Group 3 Facsimile:** The mode of black and white facsimile operation, defined in ITU-T Recommendation T.4, that incorporates means for reducing the redundant information in the signal by using a one-dimensional run-length coding scheme prior to the modulation process. *Note 1:* A 216 × 279-mm document, *i.e.*, an 8½ × 11-inch document, may be transmitted in approximately 1 minute or less over a telephone-type circuit with twice the Group 2 horizontal resolution. The vertical resolution may also be doubled. *Note 2:* Group 3 Facsimile machines have integral digital modems. *Note 3:* An optional two-dimensional bandwidth compression scheme is also defined within the Group 3 Facsimile Recommendation. *Note 4:* When any CCITT or CCIR Recommendation is modified by the ITU-T, the modified document is designated as an ITU-T Recommendation.

➢**Group 3C Facsimile:** The Group 3 digital mode of facsimile operation defined in CCITT Recommendation T.30. *Note:* Group 3C is also referred to as Group 3 Option C or as Group 3-64 kb/s.

➢ **Group 4 Facsimile:** The mode of black and white facsimile operation defined in ITU-T Recommendation T.563 and CCITT Recommendation T.6. *Note 1:* Group 4 Facsimile uses bandwidth compression techniques to transmit, essentially without errors, a 216 × 279-mm document, *i.e.*, an 8½ × 11-inch document, at a nominal resolution of 8 lines/mm in less than 1 minute over a public data network voice-grade circuit. *Note 2:* When any CCITT or CCIR Recommendation is modified by the ITU-T, the modified document is designated as an ITU-T Recommendation.

➢ **Type I Facsimile:** The mode of digital black and white facsimile operation defined in MIL-STD-188-161 used for transmission of bi-level information (*e.g.*, text and simple graphics). *Note:* Type I facsimile is interoperable with the black-and-white facsimile mode of STANAG 5000 and is designed for operation over noisy communications links such as tactical channels.

➢ **Type II Facsimile:** The mode of gray-scale facsimile operation defined in MIL-STD-188-161 used for transmission of multi-level information (*e.g.*, photographs). *Note:* Type II facsimile is interoperable with the black-and-white facsimile mode of Type I or STANAG 5000 equipment and is designed for operation over noisy communications links such as tactical channels.

facsimile converter: 1. In a facsimile receiver, a device that changes the signal modulation from frequency-shift keying (FSK) to amplitude modulation (AM). (188) 2. In a facsimile transmitter, a device that changes the signal modulation from amplitude modulation (AM) to frequency-shift keying (FSK). (188)

facsimile frequency shift: At any point in a frequency-shift facsimile system, the numerical difference between the frequency that corresponds to a white signal and the frequency that corresponds to a black signal. *Note:* Facsimile frequency shift is usually expressed in hertz. [From Weik '89]

facsimile picture signal: In facsimile systems, the baseband signal that results from the scanning process. (188)

facsimile receiver: In a facsimile system, the equipment that converts the facsimile picture signal into a recorded copy. (188)

facsimile recorder: In a facsimile receiver, the device that performs the final conversion of the facsimile picture signal to an image of the object, *i.e.*, makes the recorded copy. (188)

facsimile signal level: In a facsimile system, the signal level at any point in the system. (188) *Note 1:* The facsimile signal level is used to establish the operating levels. *Note 2:* The facsimile signal level is usually expressed in dB with respect to some standard value, such as 1 mW (milliwatt), *i.e.*, 0 dBm.

facsimile transceiver: In a facsimile system, the equipment that sends and receives facsimile signals. (188) *Note:* Full-duplex facsimile transceivers can send and receive at the same time; half-duplex facsimile transceivers cannot.

facsimile transmitter: In a facsimile system, the equipment that converts the baseband picture signals, *i.e.*, the baseband signals resulting from scanning the object, into signals suitable for transmission by a communications system. (188)

fade margin: 1. A design allowance that provides for sufficient system gain or sensitivity to accommodate expected fading, for the purpose of ensuring that the required quality of service is maintained. 2. The amount by which a received signal level may be reduced without causing system performance to fall below a specified threshold value. *Synonym* **fading margin.**

fading: In a received signal, the variation (with time) of the amplitude or relative phase, or both, of one or more of the frequency components of the signal. *Note:* Fading is caused by changes in the characteristics of the propagation path with time. (188)

fading distribution: The probability distribution that signal fading will exceed a given value relative to a specified reference level. (188) *Note 1:* In the case of phase interference fading, the time distribution of the instantaneous field strength usually approximates a Rayleigh distribution when several signal components of equal amplitude are present. *Note 2:* The field strength is usually measured in volts per meter. *Note 3:* The fading distribution may also be measured in terms of power level, where the unit of measure is usually watts per square meter and the expression is in dB.

fading margin: *Synonym* **fade margin.**

fail: *See* **failure, graceful degradation.**

fail safe: **1.** Of a device, the capability to fail without detriment to other devices or danger to personnel. (188) **2.** Pertaining to the automatic protection of programs and/or processing systems to maintain safety when a hardware or software failure is detected in a system. [NIS] **3.** Pertaining to the structuring of a system such that either it cannot fail to accomplish its assigned mission regardless of environmental factors or that the probability of such failure is extremely low.

fail-safe operation: **1.** Operation that ensures that a failure of equipment, process, or system does not propagate beyond the immediate environs of the failing entity. (188) **2.** A control operation or function that prevents improper system functioning or catastrophic degradation in the event of circuit malfunction or operator error.

failure: The temporary or permanent termination of the ability of an entity to perform its required function. (188)

failure access: [An] Unauthorized and usually inadvertent access to data resulting from a hardware or software failure in an AIS. [NIS]

fair queuing: The controlling of congestion in gateways by restricting every host to an equal share of gateway bandwidth. *Note:* Fair queuing does not distinguish between small and large hosts or between hosts with few active connections and those with many.

fall time: The time required for the amplitude of a pulse to decrease (fall) from a specified value (usually 90 percent of the peak value exclusive of overshoot or undershoot) to another specified value (usually 10 percent of the peak value exclusive of overshoot or undershoot). (188) *Note:* Limits on undershoot and oscillation, *i.e.*, hunting, may need to be specified when specifying fall time limits. *Synonym* **pulse decay time.**

false character: *See* **illegal character.**

false clock: A condition where a phase-locked loop controlling a clock locks on a frequency other than the correct frequency. *Note 1:* False clock can occur when there is excessive phase shift, as a function of frequency, in the loop. *Note 2:* False clock often occurs where the false frequency is a harmonic of the correct frequency.

false lock: A condition where a phase-locked loop locks to a frequency other than the correct one, or to an improper phase.

fan-beam antenna: A directional antenna producing a main beam having a large ratio of major to minor dimension at any transverse cross section.

fan out: *Synonym* **break out.**

FAQ file: *Abbreviation for* **Frequently Asked Questions file.** An online file that contains frequently asked questions with answers provided to assist new users and avoid repetitive offline inquiries. *Note:* An *FAQ file* is usually created for Internet news groups, but is also used in other applications.

Faraday effect: A magneto-optic effect in which the polarization plane of an electromagnetic wave is rotated under the influence of a magnetic field parallel to the direction of propagation. *Note:* The Faraday effect may be used to modulate a lightwave.

far-end crosstalk: Crosstalk that is propagated in a disturbed channel in the same direction as the propagation of a signal in the disturbing channel. *Note:* The terminals of the disturbed channel, at which the far-end crosstalk is present, and the energized terminals of the disturbing channel, are usually remote from each other. (188)

far field: *Synonym* **far-field region.**

far-field diffraction pattern: The diffraction pattern of a source (such as an LED, ILD, or the output end of an optical fiber) observed at an infinite distance from the source. *Note 1:* A far-field pattern exists at distances that are large compared with s^2/λ, where s is a characteristic dimension of the source and λ is the wavelength. For example, if the source is a uniformly illuminated circle, then s is the radius of the circle. *Note 2:* The far-field diffraction pattern of a source may be observed at infinity or (except for scale) in the focal plane of a well-corrected lens. The far-field pattern of a diffracting screen illuminated by a point source may be observed in the image plane of the source. *Synonym* **Fraunhofer diffraction pattern.** *Contrast with* **near-field diffraction pattern.**

far-field radiation pattern: A radiation pattern measured at the far field of an antenna or other emitter.

far-field region: The region where the angular field distribution is essentially independent of distance from the source. (188) *Note 1:* If the source has a maximum overall dimension D that is large compared to the wavelength, the far-field region is commonly taken to exist at distances greater than $2D^2/\lambda$ from the source, λ being the wavelength. *Note 2:* For a beam focused at infinity, the far-field region is sometimes referred to as the Fraunhofer region. *Synonyms* **far field, far zone, Fraunhofer region, radiation field.**

far zone: *Synonym* **far-field region.**

fast packet switching: A packet switching technique that increases the throughput by eliminating overhead. *Note 1:* Overhead reduction is accomplished by allocating flow control and error correction functions to either the user applications or the network nodes that interface with the user. *Note 2:* Cell relay and frame relay are two implementations of fast packet switching.

fast select: An optional user facility in the virtual call service of CCITT X.25 protocol that allows the inclusion of user data in the call request/connected and clear indication packets. *Note:* Fast select is an essential feature of the CCITT X.25 (1984) protocol.

fault: **1.** An accidental condition that causes a functional unit to fail to perform its required function. **2.** A defect that causes a reproducible or catastrophic malfunction. *Note:* A malfunction is considered reproducible if it occurs consistently under the same circumstances. **3.** In power systems, an unintentional short-circuit, or partial short-circuit, between energized conductors or between an energized conductor and ground. (188)

fault management: In network management, the set of functions that (a) detect, isolate, and correct malfunctions in a telecommunications network, (b) compensate for environmental changes, and (c) include maintaining and examining error logs, accepting and acting on error detection notifications, tracing and identifying faults, carrying out sequences of diagnostics tests, correcting faults, reporting error conditions, and localizing and tracing faults by examining and manipulating database information. (188)

fault protection subsystem: In a facility power distribution system, the subsystem that provides a direct path from each power sink to the earth electrode subsystem. (188) *Note:* The fault protection subsystem is usually referred to as a *"green wire."*

fault tolerance: The extent to which a functional unit will continue to operate at a defined performance level even though one or more of its components are malfunctioning.

FAX: *Acronym for* **facsimile.**

FC: *Abbreviation for* **functional component.**

FCC: The U.S. Government board of five presidential appointees that has the authority to regulate all non-Federal Government interstate telecommunications (including radio and television broadcasting) as well as all international communications that originate or terminate in the United States. *Note:* Similar authority for regulation of Federal Government telecommunications is vested in the National Telecommunications and Information Administration (NTIA).

FCC registration program: The Federal Communications Commission program and associated directives intended to assure that all connected terminal equipment and protective circuitry will not harm the public switched telephone network or certain private line services. *Note 1:* The FCC registration program requires the registering of terminal equipment and protective circuitry in accordance with

Subpart C of part 68, Title 47 of the *Code of Federal Regulations.* This includes the assignment of identification numbers to the equipment and the testing of the equipment. *Note 2:* The FCC registration program contains no requirement that accepted terminal equipment be compatible with, or function with, the network. (188)

FCS: *Abbreviation for* **frame check sequence.** *See* **cyclic redundancy check.**

FDDI: *Abbreviation for* **fiber distributed data interface.**

FDDI-2: *See* **fiber distributed data interface.**

FDHM: *See* **full width at half maximum.**

FDM: *Abbreviation for* **frequency-division multiplexing.**

FDMA: *Abbreviation for* **frequency-division multiple access.**

FDX: *Abbreviation for* **full duplex.**

FEC: *Abbreviation for* **forward error correction.**

Federal Communications Commission: *See* FCC.

Federal Telecommunications System (FTS): A switched long-distance telecommunications service formerly provided for official Federal Government use. *Note:* FTS has been replaced by **Federal Telecommunications Service 2000 (FTS2000).**

Federal Telecommunications System 2000 service: *See* **FTS2000.**

feed: 1. To supply a signal to the input of a system, subsystem, equipment, or component, such as a transmission line or antenna. **2.** A coupling device between an antenna and its transmission line. (188) *Note:* A feed may consist of a distribution network or a primary radiator. **3.** A transmission facility between (a) the point of origin of a signal, such as is generated in a radio or television studio, and (b) the head-end of a distribution facility, such as a broadcasting station in a network. **4.** Pertaining to the function of inserting one thing into another, such as in a feed horn, paper feed, card feed, and line feed. (188)

feedback: **1.** The return of a portion of the output, or processed portion of the output, of a (usually active) device to the input. (188) *Note 1:* The feedback signal will have a certain magnitude and phase relationship relative to the output signal or the input signal. This relationship can be used to influence the behavior, such as the gain and stability, of the overall circuit. *Note 2:* If the feedback is regenerative (additive), it is called "positive feedback," which increases gain and distortion, and decreases linearity and stability. *Note 3:* If the feedback is degenerative (subtractive), it is called "negative feedback," which reduces the gain and distortion, and increases linearity and stability. *Note 4:* Feedback may occur inadvertently, and be detrimental. **2.** Information returned as a response to an originating source.

feeder echo noise: Signal distortion resulting from reflected waves in a transmission line that is many wavelengths long and mismatched at both the generator and the load ends. (188)

feeder link: A radio link from an Earth station at a given location to a space station, or vice versa, conveying information for a space radiocommunication service other than for the fixed-satellite service. The given location may be at a specified fixed point, or at any fixed point within specified areas. [NTIA] [RR]

FEP: *Abbreviation for* **front-end processor.**

Fermat's principle: A principle stating that a ray of light follows the path that requires the least time to travel from one point to another, including reflections and refractions that may occur. *Synonym* **least-time principle.** [From Weik '89]

fetch protection: [An] AIS-provided restriction to prevent a program from accessing data in another user's segment of storage. [NIS]

FET photodetector: A photodetector using photogeneration of carriers in the channel region of a field-effect transistor structure to provide photodetection with current gain.

fiber: *See* **optical fiber.**

fiber amplifier: A device that amplifies an optical signal directly, without the need to convert it to an electrical signal, amplify it electrically, and reconvert it to an optical signal. *Note 1:* One type of fiber amplifier uses a doped fiber (*e.g.*, a fiber doped with erbium), which bears the communication signal, and which is optically pumped with a laser having a high-powered continuous output at an optical frequency slightly higher than that of the communication signal. The signal is intensified by Raman amplification. *Note 2:* Because neither optical-electrical conversion nor electrical amplification takes place, this type of amplifier is well suited for a wide variety of applications, both digital and analog. *Note 3:* Because this type of amplifier does not require extraordinary frequency (wavelength) control of the pumping laser, it is relatively simple. *Synonym* **Raman amplifier.**

fiber axis: The longitudinal center of symmetry of an optical fiber, *i.e.*, the locus of points that are determined by the centers of mechanical symmetry of the outside diameters of fiber cross sections sampled continuously along the length of the fiber.

fiber bandwidth: *See* **bandwidth (of an optical fiber).**

fiber buffer: *See* **buffer (def. #4).**

fiber cable: *See* **fiber optic cable.**

fiber cutoff wavelength (λ_{cf}): *See* **cutoff wavelength (def. #2).**

fiber dispersion: *See* **dispersion.**

fiber distributed data interface (FDDI): A concept, defined in ANSI standards, for an optical-fiber-based token-ring network, featuring (a) dual counter-rotating logical rings, each with a data transmission capacity of 100 Mb/s, (b) reliable data transfer, (c) active link monitoring, (d) station management, and (e) survivability features. *Note 1:* The four standards are (a) ANSI X3T9.5, containing Physical Media Dependent (PMD) specifications, (b) ANSI X3T9.5, containing the Physical (PHY) specifications, (c) ANSI X3.139, containing Media Access Control (MAC) specifications, and (d) ANSI X39.5, containing the Station Management (SMT) specifications. *Note 2:* The data rate of an FDDI ring may be doubled to 200 Mb/s, with loss of redundancy. *Note 3:* FDDI-2, a second-generation FDDI network standard, is under development.

fiber optic bus: *See* **bus.**

fiber optic cable: A telecommunications cable in which one or more optical fibers are used as the propagation medium. [After 2196] (188) *Note 1:* The optical fibers are surrounded by buffers, strength members, and jackets for protection, stiffness, and strength. *Note 2:* A fiber-optic cable may be an all-fiber cable, or contain both optical fibers and metallic conductors. One possible use for the metallic conductors is the transmission of electric power for repeaters. [After FAA] *Synonyms* **optical cable, optical fiber cable.**

fiber optic isolator: *See* **optical isolator.**

fiber optic link: A communications link that transmits signals by means of modulated light propagated in an optical fiber. (188)

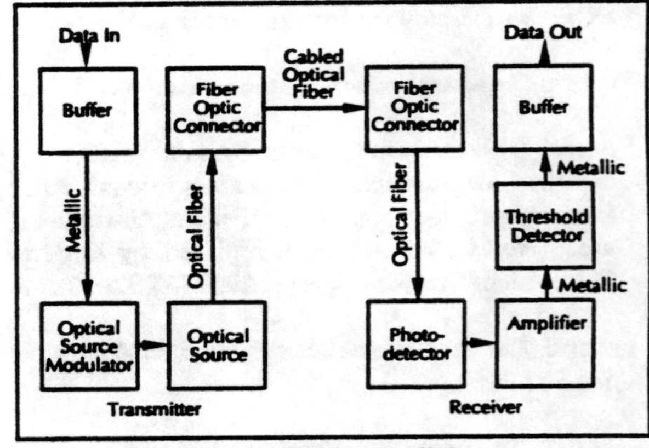

digital fiber-optic link

fiber optics (FO): The branch of optical technology concerned with the transmission of light through fibers made of transparent materials such as glasses and plastics. (188) [2196] *Note 1:* Telecommunications applications of fiber optics use flexible low-loss fibers, using a single fiber per optical path. Present-day plastic fibers have losses that are too high for telecommunications applications. *Note 2:* Various industrial and medical applications of fiber optics, such as endoscopes, use flexible fiber bundles in which individual fibers are spatially aligned, permitting optical relay of an image. *Note 3:* Some specialized industrial applications use rigid (fused) aligned fiber bundles for image transfer; such as in the fiber optics faceplates used on some cathode ray rubes (CRTs) to "flatten" the image.

fiber pigtail: *See* **pigtail (def. #1).**

fidelity: The degree to which a system, or a portion of a system, accurately reproduces, at its output, the essential characteristics of the signal impressed upon its input or the result of a prescribed operation on the signal impressed upon its input. (188)

field: 1. The volume of influence of a physical phenomenon, expressed vectorially. 2. On a data medium or in storage, a specified area used for a particular class of data, *e.g.*, a group of character positions used to enter or display wage rates on a screen. 3. Defined logical data that are part of a record. 4. The elementary unit of a record that may contain a data item, a data aggregate, a pointer, or a link. 5. In an interlaced, raster-scanned video display, a partial frame, consisting of every nth scanning line of a complete frame, where n is an integer equal to the number of fields (usually two) in a complete frame. *Note 1:* For example, in the National Television Standards Committee (NTSC) television specification used in the United States, a single frame is composed of two fields, each of which has half the number of scanning lines in a complete frame. The scanning lines of a field are separated by twice the space between the scanning lines in the full frame. The two fields are interlaced, *i.e.*, a complete frame consists of the following traces, which are listed in the order of their appearance in the complete frame, but not the order in which scanning occurs: the first line of the first field, the first line of the second field, the second line of the first field, the second line of the second field, the third line of the first field, the third line of the second field, *etc.*, until completion of the full frame. The fields are scanned alternately, one complete field at a time. Thus, the flicker rate of the display is perceived by the eye to be twice as fast as that which would result if the complete frame were to be scanned in line-by-line order. *Note 2*: Not all scanning lines are necessarily applied to user information, *i.e.*, the graphic display. Certain scanning lines, not seen under ordinary viewing conditions, are often used for transmitting test signals that indicate the quality of the displayed video.

field-disturbance sensor: A restricted radiation device which establishes a radio frequency field in its vicinity and detects changes in that field resulting from the movement of persons or objects within the radio frequency field. Examples: microwave intrusion sensors; devices that use rf energy for production line counting and sensing. [NTIA]

field intensity: The irradiance of an electromagnetic wave under specified conditions. (188) *Note:* Field intensity is usually expressed in watts per square meter.

field strength: The magnitude of an electric, magnetic, or electromagnetic field at a given point. (188) *Note:* The field strength of an electromagnetic wave is usually expressed as the rms value of the electric field, in volts per meter. The field strength of a magnetic field is usually expressed in ampere-turns per meter or in oersteds. *Synonym* **radio field intensity.**

field wire: A flexible insulated wire used in field telephone and telegraph systems. (188) *Note 1:* WD-1 and WF-16 are types of field wire. *Note 2:* Field wire usually contains conductors and high-tensile-strength strands serving as strength members.

FIFO: *Abbreviation for* **first-in first-out.**

file: 1. The largest unit of storage structure that consists of a named collection of all occurrences in a database of records of a particular record type. 2. A set of related records treated as a unit, for example, in stock control, a file could consist of a set of invoices.

file server: 1. A high-capacity disk storage device or a computer that each computer on a network can use or access and retrieve files that can be shared among attached computers. 2. A program, running on a computer, that allows different programs, running on other computers, to access the files of that computer.

file transfer, access, and management (FTAM): An application's service and protocol based on the concept of virtual file store. *Note:* FTAM allows remote access to various levels in a file structure and provides a comprehensive set of file management capabilities.

File Transfer Protocol (FTP): *See* **FTP.**

fill: *See* **bit stuffing.**

fill bit: *See* **bit stuffing.**

filled cable: A cable that has a nonhygroscopic material, usually a gel, inside the jacket or sheath. (188) *Note 1:* The nonhygroscopic material fills the spaces between the interior parts of the cable, preventing moisture from entering minor leaks in the sheath and migrating inside the cable. *Note 2:* A metallic cable, such as a coaxial cable or a metal waveguide, filled with a dielectric material, is not considered as a filled cable.

FILO: *Abbreviation for* **first-in, last-out.**

filter: In electronics, a device that transmits only part of the incident energy and may thereby change the spectral distribution of energy: (a) high-pass filters transmit energy above a certain frequency; (b) low-pass filters transmit energy below a certain frequency; (c) bandpass filters transmit energy of a certain bandwidth; (d) band-stop filters transmit energy outside a specific frequency band. [JP1]

filtered symmetric differential phase-shift keying (FSDPSK): A method of encoding information for digital transmission in which (a) a binary 0 is encoded as a +90° change in the carrier phase and a binary 1 is encoded as a –90° change in the carrier phase, and (b) abrupt phase transitions are smoothed by filtering or other functionally equivalent pulse shaping techniques.

finished call: **1.** In an information transaction, a call in which the call originator or call receiver terminates the communication and goes on hook, *i.e.*, hangs up. **2.** In an information transfer transaction, the termination of the information transfer phase.

FIP: *Acronym for* **Federal Information Processing.**

FIP equipment: In the Federal Government, any equipment or interconnected system or subsystems of equipment (as defined in 41CFR) used in the automatic acquisition, storage, manipulation, management, movement, control, display, switching, interchange, transmission, or reception of data or information.

FIP system: In the Federal Government, any organized combination of FIP equipment, software, services, support services, or related supplies.

firmware: Software that is embedded in a hardware device that allows reading and executing the software, but does not allow modification, *e.g.*, writing or deleting data by an end user. (188) *Note 1:* An example of firmware is a computer program in a read-only memory (ROM) integrated circuit chip. A hardware configuration is usually used to represent the software. *Note 2:* Another example of firmware is a program embedded in an erasable programmable read-only memory (EPROM) chip, which program may be modified by special external hardware, but not by an application program.

first-in first-out (FIFO): A queuing discipline in which entities in a queue leave the queue in the same order in which they arrive. (188) *Note 1:* Service, when available, is offered to the entity that has been in the FIFO queue the longest. *Note 2:* FIFO techniques are used in message switching.

first-in last-out (FILO): A queuing discipline in which entities in a queue leave the queue in the reverse order from that in which they arrived. (188) *Note:* An understanding of FILO techniques is important in the understanding of store-and-forward capabilities in packing switching.

first window: Of silica-based optical fibers, the transmission window at approximately 830 to 850 nm. [FAA]

FISINT: *Acronym for* **foreign instrumentation signals intelligence.**

five-hundred (500) service: A telephone service that allows individuals to receive, via a single number, telephone calls in various locations (*e.g.*, home, office, or car phone) from call originators not necessarily using the same common carrier.

fixed access: In personal communications service (PCS), terminal access to a network in which there is a set relationship between a terminal and the access interface. *Note:* A single "identifier" serves for both the access interface and the terminal. If the terminal moves to another access interface, that terminal assumes the identity of the new interface.

fixed attenuator: *See* **pad.**

fixed loop: A service feature that permits an attendant on an assisted call to retain connection through the attendant position for the duration of the call. *Note:* The attendant will usually receive a disconnect signal when the call is terminated.

fixed microwave auxiliary station: A fixed station used in connection with (a) the alignment of microwave transmitting and receiving antenna systems and equipment, (b) coordination of microwave radio survey operations, and (c) cue and contact control of television pickup station operations. [47CFR]

fixed-reference modulation: Modulation in which the significant condition for any signal element is based on a fixed reference. (188)

fixed-satellite service: A radiocommunication service between Earth stations at given positions when one or more satellites are used; the given position may be a specified fixed point or any fixed point within specified areas; in some cases this service includes satellite-to-satellite links, which may also be effected in the inter-satellite service, the fixed-satellite service may also include feeder links for other space radiocommunication services. [RR]

fixed service (FX): A radiocommunication service between specified fixed points. [NTIA] [RR]

fixed station: A station in the fixed service. [NTIA] [RR]

fixed storage: *Synonym* **read-only storage.**

fixed-tolerance-band compaction: Data compaction accomplished by storing or transmitting data only when the data fall outside prescribed limits. *Note:* An example of fixed-tolerance-band compaction in a telemetering system is the transmission of the temperature only when the temperature is above or below preestablished threshold limits. Thus, the recipient of the transmission is to assume that the value is in the prescribed range unless a signal to the contrary occurs. [From Weik '89]

flag: In data transmission or processing, an indicator, such as a signal, symbol, character, or digit, used for identification. *Note:* A flag may be a byte, word, mark, group mark, or letter that signals the occurrence of some condition or event, such as the end of a word, block, or message.

flag sequence: In data transmission or processing, a sequence of bits used to delimit, *i.e.* mark, the beginning and end of a frame. *Note 1:* An 8-bit sequence is usually used as the flag sequence; for example, the 8-bit flag sequence 01111110. *Note 2:* Flag sequences are used in bit-oriented protocols, such as Advanced Data Communication Control Procedures (ADCCP), Synchronous Data Link Control (SDLC), and High-Level Data Link Control (HDLC).

flash: A signal generated by the momentary depression of the telephone switchhook or other device. *Note:* A flash may be used to request additional services.

FLASH message: A category of precedence reserved for initial enemy contact messages or operational combat messages of extreme urgency. Brevity is mandatory. [JP1]

flat fading: Fading in which all frequency components of a received radio signal vary in the same proportion simultaneously. (188)

flat rate service: Telephone service in which a single payment permits an unlimited number of local calls to be made without further charge for a specified period of time.

flat weighting: In a noise-measuring set, a noise weighting based on an amplitude-frequency characteristic that is flat over a frequency range that must be stated. (188) *Note 1:* Flat noise power is expressed in dBrn (f_1 - f_2) or in dBm (f_1 - f_2). *Note 2: "3-kHz flat weighting"* and *"15-kHz flat weighting"* are based on amplitude-frequency characteristics that are flat between 30 Hz and the frequency indicated.

F layer: *See* **F region.**

Fleming's rule: A rule stating that if the thumb of the right hand points in the direction of an electric current, then the curled fingers point in the direction of the magnetic field that encircles the current; and further, if the curled fingers of the right hand describe the electric current in a solenoid, then the thumb points in the direction of the magnetic field inside the solenoid. *Synonym* **right-hand rule.** [From Weik '89]

flexible disk: *Synonym* **diskette.**

flip-flop: A device that may assume either one of two reversible, stable states. *Note 1:* The flip-flop is used as a basic control element in computer and communications systems. *Note 2:* In a flip-flop, the transition from one stable state to the other is unstable, *i.e.*, for the very short period during which the transition takes place, both outputs may assume the same state, which state may be unpredictable. *Synonyms* **bistable circuit, bistable multivibrator, bistable trigger circuit.**

floating-point coding compaction: Data compaction accomplished by using coefficients, a base, and exponents to specify the scale, range, or magnitude of numbers. *Note:* An example of floating-point coding compaction is using 119.8×10^6, 119.8(6), or 119.86 to represent 119,800,000. If the number is rounded to 120,000,000, it might be written as 1206 or 127 in which the last digit is the number of zeros to be appended to the preceding digits. Thus, only three positions are required instead of nine to represent the number in storage or in a message, which is only 33% of the original space and time requirement. [From Weik '89]

flooding compound: A substance surrounding the buffer tubes of a fiber-optic cable, to prevent water intrusion into the interstices in the event of a breach of the jacket. [FAA]

flood projection: In facsimile, the optical method of scanning in which the object is floodlighted and the scanning spot is defined by a masked portion of the illuminated area.

flood search routing: In a telephone network, nondeterministic routing in which a dialed number received at a switch is transmitted to all switches, *i.e.*, flooded, in the area code directly connected to that switch; if the dialed number is not an affiliated subscriber at that switch, the number is then retransmitted to all directly connected switches, and then routed through the switch that has the dialed number corresponding to the particular user end instrument affiliated with it. *Note 1:* All digits of the numbering plan are used to identify a particular subscriber. *Note 2:* Flood search routing allows subscribers to have telephone numbers independent of switch codes. *Note 3:* Flood search routing provides the highest probability that a call will go through even though a number of switches and links fail.

floppy disk: *Synonym* **diskette.**

flops: *Acronym for* **floating-point operations per second.** *Note:* For example, 15 Mflops equals 15 million floating-point arithmetic operations per second. [From Weik '89]

flowchart: A graphical representation in which symbols are used to represent such things as operations, data, flow direction, and equipment, for the definition, analysis, or solution of a problem. *Synonym* **flow diagram.**

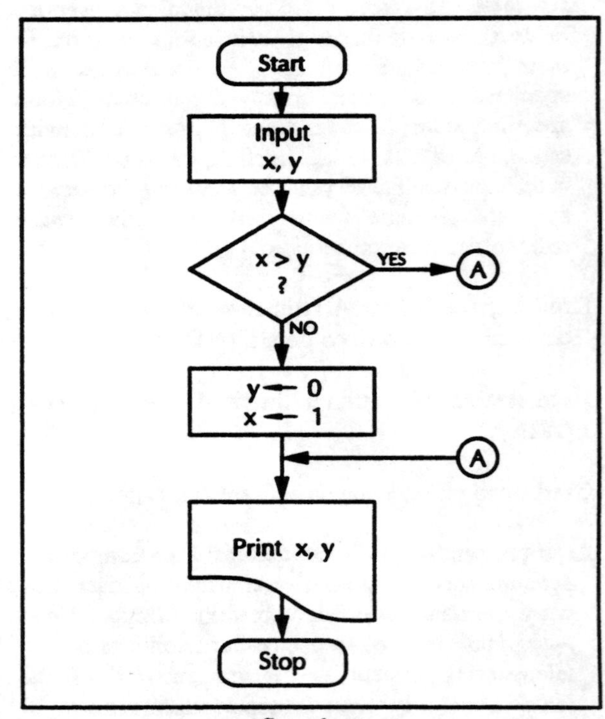

flowchart

flow control: *See* **transmit flow control.**

flow control procedure: A procedure for controlling the rate of transfer of data among elements of a network, *e.g.*, between a DTE and a data switching exchange network, to prevent overload.

flow diagram: *Synonym* **flowchart.**

flowline: On a flowchart, a line that (a) has an indicated direction, (b) represents a connection between other symbols, and (c) indicates the sequence of operations or the transfer of control.

flutter: Rapid variation of signal parameters, such as amplitude, phase, and frequency. (188) *Note:* Examples of flutter are (a) rapid variations in received signal levels, such as variations that may be caused by atmospheric disturbances, antenna movements in a high wind, or interaction with other signals, (b) in radio propagation, a phenomenon in which nearly all radio signals that are usually reflected by ionospheric layers in or above the E-region experience partial or complete absorption, (c) in radio transmission, rapidly changing signal levels, together with variable multipath time delays, caused by reflection and possible partial absorption of the signal by aircraft flying through the radio beam or common scatter volume, (d) the variation in the transmission characteristics of a loaded telephone circuit caused by the action of telegraph direct currents on the loading coils, (e) in recording and reproducing equipment, the deviation of frequency caused by irregular mechanical motion, *e.g.*, that of capstan angular velocity in a tape transport mechanism, during operation.

flux: **1.** The lines of force of a magnetic field. **2.** *Obsolete synonym for* **radiant power.**

flywheel effect: In an oscillator, the continuation of oscillations after removal of the control stimulus. (188) *Note 1:* The flywheel effect is usually caused by interacting inductive and capacitive circuits in the oscillator. *Note 2:* The flywheel effect may be desirable, such as in phase-locked loops used in synchronous systems, or undesirable, such as in voltage-controlled oscillators. *Synonym* **flywheeling.**

flywheeling: *Synonym* **flywheel effect.**

FM: *Abbreviation for* **frequency modulation.**

FM blanketing: That form of interference to the reception of other broadcast stations, which is caused by the presence of an FM broadcast signal of 115 dBμ (562 mV/m) or greater signal strength in the area adjacent to the antenna of the transmitting station. The 115-dBu contour is referred to as the *"blanketing area."* [47CFR]

FM broadcast translator: *See* **translator (def. #3).**

FM capture effect: *Synonym* **capture effect.**

FM capture ratio: *See* **capture effect.**

FM improvement factor: The quotient obtained by dividing the signal-to-noise ratio (SNR) at the output of an FM receiver by the carrier-to-noise ratio (CNR) at the input of the receiver. *Note:* When the FM improvement factor is greater than unity, the improvement in the SNR is always obtained at the expense of an increased bandwidth in the receiver and the transmission path. (188)

FM improvement threshold: The point in an FM receiver at which the peaks in the rf signal equal the peaks of the thermal noise generated in the receiver. (188) *Note:* A baseband signal-to-noise ratio of about 30 dB is typical at the improvement threshold, and this ratio improves 1 dB for each decibel of increase in the signal above the threshold.

FM threshold effect: In an FM receiver, the effect produced when the desired-signal gain begins to limit the desired signal, and thus noise limiting (suppression). (188) *Note:* FM threshold effect occurs at (and above) the point at which the FM signal-to-noise improvement is measured.

FM threshold extension: A change in the value of the FM threshold of a receiver. *Note:* FM threshold extension may be obtained by decreasing the operational bandwidth, thus decreasing the received noise power and allowing the threshold of the desired signal to occur at a lower signal input level.

FO: *Abbreviation for* **fiber optics.**

footprint: In satellite communications, that portion of the Earth's surface over which a satellite antenna delivers a specified amount of signal power under specified conditions. (188) *Note:* The limiting case of footprint area is somewhat less than one-half the Earth's surface, and depends on the altitude of the satellite.

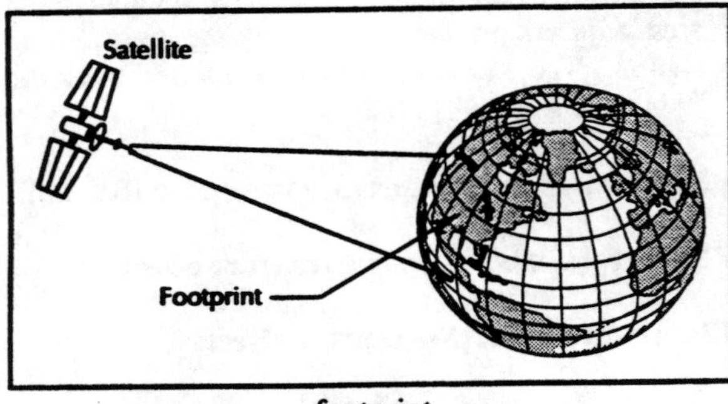

footprint

forbidden character: *Synonym* **illegal character.**

foreign exchange (FX) service: A network-provided service in which a telephone in a given local exchange area is connected, via a private line, to a central office in another, *i.e.*, *"foreign"*, exchange, rather than the local exchange area's central office. *Note:* To call originators, it appears that the subscriber having the FX service is located in the foreign exchange area.

foreign instrumentation signals intelligence (FISINT): **1.** Intelligence information derived from electromagnetic emissions associated with the testing and operational deployment of foreign aerospace, surface, and subsurface systems. **2.** Technical information and intelligence information derived from the intercept of foreign instrumentation signals by other than the intended recipients. Foreign instrumentation signals intelligence is a category of signals intelligence. *Note:* Foreign instrumentation signals include but are not limited to signals from telemetry, beaconry, electronic interrogators, tracking/fusing/arming/firing command systems, and video data links. [JP1]

format: **1.** The arrangement of bits or characters within a group, such as a word, message, or language. (188) **2.** The shape, size, and general makeup of a document. (188)

Fortran: *See* **language processor.**

fortuitous conductor: Any conductor that may provide an unintended path for signals. *Note:* Examples of fortuitous conductors are water pipes, wires, cables, and metal building and equipment structural members. (188)

fortuitous distortion: Distortion resulting from causes generally subject to laws concerning random occurrences. (188)

forward busying: In a telecommunications system, a feature in which supervisory signals are forwarded in advance of address signals in order to seize assets of the system before attempting to establish a connection. (188)

forward channel: The channel of a data circuit that transmits data from the originating user to the destination user. *Note:* The forward channel carries message traffic and some control information.

forward echo: In a transmission line, an echo propagating in the same direction as the original wave and consisting of energy reflected back by one discontinuity and then forward again by another discontinuity. (188) *Note:* Forward echoes can be supported by reflections caused by splices or other discontinuities in the transmission medium (*e.g.*, optical fiber, twisted pair, or coaxial tube). In metallic lines, they may be supported by impedance mismatches between the source or load and the characteristic impedance of the transmission medium.

forward error correction (FEC): A system of error control for data transmission wherein the receiving device has the capability to detect and correct any character or code block that contains fewer than a predetermined number of symbols in error. (188) *Note:* FEC is accomplished by adding bits to each transmitted character or code block, using a predetermined algorithm.

forward propagation ionospheric scatter (FPIS): *Synonym* **ionospheric scatter.**

forward scatter: The deflection—by diffraction, nonhomogeneous refraction, or nonspecular reflection by particulate matter of dimensions that are large with respect to the wavelength in question but small with respect to the beam diameter—of a portion of an incident electromagnetic wave, in such a manner that the energy so deflected propagates in a direction that is within 90° of the direction of propagation of the incident wave. *Note:* The scattering process may be polarization-sensitive, *i.e.*, incident waves that are identical in every respect but their polarization may be scattered differently.

forward signal: A signal sent in the direction from the calling to the called station, *i.e.*, from the original data source to the original data sink. *Note:* The forward signal is transmitted in the forward channel.

FOT: *Abbreviation for* **frequency of optimum transmission.** In the transmission of radio waves via ionospheric reflection, the highest effective, *i.e.*, working, frequency that is predicted to be usable for a specified path and time for 90% of the days of the month. (188) *Note 1:* The FOT is normally just below the value of the maximum usable frequency (MUF). In the prediction of usable frequencies, the FOT is commonly taken as 15% below the monthly median value of the MUF for the specified time and path. *Note 2:* The FOT is usually the most effective frequency for ionospheric reflection of radio waves between two specified points on Earth. *Synonyms* **frequency of optimum traffic, optimum traffic frequency, optimum transmission frequency, optimum working frequency.**

Fourier analysis: The definition of a periodic waveform of arbitrary shape as a summation of sine waves having specific amplitudes and phases, and having frequencies corresponding to the harmonics of the waveform being defined. *Note:* A Fourier analysis is particularly well suited for communications equipment design and for predicting the performance of a given design. [From Weik '89]

four-wire circuit: A two-way circuit using two paths so arranged that the respective signals are transmitted in one direction only by one path and in the other direction by the other path. *Note:* The four-wire circuit gets its name from the fact that, historically, two conductors were used in each of two directions for full-duplex operation. The name may still be applied, *e.g.*, to a communications link supported by optical fibers, even though only one fiber is required for transmission in each direction. *Contrast with* **two-wire circuit.** (188)

four-wire repeater: A repeater, consisting of two amplifiers, one associated with each direction, used in a four-wire circuit. (188)

four-wire terminating set: A balanced transformer used to perform a conversion between 4-wire and 2-wire operation. *Note 1:* For example, a 4-wire circuit may, by means of a 4-wire terminating set, be connected to a 2-wire telephone set. Also, a pair of 4-wire terminating sets may be used to introduce an intermediate 4-wire loop into a 2-wire circuit, in which loop repeaters may be situated to amplify signals in each direction without positive feedback and oscillation. *Note 2:* Four-wire terminating sets have been largely supplanted by resistance hybrids. (188)

fox message: A standard test message that includes all the alphanumerics on a teletypewriter and also function characteristics (space, figures shift, letters shift). *Note:* An example of a fox message is "THE QUICK BROWN FOX JUMPED OVER THE LAZY DOG'S BACK 1234567890."(188)

FPIS: *Abbreviation for* **forward propagation ionospheric scatter.** *See* **ionospheric scatter.**

fractional frequency fluctuation: The deviation of the frequency of an oscillator from its nominal constant frequency, normalized to the nominal frequency.

fractional T1: In the North American or Japanese hierarchies, the tariffed use of a data rate corresponding to fewer than the 24 channels served by a T1 line.

frame: **1.** In data transmission, the sequence of contiguous bits delimited by, and including, beginning and ending flag sequences. *Note 1:* A frame usually includes an information field, and usually consists of a specified number of bits between flags and contains an address field, a control field, a frame check sequence, and flags. *Note 2:* Frames usually consist of a representation of the original data to be transmitted, together with other bits which may be used for error detection or control. Additional bits may be used for routing, synchronization, or overhead information not directly associated with the original data. **2.** In the multiplex structure of pulse-code modulation (PCM) systems, a set of consecutive time slots in which the position of each digit can be identified by reference to a frame-alignment signal. (188) *Note:* The frame-alignment signal does not necessarily occur, in whole or in part, in each frame. **3.** In a time-division multiplexing (TDM) system, a repetitive group of signals resulting from a single sampling of all channels, including any required system information, such as additional synchronizing signals. (188) *Note: "In-frame"* is the condition that exists when there is a channel-to-channel and bit-to-bit correspondence, exclusive of transmission errors, between all inputs of a time-division multiplexer and the output of its associated demultiplexer. **4.** In ISDN, a block of variable length, labeled at the Data Link Layer of the Open Systems Interconnection—Reference Model. **5.** In video display, the set of all picture elements that represent one complete image. (188) *Note:* In NTSC and other television standards used throughout the world, a frame consists of two interlaced fields, each of which has half the number of scanning lines, and consequently, half the number of pixels, of one frame. **6.** In video display, one complete scanned image from a series of video images. *Note:* A video frame is usually composed of two interlaced fields.

frame alignment: In the reception of framed digital data, the extent to which a received frame is correctly aligned with respect to the clock at the receiver.

frame-alignment recovery time: *Synonym* **reframing time.**

frame alignment signal: In the transmission of data frames, a distinctive sequence of bits used to accomplish frame alignment. *Note:* A frame alignment signal may also contain additional bits for status, control, and error detection. (188)

frame-alignment time slot : A time slot starting at a particular phase or instant in each frame and allocated to the transmission of a frame-alignment signal. (188)

frame check sequence (FCS): *See* **cyclic redundancy check.**

framed interface: An interface through which information flow is partitioned into physical, periodic frames consisting of overhead information and an information payload.

frame duration: The time between the beginning of a frame and the end of that frame. *Note:* For fixed-length frames, at a fixed data rate, frame duration is constant.

frame frequency: *Synonym* **frame rate.**

frame grabber: A device that can seize and record a single frame of video information out of a sequence of many frames.

frame pitch: The distance, time, or number of bits between corresponding points, *i.e.*, significant instants, in two consecutive frames. [From Weik '89]

frame rate: The number of frames transmitted or received per unit time. (188) *Note 1:* The frame rate is usually expressed in frames per second. *Note 2:* In television transmission, the frame rate must be distinguished from the field rate, which in the NTSC and other systems, is twice the frame rate. *Synonym* **frame frequency.**

frame relay: An interface protocol for statistically multiplexed packet-switched data communications in which (a) variable-sized packets (frames) are used that completely enclose the user packets they transport, and (b) transmission rates are usually between 56 kb/s and 1.544 Mb/s (the T-1 rate). *Note 1:* In frame relay, (a) there is neither flow-control nor an error-correction capability, (b) there is information-content independence, (c) there is a correspondence only to the ISO Open systems Interconnection—Reference Model Layers 1 and 2, (d) variable-sized user packets are enclosed in larger packets (frames) that add addressing and verification information, (e) frames may vary in length up to a design limit, usually 1 kilobyte or more, (f) one frame relay packet transports one user packet, (g) implementation of fast-packet technology is used for connection-oriented frame relay services, and (h) there is a capability to handle time-delay insensitive traffic, such as LAN interworking and image transfer. *Note 2:* Frame relay is referred to as the *local management interface (LMI) standard* and is specified in *ANSI T1.617*.

frame slip: In the reception of framed data, the loss of synchronization between a received frame and the receiver clock, causing a frame misalignment event, and resulting in the loss of the data contained in the received frame. (188) *Note:* A frame slip should not be confused with a dropped frame where synchronization is not lost, *e.g.*, in the case of buffer overflow.

frame synchronization: Of a received stream of framed data, the process by which incoming frame alignment signals, *i.e.*, distinctive bit sequences, are identified, *i.e.*, distinguished from data bits, permitting the data bits within the frame to be extracted for decoding or retransmission. *Note:* The usual practice is to insert, in a dedicated time slot within the frame, a noninformation bit that is used for the actual synchronization of the incoming data with the receiver. *Synonym* **framing (def. #1).**

frame synchronization pattern: In digital communications, a prescribed recurring pattern of bits transmitted to enable the receiver to achieve frame synchronization. (188)

framing: 1. In time-division multiplexing reception, *synonym* **frame synchronization.** (188) 2. In video reception, the process of adjusting the timing of the receiver to coincide with the received video synchronization pulse. 3. In facsimile, the adjustment of the facsimile picture to a desired position in the direction of line progression. (188)

framing bit: 1. A bit used for frame synchronization. 2. In a bit stream, a bit used in determining the beginning or end of a frame. *Note 1:* The framing bit occurs at a specific position in the frame. (188) *Note 2:* In a bit stream, framing bits are noninformation bits. *Note 3:* Framing in a digital signal is usually repetitive.

framing signal: *See* **frame-alignment signal, framing bit.**

Fraunhofer diffraction pattern: *Synonym* **far-field diffraction pattern.**

Fraunhofer region: *Synonym* **far-field region.**

free net: A radio net in which any station may communicate with any other station in the net without first obtaining the permission of the net-control station. *Note:* Permission to operate as a free net is granted by the net-control station until such time as a directed net is established by the net-control station. [From Weik '89]

free routing: The routing of messages in such a manner that they are forwarded toward their destination or addressee over any available channel without dependence upon predetermined routing. [From Weik '89]

free-running capability: In a synchronized oscillator, the capability to operate in the absence of a synchronizing signal.

free space: A theoretical concept of space devoid of all matter. (188) *Note:* Free space also implies remoteness from material objects that could influence the propagation of electromagnetic waves.

free-space coupling: Coupling of magnetic, electric, or electromagnetic fields that are not confined to a conductor. (188) *Note:* Coupling by the deliberate introduction of capacitors and inductors is not considered free-space coupling.

free-space loss: The signal attenuation that would result if all absorbing, diffracting, obstructing, refracting, scattering, and reflecting influences were sufficiently removed so as to have no effect on propagation. (188) *Note:* Free-space loss is primarily caused by beam divergence, *i.e.*, signal energy spreading over larger areas at increased distances from the source.

freeze frame: A frame of visual information that is selected from a set of motion video frames, and is held in a buffer. (188) *Contrast with* **still video.**

freeze frame television: Television in which fixed ("still") images are transmitted sequentially at a rate far too slow to be perceived as continuous motion by human vision. *Note:* Transmission of an image is usually performed periodically by a processing unit that contains memory in which data representing the image are stored. For an image of specified quality, *e.g.*, resolution and color fidelity, freeze-frame television has a lower bandwidth requirement than that of full-motion television.

F region: That portion of the ionosphere existing between approximately 160 and 400 km above the surface of the Earth, consisting of layers of increased free-electron density caused by the ionizing effect of solar radiation. *Note 1:* The F region reflects normal-incident frequencies at or below the critical frequency (approximately 10 MHz) and partially absorbs waves of higher frequency. *Note 2:* The F_1 layer exists from about 160 to 250 km above the surface of the Earth and only during daylight hours. Though fairly regular in its characteristics, it is not observable everywhere or on all days. The principal reflecting layer during the summer for paths of 2,000 to 3,500 km is the F_1 layer. The F_1 layer has approximately 5×10^5 e/cm^3 (free electrons per cubic centimeter) at noontime and minimum sunspot activity, and increases to roughly 2×10^6 e/cm^3 during maximum sunspot activity. The density falls off to below 10^4 e/cm^3 at night. *Note 3:* The F_1 layer merges into the F_2 layer at night. *Note 4:* The F_2 layer exists from about 250 to 400 km above the surface of the Earth. The F_2 layer is the principal reflecting layer for HF communications during both day and night. The horizon-limited distance for one-hop F_2 propagation is usually around 4,000 km. The F_2 layer has about 10^6 e/cm^3. However, variations are usually large, irregular, and particularly pronounced during magnetic storms.

frequency: For a periodic function, the number of cycles or events per unit time. (188)

frequency accuracy: The degree of conformity to a specified value of a frequency. (188)

frequency aging: Of an oscillator, the change in frequency, over time, caused by internal changes in oscillator parameters even when external factors, such as environment and power supply characteristics, are constant.

frequency allocation: *See* **allocation (of a frequency band).**

frequency allotment: *See* **allotment (of a radio frequency or radio frequency channel).**

frequency-analysis compaction: Data compaction accomplished by using an expression composed of a number of different frequencies of different magnitudes to represent a particular curve. *Note:* An example of frequency-analysis compaction is the use of a Fourier analysis to represent an arbitrary curve, a periodic function, an aperiodic function, or a wave shape. Thus, the fundamental frequency, the amplitude of the fundamental frequency, and the amplitudes and frequencies of the harmonics are all that are needed to reconstitute the function or wave shape. The shape can thus be readily stored and transmitted in this compacted form. [From Weik '89]

frequency assignment: **1.** Authorization, given by an Administration, for a radio station to use a radio frequency or radio frequency channel under specified conditions **2.** The process of authorizing a specific frequency, group of frequencies, or frequency band to be used at a certain location under specified conditions, such as bandwidth, power, azimuth, duty cycle, or modulation. (188) *Synonym* **radio frequency channel assignment.** *See* **administration (def. #1).**

frequency assignment authority: The power granted an Administration, or its designated or delegated leader or agency via treaty or law, to specify frequencies, or frequency bands, in the electromagnetic spectrum for use in systems or equipment. *Note:* Primary frequency assignment authority for the United States is exercised by the National Telecommunications and Information Administration (NTIA) for the Federal Government and by the Federal Communications Commission (FCC) for non-Federal Government organizations. International frequency assignment authority is vested in the International Frequency Registration Board of the International Telecommunication Union. [Extracted from NTIA]

frequency averaging: **1.** The process by which the relative phases of precision clocks are compared for the purpose of defining a single time standard. **2.** A process in which network synchronization is achieved by use, at all nodes, of oscillators that adjust their frequencies to the average frequency of the digital bit streams received from connected nodes. *Note:* In frequency averaging, all oscillators are assigned equal weight in determining the ultimate network frequency. (188)

frequency band: *See* **electromagnetic spectrum.**

frequency band allocation: *See* **allocation (of a frequency band).**

frequency-change signaling: A signaling method in which one or more discrete frequencies correspond to each desired significant condition of a code. *Note 1:* The transition from one set of frequencies to the other may be a continuous or a discontinuous change in frequency or in phase. *Note 2:* Frequency-change signaling may be used in both supervisory signaling and data transmission. (188)

frequency coherence: *See* **phase coherence.**

frequency compatibility: **1.** Of an electronic device, the extent to which it will operate at its designed performance level in its intended operational environment (including the presence of interference) without causing interference to other devices. **2.** The degree to which an electrical or electronic device or devices operating on or responding to a specified frequency or frequencies is capable of functioning with other such devices.

frequency departure: An unintentional deviation from the nominal frequency value.

frequency-derived channel: A channel derived by dividing an allocated or available bandwidth over a medium into two or more portions, each usable separately. (188) *Note:* A frequency-derived channel is continuously available and may be further divided on either a frequency or time basis.

frequency deviation: **1.** The amount by which a frequency differs from a prescribed value, such as the amount an oscillator frequency drifts from its nominal frequency. **2.** In frequency modulation, the absolute difference between (a) the maximum permissible instantaneous frequency of the modulated wave or the minimum permissible instantaneous frequency of the modulated wave and (b) the carrier frequency. **3.** In frequency modulation, the maximum absolute difference, during a specified period, between the instantaneous frequency of the modulated wave and the carrier frequency. (188)

frequency dispersal: An electronic counter-countermeasure (ECCM) in which communications nets' operating frequencies are widely separated from each other, causing a requirement to spread jamming power over wider frequency bands and thus compelling a reduction of available jamming power on any single channel or frequency, or causing a requirement for more jamming power or more jamming equipment. [From Weik '89]

frequency displacement: The end-to-end shift in frequency that may result from independent frequency translation errors in a circuit. (188)

frequency distortion: *Synonym* **amplitude-vs.-frequency distortion.**

frequency diversity: Transmission and reception in which the same information signal is transmitted and received simultaneously on two or more independently fading carrier frequencies. (188)

frequency-division multiple access (FDMA): The use of frequency division to provide multiple and simultaneous transmissions to a single transponder. (188)

frequency-division multiplexing (FDM): The deriving of two or more simultaneous, continuous channels from a transmission medium by assigning a separate portion of the available frequency spectrum to each of the individual channels. (188)

frequency drift: An undesired progressive change in frequency with time. (188) *Note 1:* Causes of frequency drift include component aging and environmental changes. *Note 2:* Frequency drift may be in either direction and is not necessarily linear.

frequency-exchange signaling: Frequency-change signaling in which the change from one significant condition to another is accompanied by decay in amplitude of one or more frequencies and by buildup in amplitude of one or more other frequencies. *Note:* Frequency-exchange signaling applies to supervisory signaling and user-information transmission. (188) *Synonym* **two-source frequency keying.**

frequency fluctuation: A short-term variation, with respect to time, of the frequency of an oscillator. *Note:* Frequency fluctuation, $f(t)$, is given by

$$f(t) = \frac{1}{2\pi}\frac{d^2\theta(t)}{dt^2} ,$$

where $\theta(t)$ is the phase angle of the sinusoidal wave with respect to time, t.

frequency frogging: 1. The interchanging of the frequencies of carrier channels to accomplish specific purposes, such as to prevent feedback and oscillation, to reduce crosstalk, and to correct for a high frequency-response slope in the transmission line. (188) *Note:* Frequency frogging is accomplished by having modulators, which are integrated into specially designed repeaters, translate a low-frequency group to a high-frequency group, and vice versa. A channel will appear in the low group for one repeater section and will then be translated to the high group for the next section because of frequency frogging. This results in nearly constant attenuation with frequency over two successive repeater sections, and eliminates the need for large slope equalization and adjustments. Singing and crosstalk are minimized because the high-level output of a repeater is at a different frequency than the low-level input to other repeaters. 2. In microwave systems, the alternate use of two frequencies at repeater sites to prevent feedback and oscillation. (188)

frequency guard band: A frequency band deliberately left vacant between two channels to provide a margin of safety against mutual interference. (188)

frequency hopping: [The] repeated switching of frequencies during radio transmission according to a specified algorithm, to minimize unauthorized interception or jamming of telecommunications. [NIS] *Note:* The overall bandwidth required for frequency hopping is much wider than that required to transmit the same information using only one carrier frequency.

frequency-hopping spread spectrum: A signal structuring technique employing automatic switching of the transmitted frequency. Selection of the frequency to be transmitted is typically made in a pseudo-random manner from a set of frequencies covering a band wider than the information bandwidth. The intended receiver would frequency-hop in synchronization with the code of the transmitter in order to retrieve the desired information. [NTIA] [RR] (188) *Note:* In many cases, used as an electronic counter-countermeasure (ECCM) technique.

frequency hour: One frequency used for one hour regardless of the number of transmitters over which it is simultaneously broadcast by a station during that hour. [47CFR]

frequency instability: *See* **frequency stability.**

frequency lock: The condition in which a frequency-correcting feedback loop maintains control of an oscillator within the limits of one cycle. (188) *Note:* Frequency lock does not imply phase lock, but phase lock does imply frequency lock.

frequency modulation (FM): Modulation in which the instantaneous frequency of a sine wave carrier is caused to depart from the center frequency by an amount proportional to the instantaneous value of the modulating signal. (188) *Note 1:* In FM, the carrier frequency is called the center frequency. *Note 2:* FM is a form of angle modulation. *Note 3:* In optical communications, even if the electrical baseband signal is used to frequency-modulate an electrical carrier (an "FM" optical communications system), it is still the intensity of the lightwave that is varied (modulated) by the electrical FM carrier. In this case, the "information,"as far as the lightwave is concerned, is the electrical FM carrier. The lightwave is varied in intensity at an instantaneous rate corresponding to the instantaneous frequency of the electrical FM carrier. [After FAA]

frequency offset: The difference between the frequency of a source and a reference frequency. (188)

frequency of optimum traffic (FOT): *Synonym* **FOT.**

frequency of optimum transmission (FOT): *See* **FOT.**

frequency prediction: A prediction of the maximum usable frequency (MUF), the optimum traffic frequency, and the lowest usable frequency (LUF) for transmission between two specific locations or geographical areas during various times throughout a 24-hour period. *Note:* The prediction is usually indicated by means of a graph for each frequency plotted as a function of time. [From Weik '89]

frequency range: A continuous range or spectrum of frequencies that extends from one limiting frequency to another. *Note 1:* The frequency range for given equipment specifies the frequencies at which the equipment is operable. For example, filters pass or stop certain bands of frequencies. The frequency range for propagation indicates the frequencies at which electromagnetic wave propagation in certain modes or paths is possible over given distances. Frequency allocation, however, is made in terms of bands of frequencies. There is little, if any, conceptual difference between a range of frequencies and a band of frequencies. *Note 2: "Frequency band"* usually identifies a specific band of frequencies in the Tables of Frequency Allocations. [From Weik '89]

frequency response: *See* **insertion-loss-vs.-frequency characteristic.**

frequency response curve: A plot of the gain or attenuation of a device, such as an amplifier or a filter, as a function of frequency. *Note:* A flat curve indicates a uniform gain or attenuation over the range of frequencies for which the curve is flat. Most amplifiers have a flat frequency response over a certain band, above and below which the gain is reduced. The frequency response curve of a filter has one or more peaks or troughs. [From Weik '89]

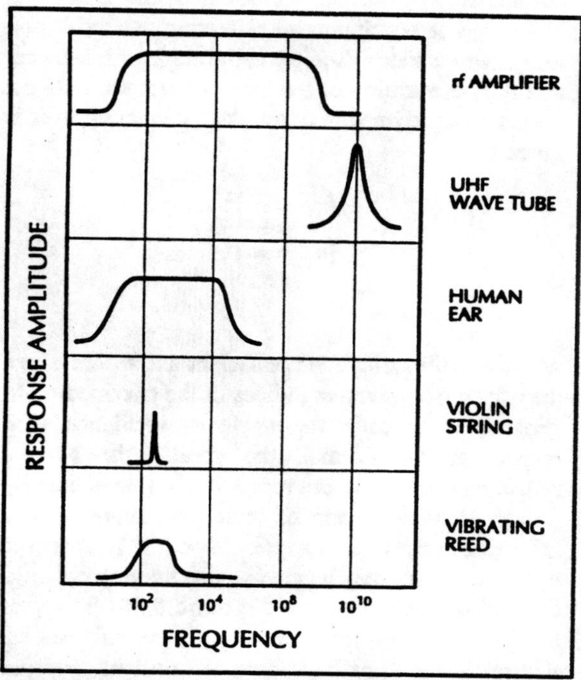

representative frequency response of various devices

frequency scanning: Conducting a search for signals over a band or range of frequencies by means of a manually or automatically tuned receiver. *Note:* The tuning rate, *i.e.,* the frequency change rate, may be fixed or variable, or it may be performed mechanically at low speed or electronically at high speed. Frequency scanning may be used to enable a radar to transmit on a clear frequency, *i.e.,* a no-interference frequency, by searching a frequency band and then tuning the system to a clear portion of that band. [From Weik '89]

frequency sharing: The assignment to or use of the same radio frequency by two or more stations that are separated geographically or that use the frequency at different times. *Note 1:* Frequency sharing reduces the potential for mutual interference where the assignment of different frequencies to each user is not practical or possible. *Note 2:* In a communications net, frequency sharing does not pertain to stations that use the same frequency.

frequency shift: 1. Any change in frequency. **2.** Any change in the frequency of a radio transmitter or oscillator. (188) *Note:* In the radio regime, frequency shift is also called rf shift. **3.** *See* **frequency-shift telegraphy. 4.** In facsimile, a frequency modulation system where one frequency represents picture black and another frequency represents picture white. Frequencies between these two limits may represent shades of gray. (188) **5.** An intentional frequency change used for modulation purposes. (188)

frequency-shift keying (FSK): Frequency modulation in which the modulating signal shifts the output frequency between predetermined values. (188) *Note 1:* Usually, the instantaneous frequency is shifted between two discrete values termed the *"mark"* and *"space"* frequencies. This is a noncoherent form of FSK. *Note 2:* Coherent forms of FSK exist in which there is no phase discontinuity in the output signal. *Synonyms* **frequency-shift modulation, frequency-shift signaling.**

frequency-shift modulation: *Synonym* **frequency-shift keying.**

frequency-shift signaling: *Synonym* **frequency-shift keying.**

frequency-shift telegraphy: Telegraphy by frequency modulation in which the telegraph signal shifts the frequency of the carrier between predetermined values. [NTIA] [RR]

frequency source: *See* **frequency standard.**

frequency spectrum: *See* **electromagnetic spectrum.**

frequency spectrum congestion: The situation that occurs when many stations transmit simultaneously using frequencies that are close together, *i.e.*, with insufficient width of frequency guard bands or channel spacing. *Note:* Frequency spectrum congestion causes (a) difficulty in discrimination by tuning, (b) overlap of (i) a sideband and an adjacent carrier, or (ii) upper and lower sidebands, respectively, of adjacent carriers, and (c) interference that occurs when frequencies shift slightly or are phase shifted by ionospheric reflection. [From Weik '89]

frequency stability: The degree to which variations of the frequency of an oscillator deviate from the mean frequency over a specified period of time. (188)

frequency standard: A stable oscillator used for frequency calibration or reference. (188) *Note 1:* A frequency standard generates a fundamental frequency with a high degree of accuracy and precision. Harmonics of this fundamental frequency are used to provide reference points. *Note 2:* Frequency standards in a network or facility are sometimes administratively designated as "primary" or "secondary." The terms *"primary"* and *"secondary,"* as used in this context, should not be confused with the respective technical meanings of these words in the discipline of precise time and frequency.

frequency synthesizer: A device that produces frequencies that are phase coherent with a reference frequency. (188) *Note:* The reference frequency may be derived from an internal or external source.

frequency tolerance: The maximum permissible departure by the center frequency of the frequency band occupied by an emission from the assigned frequency, or by the characteristic frequency of an emission from the reference frequency. Frequency tolerance is expressed in parts per 10^6 or in hertz. [NTIA] [RR] (188) *Note:* In the United States, frequency tolerance is expressed in parts per 10^n, in hertz, or in percentages. Frequency tolerance includes both the initial setting tolerance and excursions related to short- and long-term instability and aging.

frequency translation: The transfer of signals occupying a specified frequency band, such as a channel or group of channels, from one portion of the frequency spectrum to another, in such a way that the arithmetic frequency difference of signals within the band is unaltered. (188)

Fresnel diffraction pattern: *Synonym* **near-field diffraction pattern.**

Fresnel reflection: In optics, the reflection of a portion of incident light at a discrete interface between two media having different refractive indices. (188) *Note 1:* Fresnel reflection occurs at the air-glass interfaces at the entrance and exit ends of an optical fiber. Resultant transmission losses, on the order of 4% per interface, can be reduced considerably by the use of index-matching materials. *Note 2:* The coefficient of reflection depends upon the refractive index difference, the angle of incidence, and the polarization of the incident radiation. For a normal ray, the fraction of reflected incident power is given by

$$R = \frac{(n_1 - n_2)^2}{(n_1 + n_2)^2} ,$$

where R is the reflection coefficient and n_1 and n_2 are the respective refractive indices of the two media. In general, the greater the angle of incidence with respect to the normal, the greater the Fresnel reflection coefficient, but for radiation that is linearly polarized in the plane of incidence, there is zero reflection at Brewster's angle. *Note 3:* Macroscopic optical elements may be given antireflection coatings consisting of one or more dielectric thin-film layers having specific refractive indices and thicknesses. Antireflection coatings reduce overall Fresnel reflection by mutual interference of individual Fresnel reflections at the boundaries of the individual layers.

Fresnel zone: In radio communications, one of a (theoretically infinite) number of a concentric ellipsoids of revolution which define volumes in the radiation pattern of a (usually) circular aperture. *Note 1:* The cross section of the first Fresnel zone is circular. Subsequent Fresnel zones are annular in cross section, and concentric with the first. *Note 2:* Odd-numbered Fresnel zones have relatively intense field strengths, whereas even numbered Fresnel zones are nulls. *Note 3:* Fresnel zones result from diffraction by the circular aperture. (188)

front-end noise temperature: A measure of the thermal noise generated in the first stage of a receiver. (188)

front-end processor (FEP): A programmed-logic or stored-program device that interfaces data communication equipment with an input/output bus or memory of a data processing computer.

front-to-back ratio: 1. Of an antenna, the gain in a specified direction, *i.e.*, azimuth, usually that of maximum gain, compared to the gain in a direction 180° from the specified azimuth. *Note:* Front-to-back ratio is usually expressed in dB. 2. A ratio of parameters used to characterize rectifiers or other devices, in which electrical current, signal strength, resistance, or other parameters, in one direction is compared with that in the opposite direction. (188)

FSDPSK: *Abbreviation for* **filtered symmetric differential phase-shift keying.**

FSK: *Abbreviation for* **frequency-shift keying.**

FTAM: *Abbreviation for* **file transfer, access, and management.**

FTP: *Abbreviation for* **File Transfer Protocol.** The Transmission Control Protocol/Internet Protocol (TCP/IP) protocol that is (a) a standard high-level protocol for transferring files from one computer to another, (b) usually implemented as an application level program, and (c) uses the Telnet and TCP protocols. *Note:* In conjunction with the proper local software, FTP allows computers connected to the Internet to exchange files, regardless of the computer platform.

FTS: *See* **FTS2000.**

FTS2000: *Abbreviation for* **Federal Telecommunications System 2000.** A long distance telecommunications service, including services such as switched voice service for voice or data up to 4.8 kb/s, switched data at 56 kb/s and 64 kb/s, switched digital integrated service for voice, data, image, and video up to 1.544 Mb/s, packet switched service for data in packet form, video transmission for both compressed and wideband video, and dedicated point-to-point private line for voice and data. *Note 1:* Use of FTS2000 contract services is mandatory for use by U.S. Government agencies for all acquisitions subject to 40 U.S.C. 759. *Note 2:* No U.S. Government information processing equipment or customer premises equipment other than that which are required to provide an FTS2000 service are furnished. *Note 3:* The FTS2000 contractors will be required to provide service directly to an agency's terminal equipment interface. For example, the FTS2000 contractor might provide a terminal adapter to an agency location in order to connect FTS2000 ISDN services to the agency's terminal equipment. *Note 4:* GSA awarded two 10-year, fixed-price contracts covering FTS2000 services on December 7, 1988. *Note 5:* The Warner Amendment excludes the mandatory use of FTS2000 in instances related to maximum security. [FIRMR]

full carrier: A carrier that is transmitted without reduction in power, *i.e.*, a carrier that is of sufficient level to demodulate the sideband(s).

full carrier single-sideband emission: A single-sideband emission without reduction of the carrier. [NTIA] [RR]

full-duplex (FDX) circuit: A circuit that permits simultaneous transmission in both directions. (188)

full-duplex (FDX) operation: *Synonym* **duplex operation.**

full duration at half maximum (FDHM): Full width at half maximum in which the independent variable is time. *See* **full width at half maximum.**

full modulation: In an analog-to-digital converter, the condition in which the input signal amplitude has just reached the threshold at which clipping begins to occur. [From Weik '89]

full-motion operation: In television, a video frame update rate that provides the appearance of full motion without flicker or smear problems. (188) *Note:* Picture motion appears to be full at greater than 16 fps (frames per second). European television operates at 25 fps and North American television at 30 fps.

full processing: All processing functions required to recover the information bits from a received signal. (188)

full width at half maximum (FWHM): An expression of the extent of a function, given by the difference between the two extreme values of the independent variable at which the dependent variable is equal to half of its maximum value. *Note 1:* FWHM is applied to such phenomena as the duration of pulse waveforms and the spectral width of sources used for optical communications. *Note 2:* The term ***full duration at half maximum (FDHM)*** is preferred when the independent variable is time.

fully connected mesh network: *See* **network topology.**

fully connected topology: *See* **network topology.**

fully intermateable connectors: Connectors from one source that mate with complementary components from other sources without mechanical damage and with transmission properties maintained within specified limits. (188)

functional component (FC): In intelligent networks, an elemental call-processing component that directs internal network resources to perform specific actions, such as collecting dialed digits. *Note:* An FC is unique to the intelligent-network-IN/2 architecture.

functional profile: A standardization document that characterizes the requirements of a standard or group of standards, and specifies how the options and ambiguities in the standard(s) should be interpreted or implemented to (a) provide a particular information technology function, (b) provide for the development of uniform, recognized tests, and (c) promote interoperability among different network elements and terminal equipment that implement a specific profile.

functional signaling: In an integrated services digital network (ISDN), signaling in which the signaling messages are unambiguous and have clearly defined meanings that are known to both the sender and receiver of the messages. *Note:* Functional signaling is usually generated by the data terminal equipment (DTE).

functional signaling link: A combination of a communications link and the associated transfer control functions.

functional unit: An entity of hardware, software, or both, capable of accomplishing a specified purpose.

function signal: A set of signal elements that is used to transmit or represent a function-control character that actuates a control function, such as carriage return, line-feed, letters shift, or figures shift, that is to be performed by communications devices, such as teletypewriters and teleprinters. [From Weik '89]

fundamental: Of a periodic wave, the sinusoidal component, *i.e.,* Fourier component, having the lowest frequency. *Note:* Every periodic waveform may be expressed as the summation of the fundamental and its harmonics. For example, a square wave may be expressed as the summation of sine waves equal in frequency to the fundamental and all odd harmonics, each frequency having an appropriate amplitude and phase. A pure sinusoidal wave has only one component, *i.e.,* the fundamental. *Contrast with* **harmonic, overtone.**

fundamental mode: The lowest order mode of a waveguide. (188) *Note:* In optical fibers, the fundamental mode is designated LP_{01} or HE_{11}.

furcate: *Synonym* **break out.**

fuse: 1. A device that has as its critical component a metal wire or strip that will melt when heated by a prescribed (design) amperage, creating an open in the circuit of which it is a part, thereby protecting the circuit from an overcurrent condition. *Note:* Fuses are often characterized as "fast-blow" or "slow-blow," according to the time required for them to respond to an overcurrent condition. Fast-blow fuses open nearly instantaneously when exposed to an overcurrent condition. Slow-blow fuses can tolerate a transient overcurrent condition, but will open if the overcurrent condition is sustained. **2.** In optical fiber technology, to join the endfaces of a pair of optical fibers by melting, *i.e.*, welding, the endfaces together.

fused silica: *Synonym* **vitreous silica.**

fusion splice: In fiber optics, a splice created by localized heating of the ends of the two fibers to be joined. *Note:* A properly made fusion splice results in a continuous length of material with minimal discontinuities at the splice.

FWHM: *Abbreviation for* **full width at half maximum.**

FX: *Abbreviation for* **fixed service, foreign exchange service.**

(this page intentionally left blank)

gain: The ratio of output current, voltage, or power to input current, voltage, or power, respectively. (188) *Note 1:* Gain is usually expressed in dB. *Note 2:* If the ratio is less than unity, the gain, expressed in dB, will be negative, in which case there is a loss between input and output.

gain hit: *See* **hit.**

gain medium: An active medium, device, or system in which amplification of input occurs with or without feedback. *Note:* Gain media include amplifiers, lasers, and avalanche photodiodes (APDs).

gain of an antenna: *Synonym* **antenna gain.**

galactic radio noise: *Synonym* **cosmic noise.**

gap loss: 1. The power loss that occurs when an optical signal is transferred from one fiber to another that is axially aligned with it, but longitudinally separated from it. *Note:* The gap allows light from the "transmitting" fiber to spread out as it leaves the fiber endface. When it strikes the "receiving" fiber, some of the light will enter the cladding, where it is quickly lost. [After FAA] 2. An analogous form of coupling loss that occurs between an optical source, *e.g.*, an LED, and an optical fiber. *Note:* Gap loss is not usually significant at the optical detector, because the sensitive area of the detector is normally somewhat larger than the cross section of the fiber core. Unless the separation is substantial, all light emerging from the fiber, even though it diverges, will still strike the detector. *Synonym* **longitudinal offset loss.** [FAA]

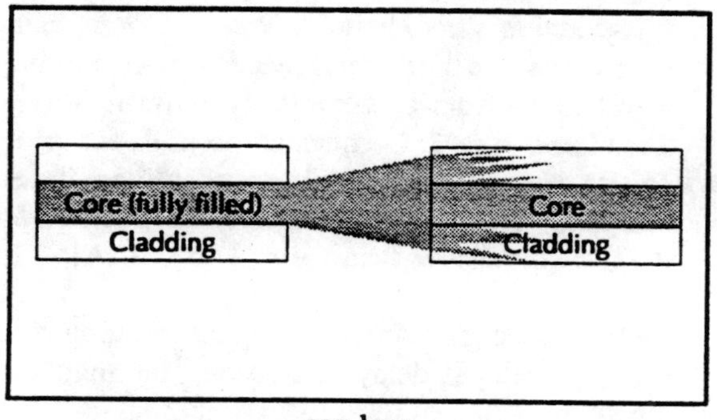

gap loss

gap-loss attenuator: An optical attenuator that exploits the principle of gap loss to reduce the optical power level when inserted in-line in the fiber path; *e.g.*, to prevent saturation of the receiver. *Note:* Gap-loss attenuators should be used in-line near the optical transmitter. [After FAA]

garble: 1. An error in transmission, reception, encryption, or decryption that changes the text of a message or any portion thereof in such a manner that it is incorrect or undecryptable. [JP1] 2. In a telephone circuit or channel, readily audible but unintelligible interference from another circuit or channel. *Note:* Garble may, for example, take place in an FDM telephone carrier system in which an interfering signal from another channel or system is demodulated in such a fashion that it has an objectionable audio power level but is nonetheless unintelligible.

gate: 1. A device having one output channel and one or more input channels, such that the output channel state is completely determined by the input channel states, except during switching transients. 2. One of many types of combinational logic elements having at least two inputs; *e.g.*, AND, OR, NAND, and NOR. (188)

gateway: 1. In a communications network, a network node equipped for interfacing with another network that uses different protocols. (188) *Note 1:* A gateway may contain devices such as protocol translators, impedance matching devices, rate converters, fault isolators, or signal translators as necessary to provide system interoperability. It also requires that mutually acceptable administrative procedures be established between the two networks. *Note 2:* A protocol translation/mapping gateway interconnects networks with different network protocol technologies by performing the required protocol conversions. 2. *Loosely,* a computer configured to perform the tasks of a gateway.

gating: 1. The process of selecting only those portions of a wave between specified time intervals or between specified amplitude limits. 2. The controlling of signals by means of combinational logic elements. (188) 3. A process in which a predetermined set of conditions, when established, permits a second process to occur. (188)

gaussian beam: A beam of light whose electric field intensity distribution is gaussian. *Note:* When such a beam is circular in cross section the intensity at distance *r* from the center, *E*(*r*), is given by

$$E(r) = E(0)\,e^{(-r/w)^2} ,$$

where *E*(0) is the electrical field strength at the beam center, *i.e.*, at *r* = 0; and *w* is the value of *r* at which the intensity is 1/e of its value on the axis.

gaussian pulse: A pulse that has a waveform described by the gaussian distribution. (188) *Note:* In the time domain, the amplitude of the waveform is given by

$$f(t) = A\,e^{(-t/\sigma)^2} ,$$

where *A* is the maximum amplitude, and σ is the pulse half-duration at the 1/e points.

GBH: *Abbreviation for* **group busy hour.**

GCT: *Abbreviation for* **Greenwich Civil Time.** *See* **Coordinated Universal Time.**

GDF: *Abbreviation for* **group distribution frame.**

gel: 1. A substance, resembling petroleum jelly in viscosity, that surrounds a fiber, or multiple fibers, enclosed in a loose buffer tube. *Note:* This gel serves to lubricate and support the fibers in the buffer tube. It also prevents water intrusion in the event the buffer tube is breached. [FAA] 2. Index-matching material in the form of a gel. [FAA] *Synonym* **index-matching gel.** *See* **index-matching material.**

general purpose computer: A computer designed to perform, or that is capable of performing, in a reasonably efficient manner, the functions required by both scientific and business applications. *Note:* A general purpose computer is often understood to be a large system, capable of supporting remote terminal operations, but it may also be a smaller computer, *e.g.*, a desktop workstation.

general purpose network: *See* **common user network.**

geometric optics: The branch of optics that describes light propagation in terms of rays. *Note 1:* Rays are bent at the interface between two dissimilar media, and may be curved in a medium in which the refractive index is a function of position. *Note 2:* The ray in geometric optics is perpendicular to the wavefront in physical optics. *Synonym* **ray optics.**

geometric spreading: *See* **inverse-square law.**

geostationary orbit: A circular orbit in the equatorial plane, any point on which revolves about the Earth in the same direction and with the same period as the Earth's rotation. (188) *Note:* An object in a geostationary orbit will remain directly above a fixed point on the equator at a distance of approximately 42,164 km from the center of the Earth, *i.e.*, approximately 35,787 km above mean sea level.

geostationary satellite: A geosynchronous satellite whose circular and direct orbit lies in the plane of the Earth's equator and which thus remains fixed relative to the Earth; by extension, a satellite that remains approximately fixed relative to the Earth. [NTIA] [RR]

geostationary satellite orbit: The orbit in which a satellite must be placed to be a geostationary satellite. [NTIA] [RR]

geosynchronous orbit: Any orbit about the Earth, which orbit has a period equal to the period of rotation of the Earth about its axis, and in the same sense, *i.e.*, direction, as the rotation of the Earth.

germanium photodiode: A germanium-based PN- or PIN-junction photodiode. *Note 1:* Germanium photodiodes are useful for direct detection of optical wavelengths from approximately 1 µm to several tens of µm. *Note 2:* Germanium-based detectors are noisier than silicon-based detectors. Silicon-based detectors are therefore usually preferred for wavelengths shorter than 1 µm. [After FAA]

ghost: A secondary image or signal resulting from echo, envelope delay distortion, or multipath reception.

gigaflop: A billion, *i.e.*, 10^9, floating point operations per second.

gigahertz (GHz): A unit of frequency denoting 10^9 Hz. (188)

glare: *Deprecated synonym for* **call collision.**

glass: 1. In the strict sense, a state of matter. [FAA] 2. In fiber-optic communication, any of a number of noncrystalline, amorphous inorganic substances, formed, by heating, from metallic or semiconductor oxides or halides, and used as the material for fibers. *Note:* The most common glasses are based on silicon dioxide (SiO_2). [After FAA]

glide slope facility: In aeronautical navigation, an instrument approach landing facility that furnishes vertical guidance information to an aircraft from its approach altitude down to the surface of the runway. (188)

global: 1. Pertaining to, or involving, the entire world. (188) 2. Pertaining to that which is defined in one subsection of an entity and used in at least one other subsection of the same entity. (188) 3. In computer, data processing, and communications systems, pertaining to what is applicable to an area beyond the immediate area of consideration. *Note:* Examples of global entities are (a) in computer programming, an entity that is defined in one subdivision of a computer program and used in at least one other subdivision of that program and (b) in personal computer systems and their software packages, a setting, definition, or condition that applies to the entire software system. [From Weik '89]

global address: In a communications network, the predefined address that is used as an address for all users of that network, and that may not be the address of an individual user, or subgroup of users, of the network.

global status: 1. The set of attributes of an entity, described at a particular time, when that set is extended to every occurrence of that entity within a prescribed boundary. (188) 2. The complete set of attributes necessary to describe an entity at a particular time. (188)

GMT: *Abbreviation for* **Greenwich Mean Time.** *Obsolete term. See* **Coordinated Universal Time.**

go-ahead message: *Synonym* **go-ahead notice.**

go-ahead notice: In a tape-relay communications system, a service message, usually sent to a relay station or to a tributary station, that contains a request to the operator to resume transmitting over a specified channel or channels. [From Weik '89] *Synonyms* **go-ahead message, start message, start notice.**

go-ahead tone: In communications systems, an audible signal transmitted by a system indicating that the system is ready to receive a message or signal. [From Weik '89]

gold code: In spread-spectrum systems, a code that is generated by summing, using modulo-two addition, the outputs of two spread-spectrum code-sequence generators. [From Weik '89]

Gopher: A menu-based information searching tool that allows users to access various types of databases, such as FTP archives and white pages databases. *Note 1:* Gopher is most often used as an Internet browser. *Note 2:* Gopher software uses the client-server model.

GOS: *Abbreviation for* **grade of service.**

GOSIP: *Acronym for* **Government Open Systems Interconnection Profile.** A definition of Federal Government functional requirements for open systems computer network products, including a common set of Open System Interconnection (OSI) data communication protocols that enables systems developed by different vendors to interoperate and enable the users of different applications on these systems to exchange information. *Note 1:* The OSI protocols were developed primarily by ISO and CCITT. *Note 2:* The GOSIP is a subset of the OSI protocols and is based on agreements reached by vendors and users of computer networks participating in the National Institute of Standards and Technology (NIST) Implementors Workshop. *Note 3:* The GOSIP is described in the latest version of FIPS PUB 146.

Government Open Systems Interconnection Profile: *See* **GOSIP.**

graceful degradation: Degradation of a system in such a manner that it continues to operate, but provides a reduced level of service rather than failing completely. (188)

graded-index fiber: An optical fiber with a core having a refractive index that decreases with increasing radial distance from the fiber axis. (188) *Note:* The most common refractive index profile for a graded-index fiber is very nearly parabolic. The parabolic profile results in continual refocusing of the rays in the core, and compensates for multimode distortion.

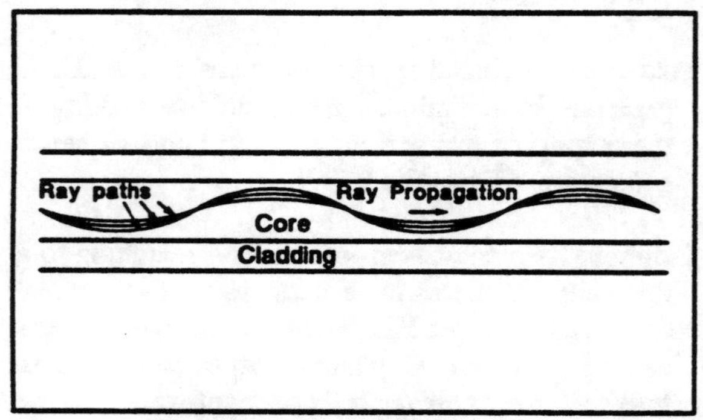

ray paths in a graded-index fiber

graded-index profile: In the core of an optical fiber, a plot of the variation of refractive index such that the refractive index decreases with increasing radial distance from the fiber axis.

grade of service (GOS): **1.** The probability of a call's being blocked or delayed more than a specified interval, expressed as a decimal fraction. (188) *Note:* Grade of service may be applied to the busy hour or to some other specified period or set of traffic conditions. Grade of service may be viewed independently from the perspective of incoming versus outgoing calls, and is not necessarily equal in each direction. **2.** In telephony, the quality of service for which a circuit is designed or conditioned to provide, *e.g.*, voice grade or program grade. *Note:* Criteria for different grades of service may include equalization for amplitude over a specified band of frequencies, or in the case of digital data transported via analog circuits, equalization for phase also.

grandfathered systems: Systems, including but not limited to, (a) PBX and key telephone systems, directly connected to the public switched telephone network on June 1, 1978, that may remain permanently connected thereto without registration unless subsequently modified, and (b) systems that are of the same type as those connected to the public switched telephone network on July 1, 1978, that were added before January 1, 1980, and that may remain permanently connected thereto without registration unless subsequently modified.

grandfathered terminal equipment: Terminal equipment (other than PBX and key telephone systems) and protective circuitry connected to the public switched telephone network before July 1, 1978, that may remain connected thereto for life without registration unless subsequently modified.

graphical user interface (GUI): A computer program or environment that displays options on the screen as icons, *i.e.*, picture symbols, by which users enter commands by selecting an icon. *Note:* Icons may be selected, *e.g.*, by pressing the <ENTER> key on the keyboard, by "clicking" a computer mouse button, or by touching the icon on a touch pad.

graphic character: **1.** A visual representation of a character, other than a control character. **2.** In the ASCII code, a character other than an alphanumeric character, intended to be written, printed, or otherwise displayed in a form that can be read by humans. *Note 1:* Graphic characters are contained in rows 2 through 7 of the ASCII code table. *Note 2:* The space and delete characters are considered to be graphic characters.

graphics: The art or science of conveying information through the use of display media, such as graphs, letters, lines, drawings, and pictures. (188) *Note:* Graphics includes the transmission of coded images such as facsimile.

Gray code: A binary code in which consecutive decimal numbers are represented by binary expressions that differ in the state of one, and only one, one bit. *Synonym* **reflected code.**

gray scale: An optical pattern consisting of discrete steps or shades of gray between black and white. (188)

great circle: A circle defined by the intersection of the surface of the Earth and any plane that passes through the center of the Earth. *Note:* On the idealized surface of the Earth, the shortest distance between two points lies along a great circle.

Greenwich Civil Time (GCT): *Synonym* **Greenwich Mean Time (GMT).** *Obsolete term. See* **Coordinated Universal Time.**

Greenwich Mean Time (GMT): Mean solar time at the meridian of Greenwich, England, formerly used as a basis for standard time throughout the world. (188) *Obsolete term. Synonym* **Greenwich Civil Time.** *See* **Coordinated Universal Time.**

ground: 1. An electrical connection to earth through an earth-electrode subsystem. (188) **2.** In an electrical circuit, a common return path that usually (a) is connected to an earth-electrode subsystem and (b) is extended throughout a facility via a facility ground system consisting of the signal reference subsystem, the fault protection subsystem, and the lightning protection subsystem. **3.** In an electrical circuit, a common return path that (a) may not necessarily be connected to earth and (b) is the zero voltage reference level for the equipment or system.

ground absorption: The dissipation of rf energy by the Earth. (188)

ground constants: The electrical parameters of earth, such as conductivity, permittivity, and magnetic permeability. *Note 1:* The values of these parameters vary with the local chemical composition and density of the earth. *Note 2:* For a propagating electromagnetic wave, such as a surface wave propagating along the surface of the Earth, these parameters vary with frequency and direction. (188)

ground current: In the presence of an electrical fault, the current that flows in the protective ground wire of a power distribution system. *Contrast with* **ground loop.**

ground loop: In an electrical system, an unwanted current that flows in a conductor connecting two points that are nominally at the same potential, *i.e.*, ground, but are actually at different potentials. *Note 1:* For example, the electrical potential at different points on the surface of the Earth can vary by hundreds of volts, primarily from the influence of the solar wind. Such an occurrence can be hazardous, *e.g.*, to personnel working on long grounded conductors such as metallic telecommunications cable pairs. *Note 2:* A ground loop can also exist in a floating ground system, *i.e.*, one not connected to an Earth ground, if the conductors that constitute the ground system have a relatively high resistance, or have, flowing through them, high currents that produce a significant voltage ("I•R") drop. *Note 3:* Ground loops can be detrimental to the operation of the electrical system. *Contrast with* **ground current.**

ground plane: An electrically conductive surface that serves as the near-field reflection point for an antenna. *Note:* A ground plane may consist of a natural (*e.g.*, Earth or sea) surface, an artificial surface of opportunity (*e.g.*, the roof of a motor vehicle), or a specially designed artificial surface (*e.g.*, the disc of a discone antenna). (188)

ground potential: The zero reference level used to apply and measure voltages in a system. *Note:* A potential difference may exist between this reference level and the ground potential of the Earth, which varies with locality, soil conditions, and meteorological phenomena.

ground-return circuit: 1. A circuit using a common return path that is at ground potential. *Note:* Earth may serve as a portion of the ground-return circuit. **2.** A circuit in which there is a common return path, whether or not connected to earth.

ground start: A method of signaling from a terminal or subscriber loop to a switch, in which method one side of a cable pair is temporarily grounded. (188)

ground wave: In radio transmission, a surface wave that propagates close to the surface of the Earth. *Note 1:* The Earth has one refractive index and the atmosphere has another, thus constituting an interface that supports surface wave transmission.

These refractive indices are subject to spatial and temporal changes. *Note 2:* Ground waves do not include ionospheric and tropospheric waves.

group: **1.** In frequency-division multiplexing, a specific number of associated voice channels, either within a supergroup or as an independent entity. *Note 1:* In wideband systems, a group usually consists of 12 voice channels and occupies the frequency band from 60 kHz to 108 kHz. *Note 2:* this is CCITT group B. *Note 3*: CCITT Basic Group A, for carrier telephone systems, consists of 12 channels occupying upper sidebands in the 12-kHz to 60-kHz band. Basic Group A is no longer mentioned in CCITT Recommendations. *Note 4:* A supergroup usually consists of 60 voice channels, *i.e.*, 5 groups of 12 voice channels each, occupying the frequency band from 312 kHz to 552 kHz. (188) *Note 5:* A mastergroup consists of 10 supergroups or 600 voice channels. (188) *Note 6:* The CCITT standard mastergroup consists of 5 supergroups. The U.S. commercial carrier standard mastergroup consists of 10 supergroups. *Note 7:* The terms "supermaster group" or "jumbo group" are sometimes used to refer to 6 mastergroups. **2.** A set of characters forming a unit for transmission or cryptographic treatment. (188)

group address: In a communications network, a predefined address used to address only a specified set of users. *Synonym* **collective address.**

group alerting and dispatching system: A service feature that (a) enables a controlling telephone to place a call to a specified number of telephones simultaneously, (b) enables the call to be recorded, (c) if any of the called lines is busy, enables the equipment to camp on until the busy line is free, and (d) rings the free line and plays the recorded message.

group busy hour (GBH): The busy hour for a given trunk group.

group delay: **1.** The rate of change of the total phase shift with respect to angular frequency, $d\theta/d\omega$, through a device or transmission medium, where θ is the total phase shift, and ω is the angular frequency equal to $2\pi f$, where f is the frequency. **2.** In an optical fiber, the transit time required for optical power, traveling at a given mode's group velocity, to travel a given distance. *Note:* For optical fiber dispersion measurement purposes, the quantity of interest is group delay per unit length, which is the reciprocal of the group velocity of a particular mode. The measured group delay of a signal through an optical fiber exhibits a wavelength dependence due to the various dispersion mechanisms present in the fiber.

group delay time: In a group of waves that have slightly different individual frequencies, the time required for any defined point on the envelope (*i.e.*, the envelope determined by the additive resultant of the group of waves) to travel through a device or transmission facility. (188)

group distribution frame (GDF): In frequency-division multiplexing, a distribution frame that provides terminating and interconnecting facilities at the group level, *i.e.*, group modulator output and group demodulator input circuits of FDM carrier equipment. *Note:* The basic spectrum of the FDM group is 60 kHz to 108 kHz. (188)

group index (*N*): In fiber optics, for a given mode propagating in a medium of refractive index η, the velocity of light in vacuum, c, divided by the group velocity of the mode. (188) *Note:* For a plane wave of wavelength λ, the group index may also be expressed,

$$N = n - \lambda \frac{dn}{d\lambda} ,$$

where n is the phase index of wavelength λ.

grouping factor: *Synonym* **blocking factor.**

group 1...4 facsimile: *See* **facsimile.**

group patch bay: *See* **patch bay.**

group velocity: **1.** The velocity of propagation of an envelope produced when an electromagnetic wave is modulated by, or mixed with, other waves of different frequencies. (188) *Note:* The group velocity is the velocity of information propagation and, loosely, of energy propagation. **2.** In optical fiber transmission, for a particular mode, the reciprocal of the rate of change of the phase constant with respect to angular frequency. *Note:* The group velocity equals the phase velocity if the phase constant is a linear function of the angular frequency, $\omega = 2\pi f$, where f is the frequency. 3. In optical-fiber transmission, the velocity of the modulated optical power.

***G/T*:** *Abbreviation for* **antenna gain-to-noise-temperature.**

guard band: *See* **frequency guard band, time guard band.**

guarded frequency: A transmission frequency that is not to be jammed or interfered with because of the value of the information being derived from it. *Note:* For example, a guarded frequency will not be jammed when the tactical, strategic, and technical information that can be obtained from the transmissions outweighs the potential operational gain achieved by jamming. [From Weik '89]

guided mode: *Synonym* **bound mode.**

guided ray: In an optical fiber, a ray that is confined primarily to the core. *Note:* A guided ray satisfies the relation given by

$$0 \le \sin\theta_r \le \sqrt{n_r^2 - n_a^2} \;,$$

where θ_r is the angle the ray makes with the fiber axis, r is the radial position, *i.e.*, radial distance, of the ray from the fiber axis, n_r is the refractive index at the radial distance r from the fiber axis, and n_a is the refractive index at the core radius, a, *i.e.*, at the core-cladding interface. Guided rays correspond to bound modes, *i.e.*, guided modes, in terms of modes rather than rays. (188) *Synonyms* **bound ray, trapped ray.**

guided wave: A wave having (a) energy concentrated near a boundary, or between substantially parallel boundaries, separating materials of different properties and (b) a direction of propagation effectively parallel to these boundaries. (188)

(this page intentionally left blank)

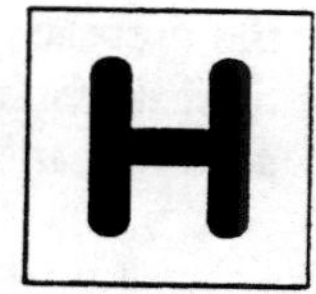

Hagelbarger code: A convolutional code that enables error bursts to be corrected provided that there are relatively long error-free intervals between the error bursts. *Note:* In the Hagelbarger code, inserted parity check bits are spread out in time so that an error burst is not likely to affect more than one of the groups in which parity is checked.

half-duplex (HDX) operation: Operation in which communication between two terminals occurs in either direction, but in only one direction at a time. (188) *Note:* Half-duplex operation may occur on a half-duplex circuit or on a duplex circuit, but it may not occur on a simplex circuit. *Synonyms* **one-way reversible operation, two-way alternate operation.**

halftone: Any photomechanical printing surface or the impression therefrom in which detail and tone values are represented by a series of evenly spaced dots in varying size and shape, varying in direct proportion to the intensity of tones they represent. [JP1]

halftone characteristic: 1. In facsimile systems, the relationship between the density of the recorded copy and the density of the object, *i.e.*, the original. (188) **2.** In facsimile systems, the relationship between the amplitude of the facsimile signal to either the density of the object or the density of the recorded copy when only a portion of the system is under consideration. *Note:* In an FM facsimile system, an appropriate parameter other than the amplitude is used.

Hamming code: An error-detecting and error-correcting binary code, used in data transmission, that can (a) detect all single- and double-bit errors and (b) correct all single-bit errors. *Note:* A Hamming code satisfies the relation $2^m \geq n+1$, where n is the total number of bits in the block, k is the number of information bits in the block, and m is the number of check bits in the block, where $m = n - k$.

Hamming distance: The number of digit positions in which the corresponding digits of two binary words of the same length are different. *Note 1:* The Hamming distance between 1011101 and 1001001 is two. *Note 2:* The concept can be extended to other notation systems. For example, the Hamming distance between 2143896 and 2233796 is three, and between "toned" and "roses" it is also three. *Synonym* **signal distance.**

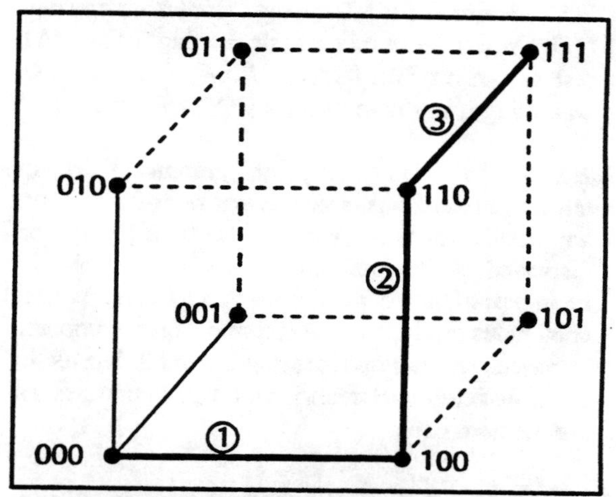

the number of digit positions in which the corresponding digits of two binary numbers or words of the same length are different

Hamming weight: The number of non-zero symbols in a symbol sequence. *Note:* For binary signaling, Hamming weight is the number of "1" bits in the binary sequence.

handoff: 1. In cellular mobile systems, the process of transferring a phone call in progress from one cell transmitter and receiver and frequency pair to another cell transmitter and receiver using a different frequency pair without interruption of the call. **2.** In satellite communications, the process of transferring ground-station control responsibility from one ground station to another without loss or interruption of service. (188)

handshaking: 1. In data communications, a sequence of events governed by hardware or software, requiring mutual agreement of the state of the operational modes prior to information exchange. **2.** The process used to establish communications parameters between two stations. (188) *Note:* Handshaking follows the establishment of a circuit between the stations and precedes information transfer. It is used to agree upon such parameters as information transfer rate, alphabet, parity, interrupt procedure, and other protocol features.

hangover: *Synonym* **tailing.**

HA1-receiver weighting: A noise weighting used in a noise measuring set to measure noise across the HA1-receiver of a 302-type or similar instrument. (188) *Note 1:* The meter scale readings of an HA1 test set are in dBa (HA1). *Note 2:* HA1 noise weighting is obsolete for new DOD applications.

hard copy: In computer graphics and in telecommunications, a permanent reproduction, on any media suitable for direct use by a person, of displayed or transmitted data. (188) *Note 1:* Examples of hard copy include teletypewriter pages, continuous printed tapes, facsimile pages, computer printouts, and radiophoto prints. *Note 2:* Magnetic tapes, diskettes, and nonprinted punched paper tapes are not hard copy.

hard disk: A flat, circular, rigid plate with a magnetizable surface on one or both sides of which data can be stored. *Note:* A hard disk is distinguished from a diskette by virtue of the fact that it is rigid. Early in the development of computer technology, hard disks, often multiple disks mounted on a common spindle, were interchangeable and removable from their drives, which were separate from the processor chassis. This technology is still in use, especially in conjunction with large mainframe computers, but physically smaller computers use hard disks that are in sealed units, along with their control electronics and read/write heads. The sealed units are usually installed permanently in the same chassis that contains the processor.

hardened: Pertaining to the condition of a facility with protective features that enable it to withstand destructive forces, such as explosions, natural disasters, or ionizing radiation. (188)

hard limiting: *See* **limiting.**

hard sectoring: In magnetic or optical disk storage, sectoring that uses a physical mark on the disk, from which mark sector locations are referenced. *Note:* Hard sectoring may be done, for example, by punching an index hole in a floppy diskette. When the presence of the index hole is recognized by an optical reader, a reference signal is generated. All sector locations can be referenced from this signal.

hardware: **1.** Physical equipment as opposed to programs, procedures, rules, and associated documentation. (188) **2.** The generic term dealing with physical items as distinguished from its capability or function such as equipment, tools, implements, instruments, devices, sets, fittings, trimmings, assemblies, subassemblies, components, and parts. *Note: Hardware* is often used in regard to the stage of development, as in the passage of a device or component from the design stage into the hardware stage as the finished object. [After JP1] **3.** In data automation, the physical equipment or devices forming a computer and peripheral components. [JP1]

hardwire: **1.** To connect equipment or components permanently in contrast to using switches, plugs, or connectors. (188) **2.** To wire in fixed logic or read-only storage that cannot be altered by program changes. (188)

harmful interference: **1.** Any emission, radiation, or induction interference that endangers the functioning or seriously degrades, obstructs, or repeatedly interrupts a communications system, such as a radio navigation service, telecommunications service, radio communications service, search and rescue service, or weather service, operating in accordance with approved standards, regulations, and procedures. (188) *Note:* To be considered harmful interference, the interference must cause serious detrimental effects, such as circuit outages and message losses, as opposed to interference that is merely a nuisance or annoyance that can be overcome by appropriate measures. **2.** Interference which endangers the functioning of a radionavigation service or of other safety services or seriously degrades, obstructs, or repeatedly interrupts a radiocommunication service operating in accordance with these *[Radio] Regulations.* [NTIA] [RR]

harmonic: **1.** Of a sinusoidal wave, an integral multiple of the frequency of the wave. *Note:* The frequency of the sine wave is called the fundamental frequency or the first harmonic, the second harmonic is twice the fundamental frequency, the third harmonic is thrice the fundamental frequency, *etc.* **2.** Of a periodic signal or other periodic phenomenon, such as an electromagnetic wave or a sound wave, a component frequency of the signal that is an integral multiple of the fundamental frequency. *Note:* The fundamental frequency is the reciprocal of the period of the periodic phenomenon. *Contrast with* **fundamental, overtone.**

harmonic distortion: In the output signal of a device, distortion caused by the presence of frequencies that are not present in the input signal. *Note:* Harmonic distortion is caused by nonlinearities within the device. (188)

hazards of electromagnetic radiation to fuel (HERF): The potential for electromagnetic radiation to cause ignition or detonation of volatile combustibles, such as aircraft fuels. (188)

hazards of electromagnetic radiation to ordnance (HERO): The potential for electromagnetic radiation to affect adversely munitions or electroexplosive devices. (188)

hazards of electromagnetic radiation to personnel (HERP): The potential for electromagnetic radiation to produce harmful biological effects in humans. (188)

H-bend: A smooth change in the direction of the axis of a waveguide, throughout which the axis remains in a plane parallel to the direction of magnetic H-field (transverse) polarization. (188) *Synonym* **H-plane bend.**

H-channel: In Integrated Services Digital Networks (ISDN), a 384-kb/s, 1472-kb/s, or 1536-kb/s channel, designated as "H_0", "H_{10}", and "H_{11}", respectively, accompanied by timing signals used to carry a wide variety of user information. (188) *Note:* Examples of types of user information representation forms include fast facsimile, video, high-speed data, high-quality audio, packet-switched data, bit streams at rates less than the respective H-channel bit rate that have been rate-adapted or multiplexed together, and packet-switched information.

HDLC: *Abbreviation for* **high-level data link control.**

HDTV: *Abbreviation for* **high-definition television.**

HDX: *Abbreviation for* **half duplex.**

head: A device that reads, writes, and/or erases data on a storage medium.

head end: **1.** A central control device required by some networks (*e.g.,* LANs or MANs) to provide such centralized functions as remodulation, retiming, message accountability, contention control, diagnostic control, and access to a gateway. **2.** A central control device, within CATV systems, that provides centralized functions such as remodulation.

header: The portion of a message that contains information used to guide the message to the correct destination. (188) *Note:* Examples of items that may be in a header are the addresses of the sender and receiver, precedence level, routing instructions, and synchronizing bits.

head-of-bus function: The function that generates management information and empty bus slots at the point on each bus where data flow begins. (188)

head-on collision: A collision that occurs on a communications channel when two or more users begin to transmit on the channel at approximately the same instant.

Heaviside layer: *Synonym* **E region.**

height gain: For a given propagation mode of an electromagnetic wave, the ratio of the field strength at a specified height to the field strength at the surface of the Earth.

helical antenna: An antenna that has the form of a helix. (188) *Note:* When the helix circumference is much smaller than one wavelength, the antenna radiates at right angles to the axis of the helix. When the helix circumference is one wavelength, maximum radiation is along the helix axis.

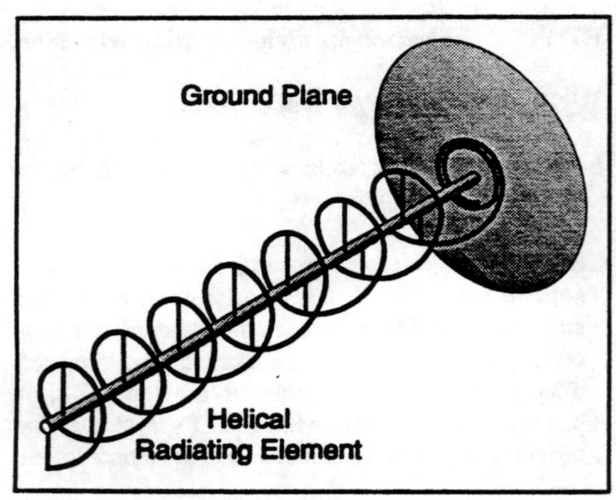

helical antenna

HEMP: *Abbreviation for* **high-altitude electromagnetic pulse.**

HE_{11} mode: Designation for the fundamental hybrid mode of an optical fiber.

HERF: *Abbreviation for* **hazards of electromagnetic radiation to fuel.**

HERO: *Abbreviation for* **hazards of electromagnetic radiation to ordnance.**

HERP: *Abbreviation for* **hazards of electromagnetic radiation to personnel.**

hertz (Hz): 1. The SI unit of frequency, equal to one cycle per second. *Note:* A periodic phenomenon that has a period of one second has a frequency of one hertz. (188) **2.** A unit of frequency which is equivalent to one cycle per second. [NTIA]

Hertzian wave: *Synonym* **radio wave.**

heterochronous: A relationship between two signals such that their corresponding significant instants do not necessarily occur at the same time. (188) *Note:* Two signals having different nominal signaling rates and not stemming from the same clock or from homochronous clocks are usually heterochronous.

heterodyne: 1. To generate new frequencies by mixing two or more signals in a nonlinear device such as a vacuum tube, transistor, or diode mixer. (188) *Note:* A superheterodyne receiver converts any selected incoming frequency by heterodyne action to a common intermediate frequency where amplification and selectivity (filtering) are provided. **2.** A frequency produced by mixing two or more signals in a nonlinear device. (188)

heterodyne repeater: In radio reception and retransmission, a repeater that converts the original band of frequencies of the received signal to a different frequency band for retransmission after amplification. *Note:* Heterodyne repeaters are used, for example, in microwave systems, to avoid undesired feedback between the receiving and transmitting antennas. (188) *Synonym* **IF repeater.**

heterogeneous multiplexing: Multiplexing in which not all the information-bearer channels operate at the same data signaling rate.

heuristic routing: Routing in which data, such as time delay, extracted from incoming messages, during specified periods and over different routes, are used to determine the optimum routing for transmitting data back to the sources. *Note:* Heuristic routing allows a measure of route optimization based on recent empirical knowledge of the state of the network.

hexadecimal: 1. Characterized by a selection, choice or condition that has sixteen possible different values or states. *Synonym* **sexadecimal. 2.** Pertaining to a fixed-radix numeration system in which the radix is sixteen.

HF: *Abbreviation for* **high frequency.**

HFDF: *Abbreviation for* **high-frequency distribution frame.**

hierarchical computer network: A computer network in which processing and control functions are performed at several levels by computers specially suited for the functions performed, such as industrial process control, inventory control, database control, or hospital automation.

hierarchically synchronized network: A mutually synchronized network in which some clocks exert more control than others, the network operating frequency being a weighted mean of the natural frequencies of the population of clocks.

hierarchical routing: Routing that is based on hierarchical addressing. *Note:* Most Transmission Control Protocol/Internet Protocol (TCP/IP) routing is based on a two-level hierarchical routing in which an IP address is divided into a network portion and a host portion. Gateways use only the network portion until an IP datagram reaches a gateway that can deliver it directly. Additional levels of hierarchical routing are introduced by the addition of subnetworks.

high-altitude electromagnetic pulse (HEMP): An electromagnetic pulse produced at an altitude effectively above the sensible atmosphere, *i.e.*, above about 120 km. (188)

high-definition television (HDTV): Television that has approximately twice the horizontal and twice the vertical emitted resolution specified by the NTSC standard. *Note 1:* In HDTV, the total number of pixels is therefore approximately four times that of the NTSC standard. *Note 2:* HDTV may include any or all improved-definition television (IDTV) and extended-television (EDTV) improvements. *Note 3:* HDTV employs a wide aspect ratio.

higher frequency ground: *Deprecated term name. See* **facility grounding system.**

high frequency (HF): Frequencies from 3 MHz to 30 MHz. (188)

high-frequency distribution frame (HFDF): A distribution frame that provides terminating and interconnecting facilities for those combined supergroup modulator output circuits and combined supergroup demodulator input circuits that contain signals occupying the baseband spectrum. (188)

high-level control: In the hierarchical structure of a primary or secondary data transmission station, the conceptual level of control or processing logic that (a) is above the Link Level and (b) controls Link Level functions, such as device control, buffer allocation, and station management.

high-level data link control (HDLC): A Link-Level protocol used to facilitate reliable point-to-point transmission of a data packet. *Note:* A subset of HDLC, known as *"LAP-B,"* is the Layer-two protocol for CCITT Recommendation X.25.

high-level language (HLL): A computer programming language that is primarily designed for, and syntactically oriented to, particular classes of problems and that is essentially independent of the structure of a specific computer or class of computers; for example, Ada®, COBOL, Fortran, Pascal. *Synonym* **high-order language.**

high-order language: *Synonym* **high-level language.**

high-pass filter: A filter that passes frequencies above a given frequency and attenuates all others. (188)

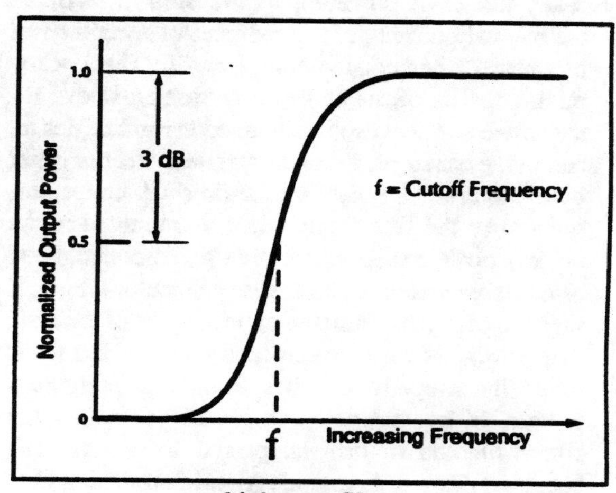

high-pass filter

high-performance equipment: Equipment that (a) has the performance characteristics required for use in trunks or links, (b) is designed primarily for use in global and tactical systems, and (c) sufficiently withstands electromagnetic interference when operating in a variety of network or point-to-point circuits. (188) *Note:* Requirements for global and tactical high-performance equipment may differ.

high-usage trunk group: A group of trunks for which an alternate route has been provided to absorb the relatively high rate of overflow traffic. (188)

highway: **1.** A digital serial-coded bit stream with time slots allotted to each call on a sequential basis. **2.** A common path or a set of parallel paths over which signals from more than one channel pass with separation achieved by time division.

hiss: Noise in the audio frequency range, having subjective characteristics analogous to prolonged sibilant sounds. (188) *Note:* Noise in which there are no pronounced low-frequency components may be considered as hiss.

hit: **1.** A transient disturbance to a communication medium. (188) **2.** A match of data to a prescribed criterion.

HLL: *Abbreviation for* **high-level language.**

hockey puck: A polishing fixture used to facilitate the manual finishing of the endfaces of certain types of optical fiber connectors. *Note 1:* The hockey puck consists of the appropriate mating sleeve for the connector in question, mounted at right angles to, and in the center of, a disk of stainless steel or other hard material. When the unfinished connector, secured to the fiber-optic cable, is mounted in the hockey puck, excess material (*e.g.*, fiber end, bead of adhesive material, and excess connector length, if present) protrudes from the opposite side of the disk. The excess is then ground away as the fixture is manually swept to and fro, usually in a figure-8 pattern, in contact with a piece of microfinishing film which is in turn supported by a rigid flat substrate. Two to four grades of microfinishing film, with abrasive particles ranging in size from 15 μm to 0.3 μm, are commonly used. [After FAA] *Note 2:* Various manufacturers use proprietary names to identify this device; however "hockey puck" has become ubiquitous.

hold-in frequency range: The range of frequencies over which a phase-locked loop can vary and still maintain frequency lock.

holding time: The total length of time that a call makes use of a trunk or channel. (188) *Note:* Holding time is usually measured in seconds.

homing: **1.** A process in which a mobile station is directed, or directs itself, toward an electromagnetic, thermal, sonic, or other source of energy, whether primary or reflected, or follows a vector force field or a gradient of a scalar force field. **2.** In radio direction finding, the locating of a moving signal source by a moving direction-finding station that has a mobile advantage. **3.** The act of approaching a source of electromagnetic radiation in which the approaching vehicle is guided by a receiver with a directional antenna. **4.** Seeking, finding, intercepting, and engaging an object, *i.e.*, a target (fixed or mobile) that may contain a signal source. [From Weik '89]

homochronous: The relationship between two signals such that their corresponding significant instants are displaced by a constant interval of time. (188)

homogeneous cladding: In an optical fiber, a cladding in which the nominal refractive index is constant throughout. *Note:* An optical fiber may have several homogeneous claddings, each having a different refractive index.

homogeneous multiplexing: Multiplexing in which all of the information-bearer channels operate at the same data signaling rate.

hop: **1.** The excursion of a radio wave from the Earth to the ionosphere and back to the Earth. (188) *Note:* The number of hops indicates the number of reflections from the ionosphere. **2.** A waveform transmitted for the duration of each relocation of the carrier frequency of a frequency-hopped system. (188) **3.** To modify a modulated waveform with constant center frequency so that it frequency hops. (188)

hop count: **1.** In a data communications network, the number of legs traversed by a packet between its source and destination. *Note:* Hop count may be used to determine the time-to-live for some packets. **2.** The number of signal regenerating devices (such as repeaters, bridges, routers, and gateways) through which data must pass to reach their destination.

horizon angle: Of a directional antenna, the angle, in a vertical plane, subtended by the lines extending (a) from the antenna to the radio horizon and (b) from the antenna in its direction of maximum radiation. (188)

horizontal redundancy check: *Synonym* **longitudinal redundancy check.**

horizontal resolution: In facsimile, the number of picture elements per unit distance in the direction of scanning or recording. (188)

horn: **1.** In radio transmission, an open-ended waveguide, of increasing cross-sectional area, which radiates directly in a desired direction or feeds a reflector that forms a desired beam. (188) *Note 1:* Horns may have one or more expansion curves, *i.e.*, longitudinal cross sections, such as elliptical, conical, hyperbolic, or parabolic curves, and not necessarily the same expansion curve in each (E-plane and H-plane) cross section. *Note 2:* A very wide range of beam patterns may be formed by controlling horn dimensions and shapes, placement of the reflector, and reflector shape and dimensions. **2.** A portion of a waveguide in which the cross section is smoothly increased along the axial direction. (188) **3.** In audio systems, a tube, usually having a rectangular transverse cross section and a linearly or exponentially increasing cross-sectional area, used for radiating or receiving acoustic waves.

horn gap switch: A switch provided with arcing horns, ordinarily used for disconnecting or breaking the charging current of overhead transmission and distribution lines. (188)

host: **1.** In packet- and message-switching communications networks, the collection of hardware and software that makes use of packet or message switching to support user-to-user, *i.e.*, end-to-end, communications, interprocess communications, and distributed data processing. [From Weik '89] **2.** *Synonym* **host computer.**

host computer: **1.** In a computer network, a computer that provides end users with services such as computation and database access and that usually performs network control functions. *Synonym* **host.** **2.** A computer on which is developed software intended to be used on another computer.

hot boot: *Synonym* **warm restart.**

hotline: A point-to-point communications link in which a call is automatically originated to the preselected destination without any additional action by the user when the end instrument goes off-hook. *Note 1:* Hotlines cannot be used to originate calls other than to preselected destinations. *Note 2:* Various priority services that require dialing are not properly termed "hotlines." *Synonyms* **automatic signaling service, off-hook service.**

hot standby: *See* **standby.**

house cable: *Deprecated term.* Communication cable within a building or a complex of buildings. *Note:* House cable owned before divestiture by the Bell System and after divestiture by the Regional Bell Operating Companies will eventually be fully depreciated and will then belong to the customer. *See* **on-premises wiring.**

housekeeping signals: *Synonym* **service signals.**

H-plane bend: *Synonym* **H-bend.**

HTML: *Abbreviation for* **Hypertext Markup Language.** An application of SGML (Standard Generalized Markup Language [ISO 8879]) implemented in conjunction with the World Wide Web to facilitate the electronic exchange and display of simple documents using the Internet.

HTTP: *Abbreviation for* **Hypertext Transfer Protocol.** In the World Wide Web, a protocol that facilitates the transfer of hypertext-based files between local and remote systems.

hub: **1.** A distribution point in a network. **2.** A device that accepts a signal from one point and redistributes it to one or more points.

Huffman coding: A coding technique used to compact data by representing the more common events with short codes and the less common events with longer codes. (188) *Note:* Huffman coding is used in Group 3 facsimile.

hundred call-seconds (CCS): *See* **call-second.**

hunting: **1.** In telephony, pertaining to the operation of a selector or other similar device to find and establish a connection with an idle circuit of a chosen group. (188) **2.** Pertaining to the failure of a device to achieve a state of equilibrium, usually by alternately overshooting and undershooting the point of equilibrium. (188)

hybrid: 1. A functional unit in which two or more different technologies are combined to satisfy a given requirement. *Note:* Examples of hybrids include (a) an electronic circuit having both vacuum tubes and transistors, (b) a mixture of thin-film and discrete integrated circuits, and (c) a computer that has both analog and digital capability. (188) **2.** A resistance hybrid. (188) **3.** A hybrid coil. (188)

hybrid balance: An expression of the degree of electrical symmetry between two impedances connected to two conjugate sides of a hybrid set or resistance hybrid. *Note 1:* Hybrid balance is usually expressed in dB. *Note 2:* If the respective impedances of the branches of the hybrid that are connected to the conjugate sides of the hybrid are known, hybrid balance may be computed by the formula for return loss. (188)

hybrid cable: An optical communications cable having two or more different types of optical fibers, *e.g.*, single-mode and multimode fibers. *Contrast with* **composite cable.**

hybrid coil: A single transformer that effectively has three windings, and which is designed to be configured as a circuit having four branches, *i.e.*, ports, that are conjugate in pairs. (188) *Note:* The primary use of a hybrid coil is to convert between 2-wire and 4-wire operation in concatenated sections of a communications circuit. Such conversion is necessary when repeaters are introduced in a 2-wire circuit. *Synonym* **bridge transformer.**

hybrid communications network: A communications network that uses a combination of line facilities, *i.e.*, trunks, loops, or links, some of which use only analog or quasi-analog signals and some of which use only digital signals. (188) *Synonym* **hybrid system.**

hybrid computer: A computer that processes both analog and digital data.

hybrid connector: A connector that contains contacts for more than one type of service. *Note:* Examples of hybrid connectors are those that have contacts for both optical fibers and twisted pairs, electric power and twisted pairs, or shielded and unshielded twisted pairs.

hybrid coupler: In an antenna system, a hybrid junction used as a directional coupler. *Note:* The loss through a hybrid coupler is usually ≈3 dB. (188)

hybrid interface structure: In integrated services digital networks (ISDN), an interface structure that uses both labeled and positioned channels.

hybrid junction: A waveguide or transmission line arranged such that (a) there are four ports, (b) each port is terminated in its characteristic impedance, and (c) energy entering any one port is transferred, usually equally, to two of the three remaining ports. (188) *Note:* Hybrid junctions are used as mixing or dividing devices.

hybrid mode: A mode consisting of components of both electrical and magnetic field vectors in the direction of propagation. (188) *Note:* In fiber optics, such modes correspond to skew (nonmeridional) rays.

hybrid network: *See* **hybrid communications network.**

hybrid routing: Routing in which numbering plans and routing tables are used to permit the collocation, in the same area code, of switches using a deterministic routing scheme with switches using a nondeterministic routing scheme, such as flood search routing. *Note:* Routing tables are constructed with no duplicate numbers, so that direct dial service can be provided to all network subscribers. This may require the use of 10-digit numbers.

hybrid set: Two or more transformers interconnected to form a network having four ports that are conjugate in pairs. *Note:* The primary use of a hybrid set is to convert between 2-wire and 4-wire operation in concatenated sections of a communications circuit. Such conversion is necessary when repeaters are introduced in a 2-wire circuit. (188)

hybrid spread spectrum: A combination of frequency hopping spread spectrum and direct-sequence spread spectrum. [NTIA]

hybrid system: *Synonym* **hybrid communications network.**

hybrid topology: *See* **network topology.**

hydroxyl ion absorption: In optical fibers, the absorption of electromagnetic waves, including the near-infrared, due to the presence of trapped hydroxyl ions remaining from water as a contaminant. *Note:* The hydroxyl (OH^-) ion can penetrate glass during or after product fabrication, resulting in significant attenuation of discrete optical wavelengths, *e.g.*, approximately 1.3 μm, used for communications via optical fibers.

hypermedia: Computer-addressable files that contain pointers for linking to multimedia information, such as text, graphics, video, or audio in the same or other documents. *Note:* The use of hypertext links is known as navigating.

hypertext: The system of coding that is used to create or navigate hypermedia in a nonsequential manner.

Hypertext Transfer Protocol: *See* **HTTP.**

Hz: *Abbreviation for* **hertz.**

(this page intentionally left blank)

I

ICA: *Abbreviation for* **International Communications Association.**

ICI: *Abbreviation for* **incoming call identification.**

ICMP: *Abbreviation for* **Internet Control Message Protocol.** An Internet protocol that reports datagram delivery errors. *Note 1:* ICMP is a key part of the TCP/IP protocol suite. *Note 2:* The packet internet gopher (ping) application is based on ICMP.

icon: In computer systems, a small, pictorial representation of an application software package, idea, or concept used in a window or a menu to represent commands, files, or options.

ICW: *Abbreviation for* **interrupted continuous wave.**

identification, friend or foe (IFF): A system using electromagnetic transmissions to which equipment carried by friendly forces automatically responds, for example, by emitting pulses, thereby distinguishing themselves from enemy forces. [JP1] *Note:* The secondary surveillance radar (SSR) system used in modern air traffic control systems is an outgrowth of the military IFF system used during World War II. The IFF equipment carried by modern military aircraft is compatible with the transponder system used for civilian air traffic control.

identification friend or foe personal identifier: The discrete identification, friend or foe code assigned to a particular aircraft, ship, or other vehicle for identification by electronic means. [JP1]

identifier (ID): **1.** In telecommunications and data processing systems, one or more characters used to identify, name, or characterize the nature, properties, or contents of a set of data elements. **2.** A string of bits or characters that names an entity, such as a program, device, or system, in order that other entities can call that entity. **3.** In programming languages, a lexical unit that names a language object, such as a variable, array, record, label, or procedure. *Note:* An identifier is placed in a label. The label is attached to, is a part of, or remains associated with, the information it identifies. If the label becomes disassociated from the information it identifies, the information may not be accessible.

IDF: *Abbreviation for* **intermediate distribution frame.**

idle-channel noise: Noise that is present in a communications channel when no signals are applied. *Note:* The channel conditions and terminations must be stated for idle-channel noise measurements to be meaningful. (188)

idle character: A control character that is transmitted when no useful information is being transmitted. [From Weik '89]

idle-line termination: A switch-controlled electrical network that maintains a desired impedance at a trunk or line terminal that is in the idle state. (188)

idle state: The telecommunications service condition that exists whenever user messages are not being transmitted but the service is immediately available for use.

idle time: A period during which a system, circuit, or component is not in use, but is available.

IDN: *Abbreviation for* **integrated digital network.**

IDTV: *Abbreviation for* **improved-definition television.**

IF: *Abbreviation for* **intermediate frequency.**

I/F: *Abbreviation for* **interface.**

IFF: *Abbreviation for* **identification, friend or foe.**

IFRB: *Abbreviation for* **International Frequency Registration Board.**

IF repeater: *See* **heterodyne repeater.**

IFS: *Abbreviation for* **ionospheric forward scatter.** *See* **ionospheric scatter.**

ILD: *Abbreviation for* **injection laser diode.**

illegal character: A character, or a combination of bits, that is not valid in a given system according to specified criteria, such as with respect to a specified alphabet, a particular pattern of bits, a rule of formation, or a check code. [From Weik '89]

Synonyms **false character, forbidden character, improper character, unallowable character, unused character.**

ILS: *Abbreviation for* **instrument landing system.**

IM: *Abbreviation for* **intensity modulation, intermodulation.**

image: In the field of image processing, a two-dimensional representation of a scene. *Synonym* **picture.**

image antenna: A hypothetical mirror-image, *i.e.*, virtual-image, of an antenna, *i.e.*, antenna element, considered to extend as far below ground, *i.e.*, the ground plane, as the actual antenna is above the ground plane. (188) *Note 1:* The image antenna is helpful in calculating electric field vectors, magnetic field vectors, and electromagnetic fields emanating from the real antenna, particularly in the vicinity of the antenna and along the ground. Each charge and current in the real antenna has its image that may also be considered as a source of radiation equal to, but differently directed from, its real counterpart. *Note 2:* An image antenna may also be considered to be on the opposite side of any equipotential plane surface, such as a metal plate acting as a ground plane, analogous to the position of a virtual optical image in a plane mirror. *Note 3:* The ground plane need not be grounded to the Earth.

image frequency: In radio reception using heterodyning in the tuning process, an undesired input frequency that is capable of producing the same intermediate frequency (IF) that the desired input frequency produces. (188) *Note:* The term *image* arises from the mirror-like symmetry of signal and image frequencies about the beating-oscillator frequency.

image frequency rejection ratio: *Synonym* **image rejection ratio.**

image rejection ratio: In reception using heterodyning in the tuning process, the ratio of (a) the intermediate-frequency (IF) signal level produced by the desired input frequency to (b) that produced by the image frequency. *Note 1:* The image rejection ratio is usually expressed in dB. (188) *Note 2:* When the image rejection ratio is measured, the input signal levels of the desired and image frequencies must be equal for the measurement to be meaningful. *Synonym* **image frequency rejection ratio.**

imagery: Collectively, the representations of objects reproduced electronically or by optical means on film, electronic display devices, or other media. [JP1]

IMD: *Abbreviation for* **intermodulation distortion.**

immediate message: A category of precedence reserved for messages relating to situations that gravely affect the security of national/allied forces or populace and that require immediate delivery to the addressee(s). [JP1]

IMP: *Abbreviation for* **interface message processor.**

impedance: The total passive opposition offered to the flow of electric current. *Note 1:* Impedance is determined by the particular combination of resistance, inductive reactance, and capacitive reactance in a given circuit. (188) *Note 2:* Impedance is a function of frequency, except in the case of purely resistive networks.

impedance matching: The connection of an additional impedance to an existing one in order to accomplish a specific effect, such as to balance a circuit or to reduce reflection in a transmission line. (188)

improper character: *Synonym* **illegal character.**

improved-definition television (IDTV): Television transmitters and receivers that (a) are built to satisfy performance requirements over and above those required by the NTSC standard and (b) remain within the general parameters of NTSC standard emissions. *Note 1:* IDTV improvements may be made at the TV transmitter or the receiver. *Note 2:* Examples of improvements include enhancements in encoding, digital filtering, scan interpolation, interlaced line scanning, and ghost cancellation. *Note 3:* IDTV improvements must allow the TV signal to be transmitted and received in the standard 4:3 aspect ratio. *Synonym* **enhanced-quality television.**

improvement threshold: *See* **FM improvement threshold.**

impulse: A short surge of electrical, magnetic, or electromagnetic energy. (188) *Synonym* **surge.**

impulse excitation: The production of oscillation in a circuit or device by impressing a stimulus (signal) for a period that is extremely short compared to the duration of the oscillation that it produces. (188) *Synonym* **shock excitation.**

impulse noise: Noise consisting of random occurrences of energy spikes having random amplitude and and spectral content. *Note:* Impulse noise in a data channel can be a definitive cause of data transmission errors. (188)

impulse response: **1.** Of a device, the mathematical function that describes the output waveform that results when the input is excited by a unit impulse. **2.** The waveform that results at the output of a device when the input is excited by a unit impulse.

IN: *Abbreviation for* **intelligent network.**

in-band noise power ratio: For multichannel equipment, the ratio of (a) the mean noise power measured in any channel, with all channels loaded with white noise, to (b) the mean noise power measured in the same channel, with all channels but the measured channel loaded with white noise. (188)

in-band signaling: Signaling that uses frequencies or time slots within the bandwidth or data stream occupied by the information channel. (188)

incidental-radiation device: A device that radiates radio frequency energy during the course of its operation although the device is not intentionally designed to generate radio frequency energy. [NTIA]

inclination of an orbit (of an Earth satellite): The angle determined by the plane containing the orbit and the plane of the Earth's equator. [NTIA] [RR]

inclined orbit: Any nonequatorial orbit of a satellite. (188) *Note:* Inclined orbits may be circular or elliptical, synchronous or asynchronous, and direct or retrograde.

inclusion: A foreign object present within, for example, an optical fiber or a crystal.

incoherent: In optics, characterized by a degree of coherence significantly less than 0.88.

incoming call identification (ICI): A switching system feature that allows an attendant to identify visually the type of service or trunk group associated with a call directed to the attendant's position.

incorrect block: A block successfully delivered to the intended destination user, but having one or more incorrect bits, additions, or deletions, in the delivered block.

incremental compaction: Data compaction accomplished by specifying only the initial value and all subsequent changes. *Note:* An example of incremental compaction is the storing or transmitting of a line voltage followed only by the deviations from the initial value. Thus, instead of transmitting the values 102, 104, 105, 103, 100, 104 and 106, only the values 102, +2, +1, –2, –3, +4, and +2, or only the values 100, +2, +4, +5, +3, 0, +4, and +6 need be sent, depending on the system used. At a given data rate, transmitting only the initial and incremental values require much less time and space than transmitting the absolute values. [From Weik '89]

incremental phase modulation (IPM): In spread-spectrum systems, phase modulation in which one binary code sequence is shifted with respect to another, usually to conduct a synchronizing search, *i.e.*, a search to discover if the two sequences are the same, and perhaps thereby enabling two data streams to be synchronized. [From Weik '89]

indefinite call sign: **1.** A call sign that represents a group of facilities, commands, authorities, activities, or units rather than one of these. **2.** In radio communications, a call sign that does not identify a station and that is used in the call-up signal or in a message that has the station call sign encrypted in the text. [From Weik '89]

independent clocks: In communication network timing subsystems, free-running precision clocks used, for synchronization purposes, at the nodes. *Note:* Variable storage buffers, installed to accommodate variations in transmission delay between nodes, are made large enough to accommodate small time (phase) departures among the nodal clocks that control transmission. Traffic may occasionally be interrupted to allow the buffers to be emptied of some or all of their stored data.

independent-sideband (ISB) transmission: Double-sideband transmission in which the information carried by each sideband is different. (188) *Note:* The carrier may be suppressed.

index dip: In an optical fiber, an undesired decrease in the refractive index at the center of the core. *Note:* An index dip is an artifact of certain manufacturing processes. *Synonym* **profile dip.**

indexing: *See* **interaction crosstalk.**

index-matching gel: *Synonym* **gel (def. #2).**

index-matching material: A substance, usually a liquid, cement (adhesive), or gel, which has an index of refraction that closely approximates that of an optical fiber, and is used to reduce Fresnel reflection at the fiber endface. (FAA) *Note 1:* An index-matching material may be used in conjunction with pairs of mated connectors, with mechanical splices, or at the ends of fibers. *Note 2:* Without the use of an index-matching material, Fresnel reflections will occur at the smooth endfaces of a fiber. These reflections may be as high as –14 dB (*i.e.*, 14 dB below the level of the incident signal). When the reflected signal returns to the transmitting end, it is reflected again and returns to the receiving end at a level that is (28 plus twice the fiber loss) dB below the direct signal. The reflected signal will also be delayed by twice the delay time introduced by the fiber. The reflected signal will have no practical effect on digital systems because of its low level relative to the direct signal; *i.e.*, it will have no practical effect on the detected signal seen at the decision point of the digital optical receiver. It may be noticeable in an analog baseband intensity-modulated video signal.

index of cooperation: **1.** In facsimile, the product of the total line length and the number of lines per unit length, divided by π. **2.** For rotating devices, the product of the drum diameter and the number of lines per unit length. (188) *Synonyms:* **diametral index of cooperation, international index of cooperation.**

index of refraction: *Synonym* **refractive index.**

index profile: *Synonym* **refractive index profile.**

indirect control: In digital data transmission, the use of a clock rate of 2^n times the modulation rate, where n is an integer greater than one. (188)

indirect wave: A wave, such as a radio wave or sound wave, that arrives at a given point by reflection or scattering from surrounding objects, rather than directly from the source. [From Weik '89]

individual line: A line that connects a single user to a switching center.

individual reception (in the broadcasting-satellite service): The reception of emissions from a space station in the broadcasting-satellite service by simple domestic installations and in particular those possessing small antennae. [NTIA] [RR]

inductive coupling: The transfer of energy from one circuit to another by virtue of the mutual inductance between the circuits. (188) *Note 1:* Inductive coupling may be deliberate and desired (as in an antenna coupler) or may be undesired (as in power line inductive coupling into telephone lines). *Note 2:* Capacitive coupling favors transfer of higher frequency components, whereas inductive coupling favors transfer of lower frequency components.

industrial, scientific, and medical (ISM) applications (of radio frequency energy): Operation of equipment or appliances designed to generate and use locally radio-frequency energy for industrial, scientific, medical, domestic or similar purposes, excluding applications in the field of telecommunications. [NTIA] [RR]

industry standard: A voluntary, industry-developed document that establishes requirements for products, practices, or operations.

information: **1.** The meaning that a human assigns to data by means of the known conventions used in their representation. [JP1] (188) **2.** In intelligence usage, unprocessed data of every description which may be used in the production of intelligence. [JP1]

information-bearer channel: **1.** A channel capable of transmitting all the information required for communication, such as user data, synchronizing sequences, and control signals. *Note:* The information-bearer channel may operate at a higher data rate than that required for user data alone. **2.** A basic communications channel with the necessary bandwidth but without enhanced or value-added services. (188)

information bit: *See* **user information bit.**

information feedback: The return of received data to the source, usually for the purpose of checking the accuracy of transmission by comparison with the original data.

information field: In data transmission, a field assigned to contain user information. *Note:* The contents of the information field are not interpreted at the link level.

information processing: *Synonym* **data processing.**

information processing center (IPC): A facility staffed and equipped for processing and distributing information. (188) *Note:* An IPC may be geographically distributed.

information security: The protection of information against unauthorized disclosure, transfer, modification, or destruction, whether accidental or intentional. (188)

information source: *Synonym* **source user.**

information superhighway: *Synonym* **National Information Infrastructure.**

information system: **1.** A system, whether automated or manual, that comprises people, machines, and/or methods organized to collect, process, transmit, and disseminate data that represent user information. (188) **2.** Any telecommunications and/or computer related equipment or interconnected system or subsystems of equipment that is used in the acquisition, storage, manipulation, management, movement, control, display, switching, interchange, transmission, or reception of voice and/or data, and includes software, firmware, and hardware. [NIS]

information systems security (INFOSEC): The protection of information systems against unauthorized access to or modification of information, whether in storage, processing or transit, and against the denial of service to authorized users or the provision of service to unauthorized users, including those measures necessary to detect, document, and counter such threats. [NIS]

information transfer: The process of moving messages containing user information from a source to a sink. (188) *Note:* The information transfer rate may or may not be equal to the transmission modulation rate.

information-transfer phase: In an information-transfer transaction, the phase during which user information blocks are transferred from the source user to a destination user.

information-transfer transaction: A coordinated sequence of user and telecommunications system actions that cause information present at a source user to become present at a destination user. *Note:* An information-transfer transaction usually consists of three consecutive phases called the access phase, the information-transfer phase, and the disengagement phase.

INFOSEC: *Acronym for* **information systems security.**

infrared (IR): The region of the electromagnetic spectrum bounded by the long-wavelength extreme of the visible spectrum (approximately 0.7 µm) and the shortest microwaves (approximately 0.1 mm).

inhibiting signal: A signal that prevents the occurrence of an event. *Note:* An inhibiting signal may be used, for example, to disable an AND gate, thus preventing any signals from passing through it as long as the inhibiting signal is present. [From Weik '89]

injection fiber: *Synonym* **launching fiber.**

injection laser diode (ILD): A laser that uses a forward-biased semiconductor junction as the active medium. *Note:* Stimulated emission of coherent light occurs at a p-n junction where electrons and holes are driven into the junction. (188) *Synonyms* **diode laser, laser diode, semiconductor laser.** *See figure on following page.*

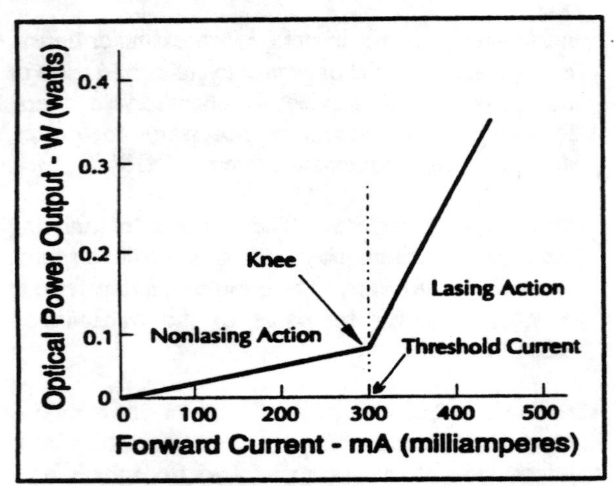

injection-laser diode: output vs. input current

ink vapor recording: Recording in which vaporized ink particles are deposited directly upon the record medium. (188)

input: **1.** In a device, process, or channel, a point that accepts data. **2.** A state, or a sequence of states, of a point that accepts data. **3.** A stimulus, such as a signal or interference, that enters a functional unit, such as a telecommunications system, a computer, or a computer program.

input data: **1.** Data being received or to be received by a device or a computer program. **2.** Data to be processed.

input-output channel: For a computer, a device that handles the transfer of data between internal memory and peripheral equipment.

input-output controller (IOC): A functional unit that controls one or more input-output channels. *Synonym* **I/O controller.**

input/output (I/O) device: A device that introduces data into or extracts data from a system. (188)

input protection: For analog input channels, protection against overvoltages that may be applied between any two input connectors or between any input connector and ground.

insertion gain: The gain resulting from the insertion of a device in a transmission line, expressed as the ratio of the signal power delivered to that part of the line following the device to the signal power delivered to that same part before insertion. (188) *Note 1:* If the resulting number is less than unity, an *"insertion loss"* is indicated. *Note 2:* Insertion gain is usually expressed in dB.

insertion loss: **1.** The loss resulting from the insertion of a device in a transmission line, expressed as the reciprocal of the ratio of the signal power delivered to that part of the line following the device to the signal power delivered to that same part before insertion. *Note:* Insertion loss is usually expressed in dB. **2.** In an optical fiber system, the total optical power loss caused by insertion of an optical component, such as a connector, splice, or coupler.

insertion-loss-vs.-frequency characteristic: Of a system or device, a plot of the amplitude as a function of frequency. (188) *Note:* The insertion-loss-vs.-frequency characteristic may be expressed as absolute gain or loss, or it may be normalized with respect to gain or loss at a specified reference frequency.

inside call: *Synonym* **internal call.**

inside plant: **1.** All the cabling and equipment installed in a telecommunications facility, including the main distribution frame (MDF) and all the equipment extending inward therefrom, such as PABX or central office equipment, MDF protectors, and grounding systems. (188) **2.** In radio and radar systems, all communications-electronics (C-E) equipment that is installed in buildings. (188)

inside wire: *See* **on-premises wiring.**

in-slot signaling: Signaling performed in the associated channel time slot.

inspection lot: A collection of produced units from which a statistically valid sample is to be drawn and inspected to determine conformance with acceptability criteria. *Note:* The inspection lot may differ from a collection of units designated as a lot for other purposes, such as for production, storage, packaging, and shipment.

instruction: In a programming language, an expression that specifies one operation and identifies its operands, if any.

instrument landing system (ILS): **1.** A radio-navigation system which provides aircraft with horizontal and vertical guidance just before and during landing and, at certain fixed points, indicates the distance to the reference point of landing. [NTIA] [RR] **2.** A system of radio navigation intended to assist aircraft in landing which provides lateral and vertical guidance, which may include indications of distance from the optimum point of landing. [JP1]

instrument landing system glide path: A system of vertical guidance embodied in the instrument landing system which indicates the vertical deviation of the aircraft from its optimum path of descent. [NTIA] [RR]

instrument landing system localizer: A system of horizontal guidance embodied in the instrument landing system which indicates the horizontal deviation of the aircraft from its optimum path of descent along the axis of the runway. [NTIA] [RR]

integrated circuit (IC): An electronic circuit that consists of many individual circuit elements, such as transistors, diodes, resistors, capacitors, inductors, and other active and passive semiconductor devices, formed on a single chip of semiconducting material and mounted on a single piece of substrate material. *Synonyms* **chip (def.#1), microcircuit.** [From Weik '89]

integrated digital network (IDN): A network that uses both digital transmission and digital switching.

integrated optical circuit (IOC): A circuit, or group of interconnected circuits, consisting of miniature solid-state optical components on semiconductor or dielectric substrates. *Note:* IOC components include light sources, optical filters, photodetectors, and thin-film optical waveguides.

integrated services digital network: *See* **ISDN.**

integrated station: A terminal device in which a telephone and one or more other devices, such as a video display unit, keyboard, or printer, are integrated and used over a single circuit.

integrated system: A telecommunication system that transfers analog and digital traffic over the same switched network. (188)

integrated voice and data terminal (IVDT): *See* **integrated station.**

integrating network: A network (circuit) that produces an output waveform that is the time integral of the input waveform. *Note:* Integrating networks are used in signal processing, such as for producing sawtooth waves from square waves.

integrity: *See* **data integrity, service integrity, and system integrity.**

intelligent network (IN): **1.** A network that allows functionality to be distributed flexibly at a variety of nodes on and off the network and allows the architecture to be modified to control the services. **2.** In North America, an advanced network concept that is envisioned to offer such things as (a) distributed call-processing capabilities across multiple network modules, (b) real-time authorization code verification, (c) one-number services, and (d) flexible private network services [including (1) reconfiguration by subscriber, (2) traffic analyses, (3) service restrictions, (4) routing control, and (5) data on call histories]. Levels of IN development are identified below:

➢ **IN/1** A proposed intelligent network targeted toward services that allow increased customer control and that can be provided by centralized switching vehicles serving a large customer base.

➢ **IN/1+** A proposed intelligent network targeted toward services that can be provided by centralized switching vehicles, *e.g.*, access tandems, serving a large customer base.

➢ **IN/2** A proposed, advanced intelligent-network concept that extends the distributed IN/1 architecture to accommodate the concept called *"service independence." Note:* Traditionally, service logic has been localized at individual switching systems. The IN/2 architecture provides flexibility in the placement of service logic, requiring the use of advanced techniques to manage the distribution of both network data and service logic across multiple IN/2 modules.

intelligent peripheral (IP): 1. A functional component that may be used most efficiently when accessed locally. 2. An intelligent-network feature that provides specialized telecommunication capabilities required by IN/2 service logic programs.

intelligibility: For voice communications, the capability of being understood. *Note:* Intelligibility does not imply the recognition of a particular voice. (188)

intelligible crosstalk: Crosstalk from which information can be derived. (188)

intensity: The square of the electric field strength of an electromagnetic wave. *Note:* Intensity is proportional to irradiance and may be used in place of the term *"irradiance"* when only relative values are important.

intensity modulation (IM): In optical communications, a form of modulation in which the optical power output of a source is varied in accordance with some characteristic of the modulating signal. (188) *Note:* In intensity modulation, there are no discrete upper and lower sidebands in the usually understood sense of these terms, because present optical sources lack sufficient coherence to produce them. The envelope of the modulated optical signal is an analog of the modulating signal in the sense that the instantaneous power of the envelope is an analog of the characteristic of interest in the modulating signal. Recovery of the modulating signal is by direct detection, not heterodyning.

interaction crosstalk: Crosstalk caused by coupling between carrier and noncarrier circuits. *Note:* If the interaction crosstalk is, in turn, coupled to another carrier circuit, that crosstalk is called "indexing." (188)

interactive data transaction: A unidirectional message, transmitted via a data channel, that requires a response in order for work to proceed logically.

interactive service: In an integrated services digital network (ISDN), a telecommunications service that facilitates a bidirectional exchange of information among users or among users and hosts. *Note:* Interactive services are grouped into conversational services, messaging services, and retrieval services.

interblock gap: On a data recording medium, an area used to indicate the end of a block or physical record. *Note:* Examples of interblock gaps are the gaps between blocks on magnetic tape and disks. (188)

intercept: 1. To stop a telephone call directed to an improper, disconnected, or restricted telephone number, and to redirect that call to an operator or a recording. (188) 2. To gain possession of communications intended for others without their consent, and, ordinarily, without delaying or preventing the transmission. (188) *Note:* An intercept may be an authorized or unauthorized action. 3. The acquisition of a transmitted signal with the intent of delaying or eliminating receipt of that signal by the intended destination user. (188)

interchangeability: A condition which exists when two or more items possess such functional and physical characteristics as to be equivalent in performance and durability, and are capable of being exchanged one for the other without alteration of the items themselves, or of adjoining items, except for adjustment, and without selection for fit and performance. [JP1]

interchangeable connectors: Connectors that share common installation geometry and have the same transmission performance. (188)

interchange circuit: A circuit that facilitates the exchange of data and signaling information between data terminal equipment (DTE) and data circuit-terminating equipment (DCE). *Note:* An interchange circuit can carry many types of signals and provide many types of service features, such as control signals, timing signals, and common return functions.

intercharacter interval: In asynchronous transmission, the time interval between the end of the stop signal of one character and the beginning of the start signal of the next character. (188) *Note:* The intercharacter interval may be of any duration. The signal sense of the intercharacter interval is always the same as the sense of the stop element, *i.e.*, "1" or "mark."

intercom: **1.** A telephone apparatus by means of which personnel can talk to each other within an aircraft, tank, ship, or activity. [JP1] **2.** A dedicated voice service within a specified user environment. (188)

interconnect facility: In a communications network, one or more communications links that (a) are used to provide local area communications service among several locations and (b) collectively form a node in the network. (188) *Note 1:* An interconnect facility may include network control and administrative circuits as well as the primary traffic circuits. *Note 2:* An interconnect facility may use any medium available and may be redundant.

interconnection: **1.** The linking together of interoperable systems. [JP1] **2.** The linkage used to join two or more communications units, such as systems, networks, links, nodes, equipment, circuits, and devices.

interexchange carrier (IXC): A communications common carrier that provides telecommunications services between LATAs or between exchanges within the same LATA. *Note:* Interexchange carriers have usually relied on local exchange carriers or competitive access providers for the local origination and termination of their traffic.

interface (I/F): **1.** In a system, a shared boundary, *i.e.*, the boundary between two subsystems or two devices. (188) **2.** A shared boundary between two functional units, defined by specific attributes, such as functional characteristics, common physical interconnection characteristics, and signal characteristics. **3.** A point of communication between two or more processes, persons, or other physical entities. **4.** A point of interconnection between user terminal equipment and commercial communications facilities. **5.** To interconnect two or more entities at a common point or shared boundary.

interface functionality: In telephony, the characteristic of interfaces that allows them to support transmission, switching, and signaling functions identical to those used in the enhanced services provided by the carrier. *Note:* As part of its comparably efficient interconnection (CEI) offering, the carrier must make available standardized hardware and software interfaces that are able to support transmission, switching, and signaling functions identical to those used in the enhanced services provided by the carrier.

interface message processor (IMP): A processor-controlled switch used in packet-switched networks to route packets to their proper destination.

interface payload: In integrated services digital networks (ISDN), the part of the bit stream through a framed interface used for telecommunications services and signaling.

interface point: *Synonym* **point of interface.**

interface standard: A standard that describes one or more functional characteristics (such as code conversion, line assignments, or protocol compliance) or physical characteristics (such as electrical, mechanical, or optical characteristics) necessary to allow the exchange of information between two or more (usually different) systems or equipment. *Note 1:* An interface standard may include operational characteristics and acceptable levels of performance. *Note 2:* In the military community, interface standards permit command and control functions to be performed using communication and computer systems.

interference: **1.** In general, extraneous energy, from natural or man-made sources, that impedes the reception of desired signals. **2.** A coherent emission having a relatively narrow spectral content, *e.g.*, a radio emission from another transmitter at approximately the same frequency, or having a harmonic frequency approximately the same as, another emission of interest to a given recipient, and which impedes reception of the desired signal by the intended recipient. *Note:* In the context of this definition, interference is distinguished from noise in that the latter is an incoherent emission from a natural source (*e.g.*, lightning) or a man-made source, of a character unlike that of the desired signal (*e.g.*, commutator noise from rotating machinery) and which usually has a broad spectral content. **3.** The effect of unwanted energy due to one or a combination of emissions, radiation, or inductions upon reception in a radiocommunication system, manifested by any performance degradation, misinterpretation, or loss of information which could be extracted in the absence of such unwanted energy. [NTIA] [RR] (188) **4.** The interaction of two or more coherent or partially coherent waves, which interaction produces a resultant wave that differs from the original waves in phase, amplitude, or both. *Note:* Interference may be constructive or destructive, *i.e.*, it may result in increased amplitude or decreased amplitude, respectively. Two waves equal in frequency and amplitude, and out of phase by 180°, will completely cancel one another. In phase, they create a resultant wave having twice the amplitude of either interfering beam. (188)

interference emission: Emission that results in an electrical signal's being propagated into, and interfering with the proper operation of, electronic or electrical equipment. *Note:* The frequency range of interference emissions may include the entire electromagnetic spectrum. (188)

interference filter: An optical filter that reflects one or more spectral bands or lines and transmits others, while maintaining a nearly zero coefficient of absorption for all wavelengths of interest. *Note 1:* An interference filter may be high-pass, low-pass, bandpass, or band-rejection. *Note 2:* An interference filter consists of multiple thin layers of dielectric material having different refractive indices. There also may be metallic layers. Interference filters are wavelength-selective by virtue of the interference effects that take place between the incident and reflected waves at the thin-film boundaries.

interferometer: An instrument that uses the principle of interference of electromagnetic waves for purposes of measurement. *Note:* Interferometers may be used to measure a variety of physical variables, such as displacement (distance), temperature, pressure, and strain.

interferometry: The branch of science devoted to the study and measurement of the interaction of waves, such as electromagnetic waves and acoustic waves. *Note 1:* The interaction of the waves can produce various spatial-, time-, and frequency-domain energy distribution patterns. [After 2196] *Note 2:* Interferometric techniques are used to measure refractive index profiles, *e.g.*, those of the preforms from which optical fibers are drawn, and to sense and measure physical variables, such as displacement (distance), temperature, pressure, and magnetic fields.

interframe time fill: In digital data transmission, a sequence of bits transmitted between consecutive frames. *Note:* Interframe time fill does not include bits stuffed within a frame.

interlaced scanning: In raster-scanned video displays, a scanning technique in which all odd-numbered scanning lines are first traced in succession, followed by the tracing of the even-numbered scanning lines in succession, each of which is traced between a pair of odd-numbered scanning lines. *Note 1:* The pattern created by tracing the odd-numbered scanning lines is called the ***"odd field"***, and the pattern created by tracing the even-numbered scanning lines is called the ***"even field"***. Each field contains half the information content, *i.e.*, pixels, of the complete video frame. *Note 2:* Image flicker is less apparent in an interlaced display than in a noninterlaced display, because the rate at which successive fields occur in an interlaced display is twice that at which successive frames would occur in a noninterlaced display containing the same number of scanning lines and having the same frame refresh rate. *Synonym* **interlacing.**

interlacing: *Synonym* **interlaced scanning.**

inter-LATA: **1.** Between local access and transport areas (LATAs). **2.** Services, revenues, and functions associated with telecommunications that originate in one LATA and that terminate in another one or that terminate outside of that LATA.

interleaving: The transmission of pulses from two or more digital sources in time-division sequence over a single path. (188)

intermediate distribution frame (IDF): In a central office or customer premises, a frame that (a) cross-connects the user cable media to individual user line circuits and (b) may serve as a distribution point for multipair cables from the main distribution frame (MDF) or combined distribution frame (CDF) to individual cables connected to equipment in areas remote from these frames.

intermediate element: In a network, a line-unit-line termination (LULT) or a line-unit-network termination (LUNT). (188)

intermediate field: *Synonym* **intermediate-field region.**

intermediate-field region: For an antenna, the transition region—lying between the near-field region and the far-field region—in which the field strength of an electromagnetic wave is dependent upon the inverse distance, inverse square of the distance, and the inverse cube of the distance from the antenna. (188) *Note:* For an antenna that is small compared to the wavelength in question, the intermediate-field region is considered to exist at all distances between 0.1 wavelength and 1.0 wavelength from the antenna. *Synonyms* **intermediate field, intermediate zone, transition zone.**

intermediate frequency (IF): A frequency to which a carrier frequency is shifted as an intermediate step in transmission or reception. (188)

intermediate language: In computer programming, a target language into which all or part of a single statement or a source program—in a source language—is translated before it is further translated or interpreted. *Note*: For a subsequent translation, an intermediate language may serve as a source language.

intermediate-level language: In computer, communications, and data processing systems, a programming language that (a) is less machine-oriented than a machine language, (b) is not so machine-independent as a common language, such as Ada®, COBOL, or Fortran, (c) contains macros that are less powerful than common-language macros, and (d) usually is the object language of a root compiler. Examples of intermediate-level languages include assembly languages, such as PL/I. [From Weik '89]

intermediate system: A system that provides an Open Systems Interconnection—Reference Model (OSI—RM) Network Layer relay function in which data received from one corresponding network entity are forwarded to another corresponding network entity.

intermediate zone: *Synonym* **intermediate-field region.**

intermodal delay distortion: *Synonym* **multimode distortion.**

intermodal dispersion: *Incorrect synonym for* **multimode distortion.**

intermodal distortion: *Synonym* **multimode distortion.**

intermodulation (IM): The production, in a nonlinear element of a system, of frequencies corresponding to the sum and difference frequencies of the fundamentals and harmonics thereof that are transmitted through the element. (188)

intermodulation distortion: Nonlinear distortion characterized by the appearance, in the output of a device, of frequencies that are linear combinations of the fundamental frequencies and all harmonics present in the input signals. (188) *Note:* Harmonic components themselves are not usually considered to characterize intermodulation distortion. When the harmonics are included as part of the distortion, a statement to that effect should be made.

intermodulation noise: In a transmission path or device, noise, generated during modulation and demodulation, that results from nonlinear characteristics in the path or device. (188)

intermodulation product: In the output of a nonlinear system, a frequency produced by intermodulation of harmonics of the frequencies present in the input signal.

internal bias: In a start-stop teletypewriter receiving mechanism, bias generated locally by the mechanism, and which has the same effect on the operating margin as bias external to the receiver, *i.e.*, applied bias. *Note:* Internal bias may be a marking bias or a spacing bias. (188)

internal call: A call placed within a private branch exchange (PBX) or local switchboard, *i.e.*, not through a central office in a public switched network. *Synonym* **inside call.** [From Weik '89]

internal memory: In a computer, all of the storage spaces that are accessible by a processor without the use of the computer input-output channels. *Note:* Internal memory usually includes several types of storage, such as main storage, cache memory, and special registers, all of which can be directly accessed by the processor. *Synonym* **internal storage.**

internal photoelectric effect: A photoconductive or photovoltaic effect.

internal storage: *Synonym* **internal memory.**

International Atomic Time (TAI): The time scale established by the International Time Bureau (BIH) on the basis of atomic clock data supplied by cooperating institutions. *Note:* The abbreviations "TAI" and "BIH" are a result of literal translation from the official international names that are written in French. (188)

international index of cooperation: *Synonym* **index of cooperation.**

International Frequency Registration Board (IFRB): *See* **Radio Regulations Board.**

International Organization for Standardization: *See* **ISO.**

International Radio Consultative Committee: *See* **CCIR, ITU-R.**

International System of Units (SI): The modern form of the metric system, which has been adopted by the United States and most other nations. *Note:* The SI is constructed from seven base units for independent physical quantities. *Tables showing these values are included on the next pages and are current as of Fall 1995.*

International Telecommunication Union (ITU): A civil international organization established to promote standardized telecommunications on a worldwide basis. (188) *Note:* The ITU-R and ITU-T are committees under the ITU. The ITU headquarters is located in Geneva, Switzerland. While older than the United Nations, it is recognized by the U.N. as the specialized agency for telecommunications.

International Telegraph Alphabet Number 5 (ITA-5): An alphabet in which (a) 128 unique 7-bit strings are used to encode upper- and lower-case letters, 10 decimal numerals, special signs and symbols, diacritical marks, data delimiters, and transmission control characters, (b) 12 of the 7-bit strings are not assigned to any letter, numeral, or control character, and (c) the unassigned bit strings are open for use in a given country that may have unique requirements, such as monetary symbols; diacritical marks, such as the tilde, umlaut, circumflex, and dieresis, and (d) a two-condition 8-bit pattern may be used that consists of seven information bits and a parity check bit. *Note:* ITA-5 is used for effecting information interchange. It is a result of a joint agreement between the International Telegraph and Telephone Consultative Committee (CCITT), now ITU-T, of the International Telecommunication Union (ITU) and the International Organization for Standardization (ISO). It is published as CCITT Recommendation V.3 and as ISO 646. It has also been adopted by NATO for military use. The United States adaptation of ITA-5 is ASCII (American Standard Code for Information Interchange) published by the American National Standards Institute (ANSI).

International Telegraph and Telephone Consultative Committee: *See* **CCITT, ITU-T.**

International Time Bureau (BIH): *See* **International Atomic Time.**

SI Prefixes. The common metric prefixes are:

Multiplication Factor		Prefix Name	Prefix	Symbol
1 000 000 000 000	=	10^{12}	tera	T
1 000 000 000	=	10^{9}	giga	G
1 000 000	=	10^{6}	mega	M
1 000	=	10^{3}	kilo	k
100	=	10^{2}	hecto	h
10	=	10^{1}	deka	da
0.1	=	10^{-1}	deci	d
0.01	=	10^{-2}	centi	c
0.001	=	10^{-3}	milli	m
0.000 001	=	10^{-6}	micro	µ
0.000 000 001	=	10^{-9}	nano	n
0.000 000 000 001	=	10^{-12}	pico	p

SI Base Units.

Quantity	Unit Name	Unit Symbol
length	meter	m
mass	kilogram	kg
time	second	s
electric current	ampere	A
thermodynamic temperature	kelvin	K
amount of substance	mole	mol
luminous intensity	candela	cd

SI derived units. Derived units are formed by combining base units and other derived units according to the algebraic relations linking the corresponding quantities. The symbols for derived units are obtained by means of the mathematical signs for multiplication, division, and use of exponents. For example, the SI unit for velocity is the *meter per second* (m/s or m•s^{-1}), and that for angular velocity is the *radian per second* (rad/s or rad•s^{-1}). Some derived SI units have been given special names and symbols, as listed in this table.

Quantity	Unit Name	Unit Symbol	Expression in Terms of Other SI Units
Absorbed dose, specific energy imparted, kerma, absorbed dose index	gray	Gy	J/kg
Activity (of a radionuclide)	becquerel	Bq	1/s
Celsius temperature	degree Celsius	°C	K
Dose equivalent	sievert	Sv	J/kg
Electric capacitance	farad	F	C/V
Electric charge, quantity of electricity	coulomb	C	A•s
Electric conductance	siemens	S	A/V
Electric inductance	henry	H	Wb/A
Electric potential, potential difference, electromotive force	volt	V	W/A
Electric resistance	ohm	Ω	V/A
Energy, work, quantity of heat	joule	J	N•m
Force	newton	N	kg•m/s^2
Frequency (of a periodic phenomenon)	hertz	Hz	1/s
Illuminance	lux	lx	lm/m^2
Luminous flux	lumen	lm	cd•sr
Magnetic flux	weber	Wb	V•s
Magnetic flux density	tesla	T	Wb/m^2
Plane angle	radian	rad	m/m
Power, radiant flux	watt	W	J/s
Pressure, stress	pascal	Pa	N/m^2
Solid angle	steradian	sr	m^2/m^2

internet: An interconnection of networks.

[The] Internet: A worldwide interconnection of individual networks operated by government, industry, academia, and private parties. *Note:* The Internet originally served to interconnect laboratories engaged in government research, and has now been expanded to serve millions of users and a multitude of purposes.

Internet protocol (IP): A DOD standard protocol designed for use in interconnected systems of packet-switched computer communication networks. *Note:* The internet protocol provides for transmitting blocks of data called *datagrams* from sources to destinations, where sources and destinations are hosts identified by fixed-length addresses. The internet protocol also provides for fragmentation and reassembly of long datagrams, if necessary, for transmission through small-packet networks.

Internet protocol (IP) spoofing: 1. The creation of IP packets with counterfeit (spoofed) IP source addresses. **2.** A method of attack used by network intruders to defeat network security measures such as authentication based on IP addresses. *Note 1:* An attack using IP spoofing may lead to unauthorized user access, and possibly root access, on the targeted system. *Note 2:* A packet-filtering-router firewall may not provide adequate protection against IP spoofing attacks. It is possible to route packets through this type of firewall if the router is not configured to filter incoming packets having source addresses on the local domain. *Note 3:* IP spoofing is possible even if no reply packets can reach the attacker. *Note 4:* A method for preventing IP spoofing problems is to install a filtering router that does not allow incoming packets to have a source address different from the local domain. In addition, outgoing packets should not be allowed to contain a source address different from the local domain, in order to prevent an IP spoofing attack from originating from the local network.

internetwork connection: *See* **gateway.**

internetworking: The process of interconnecting two or more individual networks to facilitate communications among their respective nodes. *Note:* The interconnected networks may be different types. Each network is distinct, with its own addresses, internal protocols, access methods, and administration.

interoffice trunk: A single direct transmission channel, *e.g.*, voice-frequency circuit, between central offices.

interoperability: 1. The ability of systems, units, or forces to provide services to and accept services from other systems, units or forces and to use the services so exchanged to enable them to operate effectively together. [JP1] **2.** The condition achieved among communications-electronics systems or items of communications-electronics equipment when information or services can be exchanged directly and satisfactorily between them and/or their users. The degree of interoperability should be defined when referring to specific cases. [JP1] (188)

interoperability standard: A document that establishes engineering and technical requirements that are necessary to be employed in the design of systems, units, or forces and to use the services so exchanged to enable them to operate effectively together.

interoperation: The use of interoperable systems, units, or forces. [JP1]

interposition trunk: 1. A single direct transmission channel, *e.g.*, voice-frequency circuit, between two positions of a large switchboard to facilitate the interconnection of other circuits appearing at the respective switchboard positions. **2.** Within a technical control facility, a single direct transmission circuit, between positions in a testboard or patch bay, which circuit facilitates testing or patching between the respective positions. (188)

interpret: To translate and to execute each source language statement of a computer program before translating and executing the next statement.

interrogation: 1. The transmission of a signal or combination of signals intended to trigger a response. **2.** The process whereby a station or device requests another station or device to identify itself or to give its status. (188)

interrupt: A suspension of a process, such as the execution of a computer program, caused by an event external to that process, and performed in such a way that the process can be resumed. *Synonym* **interruption.**

interrupted continuous wave (ICW): Modulation in which there is on-off keying of a continuous wave. (188)

interrupted isochronous transmission: *Synonym* **isochronous burst transmission.**

interruption: *Synonym* **interrupt.**

inter-satellite service: A radiocommunication service providing links between artificial Earth satellites. [NTIA] [RR]

interswitch trunk: A single direct transmission channel, *e.g.*, voice-frequency circuit, between switching nodes. (188)

intersymbol interference: **1.** In a digital transmission system, distortion of the received signal, which distortion is manifested in the temporal spreading and consequent overlap of individual pulses to the degree that the receiver cannot reliably distinguish between changes of state, *i.e.*, between individual signal elements. *Note 1:* At a certain threshold, intersymbol interference will compromise the integrity of the received data. *Note 2:* Intersymbol interference attributable to the statistical nature of quantum mechanisms sets the fundamental limit to receiver sensitivity. *Note 3:* Intersymbol interference may be measured by eye patterns. **2.** Extraneous energy from the signal in one or more keying intervals that interferes with the reception of the signal in another keying interval. (188) **3.** The disturbance caused by extraneous energy from the signal in one or more keying intervals that interferes with the reception of the signal in another keying interval. (188)

intertoll trunk: A single direct transmission channel, *e.g.*, voice-frequency circuit, between two toll offices.

interworking functions: Mechanisms that mask differences in physical, link, and network technologies by converting (or mapping) states and protocols into consistent network and user services.

intra-LATA: Within the boundaries of a local access and transport area (LATA).

intramodal distortion: In an optical fiber, distortion caused by dispersion, such as material or profile dispersion, of a given propagating mode. (188) [After 2196]

intraoffice trunk: A single direct transmission channel, *e.g.*, voice-frequency circuit, within a given switching center.

intrinsic joint loss: Of nonidentical optical fibers joined by a splice or a mated pair of connectors, the power loss attributable to manufacturing variations, in such parameters as physical dimensions, differences in refractive index (including profile parameter), numerical aperture, and mode field diameter.

intrinsic noise: In a transmission path or device, that noise inherent to the path or device and not contingent upon modulation. (188)

inverse multiplexer: A functional unit capable of accessing and combining two or more low-speed circuits into a virtual broadband circuit, up to and including an aggregate equal to a T1 rate.

inverse-square law: The physical law stating that irradiance, *i.e.*, the power per unit area in the direction of propagation, of a spherical wavefront varies inversely as the square of the distance from the source, assuming there are no losses caused by absorption or scattering. *Note:* For example, the power radiated from a point source, *e.g.*, an omnidirectional isotropic antenna, or from any source at very large distances from the source compared to the size of the source, must spread itself over larger and larger spherical surfaces as the distance from the source increases. Diffuse and incoherent radiation are similarly affected.

inverted position: In frequency-division multiplexing, a position of a translated channel in which an increasing signal frequency in the untranslated channel causes a decreasing signal frequency in the translated channel. (188)

inverter: **1.** In electrical engineering, a device for converting direct current into alternating current.

[JP1] (188) **2.** In computers, a device or circuit that inverts the polarity of a signal or pulse. *Deprecated synonym* **negation circuit.**

Inward Wide-Area Telephone Service (INWATS): *See* **eight-hundred (800) service.**

INWATS: *Acronym for* **Inward Wide-Area Telephone Service.** *See* **eight-hundred (800) service.**

I/O: *Abbreviation for* **input/output.**

IOC: *Abbreviation for* **input-output controller, integrated optical circuit.**

I/O controller: *Synonym* **input-output controller.**

ionosphere: That part of the atmosphere, extending from about 70 to 500 kilometers, in which ions and free electrons exist in sufficient quantities to reflect and/or refract electromagnetic waves. [After JP1]

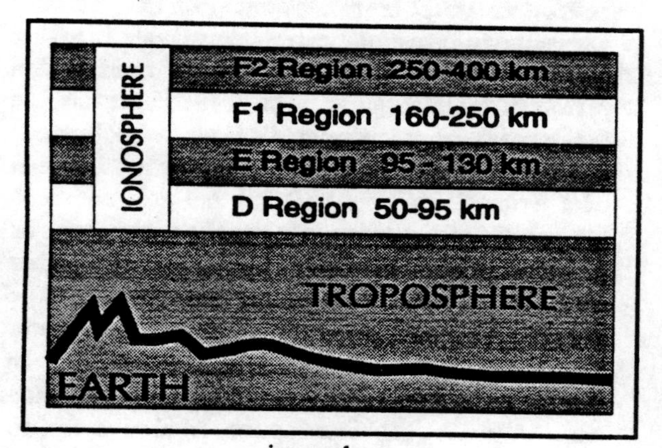

ionosphere

ionosphere sounder: A device that transmits signals for the purpose of determining ionospheric conditions. [NTIA] [RR]

ionospheric absorption: Absorption occurring as a result of interaction between an electromagnetic wave and free electrons in the ionosphere. (188)

ionospheric disturbance: An increase in the ionization of the ionosphere, caused by solar activity, which results in greatly increased radio wave absorption. (188)

ionospheric forward scatter (IFS): *Synonym* **ionospheric scatter.**

ionospheric reflection: Of electromagnetic waves propagating in the ionosphere, a redirection, *i.e.,* bending—by a complex processing involving reflection and refraction—of the waves back toward the Earth. *Note:* The amount of bending depends on the extent of penetration (which is a function of frequency), the angle of incidence, polarization of the wave, and ionospheric conditions, such as the ionization density.

ionospheric scatter: The propagation of radio waves by scattering as a result of irregularities or discontinuities in the ionization of the ionosphere. [NTIA] [RR] (188) *Synonym* **forward propagation ionospheric scatter.**

ionospheric sounding: A technique that provides real-time data on high-frequency ionospheric-dependent radio propagation, using a basic system consisting of a synchronized transmitter and receiver. *Note:* The time delay between transmission and reception is translated into effective ionospheric layer altitude. Vertical incident sounding uses a collocated transmitter and receiver and involves directing a range of frequencies vertically to the ionosphere and measuring the values of the reflected returned signals to determine the effective ionosphere layer altitude. This technique is also used to determine the critical frequency. Oblique sounders use a transmitter at one end of a given propagation path, and a synchronized receiver, usually with an oscilloscope-type display (ionogram), at the other end. The transmitter emits a stepped- or swept-frequency signal which is displayed or measured at the receiver. The measurement converts time delay to effective altitude of the ionospheric layer. The ionogram display shows the effective altitude of the ionospheric layer as a function of frequency.

ionospheric turbulence: Ongoing disturbances of the ionosphere that scatter incident electromagnetic waves. *Note:* Ionospheric turbulence results in irregularities in the composition of the ionosphere that change with time. This causes changes in reflection properties. These, in turn, cause changes in skip distance, fading, local intensification, and distortion of the incident waves. [From Weik '89]

IP: *Abbreviation for* **intelligent peripheral, Internet protocol.**

IPC: *Abbreviation for* **information processing center.**

IPX: *Abbreviation for* **Internetwork Packet Exchange.** A proprietary LAN protocol.

IR: *Abbreviation for* **infrared.**

irradiance: Radiant power incident per unit area upon a surface. *Note:* Irradiance is usually expressed in watts per square meter, but may also be expressed in joules per square meter. (188) *Deprecated synonym* **power density.**

irradiation: The product of irradiance and time, *i.e.*, the time integral of irradiance. *Note:* For example, an irradiation of 100 J/m^2 (joules per square meter) is obtained when an irradiance of 25 W/m^2 (watts per square meter) is continuously incident for 4 seconds.

ISB: *Abbreviation for* **independent sideband.** *See* **independent-sideband transmission.**

ISDN: *Abbreviation for* **integrated services digital network.** An integrated digital network in which the same time-division switches and digital transmission paths are used to establish connections for different services. *Note 1:* ISDN services include telephone, data, electronic mail, and facsimile. *Note 2:* The method used to accomplish a connection is often specified: for example, switched connection, nonswitched connection, exchange connection, ISDN connection.

ISM: *Abbreviation for* **industrial, scientific, and medical applications (of radio frequency energy).**

ISO: *Abbreviation for* **International Organization for Standardization.** An international organization that (a) consists of member bodies that are the national standards bodies of most of the countries of the world, (b) is responsible for the development and publication of international standards in various technical fields, after developing a suitable consensus, (c) is affiliated with the United Nations, and (d) has its headquarters at 1, rue de Varembé, Geneva, Switzerland. *Note:* Member bodies of ISO include, among others, the American National Standards Institute (ANSI), the Association Française de Normalisation (AFNOR), the British Standards Institution (BSI), and the Deutsche Institut für Normung (DIN).

isochrone: A line on a map or chart joining points associated with a constant time difference from the transmitter to receiver of electromagnetic waves, such as radio waves, at all points along the line. [From Weik '89]

isochronous: **1.** Of a periodic signal, pertaining to transmission in which the time interval separating any two corresponding transitions is equal to the unit interval or to a multiple of the unit interval. (188) **2.** Pertaining to data transmission in which corresponding significant instants of two or more sequential signals have a constant phase relationship. *Note:* "Isochronous" and "anisochronous" are characteristics, while "synchronous" and "asynchronous" are relationships.

isochronous burst transmission: In a data network where the information-bearer channel rate is higher than the input data signaling rate, transmission performed by interrupting, at controlled intervals, the data stream being transmitted. *Note 1:* Isochronous burst transmission enables communication between data terminal equipment (DTE) and data networks that operate at dissimilar data signaling rates, such as when the information-bearer channel rate is higher than the DTE output data signaling rate. *Note 2:* The binary digits are transferred at the information-bearer channel rate. The transfer is interrupted at intervals in order to produce the required average data signaling rate. *Note 3:* The interruption is always for an integral number of unit intervals. (188) *Note 4:* Isochronous burst transmission has particular application where envelopes are being transferred between data circuit-terminating equipment (DCE) and only the bytes contained within the envelopes are being transferred between the DCE and the DTE. *Synonyms* **burst isochronous** *(deprecated)*, **interrupted isochronous transmission.**

isochronous demodulation: Demodulation in which the time interval separating any two significant instants is equal to the unit interval or a multiple of the unit interval. (188)

isochronous distortion: The difference between the measured modulation rate and the theoretical modulation rate in a digital system. (188)

isochronous modulation: Modulation in which the time interval separating any two significant instants is equal to the unit interval or a multiple of the unit interval. (188)

isochronous signal: A signal in which the time interval separating any two significant instants is equal to the unit interval or a multiple of the unit interval. (188) *Note 1:* Variations in the time intervals are constrained within specified limits. *Note 2:* "Isochronous" is a characteristic, while "synchronous" indicates a relationship.

isolator: *See* **optical isolator.**

isotropic: 1. Pertaining to a material with properties, such as density, electrical conductivity, electric permittivity, magnetic permeability, or refractive index that do not vary with distance or direction. 2. Pertaining to a material with magnetic, electrical, or electromagnetic properties that do not vary with the direction of static or propagating magnetic, electrical, or electromagnetic fields within the material. (188)

isotropic antenna: A hypothetical antenna that radiates or receives equally in all directions. (188) *Note:* Isotropic antennas do not exist physically but represent convenient reference antennas for expressing directional properties of physical antennas.

isotropic gain: *Synonym* **absolute gain (def. #1).**

iterative impedance: In electrical circuits, for a four-terminal network, the impedance that, if connected across one pair of terminals, will match the impedance across the other pair of terminals. (188) *Note:* The iterative impedance of a uniform line is the same as its characteristic impedance.

ITU: *Abbreviation for* **International Telecommunication Union.**

ITU-R: The Radiocommunications Sector of the ITU; responsible for studying technical issues related to radiocommunications, and having some regulatory powers. *Note:* A predecessor organization was the CCIR.

ITU-T: *Abbreviation for* **International Telecommunication Union—Telecommunication Standardization Bureau.** The Telecommunications Standardization Sector of the International Telecommunication Union (ITU). *Note 1:* ITU-T is responsible for studying technical, operating, and tariff Questions and issuing Recommendations on them, with the goal of standardizing telecommunications worldwide. *Note 2:* The ITU-T combines the standards-setting activities of the predecessor organizations formerly called the International Telegraph and Telephone Consultative Committee (CCITT) and the International Radio Consultative Committee (CCIR).

IVDT: *Abbreviation for* **integrated voice data terminal.** *See* **integrated station.**

IXC: *Abbreviation for* **interexchange carrier.**

(this page intentionally left blank)

jabber: In local area networks, transmission by a data station beyond the time interval allowed by the protocol.

jacket: *Synonym* **sheath.**

jamming margin: The level of interference (jamming) that a system is able to accept and still maintain a specified level of performance, such as maintain a specified bit-error ratio even though the signal-to-noise ratio is decreasing. [From Weik '89]

jamming to signal ratio (J/S): The ratio, usually expressed in dB, of the power of a jamming signal to that of a desired signal at a given point such as the antenna terminals of a receiver.

jam signal: A signal that carries a bit pattern sent by a data station to inform the other stations that they must not transmit. *Note 1:* In carrier-sense multiple access with collision detection (CSMA/CD) networks, the jam signal indicates that a collision has occurred. *Note 2:* In carrier-sense multiple access with collision avoidance (CSMA/CA) networks, the jam signal indicates that the sending station intends to transmit. *Note 3: "Jam signal"* should not be confused with *"electronic jamming."*

jerkiness: In a video display, the perception, by human vision faculties, of originally continuous motion as a sequence of distinct "snapshots." *Note 1:* The perception of continuous motion by human vision faculties is a manifestation of complex functions, *i.e.,* characteristics, of the eyes and brain. When presented with a sequence of fixed, *i.e.,* still, images of sufficient continuity and at a sufficiently frequent update rate, the brain interpolates intermediate images, and the observer subjectively appears to see continuous motion that in reality does not exist. *Note 2:* For example, the update rate of NTSC television displays is 30 frames (60 fields) per second.

jitter: Abrupt and unwanted variations of one or more signal characteristics, such as the interval between successive pulses, the amplitude of successive cycles, or the frequency or phase of successive cycles. (188) *Note 1:* Jitter must be specified in qualitative terms (*e.g.,* amplitude, phase, pulse width or pulse position) and in quantitative terms (*e.g.,* average, RMS, or peak-to-peak). *Note 2:* The low-frequency cutoff for jitter is usually specified at 1 Hz. *Contrast with* **drift, wander.**

job: In computing, a unit of work that is defined by a user and that is to be accomplished by a computer. *Note:* A job is identified by a label and usually includes a set of computer programs, files, and control statements to the computer operating system.

job-recovery control file: *Synonym* **backup file.**

Johnson noise: *Synonym* **thermal noise.**

joint: For optical fibers, a splice or connector.

joint multichannel trunking and switching system: That composite multichannel trunking and switching system formed from assets of the Services, the Defense Communications System, other available systems, and/or assets controlled by the Joint Chiefs of Staff to provide an operationally responsive, survivable communication system, preferably in a mobile/transportable/recoverable configuration, for the joint force commander in an area of operations. [JP1]

Joint Spectrum Center (JSC): A DOD center that consolidates the former Electromagnetic Compatibility Analysis Center (ECAC) and functions of the military Department Spectrum centers.

Joint Tactical Information Distribution System (JTIDS): An advanced information distribution system that provides secure integrated communication, navigation, and identification (ICNI) capability for application to military tactical operations.

Joint Telecommunications Resources Board (JTRB): The body required to be established by Section 2(b) (3) of Executive Order No. 12472 to assist the Director of the Office of Science and Technology Policy in the exercise of assigned nonwartime emergency telecommunications functions.

journal: **1.** A chronological record of data processing operations that may be used to reconstruct a previous or an updated version of a file. *Synonym* **log.** **2.** In database management systems, the record of all stored data items that have values changed as a result of processing and manipulation of the data.

joy stick: In computer graphics, a lever (with at least two degrees of freedom) that is used as an input unit, normally as a locator.

JSC: *Abbreviation for* **Joint Spectrum Center.**

JTIDS: *Acronym for* **Joint Tactical Information Distribution System.**

JTRB: *Abbreviation for* **Joint Telecommunications Resources Board.**

Julian date: **1.** The sequential day count reckoned consecutively beginning January 1, 4713 B.C. *Note:* The Julian date on January 1, 1990, was 2,446,892. **2.** The sequential day count of the days of a year, reckoned consecutively from the first day of January. *Note:* In modern times, the definition of Julian date has been corrupted to use the first day of the year as the point of reference. To avoid ambiguity with the traditional meaning, *"day of year"* rather than *"Julian date"* should be used for this purpose. (188)

jumper: *Synonym* **cross-connection.**

junction point: *Synonym* **node (def. #1).**

justification: *See* **bit stuffing, de-stuffing, justify.**

justify: **1.** To shift the contents of a register or a field so that the significant character at the specified end of the data is at a particular position. **2.** To align text horizontally or vertically so that the first and last characters of every line, or the first and last line of the text, are aligned with their corresponding margins. *Note 1:* In English, text may be justified left, right, or both. Left justification is the most common. *Note 2:* The last line of a paragraph is usually only left justified. **3.** To align data on a designated character position.

k: *Abbreviation for* **kilo** (SI prefix for 10^3). *See* **metric system.**

K: 1. *Abbreviation for* **kelvin(s).** *See* **thermodynamic temperature.** 2. When referring to data storage capacity, 2^{10}, or 1024 in decimal notation; however this usage of an upper case K is deprecated.

Kalman filter: A computational algorithm that processes measurements to deduce an optimum estimate of the past, present, or future state of a linear system by using a time sequence of measurements of the system behavior, plus a statistical model that characterizes the system and measurement errors, plus initial condition information.

KDC: *Abbreviation for* **key distribution center.**

kelvin (K): A unit of thermodynamic temperature, taken as one of the base units of the International System of Units (SI). The kelvin is defined by setting the thermodynamic temperature of the triple point of water at 273.16 K. *Note 1:* The kelvin was formerly called *"degree Kelvin."* The term *"degree Kelvin"* is now obsolete. No degree symbol is written with K, the symbol for kelvin(s). *Note 2:* In measuring temperature intervals, the degree Celsius is equal to the kelvin. The Celsius temperature scale is defined by setting 0 °C equal to 273.16 K.

kelvin temperature scale: *See* **thermodynamic temperature.**

Kendall effect: A spurious pattern or other distortion in a facsimile record copy caused by unwanted modulation products arising from the transmission of a carrier signal, and appearing in the form of a rectified baseband that interferes with the lower sideband of the carrier. (188) *Note:* The Kendall effect occurs principally when the single-sideband width is greater than half of the facsimile carrier frequency.

Kennelly-Heaviside layer: *Synonym* **E region.**

kernel: A module of a program that forms a logical entity or performs a unit function. *Note:* The most vulnerable portion of code in a secure operating system is a special case of a kernel.

Kerr electro-optic effect: The creation of birefringence in a liquid that is not otherwise birefringent, by subjecting the liquid to an electric field. *Note 1:* The degree of birefringence, which is manifested as a difference in refractive indices for light of orthogonal linear polarizations, one of which is parallel to the induced optical axis, is directly proportional to the square of the applied electric field strength. *Note 2:* In the general case, the birefringence produced by the applied electric field can be used in conjunction with polarizers to modulate light. Devices that use this principle are called *Kerr cells.*

key: Information (usually a sequence of random or pseudorandom binary digits) used initially to set up and periodically change the operations performed in crypto-equipment for the purpose of encrypting or decrypting electronic signals, for determining electronic counter-countermeasures patterns (*e.g.*, frequency hopping or spread spectrum), or for producing other key. *Note:* *"Key"* has replaced the terms *"variable," "key(ing) variable,"* and *"crypto-variable."* [NIS]

keyboard: An input device used to enter data by manual depression of keys, which causes the generation of the selected code element. (188)

keyboard punch: *Synonym* **keypunch.**

key distribution center (KDC): A COMSEC facility that generates and distributes key in electrical form. (188)

keying: The generating of signals by the interruption or modulation of a steady signal or carrier. (188) *See* **chroma keying.**

keying variable: *Deprecated synonym for* **key.**

key management: [The] Process by which key is generated, stored, protected, transferred, loaded, used, and destroyed. [NIS]

key pulsing: A system of sending telephone calling signals in which the digits are transmitted by operation of a pushbutton key set. (188) *Note:* The type of key pulsing commonly used by users and PBX operators is dual-tone multifrequency signaling. Each pushbutton causes generation of a unique pair of tones. In military systems, pushbuttons are also provided for additional signals, such as precedence. *Synonym* **pulsing.**

keypunch: A keyboard-actuated punch that punches holes in a data medium. *Synonym* **keyboard punch.**

key set: A multiline or multifunction user terminal device. (188)

key stream: [A] sequence of symbols (or their electrical or mechanical equivalents) produced in a machine or auto-manual cryptosystem to combine with plain text to produce cipher text, control transmission security processes, or produce key. [NIS]

key telephone system (KTS): In a local environment, terminals and equipment that provide immediate access from all terminals to a variety of telephone services without attendant assistance. (188) *Note:* A KTS may interface with the public switched telephone network.

key variable: *Deprecated synonym for* **key.**

k-factor: **1.** In tropospheric radio propagation, the ratio of the effective Earth radius to the actual Earth radius. (188) *Note:* The k-factor is approximately 4/3. **2.** In ionospheric radio propagation, a correction factor that (a) is applied in calculations related to curved layers, and (b) is a function of distance and the real height of ionospheric reflection.

kHz: *Abbreviation for* **kilohertz.**

kilohertz (kHz): A unit of frequency denoting one thousand (10^3) Hz.

kilometer: A unit of distance corresponding to 1000m.

km: *Abbreviation for* **kilometer.**

knife-edge effect: In electromagnetic wave propagation, a redirection by diffraction of a portion of the incident radiation that strikes a well-defined obstacle such as a mountain range or the edge of a building. (188) *Note:* The knife-edge effect is explained by Huygens' principle, which states that a well-defined obstruction to an electromagnetic wave acts as a secondary source, and creates a new wavefront. This new wavefront propagates into the geometric shadow area of the obstacle.

kT: *See* **noise-power density.**

KTS: *Abbreviation for* **key telephone system.**

K-type patch bay: A patching facility designed for patching and monitoring of balanced digital data circuits that support data rates up to 1 Mb/s. (188)

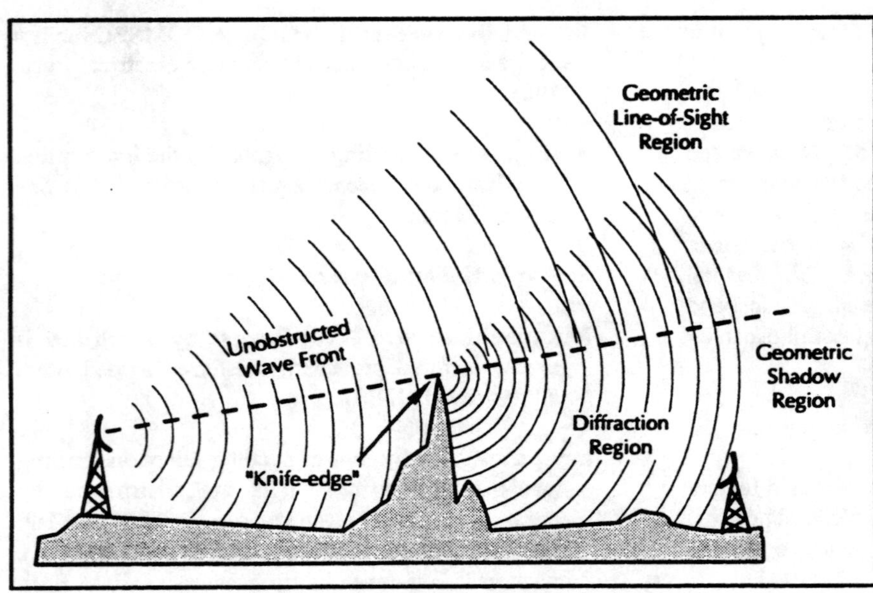

knife-edge effect

label: 1. An identifier within or attached to a set of data elements. 2. One or more characters that (a) are within or attached to a set of data elements and (b) represent information about the set, including its identification. 3. In communications, information within a message that is used to identify specific system parameters, such as the particular circuit with which the message is associated. *Note:* Messages that do not relate to call control should not contain a label. 4. In programming languages, an identifier that names a statement. 5. An identifier that indicates the sensitivity of the attached information. 6. For classified information, an identifier that indicates (a) the security level of the attached information or (b) the specific category in which the attached information belongs.

labeled channel: In integrated services digital networks, (ISDN), a time-ordered set of all block payloads that have labels containing the same information, *i.e.*, containing the same identifiers.

labeled interface structure: In integrated services digital networks (ISDN), an interface structure that provides telecommunications services and signaling by means of labeled channels.

labeled multiplexing: In integrated services digital networks (ISDN), multiplexing by concatenation of the blocks of the channels that have different identifiers in their labels.

labeled statistical channel: In integrated services digital networks (ISDN), a labeled channel in which the block payloads or the duration of each successive block is random.

Lambertian radiator: *See* **Lambert's cosine law.**

Lambertian reflector: *See* **Lambert's cosine law.**

Lambertian source: An optical source that obeys Lambert's cosine law. (188) *Note:* Conventional (surface-emitting) LEDs approximate a Lambertian source. They have a large beam divergence, and a radiation pattern that approximates a sphere. Thus, most of their total optical output is not coupled into communications fibers.

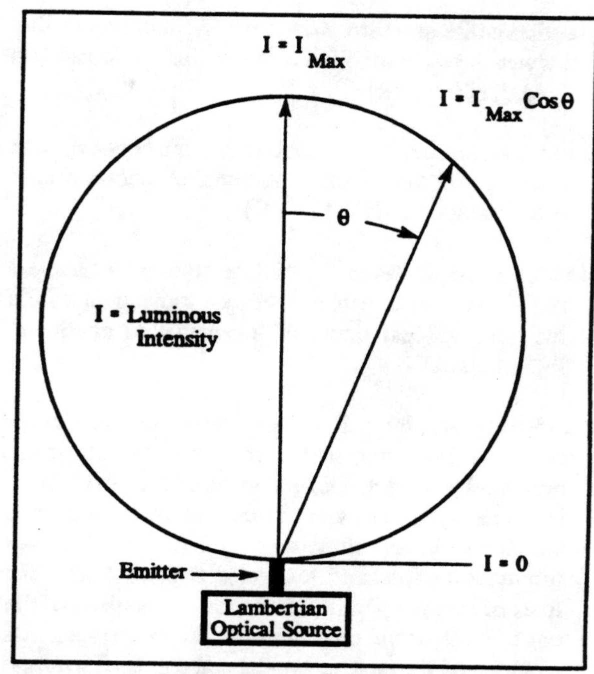

Lambertian optical source

Lambert's cosine law: The mathematical statement that the radiance of certain idealized optical sources is directly proportional to the cosine of the angle—with respect to the direction of maximum radiance—from which the source is viewed. *Note:* Lambert's cosine law may also apply to certain idealized diffuse reflectors. *Synonyms* **cosine emission law, Lambert's emission law.**

Lambert's emission law: *Synonym* **Lambert's cosine law.**

LAN: *Acronym for* **local area network.**

LAN application (software): An application software package specifically designed to operate in a local-area-network environment.

land line: A colloquial name for conventional telephone facilities. *Note:* Land lines include conventional twisted-pair lines, carrier facilities, and microwave radio facilities for supporting a conventional telephone channel, but do not include satellite links or mobile telephone links using radio transmissions.

land mobile-satellite service: A mobile-satellite service in which mobile Earth stations are located on land. [NTIA] [RR]

land mobile service: A mobile service between base stations and land mobile stations, or between land mobile stations. [NTIA] [RR]

land mobile station: A mobile station in the land mobile service capable of surface movement within the geographical limits of a country or continent. [NTIA] [RR]

landscape mode: **1.** In facsimile, the mode for scanning lines across the longer dimension of a rectangular object, *i.e.*, rectangular original. **2.** In computer graphics, the orientation of an image in which the longer dimension is horizontal. **3.** An orientation of printed text on a page such that the lines of text are parallel to the long dimension of the page. *Note:* If the page contains an image, such as a picture, and the page is viewed in the normal manner, the long dimension of the page would be parallel to the line that joins the eyes of the viewer.

land station: A station in the mobile service not intended to be used while in motion. [NTIA] [RR]

language: A set of characters, conventions, and rules that is used for conveying information. (188)

language processor: A program that performs tasks, such as translating and interpreting, required for processing a specified programming language. *Note:* Examples of language processors include a Fortran processor and a COBOL processor.

LAN operating system: *See* **network operating system.**

LAP-B: The Data Link Layer protocol as specified by CCITT Recommendation X.25 (1989).

LAP-D: *Abbreviation for* **link access procedure D.** A link protocol used in ISDN.

laser: *Acronym for* **light amplification by stimulated emission of radiation.** A device that produces a coherent beam of optical radiation by stimulating electronic, ionic, or molecular transitions to higher energy levels so that when they return to lower energy levels they emit energy. *Note 1:* Laser radiation may be either temporally coherent, spatially coherent, or both. *Note 2:* The degree of coherence of laser radiation exceeds 0.88. (188)

laser chirp: An abrupt change of the center wavelength of a laser, caused by laser instability.

laser diode: *Synonym* **injection laser diode.**

laser disk: *See* **optical disk.**

laser intelligence (LASINT): Technical and geolocation intelligence derived from laser systems; a subcategory of electro-optical intelligence. [JP1]

laser medium: *Synonym* **active laser medium.**

lasing: *See* **laser.**

lasing threshold: The lowest excitation level at which laser output is dominated by stimulated emission rather than by spontaneous emission.

LASINT: *Acronym for* **laser intelligence.**

last-in first-out (LIFO): A queuing discipline in which entities in a queue leave in the reverse order of the sequence in which they arrive. *Note:* Service, when available, is offered to the entity that has been in the LIFO queue the shortest time.

LATA: *Acronym for* **local access and transport area.**

lateral offset loss: **1.** In fiber optics, a loss of optical power at a splice or connector, caused by a lateral, *i.e.*, transverse, offset of the mating fiber cores, which offset causes an imperfect transfer of the optical signal from the "transmitting" fiber to the "receiving" fiber. *Note:* The effect of a given amount of lateral offset will depend on other parameters such as the relative diameters of the respective cores. For example: if, because of manufacturing tolerances, the "transmitting" core is smaller than the "receiving" core, the effect will be less than if both cores were the same size. [After FAA] **2.** An analogous loss of optical power caused by lateral misalignment of the fiber and optical source. [FAA] *Synonym* **transverse offset loss.**

launch angle: **1.** The angle, with respect to the normal, at which a light ray emerges from a surface. **2.** The beam divergence at an emitting surface, such as that of a light-emitting diode (LED), laser, lens, prism, or optical fiber end face. **3.** At an end face of an optical fiber, the angle between an input ray and the fiber axis. (188) *Note:* If the end face of the fiber is perpendicular to the fiber axis, the launch angle is equal to the incidence angle when the ray is external to the fiber and the refraction angle when initially inside the fiber.

launching fiber: An optical fiber used in conjunction with a source to excite the modes of another fiber in a particular fashion. *Note:* Launching fibers are most often used in test systems to improve the precision of measurements. *Synonym* **injection fiber.**

launch numerical aperture (LNA): The numerical aperture of an optical system used to couple (launch) power into an optical fiber. (188) *Note 1:* LNA may differ from the stated NA of a final focusing element if, for example, that element is underfilled or the focus is other than that for which the element is specified. *Note 2:* LNA is one of the parameters that determine the initial distribution of power among the modes of an optical fiber.

layer: **1.** In radio wave propagation, *see* **F region. 2.** In telecommunications networks and open systems architecture, a group of related functions that are performed in a given level in a hierarchy of groups of related functions. *Note:* In specifying the functions for a given layer, the assumption is made that the specified functions for the layers below are performed, except for the lowest layer.

layered system: A system in which components are grouped, *i.e.*, layered, in a hierarchical arrangement, such that lower layers provide functions and services that support the functions and services of higher layers. *Note:* Systems of ever-increasing complexity and capability can be built by adding or changing the layers to improve overall system capability while using the components that are still in place.

lay length: In communications cables—including fiber-optic cables—having the transmission media wrapped helically around a central member, the longitudinal distance along the cable required for one complete helical wrap; *i.e.*, the total cable length divided by the total number of wraps. *Note 1:* In many fiber-optic cable designs, the lay length is shorter than in metallic cables of similar diameter, to avoid overstressing the fibers during the pulling associated with the installation operation. *Note 2:* The wraps, *i.e.*, turns, that are referred to should not be confused with the twists given twisted metallic pairs, *i.e.*, wires, to reduce electromagnetic coupling. Pairs of optical fibers are not given such twists. [After FAA] *Synonym* **pitch.**

LBO: *Abbreviation for* **line buildout.** *Synonym* **building out.**

LCD: *Abbreviation for* **liquid crystal display.**

LDM: *Abbreviation for* **limited distance modem.**

leaky bucket counter: A counter that is incremented by unity each time an event occurs and that is periodically decremented by a fixed value. (188)

leaky mode: In an optical fiber, a mode having a field that decays monotonically for a finite distance in the transverse direction but becomes oscillatory everywhere beyond that finite distance. *Note:* Leaky modes correspond to leaky rays in the terminology of geometric optics. Leaky modes experience attenuation, even if the waveguide is perfect in every respect. (188) *Synonym* **tunneling mode.**

leaky ray: In an optical fiber, a ray for which geometric optics would predict total internal reflection at the boundary between the core and the cladding, but which suffers loss by virtue of the curved core boundary. *Note:* Leaky rays correspond to leaky (*i.e.*, tunneling) modes in the terminology of mode descriptors. *Synonym* **tunneling ray.**

leap second: An occasional adjustment of one second, added to, or subtracted from, Coordinated Universal Time (UTC) to bring it into approximate synchronism with UT-1, which is the time scale based on the rotation of the Earth. (188) *Note 1:* Adjustments, when required, are made at the end of June 30, or preferably, December 31, Universal Time, so that UTC never deviates from UT-1 by more than 0.9 second. *Note 2:* The last minute of the day on which an adjustment is made has 61 or 59 seconds.

leased circuit: Dedicated common-carrier facilities and channel equipment used by a network to furnish exclusive private line service to a specific user or group of users.

least-time principle: *Synonym* **Fermat's principle.**

LEC: *Abbreviation for* **local exchange carrier.**

LED: *Abbreviation for* **light-emitting diode.**

left-hand (anti-clockwise) polarized wave: An elliptically or circularly polarized wave, in which the electric field vector, observed in the fixed plane, normal to the direction of propagation, whilst looking in the direction of propagation, rotates with time in a left-hand or anticlockwise direction. [NTIA] [RR]

leg: **1.** A segment of an end-to-end route or path, such as a path from user to user via several networks and nodes within networks. *Note:* Examples of legs are several sequential microwave links between two switching centers and a transoceanic cable between two shore communications facilities, each connected to a node in a national network. **2.** A connection from a specific node to an addressable entity, such as communication link from a computer workstation to a hub.

level: **1.** The absolute or relative voltage, current, or power at a particular point in a circuit or system. (188) **2.** A tier or layer of a hierarchical system, *e.g.*, the Link-Level protocol, high-level computer language. (188)

level alignment: The adjustment of transmission levels of single links and of links in tandem to prevent problems such as overloading of transmission subsystems. (188)

LF: *Abbreviation for* **low frequency.**

LIFO: *Acronym for* **last-in first-out.**

light: In a strict sense, the region of the electromagnetic spectrum that can be perceived by human vision, *i.e.*, the visible spectrum, which is approximately the wavelength range of 0.4 µm to 0.7 µm. (188) *Note 1:* In the laser and optical communications fields, custom and practice have extended usage of the term *light* to include the much broader portion of the electromagnetic spectrum that can be handled by the basic optical techniques used for the visible spectrum. *Note 2:* The region embraced by the term *light* has not been clearly defined, but by convention and usage, is considered to extend from the near-ultraviolet region of approximately 0.3 µm, through the visible region, and into the mid-infrared region to approximately 30 µm.

light-emitting diode (LED): A semiconductor device that emits incoherent optical radiation when biased in the forward direction. (188)

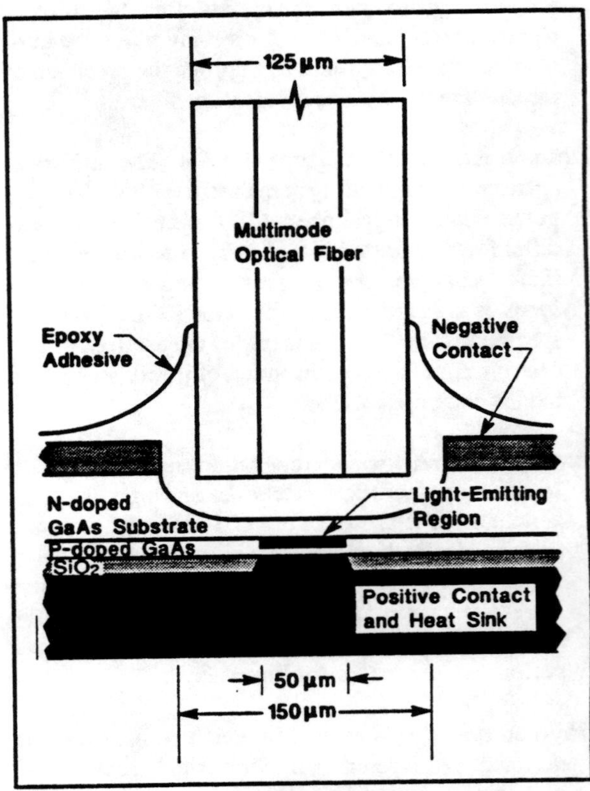

surface-emitting LED

lightguide: *See* **optical fiber.**

lightning down-conductor: In a lightning protection subsystem, the conductor connecting the air terminal or overhead ground wire to the earth electrode subsystem. (188)

lightning protection subsystem: All of the components used to protect a facility from the effects of lightning. (188) *Note:* The lightning protection subsystem includes air terminals, lightning down-conductors, the earth electrode subsystem, air gaps, arresters, and their interconnections.

light pen: A stylus, usually hand-held, that contains a photodetector or light source, and that allows interaction with a computer through a specially designed monitor screen.

light valve: *Synonym* **optical switch.**

limited protection: A form of short-term communications security applied to the electromagnetic or acoustic transmission of unclassified information that warrants a degree of protection against simple analysis and easy exploitation but that does not warrant protection to the extent needed for security of classified information. (188)

limited-protection voice equipment: Equipment that provides limited security for unclassified voice communications. (188)

limiter: A device in which the voltage or some other characteristic of the output signal is automatically prevented from exceeding a specified value. (188)

limiter circuit: A circuit of nonlinear elements that restricts the electrical excursion of a variable in accordance with specified criteria. (188)

limiting: Any process by which a specified characteristic (usually amplitude) of the output of a device is prevented from exceeding a predetermined value. (188) *Note 1: Hard limiting ("clipping")* is a limiting action in which there is (a) over the permitted dynamic range, negligible variation in the expected characteristic of the output signal, and (b) a steady-state signal, at the maximum permitted level, for the duration of each period when the output would otherwise be required to exceed the permitted dynamic range in order to correspond to the transfer function of the device. *Note 2: Soft limiting* is limiting in which the transfer function of a device is a function of its instantaneous or integrated output level. The output waveform is therefore distorted, but not clipped.

limits of interference: In radio transmission, the maximum permissible interference as specified in recommendations of the International Special Committee on Radio Interference or other recognized authority. (188)

line: **1.** A physical medium for transferring electrical or electromagnetic energy from one point to another for purposes of communications. (188) **2.** A land line. **3.** A metallic medium used for the transmission of electrical power. **4.** *See* **scanning line.**

line adapter circuit: *See* **four-wire terminating set.**

linear analog control: *Synonym* **linear analog synchronization.**

linear analog synchronization: Synchronization in which the functional relationships used to obtain synchronization are of simple proportionality. *Synonym* **linear analog control.**

linear combiner: A diversity combiner in which the combining consists of simple addition of two or more signals. (188)

linear device: A device for which the output is, within a given dynamic range, linearly proportional to the input.

linearity: The property of a system in which, if input signals X and Y result in system output $S(X)$ and $S(Y)$ respectively, the input signal $aX + bY$ will result in the output $aS(X) + bS(Y)$, where S is the system transfer function and a and b are scalars.

linearly polarized (LP) mode: A mode for which the field components in the direction of propagation are small compared to components perpendicular to that direction. *Note:* The LP description is an approximation that is valid for a weakly guiding optical fiber, including typical telecommunications grade fibers.

linear network: *See* **network topology.**

linear optimization: *Synonym* **linear programming.**

linear polarization: Of an electromagnetic wave, confinement of the E-field vector or H-field vector to a given plane. *Note:* Historically, the orientation of a polarized electromagnetic wave has been defined in the optical regime by the orientation of the electric vector, and in the radio regime, by the orientation of the magnetic vector. *Synonym* **plane polarization.**

linear predictive coding (LPC): A method of digitally encoding analog signals, which method uses a single-level or multilevel sampling system in which the value of the signal at each sample time is predicted to be a linear function of the past values of the quantized signal. *Note:* LPC is related to adaptive predictive coding (APC) in that both use adaptive predictors. However, LPC uses more prediction coefficients to permit use of a lower information bit rate than APC, and thus requires a more complex processor.

linear programming (LP): In operations research, a procedure for locating the maximum or minimum of a linear function of variables that are subject to linear constraints. *Synonym* **linear optimization.**

linear topology: *See* **network topology.**

line balance: The degree of electrical similarity of the two conductors of a transmission line. (188) *Note:* A high degree of line balance reduces pickup of extraneous disturbances of all kinds, including crosstalk.

line buildout (LBO): *Synonym* **building out.**

line code: A code chosen for use within a communications system for transmission purposes. (188) *Note 1:* A line code may differ from the code generated at a user terminal, and thus may require translation. *Note 2:* A line code may, for example, reflect a requirement of the transmission medium, *e.g.*, optical fiber versus shielded twisted pair.

line driver: An amplifier used to enhance the transmission reliability of a usually digital intrafacility metallic transmission line, over extended distances, by driving the input to the transmission line with a higher than normal signal level. *Note:* An example of a line driver is an amplifier used to extend the range of an RS-232C digital signal beyond 50 feet (~15 m) while maintaining a specified bit-error ratio.

line filter balance: A network designed to maintain phantom group balance when one side of the group is equipped with a carrier system. (188) *Note:* Since it must balance the phantom group for only voice frequencies, the line filter balance configuration is usually simple compared with the filter that it balances.

line hit: *See* **hit.**

line load control: A network-provided service feature that allows selective denial of call origination to certain lines when excessive demands for service are required of a switching center. (188)

line loop: *See* **loop.**

line-of-sight (LOS) propagation: Of an electromagnetic wave, propagation in which the direct ray from the transmitter to the receiver is unobstructed, *i.e.*, the transmission path is not established by or dependent upon reflection or diffraction. *Note:* The need for LOS propagation is most critical at VHF and higher frequencies.

line-route map: A map or overlay for signal communications operations that shows the actual routes and types of construction of wire circuits in the field. It also gives the locations of switchboards and telegraph stations. [JP1]

line side: The portion of a device that is connected to external, *i.e.*, outside plant, facilities such as trunks, local loops, and channels.

line source: **1.** In spectroscopy, an optical source that emits one or more spectrally narrow lines as opposed to a continuous spectrum. **2.** In the geometric sense, an optical source having an emitting area in the form of a spatially narrow line, *e.g.*, a slit. *Synonym* **slit source.**

line spectrum: In optics, an emission or absorption spectrum consisting of one or more narrow spectral lines, as opposed to a continuous spectrum.

line speed: *See* **modulation rate.**

line-to-line correlation: In facsimile, the correlation of object information from scanning line to scanning line. *Note:* Line-to-line correlation is used in two-dimensional encoding.

line traffic coordinator (LTC): In a DDN switching center, the processor that controls traffic on a line. (188)

line verification: *See* **busy verification.**

linewidth: *See* **spectral width.**

link: 1. The communications facilities between adjacent nodes of a network. (188) 2. A portion of a circuit connected in tandem with, *i.e.*, in series with, other portions. 3. A radio path between two points, called a radio link. (188) 4. In communications, a general term used to indicate the existence of communications facilities between two points. [JP1] 5. A conceptual circuit, *i.e.*, logical circuit, between two users of a network, that enables the users to communicate, even when different physical paths are used. *Note 1:* In all cases, the type of link, such as data link, downlink, duplex link, fiber optic link, line-of-sight link, point-to-point link, radio link and satellite link, should be identified. *Note 2:* A link may be simplex, half-duplex, or duplex. 6. In a computer program, a part, such as a single instruction or address, that passes control and parameters between separate portions of the program. 7. In hypertext, the logical connection between discrete units of data.

link encryption: The application of on-line crypto-operation to a link of a communications system so that all information passing over the link is encrypted in its entirety. [JP1]

linking protection (LP): In adaptive high-frequency (HF) radio, protection intended to prevent the establishment of unauthorized links or the unauthorized manipulation of legitimate links, and which are administered through an authorization process. [After FED-STD-1049/1]

Link Layer: *Deprecated term for* **Data Link Layer.** *See* **Open Systems Interconnection—Reference Model.**

link level: In the hierarchical structure of a primary or secondary station, the conceptual level of control or data processing logic that controls the data link. *Note:* Link-level functions provide an interface between the station high-level logic and the data link. Link-level functions include (a) transmit bit injection and receive bit extraction, (b) address and control field interpretation, (c) command response generation, transmission and interpretation, and (d) frame check sequence computation and interpretation.

link orderwire: A voice or data communications circuit that (a) serves as a transmission link between adjacent communications facilities that are interconnected by a transmission link and (b) is used only for coordination and control of link activities, such as traffic monitoring and traffic control. (188)

link protocol: A set of rules relating to data communications over a data link. *Note:* Link protocols define data link parameters, such as transmission code, transmission mode, control procedures, and recovery procedures.

link quality analysis (LQA): In adaptive high-frequency (HF) radio, the overall process by which measurements of signal quality are made, assessed, and analyzed. *Note 1:* In LQA, signal quality is determined by measuring, assessing, and analyzing link parameters, such as bit error ratio (BER), and the levels of the ratio of signal-plus-noise-plus-distortion to noise-plus-distortion (SINAD). Measurements are stored at—and exchanged between—stations, for use in making decisions about link establishment. *Note 2:* For adaptive HF radio, LQA is automatically performed and is usually based on analyses of pseudo-BERs and SINAD readings.

lip synchronization: The synchronization of audio and corresponding video signals so that there is no noticeable lack of simultaneity between them. (188) *Note:* An example of a lip synchronization problem is the case in which television video and audio signals are transported via different facilities (*e.g.*, a geosynchronous satellite link and a landline) that have significantly differently delay times, respectively. In such cases it is necessary to delay the audio electronically to allow for the difference in propagation times.

liquid crystal display (LCD): A display device that creates characters by means of the action of electrical signals on a matrix of liquid cells that become opaque when energized. *Note:* A liquid crystal display may be designed to be viewed by reflected or transmitted light.

LLC: *Abbreviation for* **logical link control.** *See* **logical link control sublayer.**

LNA: *Abbreviation for* **launch numerical aperture.**

load: **1.** The power consumed by a device or circuit in performing its function. (188) **2.** A power-consuming device connected to a circuit. (188) **3.** To enter data or programs into storage or working registers. (188) **4.** To insert data values into a database that previously contained no occurrences of data. **5.** To place a magnetic tape reel on a tape drive, or to place cards into the card hopper of a card punch or reader. (188)

load capacity: In pulse-code modulation (PCM), the level of a sinusoidal signal that has positive and negative peaks that coincide with the positive and negative virtual decision values of the encoder. *Note:* Load capacity is usually expressed in dBm0. *Synonym* **overload point.**

loader: A routine that reads data into main storage.

load factor: The ratio of the average load over a designated period of time to the peak load occurring during that period. (188)

loading: **1.** The insertion of impedance into a circuit to change the characteristics of the circuit. (188) **2.** In multichannel communications systems, the insertion of white noise or equivalent dummy traffic at a specified level to simulate system traffic and thus enable analysis of system performance. (188) **3.** In telephone systems, the load, *i.e.*, power level, imposed by the busy hour traffic. *Note 1:* The loading may be expressed as (a) the equivalent mean power and the peak power as a function of the number of voice channels or (b) the equivalent power of a multichannel complex or signal composite referred to zero transmission level point (0TLP). *Note 2:* Loading is a function of the number of channels and the specified voice channel mean power. (188)

loading characteristic: In multichannel telephone systems, a plot, for the busy hour, of the equivalent mean power and the peak power as a function of the number of voice channels. (188) *Note:* The equivalent power of a multichannel signal referred to the zero transmission level point is a function of the number of channels and has for its basis a specified voice channel mean power.

loading coil: A coil that does not provide coupling to any other circuit, but is inserted in a circuit to increase its inductance. (188) *Note 1:* Loading coils inserted periodically in a pair of wires reduce the attenuation at the higher voice frequencies up to the cutoff frequency of the low-pass filter formed by (a) the inductance of the coils and distributed inductance of the wires, and (b) the distributed capacitance between the wires. Above the cutoff frequency, attenuation increases rapidly. *Note 2:* A common application of loading coils is to improve the voice-frequency amplitude response characteristics of twisted cable pairs. When connected across a twisted pair at regular intervals, loading coils, in concert with the distributed resistance and capacitance of the pair, form an audio-frequency filter that improves the high-frequency audio response of the pair. *Note 3:* When loading coils are in place, signal attenuation increases rapidly for frequencies above the audio cutoff frequency. Thus, when a pair is used to support applications that require higher frequencies, such as carrier systems, loading coils must be absent.

lobe: **1.** An identifiable segment of an antenna radiation pattern. *Note:* A lobe is characterized by a localized maximum bounded by identifiable nulls. **2.** A pair of channels between a data station and a lobe attaching unit, one channel for sending and one for receiving, as seen from the point of view of the attached data station.

lobe attaching unit: In a ring network, a functional unit used to connect and disconnect data stations to and from the ring without disrupting network operations.

local access and transport area (LATA): Under the terms of the Modification of Final Judgment (MFJ), a geographical area within which a divested Bell Operating Company (BOC) is permitted to offer exchange telecommunications and exchange access services. *Note:* Under the terms of the MFJ, the BOCs are generally prohibited from providing services that originate in one LATA and terminate in another.

local area network (LAN): A data communications system that (a) lies within a limited spatial area, (b) has a specific user group, (c) has a specific topology, and (d) is not a public switched telecommunications network, but may be connected to one. (188) *Note 1:* LANs are usually restricted to relatively small areas, such as rooms, buildings, ships, and aircraft. *Note 2:* An interconnection of LANs within a limited geographical area, such as a military base, is commonly referred to as a campus area network. An interconnection of LANs over a city-wide geographical area is commonly called a metropolitan area network (MAN). An interconnection of LANs over large geographical areas, such as nationwide, is commonly called a wide area network (WAN). *Note 3:* LANs are not subject to public telecommunications regulations.

local battery: **1.** In telegraphy, the source of power that actuates the telegraphic station recording instruments, as distinguished from the source of power that furnishes current to the line. (188) **2.** In telephony, a system in which each telephone instrument has its own source of power, as opposed to being powered from the central office. (188) **3.** A source of local power for a telephone instrument.

local call: **1.** Any call using a single switching facility. (188) **2.** Any call for which an additional charge, *i.e.*, toll charge, is not made to the calling or called party. *Note:* Calls such as those via "800" numbers do not qualify as local calls, because the called party is charged.

local central office: *Synonym* **central office.**

local clock: A source of timing located in close proximity to an associated facility, such as a communications station, central office, or node. *Note:* The same clock might be a remote clock relative to some other facility.

local exchange: *Synonym* **central office.**

local exchange carrier (LEC): A local telephone company, *i.e.*, a communications common carrier that provides ordinary local voice-grade telecommunications service under regulation within a specified service area.

local exchange loop: An interconnection between customer premises equipment and telephone central office.

local line: *See* **loop.**

local loop: *Synonym* **loop.**

local measured service: *See* **measured-rate service.**

local office: *Synonym* **central office.**

local orderwire: A communications circuit between a technical control facility and selected terminal or repeater locations within the communications complex. (188) *Note:* In multichannel radio systems, the local orderwire is usually a handset connection at the radio location.

local side: The portion of a device that is connected to internal facilities, such as switches, patch panels, test bays and supervisory equipment. (188)

lock-in frequency: A frequency at which a closed-loop system can acquire and track a signal. *See* **lock-in range.**

lock-in range: **1.** The range of frequencies within which a closed-loop system can aquire and track a signal. (188) **2.** The dynamic range within which a closed-loop system can aquire and track a signal.

lockout: **1.** In telephone systems, treatment of a user's line or trunk that is in trouble, or in a permanent off-hook condition, by automatically disconnecting the line from the switching equipment. (188) **2.** In public telephone systems, a process that denies an attendant or other users the ability to reenter an established connection. **3.** In a telephone circuit controlled by two voice-operated devices, the inability of one or both users to get through, either because of excessive local circuit noise or because of continuous speech from either or both users. (188) **4.** In mobile communications, an arrangement of control circuits whereby only one receiver can feed the system at a time. (188) *Synonym* **receiver lockout system.**

5. An arrangement for restricting access to use of all, or part of, a computer system. *Synonym* **protection.**

log: *Synonym* **journal.**

logical circuit: *Synonym* **virtual circuit.**

logical link control (LLC) sublayer: In a local-area-network/metropolitan-area-network (LAN/MAN) system, the part of the link level that (a) supports medium-independent data link functions and (b) uses the services of the medium access control sublayer to provide services to the network layer.

logical route: *Synonym* **virtual circuit.**

logical signaling channel: A logical channel that provides a signaling path within an information channel or within a physical signaling channel.

logical topology: Of a network, the schematic configuration that reflects the network's function, use, or implementation without regard to the physical interconnection of network elements.

log in: To perform a login procedure. *Synonym* **log on.**

login: The procedure that is followed by a user in beginning a session, *e.g.*, a period of terminal operation. *Synonym* **logon.**

log off: To perform a log-off procedure. *Synonym* **log out.**

log-off: The procedure that is followed by a user in closing a session, *e.g.*, a period of terminal operation. *Synonym* **log-out.**

logon: *Synonym* **login.**

log on: *Synonym* **log in.**

log out: *Synonym* **log off.**

log-periodic (LP) antenna: A broadband, multielement, unidirectional, narrow-beam antenna that has impedance and radiation characteristics that are regularly repetitive as a logarithmic function of the excitation frequency. *Note:* The length and spacing of the elements of a log-periodic antenna increase logarithmically from one end to the other. *Synonym* **log-periodic array.** (188)

log-periodic (LP) array: *Synonym* **log-periodic antenna.**

long-distance call: Any telephone call to a destination outside the local service area of the calling station, whether inter-LATA or intra-LATA, and for which there is a charge beyond that for basic service. *Synonym* **toll call.**

long-haul communications: **1.** In public switched networks, pertaining to circuits that span large distances, such as the circuits in inter-LATA, interstate, and international communications. **2.** In the military community, communications among users on a national or worldwide basis. *Note 1:* Compared to tactical communications, long-haul communications are characterized by (a) higher levels of users, such as the National Command Authority, (b) more stringent performance requirements, such as higher quality circuits, (c) longer distances between users, including world wide distances, (d) higher traffic volumes and densities, (e) larger switches and trunk cross sections, and (f) fixed and recoverable assets. *Note 2: "Long-haul communications"* usually pertains to the U.S. Defense Communications System. (188)

longitudinal balance: **1.** The electrical symmetry, with respect to ground, of the two wires of a pair (188) **2.** An expression of the difference in impedance of the two sides of a circuit.

longitudinal offset loss: *Synonym* **gap loss.**

longitudinal redundancy check (LRC): A system of error control based on the formation of a block check following preset rules. *Note 1:* The block check formation rules are applied in the same manner to each character. (188) *Note 2:* A combination of longitudinal and vertical redundancy check allows the detection and correction of single bit errors. *Synonym* **horizontal redundancy check.**

longitudinal voltage: A voltage induced or appearing along the length of a transmission medium. *Note 1:* Longitudinal voltage may be effectively eliminated by using differential amplifiers or receivers that respond only to voltage differences, *e.g.*, those between the wires that constitute a pair. *Note 2:* Induced longitudinal voltages at low (power-line) frequencies can be greatly reduced by twisting parallel wires to create what are referred to as *"twisted wire pairs."* *Synonym* **common-mode voltage.**

long line: A transmission line in a long-distance communications network. (188) *Note:* Examples of long lines are TDM and FDM carrier systems, microwave radio links, geosynchronous satellite links, underground cables, aerial cables and open wire, and submarine cables.

long-range aid to navigation (loran) system: *See* **loran.**

long-range radio aid to navigation system: *See* **loran.**

long-term stability: Of an oscillator, the degree of uniformity of frequency over time, when the frequency is measured under identical environmental conditions, such as supply voltage, load, and temperature. *Note:* Long-term frequency changes are caused by changes in the oscillator elements that determine frequency, such as crystal drift, inductance changes, and capacitance changes. (188)

long wavelength: In fiber optic communications, pertaining to optical wavelengths greater than ≈1 µm.

look-ahead-for-busy (LFB) information: Information concerning network resources available to support higher precedence calls. *Note 1:* Available resources include idle circuits and circuits used for lower precedence calls. *Note 2:* LFB information may be used to make call-path reservations.

loop: **1.** A communications channel from a switching center or an individual message distribution point to the user terminal. (188) *Synonym* **subscriber line. 2.** In telephone systems, a pair of wires from a central office to a subscriber's telephone. (188) *Synonyms* **local loop, user line. 3.** Go-and-return conductors of an electric circuit; a closed circuit. **4.** A closed path under measurement in a resistance test. **5.** A type of antenna, in the form of a circle or rectangle, usually used in direction-finding equipment and in UHF reception. (188) **6.** A sequence of instructions that may be executed iteratively while a certain condition prevails until the loop has been executed once.

loop-back: **1.** A method of performing transmission tests of access lines from the serving switching center, which method usually does not require the assistance of personnel at the served terminal. (188) **2.** A method of testing between stations (not necessarily adjacent) wherein two lines are used, with the testing being done at one station and the two lines interconnected at the distant station. (188) **3.** A patch, applied manually or automatically, remotely or locally, that facilitates a loop-back test.

loop check: *Synonym* **echo check.**

looped dual bus: A distributed-queue dual-bus (DQDB) scheme in which the head-of-bus functions for both buses are at the same location.

loop filter: In a phase-locked loop, a filter located between the phase detector (or time discriminator) and the voltage controlled oscillator (or phase shifter).

loop gain: **1.** The total usable power gain of a carrier terminal or two-wire repeater. *Note:* The maximum usable gain is determined by, and may not exceed, the losses in the closed path. **2.** The sum of the gains, expressed in dB, acting on a signal passing around a closed path, *i.e.*, a loop. (188)

loop noise: The noise contributed by one or both loops of a telephone circuit to the total circuit noise. (188) *Note:* In a given case, it should be stated whether the loop noise is for one or both loops.

loop start: A supervisory signal given by a telephone or PBX in response to completing the loop path.

loop test: A test that uses a closed circuit, *i.e.*, loop, to detect and locate faults. (188)

loop transmission: Multipoint transmission in which (a) all the stations in a network are serially connected in one closed loop, (b) there are no cross-connections, (c) the stations serve as regenerative repeaters, forwarding messages around the loop until they arrive at their destination stations, and (d) any station can introduce a message into the loop by interleaving it with other messages. [From Weik '89]

loose buffer: *See* **buffer.**

loran: *Acronym for* **long-range radio navigation.** A long-range radio navigation position-fixing system consisting of an array of fixed stations that transmit precisely synchronized signals to mobile receivers. *Note:* A loran receiver measures differences in the times of arrival of the signals from the various stations. A fixed difference in the time of arrival of the signals from any two stations will define a hyperbolic arc on which the receiver must lie. Three or more stations are needed to remove ambiguities in the position of the receiver. *Synonyms* **long-range aid to navigation system, long-range radio aid to navigation system.**

loran station: A long distance radionavigation land station transmitting synchronized pulses. Hyperbolic lines of position are determined by the measurement of the difference in the time of arrival of these pulses. [NTIA]

LOS: *Abbreviation for* **line of sight.** *See* **line-of-sight propagation.**

loss: **1.** The diminution, usually expressed in dB, of signal level in a communications medium. (188) **2.** The power, usually expressed in watts, consumed by a circuit or component. (188) **3.** The energy dissipated without accomplishing useful work or purpose.

lossy medium: A medium in which a significant amount of the energy of a propagating electromagnetic wave is absorbed per unit distance traveled by the wave. [After 2196]

lost block: A block not delivered to the user within a specified maximum end-to-end block transfer time.

lost call: A call that has not been completed for any reason other than cases where the call receiver (termination) is busy. (188)

lower frequency ground: *Deprecated term. See* **facility grounding system.**

lowest usable high frequency (LUF): The lowest frequency in the HF band at which the received field intensity is sufficient to provide the required signal-to-noise ratio for a specified time period, *e.g.*, 0100 to 0200 UTC, on 90% of the undisturbed days of the month. (188)

low frequency (LF): Any frequency in the band from 30 kHz to 300 kHz. (188)

low-level keying: *Synonym* **low-level signaling.**

low-level language: *Synonym* **computer-oriented language.**

low-level modulation: Modulation of a signal, *e.g.*, a carrier, at a point in a system or device, such as a radio transmitter, where the power level is low compared to the final output power. (188)

low-level signaling: The use on signal lines of voltage levels that are between the limits of positive and negative 6 volts. (188) *Synonym* **low-level keying.**

low-pass filter: A filter network that passes all frequencies below a specified frequency with little or no loss, but strongly attenuates higher frequencies. (188)

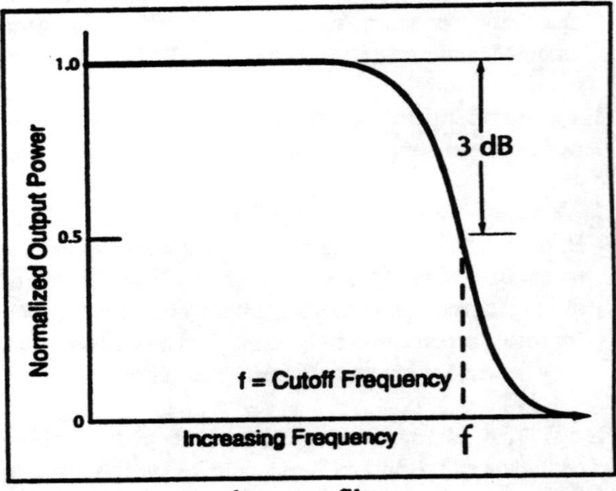

low pass filter

low-performance equipment: **1.** Equipment that has imprecise characteristics that do not meet system reliability requirements. **2.** In military communications, equipment that has insufficiently exacting characteristics to permit its use in trunks or links. (188) *Note:* Low-performance equipment may be used in loops if it meets loop performance requirements. **3.** Tactical ground and airborne equipment that (a) has size, weight, and complexity characteristics that must be kept to a minimum and (b) is used in systems that have components with similar minimum performance characteristics. (188)

low-power communication device: A restricted radiation device, exclusive of those employing conducted or guided radio frequency techniques, used for the transmission of signs, signals (including control signals), writing, images and sounds or intelligence of any nature by radiation of electromagnetic energy. Examples: Wireless microphone, phonograph oscillator, radio-controlled garage door opener, and radio-controlled models. [NTIA]

LP: *Abbreviation for* **linear programming, linking protection.**

L-pad: A pad composed of two discrete components, one series component and one shunt component. *Note:* In schematic representation, the components resemble the upper-case letter "L,"hence the name. (188)

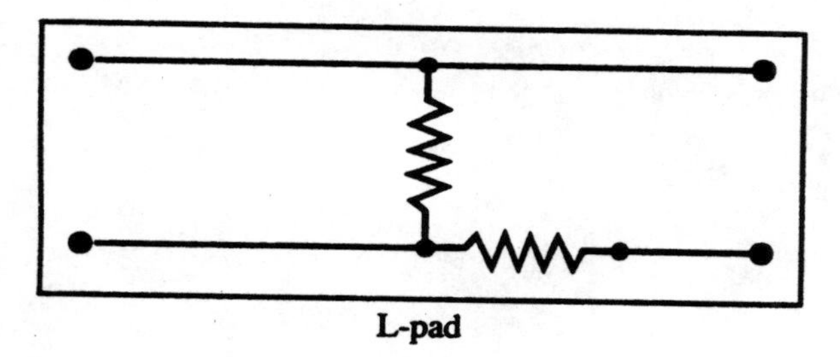

L-pad

LPC: *Abbreviation for* **linear predictive coding.**

LP mode: *Abbreviation for* **linearly polarized mode.**

LP_{01} mode: Designation of the fundamental LP mode. *See* **fundamental mode.**

LQA: *Abbreviation for* **link quality analysis.**

LRC: *Abbreviation for* **longitudinal redundancy check.**

LSB: *Abbreviation for* **least significant bit, lower sideband.**

LTC: *Abbreviation for* **line traffic coordinator.**

LUF: *Acronym for* **lowest usable high frequency.**

luminescent diode: *See* **superluminescent LED.**

lynx: A World Wide Web browser that provides a character-based user interface to hypertext-based information. *Note:* Lynx can display only character-based portions of the hypertext-based information.

(this page intentionally left blank)

MAC: *Abbreviation for* **medium access control.** *See* **medium access control sublayer.**

machine-independent: In telecommunications, computer, and data processing systems, pertaining to operations, procedures, computer programs, and processing that do not depend upon specific hardware for their successful execution. [From Weik '89]

machine instruction: An instruction that is written in a machine language and can be executed directly by the processor for which it was designed without translation or interpretation.

machine language: A language that need not be modified, translated, or interpreted before it can be used by the processor for which it was designed. (188) *Note 1:* The operation codes and addresses used in instructions written in machine language can be directly sensed by the arithmetic and control unit circuits of the processor for which the language is designed. *Note 2:* Instructions written in an assembly language or a high-level language must be translated into machine language before they can be executed by a processor. *Note 3:* Machine languages are usually used by computer designers rather than computer users.

machine learning: The ability of a device to improve its performance based on its past performance.

machine-oriented language: *Synonym* **computer-oriented language.**

machine-readable medium: A medium capable of storing data in a form that can be accessed by an automated sensing device. *Note:* Examples of machine-readable media include (a) magnetic disks, cards, tapes, and drums, (b) punched cards and paper tapes, (c) optical disks, and (d) magnetic ink characters. *Synonym* **automated data medium.**

machine word: *Synonym* **computer word.**

macrobend: A relatively large-radius bend in an optical fiber, such as might be found in a splice organizer tray or a fiber-optic cable that has been bent. *Note:* A macrobend will result in no significant radiation loss if it is of sufficiently large radius. The definition of "sufficiently large" depends on the type of fiber. Single-mode fibers have a low numerical aperture, typically less than 0.15, and are therefore are more susceptible to bend losses than other types. Normally, they will not tolerate a minimum bend radius of less than 6.5 to 7.5 cm (2.5 to 3 inches). Certain specialized types of single-mode fibers, however, can tolerate a far shorter minimum bend radius without appreciable loss. A graded-index multimode fiber having a core diameter of 50 µm and a numerical aperture of 0.20 will typically tolerate a minimum bend radius of not less than 3.8 cm (1.5 inches). The fibers commonly used in customer-premises applications (62.5-µm core) typically have a relatively high numerical aperture, (approximately 0.27), and can tolerate a bend radius of less than an inch (2.5 cm). [After FAA]

macrobend loss: In an optical fiber, that loss attributable to macrobending. *Synonym* **curvature loss.**

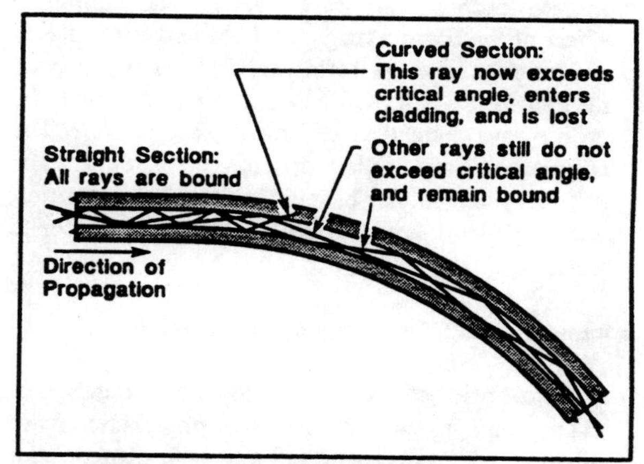

macrobend loss

magnetic card: A card with a magnetizable surface on which data can be stored and retrieved. (188)

magnetic circuit: **1.** The complete closed path taken by magnetic flux. **2.** A region of ferromagnetic material, such as the core of a transformer or solenoid, that contains essentially all of the magnetic flux.

magnetic core storage: In computer technology, a storage device that uses ferromagnetic materials such as iron, iron oxide, or ferrite and in such shapes as wires, toroids, and rods.

magnetic disk: *See* **diskette, hard disk.**

magnetic drum: A right circular cylinder with a magnetizable surface on which digital data can be stored and retrieved.

magnetic storm: A perturbation of the Earth's magnetic field, caused by solar disturbances, usually lasting for a brief period (several days) and characterized by large deviations from the usual value of at least one component of the field. *Note:* Magnetic storms can affect radio propagation because they disturb the ionosphere.

magnetic tape: **1.** A tape with a magnetizable surface on which data can be stored and retrieved. **2.** A tape or ribbon of any material impregnated or coated with magnetic or other material on which information may be placed in the form of magnetically polarized spots. [JP1]

magneto-ionic double refraction: The combined effect of the Earth's magnetic field and atmospheric ionization, whereby a linearly polarized wave entering the ionosphere is split into two components called the ordinary wave and the extraordinary wave. *Note:* The component waves follow different paths, experience different attenuations, have different phase velocities, and, in general, are elliptically polarized in opposite senses.

magneto-optic: *See* **magneto-optic effect.**

magneto-optic effect: Any one of a number of phenomena in which an electromagnetic wave interacts with a magnetic field, or with matter under the influence of a magnetic field. (188) *Note:* The most important magneto-optic effect having application to optical communication is the Faraday effect, in which the plane of polarization is rotated under the influence of a magnetic field parallel to the direction of propagation. This effect may be used to modulate a lightwave.

mailbox-type facility: A facility in which a message from an originating user is stored until the destination user requests delivery of that message.

main beam: *Synonym* **main lobe.**

main distribution frame (MDF): A distribution frame on one part of which the external trunk cables entering a facility terminate, and on another part of which the internal user subscriber lines and trunk cabling to any intermediate distribution frames terminate. *Note 1:* The MDF is used to cross-connect any outside line with any desired terminal of the multiple cabling or any other outside line. (188) *Note 2:* The MDF usually holds central office protective devices and functions as a test point between a line and the office. *Note 3:* The MDF in a private exchange performs functions similar to those performed by the MDF in a central office. *Synonym (in telephony)* **main frame.**

main frame: *Synonym (in telephony)* **main distribution frame.**

mainframe: A large computer, usually one to which other computers and/or terminals are connected to share its resources and computing power.

main lobe: Of an antenna radiation pattern, the lobe containing the maximum power (exhibiting the greatest field strength). (188) *Note:* The horizontal radiation pattern, *i.e.*, that which is plotted as a function of azimuth about the antenna, is usually specified. The width of the main lobe is usually specified as the angle encompassed between the points where the power has fallen 3 dB below the maximum value. The vertical radiation pattern, *i.e.*, that which is plotted as a function of elevation from a specified azimuth, is also of interest and may be similarly specified. (188) *Synonym* **main beam.**

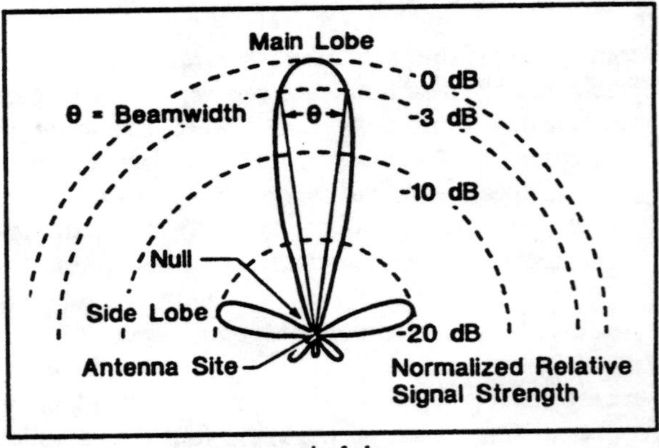

main lobe

main station: A user instrument, *e.g.*, telephone set or terminal, with a distinct call number designation, connected to a local loop, used for originating calls, and on which incoming calls from the exchange are answered.

main storage: In a computer, program-addressable storage from which instructions and other data may be loaded directly into registers for subsequent execution or processing. *Note 1:* Main storage includes the total program-addressable execution space that may include one or more storage devices. *Note 2: "Main storage"* usually refers to large and intermediate computers, whereas *"memory"* usually refers to microcomputers, minicomputers, and calculators.

maintainability: **1.** A characteristic of design and installation, expressed as the probability that an item will be retained in or restored to a specified condition within a given period of time, when the maintenance is performed in accordance with prescribed procedures and resources. (188) **2.** The ease with which maintenance of a functional unit can be performed in accordance with prescribed requirements.

maintenance: **1.** Any activity, such as tests, measurements, replacements, adjustments and repairs, intended to restore or retain a functional unit in a specified state in which the unit can perform its required functions. (188) **2.** [For materiel], All action taken to retain materiel in a serviceable condition or to restore it to serviceability. It includes inspection, testing, servicing, classification as to serviceability, repair, rebuilding, and reclamation. [JP1] **3.** [For materiel], All supply and repair action taken to keep a force in condition to carry out its mission. [JP1] **4.** [For materiel], The routine recurring work required to keep a facility (plant, building, structure, ground facility, utility system, or other real property) in such condition that it may be continuously used, at its original or designed capacity and efficiency for its intended purpose. [JP1]

maintenance control circuit (MCC): In a communications link, a circuit used by maintenance personnel for coordination. *Note:* An MCC is not available to operations or technical control personnel. (188)

major lobe: *See* **main lobe.**

MAN: *Acronym for* **metropolitan area network.**

managed object: **1.** In a network, an abstract representation of network resources that are managed. (188) *Note:* A managed object may represent a physical entity, a network service, or an abstraction of a resource that exists independently of its use in management. **2.** In telecommunications management, a resource within the telecommunications environment that may be managed through the use of operation, administration, maintenance, and provisioning application protocols. (188)

management information system (MIS): An organized assembly of resources and procedures required to collect, process, and distribute data for use in decision making. (188)

Manchester code: A code in which (a) data and clock signals are combined to form a single self-synchronizing data stream, (b) each encoded bit contains a transition at the midpoint of a bit period, (c) the direction of transition determines whether the bit is a "0" or a "1," and (d) the first half is the true bit value and the second half is the complement of the true bit value. (188) *Contrast with* **non-return-to-zero.**

mandrel wrapping: In multimode fiber optics, a technique used to modify the modal distribution of a propagating optical signal. *Note:* A cylindrical rod wrap consists of a specified number turns of fiber on a mandrel of specified size, depending on the fiber characteristics and the desired modal distribution. It has application in optical transmission performance tests, to simulate, *i.e.*, establish, equilibrium mode distribution in a launch fiber (a fiber used to inject a test signal in another fiber that is under test). If the launch fiber is fully filled ahead of the mandrel wrap, the higher-order modes will be stripped off, leaving only lower-order modes. If the launch fiber is underfilled, *e.g.*, as a consequence of being energized by a laser diode or edge-emitting LED, there will be a redistribution to higher-order modes until modal equilibrium is reached.

man-machine system: A system in which the functions of a human operator and a machine are integrated.

margin: **1.** In communications systems, the maximum degree of signal distortion that can be tolerated without affecting the restitution, *i.e.*, without its being interpreted incorrectly by the decision circuit. **2.** The allowable error rate, deviation from normal, or degradation of the performance of, a system or device.

marine broadcast station: A coast station which makes scheduled broadcasts of time, meteorological, and hydrographic information. [NTIA]

marine utility station: A station in the maritime mobile service consisting of one or more hand-held radiotelephone units licensed under a single authorization. Each unit is capable of operation while being hand-carried by an individual.

maritime air communications: Communications systems, procedures, operations, and equipment that are used for message traffic between aircraft stations and ship stations in the maritime service. *Note:* Commercial, private, naval, and other ships are included in maritime air communications.

maritime broadcast communications net: A communications net that is used for international distress calling, including international lifeboat, lifecraft, and survival-craft high-frequency (HF); aeronautical emergency very high-frequency (VHF); survival ultra high-frequency (UHF); international calling and safety very high-frequency (VHF); combined scene-of-search-and-rescue; and other similar and related purposes. *Note:* Basic international distress calling is performed at either medium frequency (MF) or at high frequency (HF).

maritime mobile-satellite service: A mobile-satellite service in which mobile Earth stations are located on board ships; survival craft stations and emergency position-indicating radiobeacon stations may also participate in this service. [NTIA] [RR]

maritime mobile service: A mobile service between coast stations and ship stations, or between ship stations, or between associated on-board communication stations; survival craft stations and emergency position-indicating radiobeacon stations may also participate in this service. [NTIA] [RR]

maritime radionavigation-satellite service: A radionavigation-satellite service in which Earth stations are located on board ships. [NTIA] [RR]

maritime radionavigation service: A radionavigation service intended for the benefit and for the safe operation of ships. [NTIA] [RR]

mark: **1.** In telegraphy, one of the two significant conditions of encoding. (188) *Note 1:* The complementary significant condition is called a *"space." Note 2:* In modern digital communications, the two corresponding significant conditions of encoding are called "1" and "0." *Synonyms* **marking pulse, marking signal. 2.** A symbol or symbols that indicate the beginning or the end of a field, of a word, or of a data item in a file, record, or block.

marker beacon: A transmitter in the aeronautical radionavigation service which radiates vertically a distinctive pattern for providing position information to aircraft. [NTIA] [RR]

marking bias: The uniform lengthening of all marking signal pulse widths at the expense of the pulse widths of all spacing pulses. (188)

marking end distortion: *See* **end distortion.**

marking pulse: *Synonym* **mark.**

marking signal: *Synonym* **mark.**

***m*-ary code:** *See* ***n*-ary code.**

***m*-ary signaling:** *See* ***n*-ary code.**

maser: *Acronym for* **microwave amplification by stimulated emission of radiation.** A member of the general class of microwave oscillators based on molecular interaction with electromagnetic radiation.

mask: **1.** In communications systems, to obscure, hide, or otherwise prevent information from being derived from a signal. *Note 1:* Masking is usually the result of interaction with another signal, such as noise, static, jamming, or other forms of interference. *Note 2:* Masking is not synonymous with erasing or deleting. **2.** In computing and data processing systems, a pattern of bits that can be used to retain or suppress segments of another pattern of bits. [From Weik '89]

masked threshold: The level at which an indistinguishable signal of interest becomes distinguishable from other signals or noise. *Note:* In acoustics, the masked threshold is usually expressed in dB.

master clock: A device that generates periodic, accurately spaced signals that are used for such purposes as timing, regulation of the operations of a processor, or generation of interrupts.

master frequency generator: In frequency-division multiplexing (FDM), equipment used to provide system end-to-end carrier frequency synchronization and frequency accuracy of tones. (188) *Note:* The following types of oscillators are used in the Defense Communications System FDM systems:
Type 1 - A master carrier oscillator as an integral part of the multiplexer set.
Type 2 - A submaster oscillator equipment or slave oscillator equipment as an integral part of the multiplexer set.
Type 3 - An external master oscillator equipment that has extremely accurate and stable characteristics.
Synonym **master oscillator.**

mastergroup: *See* **group.**

master oscillator: *Synonym* **master frequency generator.**

master-slave timing: Timing in which one station or node supplies the timing reference for all other interconnected stations or nodes.

master station: **1.** In a data network, the station that is designated by the control station to ensure data transfer to one or more slave stations. *Note:* A master station controls one or more data links of the data communications network at any given instant. The assignment of master status to a given station is temporary and is controlled by the control station according to the procedures set forth in the operational protocol. Master status is normally conferred upon a station so that it may transmit a message, but a station need not have a message to send to be designated the master station. **2.** In navigation systems using precise time dissemination, a station that has the clock used to synchronize the clocks of subordinate stations. **3.** In basic mode link control, the data station that has accepted an invitation to ensure a data transfer to one or more slave stations. *Note:* At a given instant, there can be only one master station on a data link.

matched junction: A waveguide component having four or more ports, and so arranged that if all ports except one are terminated in the correct impedance, there will be no reflection of energy from the junction when the fourth port is driven by a transmission line having a matching impedance. (188)

matching gel: *See* **gel.**

material absorption: *See* **absorption.**

material dispersion: *See* **dispersion.**

material dispersion coefficient [$M(\lambda)$]: In an optical fiber, pulse broadening per unit length of fiber and per unit of spectral width, usually expressed in picoseconds per (nanometer•kilometer). *Note 1:* For many optical fiber materials, $M(\lambda)$ approaches zero at a specific wavelength λ_0 between 1.3 µm and 1.5 µm. At wavelengths shorter than λ_0, $M(\lambda)$ is negative and increases with wavelength; at wavelengths longer than λ_0, $M(\lambda)$ is positive and decreases with wavelength. *Note 2:* Pulse broadening caused by material dispersion in a unit length of optical fiber is given by the product of $M(\lambda)$ and spectral width ($\Delta\lambda$).

$$M(\lambda) = \frac{1}{c}\frac{dN}{d\lambda} = -\frac{1}{c}\frac{d^2n}{d\lambda^2} ,$$

where n is the refractive index of the material, N is the group index expressed as

$$N = n - \lambda\frac{dn}{d\lambda} ,$$

λ is the wavelength of interest, and c is the velocity of light in vacuum.

material scattering: Of an electromagnetic wave, scattering that is attributable to the intrinsic properties of the material through which the wave is propagating. *Note 1:* Ionospheric scattering and Rayleigh scattering are examples of material scattering. *Note 2:* In an optical fiber, material scattering is caused by micro-inhomogeneities in the refractive indices of the materials used to fabricate the fiber, including the dopants used to modify the refractive index profile.

maximal-ratio combiner: A diversity combiner in which (a) the signals from each channel are added together, (b) the gain of each channel is made proportional to the rms signal level and inversely proportional to the mean square noise level in that channel, and (c) the same proportionality constant is used for all channels. (188) *Synonyms* **ratio-squared combiner, post-detection combiner, predetection combining, selective combiner.**

maximum block transfer time: The maximum allowable waiting time between initiation of a block transfer attempt and completion of a successful block transfer.

maximum calling area: Geographic calling limits permitted to a particular access line based on requirements for the particular line. *Note:* Maximum calling area restrictions are imposed for network control purposes.

maximum disengagement time: The maximum allowable waiting time between initiation of a disengagement attempt and successful disengagement.

maximum justification rate: *Synonym* **maximum stuffing rate.**

maximum keying frequency: In facsimile systems, the frequency in hertz numerically equal to the spot speed divided by twice the X-dimension of the scanning spot. (188)

maximum modulating frequency: In a facsimile transmission system, the highest picture frequency that is required. (188) *Note:* The maximum modulating frequency and the maximum keying frequency are not necessarily equal.

maximum stuffing rate: In a bit stream, the maximum rate at which stuffing bits can be inserted into the stream. (188) *Synonym* **maximum justification rate.**

maximum usable frequency (MUF): In radio transmission using reflection from the regular ionized layers of the ionosphere, the upper frequency limit that can be used for transmission between two points at a specified time. (188) *Note:* MUF is a median frequency applicable to 50% of the days of a month, as opposed to 90% cited for the lowest usable high frequency (LUF) and the optimum traffic frequency (FOT).

maximum user signaling rate: The maximum rate, in bits per second, at which binary information can be transferred in a given direction between users over the telecommunications system facilities dedicated to a particular information transfer transaction, under conditions of continuous transmission and no overhead information. *Note 1:* For a single channel, the signaling rate is given by

$$SCSR = \frac{\log_2 n}{T} \ ,$$

where *SCSR* is the single-channel signaling rate in bits per second, *T* is the minimum time interval in seconds for which each level must be maintained, and n is the number of significant conditions of modulation of the channel. *Note 2:* In the case where an individual end-to-end telecommunications service is provided by parallel channels, the parallel-channel signaling rate is given by

$$PCSR = \sum_{i=1}^{m} \frac{\log_2 n_i}{T_i} \ ,$$

where *PCSR* is the total signaling rate for *m* channels, *m* is the number of parallel channels, T_i is the minimum interval between significant instants for the *I*-th channel, and n_i is the number of significant conditions of modulation for the *I*-th channel. *Note 3:* In the case where an end-to-end telecommunications service is provided by tandem channels, the end-to-end signaling rate is the lowest signaling rate among the component channels.

Maxwell's equations: A set of partial differential equations that describe and predict the behavior of electromagnetic waves in free space, in dielectrics, and at conductor-dielectric boundaries. *Note:*

Maxwell's equations expand upon and unify the laws of Ampere, Faraday, and Gauss, and form the foundation of modern electromagnetic theory.

MCC: *Abbreviation for* **maintenance control circuit.**

MCM: *Abbreviation for* **multicarrier modulation.**

MDF: *Abbreviation for* **main distribution frame.**

meaconing: A system of receiving radio beacon signals and rebroadcasting them on the same frequency to confuse navigation. The meaconing stations cause inaccurate bearings to be obtained by aircraft or ground stations. [JP1]

mean power (of a radio transmitter): The average power supplied to the antenna transmission line by a transmitter during an interval of time sufficiently long compared with the lowest frequency encountered in the modulation taken under normal operating conditions. [NTIA] [RR] (188) *Note:* Normally, a time of 0.1 second, during which the mean power is greatest, will be selected.

mean time between failures (MTBF): **1.** An indicator of expected system reliability calculated on a statistical basis from the known failure rates of various components of the system. *Note:* MTBF is usually expressed in hours. **2.** Of a system, over a long performance measurement period, the measurement period divided by the number of failures that have occurred during the measurement period. **3.** For population of items, during a measurement period, the total functioning life of the population of items divided by the total number of failures within the population during the measurement period. *Note 1:* The total functioning life of the population may be calculated as the summation of the operating life of every item in the population over the measurement period. When computing the MTBF, any measure of operating life may be used, such as time, cycles, kilometers, or events. *Note 2:* For example, if a total of 1,000 events, such as data transfers, radio transmissions, or system boots, occurs in a population of items during a measurement period of 100 hours and there are a total of 10 failures among the entire population, the MTBF for each item is (1000)(100)/10 = 10^4 hours.

mean time between outages (*MTBO*): In a system, the mean time between equipment failures that result in loss of system continuity or unacceptable degradation. *Note:* The *MTBO* is calculated by the equation,

$$MTBO = \frac{MTBF}{1 - FFAS},$$

where *MTBF* is the nonredundant mean time between failures and *FFAS* is the fraction of failures for which the failed equipment is automatically bypassed. (188)

mean time to repair (MTTR): The total corrective maintenance time divided by the total number of corrective maintenance actions during a given period of time. (188)

mean time to service restoral (MTSR): The mean time to restore service following system failures that result in a service outage. *Note:* The time to restore includes all time from the occurrence of the failure until the restoral of service. (188)

measured-rate service: Telephone service for which charges are made in accordance with the total connection time of the line.

measurement period: *See* **performance measurement period.**

mechanically induced modulation: Optical signal modulation induced by mechanical means. *Note:* An example of deleterious mechanically induced modulation is speckle noise created in a multimode fiber by an imperfect splice or imperfectly mated connectors. Mechanical disturbance of the fiber ahead of the joint will introduce changes in the modal structure, resulting in variations of joint loss.

mechanically intermateable connectors: Connectors that are mechanically mateable, without creating mechanical damage, and without regard to attenuation properties.

mechanical splice: Of optical fibers, a splice, *i.e.*, permanent joint, accomplished by aligning the mating fibers in some kind of mechanical fixture. *Note 1:* The fibers may be secured by mechanical means or with an optical adhesive. *Note 2:* When the fibers are secured by mechanical means, the gap between them is usually filled with an index-matching gel to reduce Fresnel reflection. Likewise, the optical adhesives that are used in conjunction with mechanical splices are formulated to have a refractive index that approximates that of the glass, and also serve to reduce Fresnel reflection. [After FAA]

mediation function: In telecommunications network management, a function that routes or acts on information passing between network elements and network operations. (188) *Note 1:* Examples of mediation functions are communications control, protocol conversion, data handling, communications of primitives, processing that includes decision-making, and data storage. *Note 2:* Mediation functions can be shared among network elements, mediation devices, and network operation centers.

medium: **1.** In telecommunications, the transmission path along which a signal propagates, such as a wire pair, coaxial cable, waveguide, optical fiber, or radio path. (188) **2.** The material on which data are or may be recorded, such as plain paper, paper tapes, punched cards, magnetic tapes, magnetic disks, or optical disks. (188)

medium access control (MAC) sublayer: In a communications network, the part of the data link layer that supports topology-dependent functions and uses the services of the physical layer to provide services to the logical link control sublayer.

medium access unit (MAU): In a communications system, the equipment that adapts or formats the signal for transmittal over the communication medium. *Note:* An example of a MAU is an optical transmitter, which accepts an electrical signal at its input port and converts it to an optical signal accessible at its output port.

medium frequency (MF): Frequencies from 300 kHz to 3000 kHz. (188)

medium interface connector (MIC): In communications systems, the connector at the interface point between the bus interface unit and the terminal, *i.e.*, the medium interface point.

medium interface point (MIP): In communication systems, the location at which the standards for the interface parameters between a terminal and the line facility are implemented.

medium-power talker: A hypothetical talker, within a log-normal distribution of talkers, whose volume lies at the medium power of all talkers determining the volume distribution at the point of interest. (188) *Note:* When the distribution follows a log-normal curve (values expressed in decibels), the mean and standard deviation can be used to compute the medium-power talker. The talker volume distribution follows a log-normal curve and the medium-power talker is uniquely determined by the average talker volume. The medium-power talker volume, V, is given by $V = V_o + 0.115\sigma^2$, where V_o is the average of the talker volume distribution in volume units (vu), and σ^2 is the variance of the distribution.

megahertz (MHz): A unit of frequency denoting one million (10^6) Hz. (188)

memory: **1.** All of the addressable storage space in a processing unit and other internal memory that is used to execute instructions. **2.** *Loosely*, the volatile, main storage in computers. *See* **random access memory.** *Contrast with* **hard disk.**

menu: A displayed list of options from which a user selects actions to be performed.

MERCAST: *Acronym for* **merchant-ship broadcast system.**

merchant-ship broadcast system (MERCAST): A maritime shore-to-ship broadcast system in which the ocean areas are divided into primary broadcast areas each covered by a high-powered shore radio station that broadcasts simultaneously on one medium frequency (MF) and one or more high frequencies (HF) for routing messages to ocean-going ships. *Note:* In some instances, coast stations may repeat the messages. [From Weik '89]

meridional ray: In fiber optics, a ray that passes through the optical axis of an optical fiber (in contrast with a skew ray, which does not).

mesh network: *See* **network topology.**

mesh topology: *See* **network topology.**

mesochronous: The relationship between two signals such that their corresponding significant instants occur at the same average rate. (188)

message: 1. Any thought or idea expressed briefly in a plain or secret language, prepared in a form suitable for transmission by any means of communication. [JP1] *Note:* A message may be a one-unit message or a multiunit message. 2. [In telecommunications,] Record information expressed in plain or encrypted language and prepared in a format specified for intended transmission by a telecommunications system. [JP1] 3. An arbitrary amount of information whose beginning and end are defined or implied.

message alignment indicator: In a signal message, data transmitted between the user part and the message transfer part to identify the boundaries of the signal message.

message broadcast: An electronic-mail conference capability using data terminals. *Note:* Control can be maintained by the user or by the network.

message center: *See* **communications center.**

message feedback: A method of checking the accuracy of transmission of data by sending received data back to the sending end for comparison with the original data that have been stored there for this purpose. [From Weik '89]

message format: A predetermined or prescribed spatial or time-sequential arrangement of the parts of a message that is recorded in or on a data storage medium. *Note:* Messages prepared for electrical transmission are usually composed on a printed blank form with spaces for each part of the message and for administrative entries.

message handling system (MHS): In the CCITT X.400 Recommendations, the family of services and protocols that provides the functions for global electronic-mail transfer among local mail systems.

message heading: In radio communications, the message part or parts that (a) precede the text, *i.e.,* the message body, in time or space according to established conventions and (b) may include several data items, such as address groups, routing indicators, action addressee designators, information addressee designators, exempted addressee designators, prosigns, prowords, clear indicators, date-time groups, originator designators, special instructions, and protocol symbols. *Note:* Several message heading data items may be combined into a message preamble. [From Weik '89]

message part: 1. In radio communications, one of the three major subdivisions of a message, namely the heading, the text, or the ending. *Note:* Each message part may have separate components and each component may have elements and contents. 2. In cryptosystems, text that results from the division of a long message into several shorter messages of different lengths as a transmission security measure. *Note:* Message parts are usually prepared in such a manner as to appear unrelated externally. Statements that identify the parts for assembly at reception are encrypted in the texts. [From Weik '89]

message register leads: Terminal equipment leads at the interface used solely for receiving dc message register pulses from a central office at a PBX so that message unit information normally recorded at the central office only is also recorded at the PBX. . . . [47CFR]

message service: Switched service furnished to the general public (as distinguished from private line service). Except as otherwise provided, this includes exchange switched services and all switched services provided by interexchange carriers and completed by a local telephone company's access services. . . . [47CFR] *Synonym* **message toll service.**

message switching: A method of handling message traffic through a switching center, either from local users or from other switching centers, whereby the message traffic is stored and forwarded through the system. (188)

message toll service: *Synonym* **message service.**

message transfer part: The part of a common-channel signaling system that transfers signal messages and performs associated functions, such as error control and signaling link security.

message unit: A unit of measure for charging telephone calls, based on parameters such as the length of the call, the distance called, and/or the time of day.

messaging service: In integrated services digital networks (ISDN), an interactive telecommunications service that provides for information interchange among users by means of store-and-forward, electronic mail, or message-handling functions.

metallic circuit: A circuit in which metallic conductors are used and in which the ground or earth forms no part. (188)

metallic voltage: A potential difference between metallic conductors, as opposed to a potential difference between a metallic conductor and ground.

meteor burst communications: Communications by the propagation of radio signals reflected by ionized meteor trails. [NTIA]

meteorological aids service: A radiocommunication service used for meteorological, including hydrological, observations and exploration. [NTIA] [RR]

meteorological-satellite service: An Earth exploration-satellite service for meteorological purposes. [NTIA] [RR]

metric system: A decimal system of weights and measures based on the meter as a unit of length and the kilogram as a unit of mass. *Note:* The modern form of the metric system is the International System of Units (SI). *See* **International System of Units.**

metropolitan area network (MAN): A data communications network that (a) covers an area larger than a campus area network and smaller than a wide area network (WAN), (b) interconnects two or more LANs, and (c) usually covers an entire metropolitan area, such as a large city and its suburbs. (188)

MF: *Abbreviation for* **medium frequency.**

MFD: *Abbreviation for* **mode field diameter.**

MFJ: *Abbreviation for* **Modification of Final Judgment.**

MFSK: *Abbreviation for* **multiple frequency-shift keying.**

MHS: *Abbreviation for* **message handling system.**

MHz: *Abbreviation for* **megahertz.**

MIC: *Abbreviation for* **medium interface connector.**

microbend: In an optical waveguide, sharp curvatures involving local axial displacements of a few micrometers and spatial wavelengths of a few millimeters. (188) *Note:* Microbends can result from waveguide coating, cabling, packaging, and installation. Microbending can cause significant radiative loss and mode coupling.

microbending: *See* **microbend.**

microbend loss: In an optical fiber, the optical power loss caused by a microbend. [2196]

microcircuit: *Synonym* **integrated circuit.**

microcode: A sequence of microinstructions that is fixed in storage that is not program-addressable, and that performs specific processing functions.

microcomputer: A computer (a) in which the processing unit is a microprocessor and (b) that usually consists of a microprocessor, a storage unit, an input channel, and an output channel, all of which may be on one chip.

microfinishing film: A film of dimensionally stable plastic, to which are adhered carefully graded abrasive or polishing powders, *i.e.*, particles, having dimensions in the micrometer or submicrometer range. *Note:* Microfinishing films resemble sandpaper, but have much smaller abrasive or polishing particles. They are used commercially to shape and/or polish machined parts. They are also used to finish the endfaces of certain types of optical connectors. [After FAA]

microinstruction: An instruction that controls data flow and instruction-execution sequencing in a processor at a more fundamental level than machine instructions. *Note:* A series of microinstructions is necessary to perform an individual machine instruction.

micro-mainframe link: A physical or logical connection established between a remote microprocessor and mainframe host computer for the express purpose of uploading, downloading, or viewing interactive data and databases on-line in real time. *Note:* A micro-mainframe link usually requires terminal emulation software on the microcomputer.

microprocessor: A central processing unit implemented on a single chip. (188)

microprogram: A sequence of microinstructions that are in special storage where they can be dynamically accessed to perform various functions.

microwave (mw): Loosely, an electromagnetic wave having a wavelength from 300 mm to 10 mm (1 GHz to 30 GHz). *Note:* Microwaves exhibit many of the properties usually associated with waves in the optical regime, *e.g.*, they are easily concentrated into a beam. (188)

Mie scattering: Scattering of an electromagnetic wave by particles or refractive index inhomogeneities of a size on the order of the wavelength of interest.

mileage: In telecommunications, a specified distance used in tariff calculations, *i.e.*, toll charge calculations. *Note:* Mileage is locally defined and often refers to airline distance rather than actual communication system route miles.

military common emergency frequency: A frequency that (a) is used by all military units that are equipped to operate at that frequency or in the band in which that frequency lies and (b) is also used internationally by survival-craft stations and survival-craft equipment. [From Weik '89]

millimeter wave: Loosely, an electromagnetic wave having a wavelength from 1 mm to 0.1 mm (300 GHz to 3000 GHz). *Note:* Millimeter waves exhibit many of the properties usually associated with waves in the optical regime, *e.g.*, they are easily concentrated into a beam. (188)

minicomputer: *See* **computer.**

minimize: A condition wherein normal message and telephone traffic is drastically reduced in order that messages connected with an actual or simulated emergency shall not be delayed. [JP1]

minimum bend radius: The radius below which an optical fiber or fiber-optic cable should not be bent. *Note 1:* The minimum bend radius is of particular importance in the handling of fiber-optic cables. It will vary with different cable designs. The manufacturer should specify the minimum radius to which the cable may safely be bent during installation, and for the long term. The former is somewhat shorter than the latter. *Note 2:* The minimum bend radius is in general also a function of tensile stresses, *e.g.*, during installation, while being bent around a sheave while the fiber or cable is under tension. *Note 3:* If no minimum bend radius is specified, one is usually safe in assuming a minimum long-term low-stress radius not less than 15 times the cable diameter.

minimum discernable signal (MDS): *See* **threshold.**

minimum-dispersion slope: *See* **zero-dispersion slope.**

minimum-dispersion wavelength: *Synonym* **zero-dispersion wavelength.**

minimum-dispersion window: 1. The window of an optical fiber at which material dispersion is very small. *Note 1:* In silica-based fibers, the minimum-dispersion window occurs at a wavelength of approximately 1.3 µm. *Note 2:* The minimum-dispersion window may be shifted toward the minimum-loss window, *i.e.*, 1.55 µm, by the addition of dopants during manufacture. [After FAA] 2. In a single-mode fiber, the window or, in the case of doubly or quadruply clad fibers, windows, at which material and waveguide dispersion cancel one another, resulting in extremely wide bandwidth, *i.e.*, extremely low dispersion, over a very narrow range of wavelengths. [After FAA] *Synonym* **zero dispersion window.**

minimum-loss window: Of an optical fiber, the transmission window at which the attenuation coefficient is at or near the theoretical (quantum-limited) minimum. *Note 1:* If the losses from various mechanisms are plotted on a single graph as a function of wavelength, the minimum-loss window occurs in the vicinity of the wavelength at which the Rayleigh-scattering attenuation curve and the infrared-phonon-absorption curve intersect. *Note 2:* For silica-based fibers, the minimum-loss window occurs at approximately 1.55 μm. [After FAA]

minimum picture interval: The minimum time between the television pictures that have been selected for encoding. (188) *Note:* CCITT Recommendation H.221 cites the following values for picture interval: 1/29.97, 2/29.97, 3/29.97, and 4/29.97 seconds per picture.

MIP: *Abbreviation for* **medium interface point.**

MIS: *Abbreviation for* **management information system.**

misalignment loss: *See* **angular misalignment loss, gap loss, lateral offset loss.**

misdelivered block: A block received by a user other than the one intended by the message source.

mission bit stream: *Synonym* **payload.**

mixer: A nonlinear circuit or device that accepts as its input two different frequencies and presents at its output (a) a signal equal in frequency to the sum of the frequencies of the input signals, (b) a signal equal in frequency to the difference between the frequencies of the input signals, and, if they are not filtered out, (c) the original input frequencies.

mixing: *See* **heterodyne.**

MLPP: *Abbreviation for* **multilevel precedence and preemption.**

MM patch bay: A patching facility designed for patching and monitoring of digital data circuits at rates exceeding 3 Mb/s.

mobile Earth station: An Earth station in the mobile-satellite service intended to be used while in motion or during halts at unspecified points. [NTIA] [RR]

mobile-satellite service: A radiocommunication service:

- between mobile earth stations and one or more space stations, or between space stations used by this service; or
- between mobile earth stations by means of one or more space stations. This service may also include feeder links necessary for its operation. [NTIA] [RR]

mobile service: A radiocommunication service between mobile and land stations, or between mobile stations. [NTIA] [RR]

mobile services switching center (MSC): In an automatic cellular mobile system, the interface between the radio system and the public switched telephone network. *Note:* The MSC performs all signaling functions that are necessary to establish calls to and from mobile stations.

mobile station: A station in the mobile service intended to be used while in motion or during halts at unspecified points. [NTIA] [RR]

modal dispersion: *Incorrect synonym for* **multimode distortion.**

modal distortion: *Synonym* **multimode distortion.**

modal distribution: In an optical waveguide operating at a given wavelength, the number of modes supported, and their propagation time differences. (188)

modal loss: In an open waveguide, such as an optical fiber, a loss of energy on the part of an electromagnetic wave due to obstacles outside the waveguide, abrupt changes in direction of the waveguide, or other anomalies, that cause changes in the propagation mode of the wave in the waveguide. (188)

modal noise: Noise generated in an optical fiber system by the combination of mode-dependent optical losses and fluctuation in the distribution of optical energy among the guided modes or in the relative phases of the guided modes. (188) *Synonym* **speckle noise.**

mode: **1.** In a waveguide or cavity, one of the various possible patterns of propagating or standing electromagnetic fields. (188) *Note 1:* Each mode is characterized by frequency, polarization, electric field strength, and magnetic field strength. *Note 2:* The electromagnetic field pattern of a mode depends on the frequency, refractive indices or dielectric constants, and waveguide or cavity geometry. **2.** Any electromagnetic field distribution that satisfies Maxwell's equations and the applicable boundary conditions. **3.** In data communications, a protocol used to transfer data from switch to switch or from switch to terminal. (188) **4.** In statistics, the value associated with the highest peak in a probability density function.

mode coupling: In an electromagnetic waveguide, the exchange of power among modes. (188) *Note:* In a multimode optical fiber, mode coupling reaches statistical equilibrium, *i.e.*, equilibrium mode distribution, after the equilibrium length has been traversed.

mode field diameter (MFD): An expression of distribution of the irradiance, *i.e.*, the optical power, across the end face of a single-mode fiber. *Note:* For a Gaussian power distribution in a single-mode optical fiber, the mode field diameter is that at which the electric and magnetic field strengths are reduced to 1/e of their maximum values, *i.e.*, the diameter at which power is reduced to $1/e^2$ of the maximum power, because the power is proportional to the square of the field strength.

mode filter: A device used to select, reject, or attenuate a certain mode or modes. (188)

mode [identification friend or foe]: The number or letter referring to the specific pulse spacing of the signals transmitted by an interrogator. [JP1]

modem: *Acronym for* **modulator/demodulator.** **1.** In general, a device that both modulates and demodulates signals. (188) **2.** In computer communications, a device used for converting digital, signals into, and recovering them from, quasi-analog signals suitable for transmission over analog communications channels. *Note:* Many additional functions may be added to a modem to provide for customer service and control features. *Synonym* **signal conversion equipment.** **3.** In FDM carrier systems, a device that converts the voice band to, and recovers it from, the first level of frequency translation.

mode mixer: *Synonym* **mode scrambler.**

modem patch: A method of electrically interconnecting circuits by using back-to-back modems. (188)

mode partition noise: In an optical communications link, phase jitter of the signal caused by the combined effects of mode hopping in the optical source and intramodal distortion in the fiber. *Note:* Mode hopping causes random wavelength changes which in turn affect the group velocity, *i.e.*, the propagation time. Over a long length of fiber, the cumulative effect is to create jitter, *i.e.*, mode partition noise. The variation of group velocity creates the mode partition noise.

mode scrambler: **1.** A device for inducing mode coupling in an optical fiber. (188) **2.** A device composed of one or more optical fibers in which strong mode coupling occurs. *Note:* Mode scramblers are used to provide a modal distribution that is independent of the optical source, for purposes of laboratory or field measurements or tests. *Synonym* **mode mixer.**

mode stripper: *See* **cladding mode stripper.**

mode volume: The number of bound modes that an optical fiber is capable of supporting. (188) *Note:* The mode volume M is approximately given by $V^2/2$ and $(V^2/2)[g/(g+2)]$ respectively for step-index and power-law profile fibers, where g is the profile parameter, and V is the normalized frequency greater than 5.

Modification of Final Judgment (MFJ): The 1982 antitrust suit settlement agreement ("*Consent Decree*") entered into by the United States Department of Justice and the American Telephone and Telegraph Company (AT&T) that, after modification and upon approval of the United States District Court for the District of Columbia, required the divestiture of the Bell Operating Companies from AT&T.

modified AMI code: *Abbreviation for* **modified alternate mark inversion code.** A T-carrier AMI line code in which bipolar violations may be deliberately inserted to maintain system synchronization. *Note 1:* The clock rate of an incoming T-carrier signal is extracted from its bipolar line code. T-carrier was originally developed for voice applications. When voice signals are digitized for transmission via T-carrier, there is no problem in maintaining system synchronization, because of the nature of the digitized signals. However, when used for the transmission of digital data, the conventional AMI line code may fail to have sufficient marks, *i.e.*, "1's," to permit recovery of the incoming clock, and synchronization is lost. This happens when there are too many consecutive zeros in the user data being transported. To prevent loss of synchronization when a long string of zeros is present in the user data, deliberate bipolar violations are inserted into the line code, to create a sufficient number of marks to maintain synchronization. The receive terminal equipment recognizes the bipolar violations and removes from the user data the marks attributable to the bipolar violations. *Note 2:* The exact pattern of bipolar violations that is transmitted in any given case depends on the line rate and the polarity of the last valid mark in the user data prior to the unacceptably long string of zeros. *Note 3:* The number of consecutive zeros that can be tolerated in user data depends on the data rate, *i.e.*, the level of the line code in the T-carrier hierarchy. The North American T1 line code (1.544 Mb/s) does not use bipolar violations. The European T1 line code (2.048 Mb/s) may use bipolar violations when 8 or more consecutive zeros are present. This line code is called *bipolar with eight-zero substitution (B8ZS).* (In all levels of the European T-carrier hierarchy, the patterns of bipolar violations that are used differ from those used in the North American hierarchy.) At the North American T2 rate (6.312 Mb/s), bipolar violations are inserted if 6 or more consecutive zeros occur. This line code is called *bipolar with six-zero substitution (B6ZS).* At the North American T3 rate (44.736 Mb/s), bipolar violations are inserted if 3 or more consecutive zeros occur. This line code is called *"bipolar with three-zero substitution" (B3ZS).*

modular: Pertaining to the design concept in which interchangeable units are used to create a functional end product.

modular jack: A device that conforms to the *Code of Federal Regulations,* Title 47, part 68, which defines the size and configuration of all units that are permitted for connection to the public exchange facilities.

modulation: The process, or result of the process, of varying a characteristic of a carrier, in accordance with an information-bearing signal. (188)

modulation factor: In amplitude modulation, the ratio of the peak variation actually used, to the maximum design variation in a given type of modulation. (188) *Note:* In conventional amplitude modulation, the maximum design variation is considered that for which the instantaneous amplitude of the modulated signal reaches zero. When zero is reached, the modulation is considered 100%.

modulation index: In angle modulation, the ratio of the frequency deviation of the modulated signal to the frequency of a sinusoidal modulating signal. (188) *Note:* The modulation index is numerically equal to the phase deviation in radians.

modulation rate: **1.** The rate at which a carrier is varied to represent the information in a digital signal. *Note:* Modulation rate and information transfer rate are not necessarily the same. **2.** For modulated digital signals, the reciprocal of the unit interval of the modulated signal, measured in seconds. (188)

modulation suppression: In the reception of an amplitude-modulated signal, an apparent reduction in the depth of modulation of a wanted signal, caused by the presence, at the detector, of a stronger unwanted signal. (188)

modulator: A device that imposes a signal on a carrier. (188)

modulator-demodulator (modem): *See* **modem.**

module: **1.** An interchangeable subassembly that constitutes part of, *i.e.*, is integrated into, a larger device or system. (188) **2.** In computer programming, a program unit that is discrete and identifiable with respect to compiling, combining with other modules, and loading. (188)

monitor: **1.** Software or hardware that is used to scrutinize and to display, record, supervise, control, or verify the operations of a system. *Note:* Possible uses of monitors are to indicate significant departures from the norm, or to determine levels of utilization of particular functional units. **2.** A device used for the real-time temporary display of computer output data. *Note:* Monitors usually use cathode-ray-tube or liquid-crystal technology. *Synonyms* **video display terminal, video display unit, visual display unit.**

monitoring: **1.** The act of listening, carrying out surveillance on, and/or recording the emissions of one's own or allied forces for the purpose of maintaining and improving procedural standards and security, or for reference, as applicable. [JP1] **2.** The act of listening, carrying out surveillance on, and/or recording of enemy emissions for intelligence purposes. [JP1] **3.** The act of detecting the presence of signals, such as electromagnetic radiation, sound, or visual signals, and the measurement thereof with appropriate measuring instruments. **4.** The act of detecting the presence of radiation and the measurement thereof with radiation measuring instruments. *Synonym* **radiological monitoring.** [JP1]

monitor jack: A jack used to access communications circuits to observe signal conditions without interrupting the services. (188)

monitor key: A key used to access communications circuits to observe signal conditions without interrupting the services. (188)

monochromatic: In optics, pertaining to a single wavelength of electromagnetic radiation or to a single color. *Note:* In practice, optical radiation is never perfectly monochromatic, *i.e.*, it never consists of only one wavelength. It always has a finite spectral width, albeit narrow.

monochromator: In optics, an instrument for isolating narrow portions of the spectrum.

monomode optical fiber: *Synonym* **single-mode optical fiber.**

Mosaic: A portable World Wide Web browser that provides a graphical user interface to hypertext-based information.

mosquito noise: In a video display, distortion sometimes seen around the edges of moving objects, and characterized by moving artifacts around edges and/or by blotchy noise patterns superimposed over the objects, resembling a mosquito flying around a person's head and shoulders.

motion compensation: Interframe coding that (a) is used to compress motion of video images and (b) uses an algorithm to examine a sequence of image frames to measure the difference from frame to frame in order to send motion vector information. (188)

motion response degradation: The deterioration of motion video quality, resulting in a loss of perceived spatiotemporal resolution.

motion video: In video systems, temporally varying visual imagery intended to communicate or to convey movement or change.

mouse: A hand-held computer input device that generates signals that increment, *i.e.*, slew, the position of a cursor on a video display. *Note:* A mouse is placed on a flat surface and moved manually in the direction in which it is desired to move the cursor. A mouse has momentary switches ("buttons") that may be finger-operated to trigger an event after the cursor is positioned correctly.

M-patch bay: A patching facility designed for patching and monitoring digital data circuits at data signaling rates from 1 Mb/s (megabits per second) to 3 Mb/s. (188)

m-sequence: *See* ***n*-sequence.**

MTBF: *Abbreviation for* **mean time between failures.**

MTBO: *Abbreviation for* **mean time between outages.**

MTSR: *Abbreviation for* **mean time to service restoral.**

MTTR: *Abbreviation for* **mean time to repair.**

mudbox: Equipment that is sufficiently rugged to withstand adverse environments. *Note:* A mudbox is expected to operate unsheltered on the ground.

MUF: *Abbreviation for* **maximum usable frequency.**

mu-law (μ-law): *See* **mu-law (μ-law) algorithm.**

mu-law (μ-law) algorithm: A standard analog signal compression algorithm, used in digital communications systems of the North American digital hierarchy, to optimize, *i.e.*, modify, the dynamic range of an analog signal prior to digitizing. *Note:* The wide dynamic range of speech does not lend itself well to efficient linear digital encoding. Mu-law encoding effectively reduces the dynamic range of the signal, thereby increasing the coding efficiency and resulting in a signal-to-distortion ratio that is greater than that obtained by linear encoding for a given number of bits.

muldem: *Acronym for* **multiplexer/ demultiplexer.**

multiaddress calling: A service feature that permits a user to designate more than one addressee for the same data. *Note:* Multiaddress calling may be performed sequentially or simultaneously.

multicarrier modulation (MCM): A technique of transmitting data by dividing the data into several interleaved bit streams and using these to modulate several carriers. *Note:* MCM is a form of frequency-division multiplexing.

multicast: **1.** In a network, a technique that allows data, including packet form, to be simultaneously transmitted to a selected set of destinations. *Note:* Some networks, such as Ethernet, support multicast by allowing a network interface to belong to one or more multicast groups. **2.** To transmit identical data simultaneously to a selected set of destinations in a network, usually without obtaining acknowledgement of receipt of the transmission.

multicast address: A routing address that (a) is used to address simultaneously all the computers in a group and (b) usually identifies a group of computers that share a common protocol, as opposed to a group of computers that share a common network. *Note:* Multicast address also applies to radio communications. *Synonym (in Internet protocol)* **class d address.**

multichannel: Pertaining to communications, usually full-duplex communications, on more than one channel. *Note:* Multichannel transmission may be accomplished by time-division multiplexing, frequency-division multiplexing, phase-division multiplexing, or space diversity.

multicoupler: In radio communications, a device for connecting several receivers or transmitters to one antenna in such a way that the equipment impedances are properly matched to the antenna impedance.

multi-element dipole antenna: An antenna consisting of an arrangement of multiple dipole antennas. (188) *Note:* Various directivity patterns may be obtained by varying the arrangement of the dipoles and the way they are driven.

multifiber cable: A fiber-optic cable having two or more fibers, each of which is capable of serving as an independent optical transmission channel. [After FAA]

multifiber cable assembly: *See* **cable assembly.**

multifiber joint: An optical splice or connector designed to mate two multifiber cables, providing simultaneous optical alignment of all individual optical fibers.

multiframe: In PCM systems, a set of consecutive frames in which the position of each frame can be identified by reference to a multiframe alignment signal. (188) *Note:* The multiframe alignment signal does not necessarily occur, in whole or in part, in each multiframe.

multifrequency pulsing: *Synonym* **multifrequency signaling.** *See* **dual-tone multifrequency signaling.**

multifrequency signaling: *Synonym* **dual-tone multifrequency signaling.**

multilayer filter: *See* **interference filter.**

multilevel modulation: *See* **n-ary code.**

multilevel precedence and preemption (MLPP): In military communications, a priority scheme (a) for assigning one of several precedence levels to specific calls or messages so that the system handles them in a predetermined order and time frame, (b) for gaining controlled access to network resources in which calls and messages can be preempted only by higher priority calls and messages, (c) that is recognized only within a predefined domain, and (d) in which the precedence level of a call outside the predefined domain is usually not recognized.

multilink operation: In packet-switched networks, the simultaneous use of multiple links for the transmission of different segments of the same message unit. *Note:* Use of multilink operation is intended to increase the effective rate of message transmission. Multilink operation requires special procedures for multiplexing/demultiplexing control.

multimedia: Pertaining to the processing and integrated presentation of information in more than one form, *e.g.*, video, voice, music, or data. *Contrast with* **multiple media.**

multimode dispersion: *Incorrect synonym for* **multimode distortion.**

multimode distortion: A distortion mechanism, occurring in multimode fibers, in which the signal is spread in time because the velocity of propagation of the optical signal is not the same for all modes. (188) *Note 1:* In the ray-optics analogy, multimode distortion in a step-index optical fiber may be compared to multipath propagation of a radio signal. The direct signal is distorted by the arrival of the reflected signal a short time later. In a step-index optical fiber, rays taking more direct paths through the fiber core, *i.e.*, those which undergo the fewest reflections at the core-cladding boundary, will traverse the length of the fiber sooner than those rays which undergo more reflections. This results in distortion of the signal. *Note 2:* Multimode distortion limits the bandwidth of multimode fibers. For example, a typical step-index fiber with a 50-µm core would be limited to approximately 20 MHz for a one-kilometer length, *i.e.*, a bandwidth of 20 MHz•km. *Note 3:* Multimode distortion may be considerably reduced, but never completely eliminated, by the use of a core having a graded refractive index. The bandwidth of a typical off-the-shelf graded-index multimode fiber, having a 50-µm core, may approach 1 GHz•km or more. Multimode graded-index fibers having bandwidths approaching 3 GHz•km have been produced. *Note 4:* Because of its similarity to dispersion in its effect on the optical signal, multimode distortion is sometimes incorrectly referred to as *"intermodal dispersion," "modal dispersion,"* or *"multimode dispersion."* Such usage is incorrect because multimode distortion is not a truly dispersive effect. Dispersion is a wavelength-dependant phenomenon, whereas multimode distortion may occur at a single wavelength. [After FAA] *Synonyms* **intermodal delay distortion, intermodal distortion, modal distortion.**

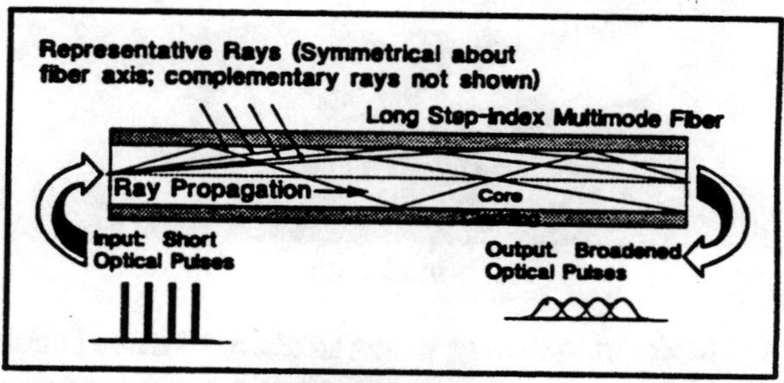

multimode distortion

multimode group delay: *Synonym* **differential mode delay.**

multimode optical fiber: An optical fiber that supports the propagation of more than one bound mode. (188) *Note:* A multimode optical fiber may be either a graded-index (GI) fiber or a step-index (SI) fiber.

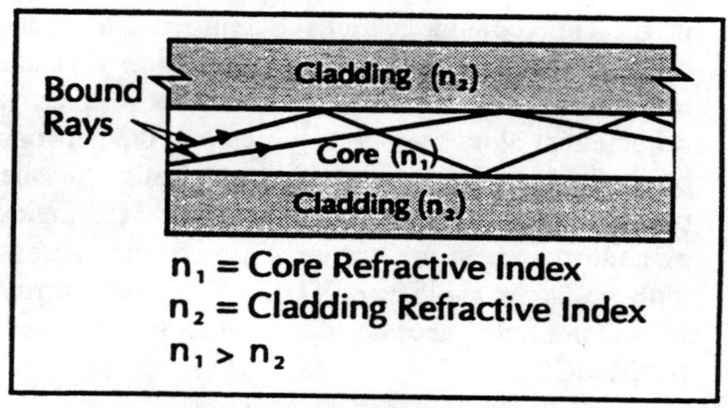

step index, multimode fiber

multiparty line: *Synonym* **party line.**

multipath: The propagation phenomenon that results in radio signals' reaching the receiving antenna by two or more paths. (188) *Note 1:* Causes of multipath include atmospheric ducting, ionospheric reflection and refraction, and reflection from terrestrial objects, such as mountains and buildings. *Note 2:* The effects of multipath include constructive and destructive interference, and phase shifting of the signal. *Note 3:* In facsimile and television transmission, multipath causes jitter and ghosting.

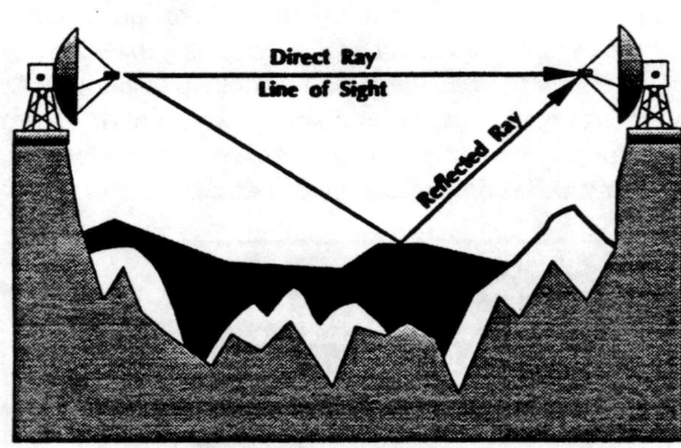

multipath

multiple: A system of wiring so arranged that a circuit, a line, or a group of lines is accessible at a number of points. (188) *Synonym* **multipoint.**

multiple access: **1.** The connection of a user to two or more switching centers by separate access lines using a single message routing indicator or telephone number. (188) **2.** In satellite communications, the capability of a communications satellite to function as a portion of a communications link between more than one pair of satellite terminals concurrently. (188) *Note:* The three types of multiple access presently used with communications satellites are code-division, frequency-division, and time-division multiple access. **3.** In computer networking, a scheme that allows temporary access to the network by individual users, on a demand basis, for the purpose of transmitting information. *Note:* Examples of multiple access are carrier sense multiple access with collision avoidance (CSMA/CA) and carrier sense multiple access with collision detection (CSMA/CD).

multiple call: *Synonym* **conference call.**

multiple frequency-shift keying (MFSK): Frequency-shift keying (FSK) in which multiple codes are used in the transmission of digital signals. (188) *Note:* In MFSK, the coding schemes use multiple frequencies that are transmitted concurrently or sequentially.

multiple homing: **1.** In telephone systems, the connection of a terminal facility so that it can be served by one or several switching centers. (188) *Note:* Multiple homing may use a single directory number. **2.** In telephone systems, the connection of a terminal facility to more than one switching center by separate access lines. *Note:* Separate directory numbers are applicable to each switching center accessed. (188)

multiple media: Transmission media using more than one type of transmission path (*e.g.*, optical fiber, radio, and copper wire) to deliver information. *Contrast with* **multimedia.**

multiple-spot scanning: In facsimile systems, scanning performed simultaneously by two or more scanning spots, each one analyzing its fraction of the total scanned area of the object. (188)

multiplex (MUX): *See* **multiplexing.**

multiplex aggregate bit rate: In a time-division multiplexer, the bit rate that is equal to the sum of (a) the input channel data signaling rates available to the user and (b) the rate of the overhead bits required. (188)

multiplex baseband: **1.** In frequency-division multiplexing, the frequency band occupied by the aggregate of the signals in the line interconnecting the multiplexing and radio or line equipment. (188) **2.** In frequency division multiplexed carrier systems, at the input to any stage of frequency translation, the frequency band occupied. *Note:* For example, the output of a group multiplexer consists of a band of frequencies from 60 kHz to 108 kHz. This is the group-level baseband that results from combining 12 voice-frequency input channels, having a bandwidth of 4 kHz each, including guard bands. In turn, 5 groups are multiplexed into a super group having a baseband of 312 kHz to 552 kHz. This baseband, however, does not represent a group-level baseband. Ten super groups are in turn multiplexed into one master group, the output of which is a baseband that may be used to modulate a microwave-frequency carrier.

multiplexer (MUX): A device that combines multiple inputs into an aggregate signal to be transported via a single transmission channel. (188) *Synonym* **multiplexing equipment.**

multiplexer/demultiplexer (muldem): A device that combines the functions of multiplexing and demultiplexing of digital signals. *Note:* The term *muldem* should not be confused with *modem.*

multiplex hierarchy: In frequency-division multiplexing, the rank of frequency bands occupied:

12 channels	group
5 groups (60 channels)	super group
5 super groups (300 channels)	master group (CCITT)
10 super groups (600 channels)	master group (U.S. standard)
6 U.S. master groups (3600 channels)	jumbo group

(188)

multiplexing (MUXing): The combining of two or more information channels onto a common transmission medium. *Note:* In electrical communications, the two basic forms of multiplexing are time-division multiplexing (TDM) and frequency-division multiplexing (FDM). In optical communications, the analog of FDM is referred to as wavelength-division multiplexing (WDM).

multiplexing equipment: *Synonym* **multiplexer.**

multiplex link encryption: Encryption in which a single cryptographic device is used to encrypt all of the data in a multiplexed link. (188)

multipoint: *Synonym* **multiple.**

multipoint access: Access in which more than one terminal is supported by a single network termination.

multipoint circuit: A circuit that interconnects three or more separate points. (188)

multipoint distribution service: A one-way domestic public radio service rendered on microwave frequencies from a fixed station transmitting (usually in an omnidirectional pattern) to multiple receiving facilities located at fixed points. [47CFR]

multipoint grounding system: Equipment bonded together and also bonded to the facility grounding system at the point nearest the equipment. (188)

multipoint link: A data communications link that interconnects three or more terminals.

multiport repeater: In digital networking, an active device, having multiple input/output (I/O) ports, in which a signal introduced at the input of any port appears at the output of every port. *Note 1:* A multiport repeater usually performs regenerative functions, *i.e.*, it reshapes the digital signals. *Note 2:* Depending on the application, a multiport repeater may be designed not to repeat a signal back to the port from which it originated.

multiprocessing: **1.** Simultaneous processing by two or more processors acting in concert. **2.** The simultaneous execution of two or more computer programs or sequences of instructions by a single processor.

multiprocessor: A computer that has two or more processors that have common access to a main storage.

multiprogramming: A mode of operation that provides for the interleaved execution of two or more computer programs by a single processor. (188)

multi-satellite link: A radio link between a transmitting Earth station and a receiving Earth station through two or more satellites, without any intermediate Earth station. A multi-satellite link comprises one uplink, one or more satellite-to-satellite links, and one downlink. [NTIA] [RR]

multitasking: The concurrent performance or interleaved execution of two or more tasks. *Synonym* **concurrent operation.**

mutually synchronized network: A network that has a synchronizing arrangement in which each clock in the network exerts a degree of control on all others.

mutual synchronization: Synchronization in which the frequency of the clock at a particular node is controlled by a weighted average of the timing on all signals received from neighboring nodes.

MUX: *Abbreviation for* **multiplex, multiplexer.** *See* **multiplexing.**

mw: *Abbreviation for* **microwave.**

NA: *Abbreviation for* **numerical aperture.**

nailed-up circuit: *Deprecated term. See* **dedicated circuit, permanent virtual circuit.**

NAK: *Acronym for* **negative-acknowledge character.**

NAK attack: In communications security systems, a security penetration technique that makes use of the negative-acknowledge transmission-control character and capitalizes on a potential weakness in a system that handles asynchronous transmission interruption in such a manner that the system is in an unprotected state against unauthorized access during certain periods. [From Weik '89]

narrative traffic: Traffic consisting of plain or encrypted messages written in a natural language and transmitted in accordance with standard formats and procedures. (188) *Note:* Examples of narrative traffic include (a) messages that are placed on paper tape and transmitted via a teletypewriter (TTY), and on reception, are converted back to a printed page on another teletypewriter or teleprinter and (b) messages printed on a sheet of paper, transmitted via optical character recognition (OCR) equipment, and on reception, converted back to a printed page on a printer.

narrowband modem: A modem whose modulated output signal has an essential frequency spectrum that is limited to that which can be wholly contained within, and faithfully transmitted through, a voice channel with a nominal 4-kHz bandwidth. (188) *Note:* High frequency (HF) modems are limited to operation over a voice channel with a nominal 3-kHz bandwidth.

narrowband radio voice frequency (NBRVF): In narrowband radio, the nominal 3-kHz bandwidth allocated for single channel radio that provides a transmission path for analog and quasi-analog signals. (188)

narrowband signal: Any analog signal or analog representation of a digital signal whose essential spectral content is limited to that which can be contained within a voice channel of nominal 4-kHz bandwidth. (188) *Note:* Narrowband radio uses a voice channel with a nominal 3-kHz bandwidth.

***n*-ary code:** A code that has *n* significant conditions, where n is a positive integer greater than 1. (188) *Note 1:* The integer substituted for *n* indicates the specific number of significant conditions, *i.e.*, quantization states, in the code. For example, an 8-ary code has eight significant conditions and can convey three bits per code symbol. *Note 2:* A prefix that indicates an integer, *e.g.*, "bi," "tern," or "quater," may be used in lieu of a numeral, to produce "binary," "ternary," or "quaternary" (2, 3, and 4 states respectively).

***n*-ary signaling:** *See* ***n*-ary code.**

NATA: *Abbreviation for* **North American Telecommunications Association.**

National Command Authorities (NCA): The President and the Secretary of Defense or their duly deputized alternates or successors. [JP1]

National Communications System (NCS): 1. The organization established by Section 1(a) of Executive Order No. 12472 to assist the President, the National Security Council, the Director of the Office of Science and Technology Policy, and the Director of the Office of Management and Budget, in the discharge of their national security emergency preparedness telecommunications functions. The NCS consists of both the telecommunications assets of the entities represented on the NCS Committee of Principals and an administrative structure consisting of the Executive Agent, the NCS Committee of Principals, and the Manager. **2.** The telecommunications system that results from the technical and operational integration of the separate telecommunications systems of the several executive branch departments and agencies having a significant telecommunications capability. [JP1]

National Coordinating Center (NCC) for Telecommunications: The joint telecommunications industry/Federal Government operation established by the National Communications System to assist in the initiation, coordination, restoration, and reconstitution of National Security or Emergency Preparedness (NS/EP) telecommunications services or facilities.

National Electric Code® (NEC): A standard that governs the use of electrical wire, cable, and fixtures, and electrical and optical communications cable installed in buildings. *Note:* The NEC was developed by the NEC Committee of the American National Standards Institute (ANSI), was sponsored by the National Fire Protection Association (NFPA), and is identified by the description ANSI/NFPA 70-XXXX, the last four digits representing the year of the NEC revision. (188)

National Information Infrastructure (NII): A proposed, advanced, seamless web of public and private communications networks, interactive services, interoperable hardware and software, computers, databases, and consumer electronics to put vast amounts of information at users' fingertips. *Note:* NII includes more than just the physical facilities (more than the cameras, scanners, keyboards, telephones, fax machines, computers, switches, compact disks, video and audio tape, cable, wire, satellites, optical fiber transmission lines, microwave nets, switches, televisions, monitors, and printers) used to transmit, store, process, and display voice, data, and images; it encompasses a wide range of interactive functions, user-tailored services, and multimedia databases that are interconnected in a technology-neutral manner that will favor no one industry over any other. *Synonym* **information superhighway.**

National Security or Emergency Preparedness telecommunications: *See* **NS/EP telecommunications.**

National Television Standards Committee standard: *See* **NTSC standard.**

natural frequency: Of an antenna, the lowest frequency at which the antenna resonates without the addition of any inductance or capacitance. (188)

nautical mile (nmi): A unit of distance used in navigation and based on the length of one minute of arc taken along a great circle. *Note 1:* Because the Earth is not a perfect sphere, various values have been assigned to the nautical mile. The value 1852 meters (6076.1 ft.) has been adopted internationally. *Note 2:* The nautical mile is frequently confused with the geographical mile, which is equal to 1 min of arc on the Earth's equator (6087.15 ft.).

NBH: *Abbreviation for* **network busy hour.** *See* **busy hour.**

NBRVF: *Abbreviation for* **narrowband radio voice frequency.**

NCA: *Abbreviation for* **National Command Authorities.**

NCC: *Abbreviation for* **National Coordinating Center for Telecommunications.**

NCS: *Abbreviation for* **National Communications System, net control station.**

near-end crosstalk: Crosstalk that is propagated in a disturbed channel in the direction opposite to the direction of propagation of a signal in the disturbing channel. *Note:* The terminals of the disturbed channel, at which the near-end crosstalk is present, and the energized terminal of the disturbing channel, are usually near each other. (188)

near field: *Synonym* **near-field region (def. #1).**

near-field diffraction pattern: The diffraction pattern of an electromagnetic wave, which pattern is observed close to a source or aperture, as distinguished from a far-field diffraction pattern. *Note:* The pattern in the output plane is called the near-field radiation pattern. *Synonym* **Fresnel diffraction pattern.** *Contrast with* **far-field diffraction pattern.**

near-field region: **1.** The close-in region of an antenna wherein the angular field distribution is dependent upon distance from the antenna. (188) *Synonyms* **near field, near zone.** **2.** In optical fiber communications, the region close to a source or aperture. *Note:* The diffraction pattern in this region typically differs significantly from that observed at infinity and varies with distance from the source.

near-field scanning: A technique for measuring the refractive-index profile of an optical fiber by using an extended source to illuminate an endface and measuring the point-by-point radiance at the exit face.

near real time: **1.** Pertaining to the delay introduced, by automated data processing, between the occurrence of an event and the use of the processed data, *e.g.,* for display or feedback and control purposes. *Note 1:* For example, a near-real-time display depicts an event or situation as it existed at the current time less the processing time. *Note 2:* The distinction between near real time and real time is somewhat nebulous and must be defined for the situation at hand. *Contrast with* **real time.** **2.** Pertaining to the timeliness of data or information which has been delayed by the time required for electronic communication and automatic data processing. This implies that there are no significant delays. [JP1]

near-vertical-incidence skywave: In radio propagation, a wave that is reflected from the ionosphere at a nearly vertical angle and that is used in short-range communications to reduce the area of the skip zone and thereby improve reception beyond the limits of the ground wave. (188)

near zone: *Synonym* **near-field region (def. #1).**

NEC: *Abbreviation for* **National Electric Code®.**

necessary bandwidth: For a given class of emission, the width of the frequency band which is just sufficient to ensure the transmission of information at the rate and with the quality required under specified conditions. [NTIA] [RR] (188) *Note:* Emissions useful for the adequate functioning of the receiving equipment, *e.g.*, the emission corresponding to the carrier of reduced carrier systems, must be included in the necessary bandwidth. (188) (*See* Annex J of *NTIA Manual of Regulations and Procedures for Federal Radio Frequency Management* for formulas used to calculate necessary bandwidth.)

negation circuit: *Deprecated synonym for* **inverter.**

negative-acknowledge character (NAK): A transmission control character sent by a station as a negative response to the station with which the connection has been set up. (188) *Note 1:* In binary synchronous communication protocol, the NAK is used to indicate that an error was detected in the previously received block and that the receiver is ready to accept retransmission of that block. *Note 2:* In multipoint systems, the NAK is used as the not-ready reply to a poll.

negative feedback: *See* **feedback (def. #1).**

negative justification: *Synonym* **de-stuffing.**

negative pulse stuffing: *Synonym* **de-stuffing.**

***n*-entity:** An active element in the *n*-th layer of the Open Systems Interconnection—Reference Model (OSI-RM) that (a) interacts directly with elements, *i.e.*, entities, of the layer immediately above or below the *n*-th layer, (b) is defined by a unique set of rules, *i.e.*, syntax, and information formats, including data and control formats, and (c) performs a defined set of functions. *Note 1:* The *n* refers to any one of the 7 layers of the OSI-RM. *Note 2:* In an existing layered open system, the *n* may refer to any given layer in the system. *Note 3:* Layers are conventionally numbered from the lowest, *i.e.*, the physical layer, to the highest, so that the $n + 1$ layer is above the *n*-th layer and the $n - 1$ layer is below.

NEP: *Abbreviation for* **noise equivalent power.**

neper (Np): A unit used to express ratios, such as gain, loss, and relative values. *Note 1:* The neper is analogous to the decibel, except that the Naperian base 2.718281828... is used in computing the ratio in nepers. *Note 2:* The value in nepers, *Np*, is given by $Np = \ln(x_1/x_2)$, where x_1 and x_2 are the values of interest, and ln is the natural logarithm, *i.e.*, logarithm to the base e. (188) *Note 3:* One neper (Np) = 8.686 dB, where 8.686 = 20/(ln 10). *Note 4:* The neper is often used to express voltage and current ratios, whereas the decibel is usually used to express power ratios. *Note 5:* Like the dB, the Np is a dimensionless unit. *Note 6:* The ITU recognizes both units.

nested command menu: A command menu within another command menu. *See* **command menu.**

net: *Synonym* **communications net.**

net control station (NCS): **1.** A radio station that performs net control functions, such as controlling traffic and enforcing operational discipline. (188) [From Weik '89] **2.** [A] terminal in a secure telecommunications net responsible for distributing key in electronic form to the members of the net. [NIS]

net gain: The overall gain of a transmission circuit. (188) *Note 1:* Net gain is measured by applying a test signal at an appropriate power level *(see Note 5)* at the input port of a circuit and measuring the power delivered at the output port. The net gain in dB is calculated by taking 10 times the logarithm of the ratio of the output power to the input power. *Note 2:* The net gain expressed in dB may be positive or negative. *Note 3:* If the net gain expressed in dB is negative, it is also called the "net loss." *Note 4:* If the net gain is expressed as a ratio, and the ratio is less than unity, a net loss is indicated. *Note 5:* The test signal must be chosen so that its power level is within the usual operating range of the circuit being tested.

net loss: The overall loss of a transmission circuit. (188)

net loss variation: The maximum change in net loss occurring in a specified portion of a communication system during a specified period. (188)

net operation: The operation of an organization of stations capable of direct communication on a common channel or frequency. *Note:* Net operations (a) allow participants to conduct ordered conferences among participants who usually have common information needs or related functions to perform, (b) are characterized by adherence to standard formats and procedures, and (c) are responsive to a common supervisory station, called the *"net control station,"* which permits access to the net and maintains net operational discipline.

net radio interface (NRI): An interface between a single-channel radio station (usually in a radio net) and switched communications systems. (188)

network: **1.** An interconnection of three or more communicating entities. **2.** An interconnection of usually passive electronic components that performs a specific function (which is usually limited in scope), *e.g.*, to simulate a transmission line or to perform a mathematical function such as integration or differentiation. *Note:* A network may be part of a larger circuit. (188)

network administration: A group of network management functions that (a) provide support services, (b) ensure that the network is used efficiently, and (c) ensure prescribed service-quality objectives are met. (188) *Note:* Network administration may include activities such as network address assignment, assignment of routing protocols and routing table configuration, and directory service configuration.

network architecture: **1.** The design principles, physical configuration, functional organization, operational procedures, and data formats used as the bases for the design, construction, modification, and operation of a communications network. (188) **2.** The structure of an existing communications network, including the physical configuration, facilities, operational structure, operational procedures, and the data formats in use. (188)

network busy hour (NBH): *See* **busy hour.**

network connectivity: The topological description of a network that specifies, in terms of circuit termination locations and quantities, the interconnection of the transmission nodes. (188)

network control program (NCP): In a switch or network node, software designed to store and forward frames between nodes. *Note:* An NCP may be used in local area networks or larger networks.

network element (NE): In integrated services digital networks, a piece of telecommunications equipment that provides support or services to the user. (188)

network engineering: **1.** In telephony, the discipline concerned with (a) determining internetworking service requirements for switched networks, and (b) developing and implementing hardware and software to meet them. **2.** In computer science, the discipline of hardware and software engineering to accomplish the design goals of a computer network. **3.** In radio communications, the discipline concerned with developing network topologies.

Network File System: *See* **NFS.**

network interface: **1.** The point of interconnection between a user terminal and a private or public network. **2.** The point of interconnection between the public switched network and a privately owned terminal. (188) *Note: Code of Federal Regulations,* Title 47, part 68, stipulates the interface parameters. **3.** The point of interconnection between one network and another network. (188)

network interface card (NIC): A network interface device (NID) in the form of circuit card that is installed in an expansion slot of a computer, to provide network access. *Note:* Examples of NICs are cards that interface a computer with an Ethernet LAN and cards that interface a computer with an FDDI ring network.

network interface device (NID): **1.** A device that performs interface functions, such as code conversion, protocol conversion, and buffering, required for communications to and from a network. **2.** A device used primarily within a local area network (LAN) to allow a number of independent devices, with varying protocols, to communicate with each other. *Note 1:* An NID converts each device protocol into a common transmission protocol. *Note 2:* The transmission protocol may be chosen to accommodate directly a number of the devices used within the network without the need for protocol conversion for those devices by the NID. *Synonym* **network interface unit.**

network interface unit (NIU): *Synonym* **network interface device.**

network inward dialing (NID): *Synonym* **direct inward dialing.**

Network Layer: *See* **Open Systems Interconnection—Reference Model.**

network management: The execution of the set of functions required for controlling, planning, allocating, deploying, coordinating, and monitoring the resources of a telecommunications network, including performing functions such as initial network planning, frequency allocation, predetermined traffic routing to support load balancing, cryptographic key distribution authorization, configuration management, fault management, security management, performance management, and accounting management. *Note:* Network management does not include user terminal equipment. (188)

network manager: In network management, the entity that initiates requests for management information from managed systems or receives spontaneous management-related notifications from managed systems. (188)

network operating system (NOS): Software that (a) controls a network and its message (*e.g.*, packet) traffic, and queues, (b) controls access by multiple users to network resources such as files, and (c) provides for certain administrative functions, including security. *Note 1:* A network operating system is most frequently used with local area networks and wide area networks, but could also have application to larger network systems. *Note 2:* The upper 5 layers of the OSI—Reference Model provide the foundation upon which many network operating systems are based.

network outward dialing (NOD): *Synonym* **direct outward dialing.**

network terminal number (NTN): In the CCITT International X.121 format, the sets of digits that comprise the complete address of the data terminal end point. *Note:* For an NTN that is not part of a national integrated numbering format, the NTN is the 10 digits of the CCITT X.25 14-digit address that follow the Data Network Identification Code (DNIC). When part of a national integrated numbering format, the NTN is the 11 digits of the CCITT X.25 14-digit address that follow the DNIC.

network terminating interface (NTI): *Synonym for* **demarcation point.**

network termination: Network equipment that provides functions necessary for network operation of ISDN access protocols. *Note:* Network termination provides functions essential for transmission services.

network termination 1 (NT1): In Integrated Services Digital Networks (ISDN), a functional grouping of customer-premises equipment that includes functions that may be regarded as belonging to OSI Layer 1, *i.e.*, functions associated with ISDN electrical and physical terminations on the user premises. *Note:* The NT1 forms a boundary to the network and may be controlled by the provider of the ISDN services.

network termination 2 (NT2): In Integrated Services Digital Networks (ISDN), an intelligent device that may include functionality for OSI Layers 1 through 3 (dependent on individual systems requirements).

network topology: The specific physical, *i.e.*, real, or logical, *i.e.*, virtual, arrangement of the elements of a

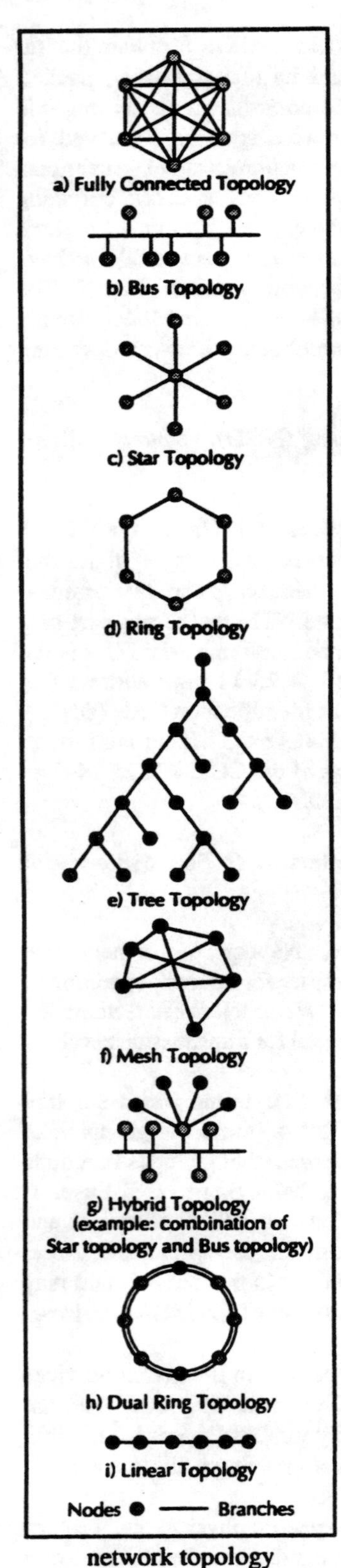

network topology

network. *Note 1:* Two networks have the same topology if the connection configuration is the same, although the networks may differ in physical interconnections, distances between nodes, transmission rates, and/or signal types. *Note 2:* The common types of network topology are illustrated *[refer to the figure on this page]* and defined in alphabetical order below:

➢ **bus topology:** A network topology in which all nodes, *i.e.,* stations, are connected together by a single bus.

➢ **fully connected topology:** A network topology in which there is a direct path (branch) between any two nodes. *Note:* In a fully connected network with n nodes, there are n(n–1)/2 direct paths, *i.e.,* branches. *Synonym* **fully connected mesh network.**

➢ **hybrid topology:** A combination of any two or more network topologies. *Note 1:* Instances can occur where two basic network topologies, when connected together, can still retain the basic network character, and therefore not be a hybrid network. For example, a tree network connected to a tree network is still a tree network. Therefore, a hybrid network accrues only when two basic networks are connected and the resulting network topology fails to meet one of the basic topology definitions. For example, two star networks connected together exhibit hybrid network topologies. *Note 2:* A hybrid topology always accrues when two different basic network topologies are connected.

➢ **linear topology:** *See* **bus topology.**

➢ **mesh topology:** A network topology in which there are at least two nodes with two or more paths between them.

➢ **ring topology:** A network topology in which every node has exactly two branches connected to it.

➢ **star topology:** A network topology in which peripheral nodes are connected to a central node, which rebroadcasts all transmissions received from any peripheral node to all peripheral nodes on the network, including the originating node. *Note 1:* All peripheral nodes may thus communicate with all others by transmitting to, and receiving from, the central node only. *Note 2:* The failure of a transmission line, *i.e.,* channel, linking any peripheral node to the central node will result in the isolation of that peripheral node from all others. *Note 3:* If the star central node is passive, the originating node must be able to tolerate the reception of an echo of its own transmission, delayed by the two-way transmission time, *i.e.,* to and from the central node, plus any delay generated in the central node. An active star network has an active central node that usually has the means to prevent echo-related problems. (188)

➢ **tree topology:** A network topology that, from a purely topologic viewpoint, resembles an interconnection of star networks in that individual peripheral nodes are required to transmit to and receive from one other node only, toward a central node, and are not required to act as repeaters or regenerators. (188) *Note 1:* The function of the central node may be distributed. *Note 2:* As in the conventional star network, individual nodes may thus still be isolated from the network by a single-point failure of a transmission path to the node. *Note 3:* A single-point failure of a transmission path within a distributed node will result in partitioning two or more stations from the rest of the network.

network utility: An internetwork administrative signaling mechanism in the call control procedure between packet switching public data networks.

neutral: **1.** In ac power distribution, the conductor that (a) is intentionally grounded on the supply side of the service disconnect and (b) provides a current return path for ac power currents. (188) **2.** In three-phase ac "Y," *i.e.*, wye, power distribution, the low-potential fourth wire that conducts only that current required to achieve electrical balance, *i.e.*, to provide a return path for any current imbalance among the three phases.

neutral direct-current telegraph system: A telegraph system in which (a) current flows during marking intervals and no current flows during spacing intervals for the transmission of signals over a line, and (b) the direction of current flow is immaterial. (188) *Synonyms* **single-current system, single-current transmission system, single-Morse system.**

neutral ground: An intentional ground applied to the neutral conductor or neutral point of a circuit, transformer, machine, apparatus, or system. (188)

neutral operation: A method of teletypewriter operation in which marking signals are formed by current pulses of one polarity, either positive or negative, and spacing signals are formed by reducing the current to zero or nearly zero. (188)

neutral relay: A relay in which the direction of movement of the armature does not depend upon the direction of the current in the circuit controlling the armature. (188)

new customer premises equipment: All customer premises equipment not in service or in the inventory of a regulated telephone utility as of December 31, 1982.

NF: *Abbreviation for* **noise figure.**

NFS: *Abbreviation for* **Network File System.** A proprietary distributed file system that is widely used by TCP/IP vendors. *Note:* NFS allows different computer systems to share files, and uses user datagram protocol (UDP) for data transfer.

***n*-function:** A defined action performed by an *n*-entity. *Note:* An *n*-function may be (a) a single action, *i.e.*, a primitive function, or (b) a set of actions.

nibble: Part of a byte, usually half of a byte. *(Obsolete)*

NIC: *Abbreviation for* **network interface card.**

NID: *Acronym for* **network interface device, network inward dialing.**

NII: *Abbreviation for* **National Information Infrastructure.**

nine-hundred (900) service: A telephone service via which the caller may access information on a charge-per-call or charge-per-time basis.

NIU: *Abbreviation for* **network interface unit.** *See* **network interface device.**

nmi: *Abbreviation for* **nautical mile.**

NOD: *Acronym for* **network outward dialing.**

nodal clock: The principal clock or alternate clock located at a particular node that provides the timing reference for all major functions at that node.

nodal point: *Synonym* **node (def. #1).**

node: **1.** In network topology, a terminal of any branch of a network or an interconnection common to two or more branches of a network. (188) *Synonyms* **junction point, nodal point.** **2.** In a switched network, one of the switches forming the network backbone. *Note:* A node may also include patching and control facilities. (188) **3.** A technical control facility (TCF). (188) **4.** A point in a standing or stationary wave at which the amplitude is a minimum. (188) *In this sense, synonym* **null.**

noise: **1.** An undesired disturbance within the frequency band of interest; the summation of unwanted or disturbing energy introduced into a communications system from man-made and natural sources. (188) **2.** A disturbance that affects a signal and that may distort the information carried by the signal. **3.** Random variations of one or more characteristics of any entity such as voltage, current, or data. **4.** A random signal of known statistical properties of amplitude, distribution, and spectral density. **5.** *Loosely,* any disturbance tending to interfere with the normal operation of a device or system.

noise current: **1.** Interfering and unwanted electrical currents in a device or system. **2.** In optical communications, the rms component of the optical detector output electrical current with no incoming signal present.

noise equivalent power (NEP): At a given data-signaling rate or modulation frequency, operating wavelength, and effective noise bandwidth, the radiant power that produces a signal-to-noise ratio of unity at the output of a given optical detector. (188) *Note 1:* Some manufacturers and authors define NEP as the minimum detectable power per square root bandwidth. When defined this way, NEP has the units of watts per (hertz)$^{1/2}$. Therefore, the term is a misnomer, because the units of power are watts. *Note 2:* Some manufacturers define NEP as the radiant power that produces a signal-to-dark-current noise ratio of unity. The NEP measurement is valid only if the dark-current noise dominates the noise level.

noise factor: *Synonym* **noise figure.**

noise figure (NF): The ratio of the output noise power of a device to the portion thereof attributable to thermal noise in the input termination at standard noise temperature (usually 290 K). (188) *Note:* The noise figure is thus the ratio of actual output noise to that which would remain if the device itself did not introduce noise. In heterodyne systems, output noise power includes spurious contributions from image-frequency transformation, but the portion attributable to thermal noise in the input termination at standard noise temperature includes only that which appears in the output via the principal frequency transformation of the system, and excludes that which appears via the image frequency transformation. *Synonym* **noise factor.**

noise level: The noise power, usually relative to a reference. (188) *Note:* Noise level is usually measured in dB for relative power or picowatts for absolute power. A suffix is added to denote a particular reference base or specific qualities of the measurement. Examples of noise-level measurement units are dBa, dBa(F1A), dBa(HA1), dBa0, dBm, dBm(psoph), dBm0, dBm0P, dBrn, dBrnC, dBrn(f_1-f_2), dBrn(144-line), pW, pWp, and pWp0.

noise power: **1.** The power generated by a random electromagnetic process. (188) **2.** Interfering and unwanted power in an electrical device or system. **3.** In the acceptance testing of radio transmitters, the mean power supplied to the antenna transmission line by a radio transmitter when loaded with noise having a Gaussian amplitude-vs.-frequency distribution. (188)

noise power density: The noise power in a bandwidth of 1 Hz, *i.e.*, the noise power per hertz at a point in a noise spectrum. *Note:* The noise-power density of the internal noise that is contributed by a receiving system to an incoming signal is expressed as the product of Boltzmann's constant, k, and the equivalent noise temperature, T_n. Thus, the noise-power density is often expressed simply as k*T*. *Synonym* **k*T*.** [From Weik '89]

noise suppression: **1.** Reduction of the noise power level in electrical circuits. **2.** The process of automatically reducing the noise output of a receiver during periods when no carrier is being received. (188) *Contrast with* **squelch.**

noise temperature: At a pair of terminals, the temperature of a passive system having an available noise power per unit bandwidth at a specified frequency equal to that of the actual terminals of a network. *Note:* The noise temperature of a simple resistor is the actual temperature of that resistor. The noise temperature of a diode may be many times the actual temperature of the diode.

noise voltage: **1.** Interfering and unwanted voltage in an electronic device or system. **2.** In optical communications, the rms component of the optical detector output electrical voltage with no incoming signal present.

noise weighting: A specific amplitude-vs.-frequency characteristic that permits a measuring set to give numerical readings that approximate the interfering effects to any listener using a particular class of telephone instrument. (188) *Note 1:* Noise weighting measurements are made in lines terminated either by the measuring set or the class of instrument. *Note 2:* The most widely used noise weightings were established by agencies concerned with public telephone service, and are based on characteristics of specific commercial telephone instruments, representing successive stages of technological development. The coding of commercial apparatus appears in the nomenclature of certain weightings. The same weighting nomenclature and units are used in military versions of commercial noise measuring sets.

noise window: A notch, *i.e.*, a dip, in the noise frequency spectrum characteristic of a device, such as a transmitter, receiver, channel, or amplifier, from external sources or internal sources. *Note:* The noise window is usually represented as a band of lower amplitude noise in a wider band of higher amplitude noise. [From Weik '89]

noisy black: **1.** In facsimile or display systems, such as television, a nonuniformity in the black area of the image, *i.e.*, document or picture, caused by the presence of noise in the received signal. **2.** A signal or signal level that is supposed to represent a black area on the object, but has a noise content sufficient to cause the creation of noticeable white spots on the display surface or record medium.

noisy white: **1.** In facsimile or display systems, such as television, a nonuniformity in the white area of the image, *i.e.*,, document or picture, caused by the presence of noise in the received signal. **2.** A signal or signal level that is supposed to represent a white area on the object, but has a noise content sufficient to cause the creation of noticeable black spots on the display surface or record medium.

nominal bandwidth: The widest band of frequencies, inclusive of guard bands, assigned to a channel. (188) *Note:* Nominal bandwidth should not be confused with the terms *"necessary bandwidth," "occupied bandwidth,"* or *"rf bandwidth."*

nominal bit stuffing rate: The rate at which stuffing bits are inserted when both the input and output bit rates are at their nominal values. (188)

nominal linewidth: In facsimile systems, the average separation between centers of adjacent scanning or recording lines. (188)

nonassociated common-channel signaling: A form of common-channel signaling where the signaling channel serves one or more trunk groups, at least one of which terminates at a point other than the signal transfer point at which the signaling channel terminates. (188)

nonblocking switch: A switch that has enough paths across it that an originated call can always reach an available line without encountering a busy condition. (188)

non-call associated signaling (NCAS): Signaling that is independent of an end-to-end bearer connection, including support for the functions of registration, authentication, and validation.

noncentralized operation: Operation that uses a control discipline for multipoint data communication links in which transmission may be between tributary stations or between the control station and tributary stations.

noncircularity: *Synonym* **ovality.**

noncritical technical load: Of the total technical load at a facility during normal operation, the part that is not required for synchronous operation. (188)

nonerasable storage: *Synonym* **read-only memory.**

non-fixed access: In personal communications service (PCS), terminal access to a network in which there is no set relationship between a terminal and the access interface. *Note:* The access interface and the terminal each has its own separate "identifiers." The terminal may be moved from one access interface to another while maintaining the terminal's unique identity.

nonlinear distortion: Distortion caused by a deviation from a linear relationship between specified input and output parameters of a system or component. (188)

nonlinear scattering: Direct conversion of a photon from one wavelength to one or more other wavelengths. *Note 1:* In an optical fiber, nonlinear scattering is usually not important below the threshold irradiance for stimulated nonlinear scattering. *Note 2:* Examples of nonlinear scattering are Raman and Brillouin scattering.

nonloaded twisted pair: A twisted pair that has no intentionally added inductance.

nonoperational load: Administrative, support, and housing power requirements. (188) *Synonym* **utility load.**

nonresonant antenna: *Synonym* **aperiodic antenna.**

non-return-to-zero (NRZ): A code in which "1s" are represented by one significant condition and "0s" are represented by another, with no neutral or rest condition, such as a zero amplitude in amplitude modulation (AM), zero phase shift in phase-shift keying (PSK), or mid-frequency in frequency-shift keying (FSK). (188) *Note 1: Contrast with* **Manchester code, return-to-zero.** *Note 2:* For a given data signaling rate, *i.e.,* bit rate, the NRZ code requires only one-half the bandwidth required by the Manchester code.

non-return-to-zero change-on-ones (NRZ1): A code in which "1s" are represented by a change in a significant condition and "0s" are represented by no change. (188)

non-return-to-zero mark (NRZ-M): A binary encoding scheme in which a signal parameter, such as electric current or voltage, undergoes a change in a significant condition or level every time that a "one" occurs, but when a "zero" occurs, it remains the same, *i.e.,* no transition occurs. *Note 1:* The transitions could also occur only when "zeros" occur and not when "ones" occur. If the significant condition transition occurs on each "zero," the encoding scheme is called "non-return-to-zero space" (NRZ-S). *Note 2:* NRZ-M and NRZ-S signals are technically interchangeable; *i.e.,* one is the logical "NOT" (inverse) of the other. It is necessary for the receiver to have prior knowledge of which scheme is being used. Without such knowledge, it is impossible for the receiver to interpret the data stream correctly; *i.e.,* its output may be the correct data stream or the logical inverse of the correct data stream. [From Weik '89] *Contrast with* **non-return-to-zero space.** *Synonyms* **conditioned baseband representation, differentially encoded baseband, non-return-to-zero one (NRZ-1), NRZ-B.**

non-return-to-zero one (NRZ-1): *Synonym* **non-return-to-zero mark.**

non-return-to-zero space (NRZ-S): A binary encoding scheme in which a signal parameter, such as electric current or voltage, undergoes a change in a significant condition or level every time that a "zero" occurs, but when a "one" occurs, it remains the same, *i.e.,* no transition occurs. *Note 1:* The transitions could also occur only when "ones" occur and not when "zeros" occur. If the significant condition transition occurs on each "one," the encoding scheme is called "non-return to zero mark" (NRZ-M). *Note 2:* NRZ-S and NRZ-M signals are technically interchangeable; *i.e.,* one is the logical "NOT" (inverse) of the other. It is necessary for the receiver to have prior knowledge of which scheme is being used. Without such knowledge, it is impossible for the receiver to interpret the data stream correctly; *i.e.,* its output may be the correct data stream or the logical inverse of the correct data stream. [From Weik '89] *Contrast with* **non-return-to-zero mark.** *Synonym* **non-return-to-zero.**

nonshifted fiber: *Synonym* **dispersion-unshifted fiber.**

nonsynchronous data transmission channel: A data transmission channel in which separate timing information is not transferred between the data terminal equipment (DTE) and the data circuit terminating equipment (DCE). (188)

nonsynchronous network: *Synonym* **asynchronous network.**

nonsynchronous system: *See* **asynchronous transmission.**

nonsynchronous transmission: *See* **asynchronous transmission.**

nontechnical load: Of the total operational load at a facility during normal operation, the part used for support purposes, such as general lighting, heating, air-conditioning, and ventilating equipment. (188)

nontransparent mode: A mode of operating a data transmission system in which control characters are treated and interpreted as such, rather than simply as data or text bits in a bit. [From Weik '89]

normalized frequency (*V*): **1.** In an optical fiber, a dimensionless quantity, *V*, given by

$$V = \frac{2\pi a}{\lambda}\sqrt{n_1^2 - n_2^2} \ ,$$

where a is the core radius, λ is the wavelength in vacuum, n_1 is the maximum refractive index of the core, and n_2 is the refractive index of the homogeneous cladding. *Note 1:* In multimode operation of an optical fiber having a power-law refractive index profile, the approximate number of bound modes, *i.e.*, the mode volume, is given by

$$\frac{V^2}{2}\left(\frac{g}{g+2}\right) \ ,$$

where V is the normalized frequency greater than 5 and g is the profile parameter. *Note 2:* For a step index fiber, the mode volume is given by $V^2/2$. For single-mode operation, $V < 2.405$. *Synonym* **V number.** **2.** The ratio between an actual frequency and a reference value. **3.** The ratio between an actual frequency and its nominal value. (188)

notch: In a relatively wide band of frequencies, not necessarily of uniform amplitude, a narrow band of frequencies having relatively low amplitudes.

notched filter: *Synonym* **band-stop filter.**

notched noise: Noise from which a narrow band of frequencies has been removed. (188) *Note:* Notched noise is usually used for testing devices or circuits.

not-ready condition: At the data terminal equipment/data circuit-terminating equipment (DTE/DCE) interface, a steady-state condition that indicates that the DCE is not ready to accept a call-request signal or that the DTE is not ready to accept an incoming call. (188)

Np: *Abbreviation for* **neper.**

NRI: *Abbreviation for* **net radio interface.**

NRZ: *Abbreviation for* **non-return-to-zero.**

NRZ-M: *Synonym* **non-return-to-zero mark.**

NS/EP telecommunications: *Abbreviation for* **National Security or Emergency Preparedness telecommunications.** Telecommunications services that are used to maintain a state of readiness or to respond to and manage any event or crisis (local, national, or international) that causes or could cause injury or harm to the population, damage to or loss of property, or degrade or threaten the national security or emergency preparedness posture of the United States.

***n*-sequence:** A pseudorandom binary sequence of n bits that (a) is the output of a linear shift register and (b) has the property that, if the shift register is set to any nonzero state and then cycled, a pseudorandom binary sequence of a maximum of $n = 2^m-1$ bits will be generated, where m is the number of stages, *i.e.*, the number of bit positions in the register, before the shift register returns to its original state and the *n*-bit output sequence repeats. *Note:* The register may be used to control the sequence of frequencies for a frequency-hopping spread spectrum transmission system.

NTI: *Abbreviation for* **network terminating interface.**

NTN: *Abbreviation for* **network terminal number.**

NTSC standard: *Abbreviation for* **National Television Standards Committee standard.** The North American standard (525-line interlaced raster-scanned video) for the generation, transmission, and reception of television signals. *Note 1:* In the NTSC standard, picture information is transmitted in vestigial-sideband AM and sound information is transmitted in FM. *Note 2:* In addition to North America, the NTSC standard is used in Central America, a number of South American countries, and some Asian countries, including Japan. *Contrast with* **PAL, PAL-M, SECAM.**

nuclear hardness: 1. An expression of the extent to which the performance of a system, facility, or device is expected to degrade in a given nuclear environment. **2.** The physical attributes of a system or component that will allow survival in an environment that includes nuclear radiation and electromagnetic impulses (EMI). *Note 1:* Nuclear hardness may be expressed in terms of either susceptibility or vulnerability. *Note 2:* The extent of expected performance degradation (*e.g.,* outage time, data lost, and equipment damage) must be defined or specified. The environment (*e.g.,* radiation levels, overpressure, peak velocities, energy absorbed, and electrical stress) must be defined or specified. **3.** The physical attributes of a system or component that will allow a defined degree of survivability in a given environment created by a nuclear weapon. *Note:* Nuclear hardness is determined for specified or actual quantified environmental conditions and physical parameters, such as peak radiation levels, overpressure, velocities, energy absorbed, and electrical stress. It is achieved through design specifications and is verified by test and analysis techniques.

null: 1. In an antenna radiation pattern, a zone in which the effective radiated power is at a minimum relative to the maximum effective radiated power of the main beam. *Note 1:* A null often has a narrow directivity angle compared to that of the main beam. Thus, the null is useful for several purposes, such as radio navigation and suppression of interfering signals in a given direction. *Note 2:* Because there is reciprocity between the transmitting and receiving characteristics of an antenna, there will be corresponding nulls for both the transmitting and receiving functions. **2.** A dummy letter, letter symbol, or code group inserted in an encrypted message to delay or prevent its solution, or to complete encrypted groups for transmission or transmission security purposes. [NIS] **3.** In database management systems, a special value assigned to a row or a column indicating either unknown values or inapplicable usage. **4.** *Synonym* **node (def. #4).**

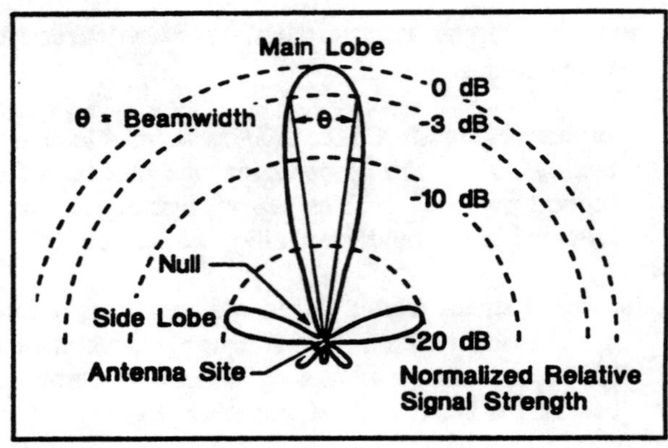

null

null character (NUL): In transmission systems, a control character (a) that is used to accomplish media-fill stuffing or a time-fill stuffing in storage device or in a data transmission line and (b) that may be inserted and removed from a series of characters without affecting the meaning of the series. *Note:* The null character may affect the control of equipment or the format of messages. [From Weik '89]

numerical aperture (*NA*): 1. The sine of the vertex angle of the largest cone of meridional rays that can enter or leave an optical system or element, multiplied by the refractive index of the medium in which the vertex of the cone is located. *Note:* The *NA* is generally measured with respect to an object or image point and will vary as that point is moved. **2.** For an optical fiber in which the refractive index decreases monotonically from n_1 on the axis to n_2 in the cladding, an expression of the extent of the fiber's ability to accept, in its bound modes, non-normal incident rays, given by $NA = (n_1^2 - n_2^2)^{1/2}$. *Note:* In multimode fibers, the term *equilibrium numerical aperture* is sometimes used. This refers to the numerical aperture with respect to the extreme exit angle of a ray emerging from a fiber in which equilibrium mode distribution has been established. (188) **3.** *Colloquially,* the sine of the radiation or acceptance angle of an optical fiber, multiplied by the refractive index of the material in contact with the exit or entrance face. *Note:* This usage is approximate and imprecise, but is often encountered.

numerical aperture loss: A loss of optical power that occurs at a splice or a pair of mated connectors when the numerical aperture of the "transmitting" fiber exceeds that of the "receiving" fiber, even if the cores are precisely the same diameter and are perfectly aligned. [FAA] *Note 1:* The higher numerical aperture of the transmitting fiber means that it emits a larger cone of light than the receiving fiber is capable of accepting, resulting in a coupling loss. [FAA] *Note 2:* In the opposite case of numerical aperture mismatch, where the transmitting fiber has the lower numerical aperture, no numerical aperture loss occurs, because the receiving fiber is capable of accepting light from any bound mode of the transmitting fiber. [After FAA]

***n*-unit code:** A code in which the signals or groups of digits that represent coded items, such as characters, have the same number of signal elements or digits, namely *n* elements or digits, where *n* may be any positive integer. *Note:* An example of an *n*-unit code is the 7-unit code (8-unit with parity) ASCII code. Each character is represented by a pattern of 7 binary digits. The units may also be characters or other special signs. [From Weik '89]

***n*-user:** In the ISO Open Systems Interconnection—Reference Model (OSI—RM), an *n*+1 entity that uses the services of the *n*-layer, and below, to communicate with another *n*+1 entity. *Note:* If *n* identifies a specific or a reference level, the *n*+1 layer is the layer above the n layer and the *n*–1 layer is the layer below the *n* layer. Thus, the *n*+2 layer is two layers above the *n* layer.

NVIS: *Abbreviation for* **near vertical-incidence skywave.**

NXX code: In the North American direct distance dialing numbering plan, a central office code of three digits that designates a particular central office or a given 10,000-line unit of subscriber lines; "N" is any number from 2 to 9, and "X" is any number from 0 to 9.

Nyquist interval: The maximum time interval between equally spaced samples of a signal that will enable the signal waveform to be completely determined. (188) *Note 1:* The Nyquist interval is equal to the reciprocal of twice the highest frequency component of the sampled signal. *Note 2:* In practice, when analog signals are sampled for the purpose of digital transmission or other processing, the sampling rate must be more frequent than that defined by Nyquist's theorem, because of quantization error introduced by the digitizing process. The required sampling rate is determined by the accuracy of the digitizing process.

Nyquist rate: The reciprocal of the Nyquist interval, *i.e.,*the minimum theoretical sampling rate that fully describes a given signal, *i.e.,* enables its faithful reconstruction from the samples. *Note:* The actual sampling rate required to reconstruct the original signal will be somewhat higher than the Nyquist rate, because of quantization errors introduced by the sampling process.

Nyquist's theorem: A theorem, developed by H. Nyquist, which states that an analog signal waveform may be uniquely reconstructed, without error, from samples taken at equal time intervals. The sampling rate must be equal to, or greater than, twice the highest frequency component in the analog signal. *Synonym* **sampling theorem.**

(this page intentionally left blank)

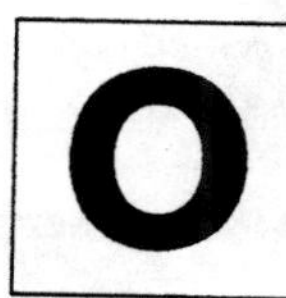

object: 1. In image processing, a sub-region of an image that is perceived as a single entity. *Note:* An image can contain more than one object. **2.** In facsimile systems, the image, the likeness of which is to be transmitted.

object persistence: In a video display, distortion wherein the entirety of some object (or objects) that appeared in a previous frame (and that should no longer appear) remain in the current frame and in subsequent frames as a faded image or as an outline.

object retention: In a video display, distortion in which a fragment of an object that appeared in a previous frame (and should no longer appear) remains in the current and subsequent video frames.

OCC: *Abbreviation for* **other common carrier.**

occupancy: For equipment, such as a circuit or a switch, the ratio of the actual time in use to the available time during a 1-hour period. *Note 1:* Occupancy is usually expressed in percent. *Note 2:* Occupancy may be plotted versus time of day. *Synonym* **usage.**

occupied bandwidth: The width of a frequency band such that, below the lower and above the upper frequency limits, the mean powers emitted are each equal to a specified percentage *B*/2 of the total mean power of a given emission. Unless otherwise specified by the CCIR for the appropriate class of emission, the value of *B*/2 should be taken as 0.5%. [NTIA] [RR] (188) *Note 1:* The percentage of the total power outside the occupied bandwidth is represented by *B*. *Note 2:* In some cases, *e.g.*, multichannel frequency-division multiplexing systems, use of the 0.5% limits may lead to certain difficulties in the practical application of the definition of occupied and necessary bandwidth; in such cases, a different percentage may prove useful.

oceanographic data interrogating station: A station in the maritime mobile service the emissions of which are used to initiate, modify or terminate functions of equipment directly associated with an oceanographic data station, including the station itself. [NTIA]

oceanographic data station: A station in the maritime mobile service located on a ship, buoy, or other sensor platform the emissions of which are used for transmission of oceanographic data. [NTIA]

OCR: *Abbreviation for* **optical character reader, optical character recognition.**

octet: A byte of eight binary digits usually operated upon as an entity. (188)

octet alignment: The configuration of a field composed of an integral number of octets. *Note:* If the field is not divisible by eight, bits (usually zeros) are added to either the first octet (left justification) or the last octet (right justification).

OD: *Abbreviation for* **optical density.**

odd-even check: *Synonym* **parity check.**

odd parity: *See* **parity, parity check.**

OFC: *Abbreviation for* **optical fiber, conductive.** *Note:* OFC is the designation given by the National Fire Protection Association (NFPA) to interior fiber-optic cables which contain at least one electrically conductive, non-current-carrying component, such as a metallic strength member or vapor barrier, and which are not certified for use in plenum or riser applications. [After FAA]

OFCP: *Abbreviation for* **optical fiber, conductive, plenum**. *Note:* OFCP is the designation given by the National Fire Protection Association (NFPA) to interior fiber-optic cables which contain at least one electrically conductive, non-current-carrying component such as a metallic strength member or vapor barrier, and which are certified for use in plenum applications. [After FAA]

OFCR: *Abbreviation for* **optical fiber, conductive, riser**. *Note:* OFCR is the designation given by the National Fire Protection Association (NFPA) to interior fiber-optic cables which contain at least one electrically conductive, non-current-carrying component such as a metallic strength member or vapor barrier, and which are certified for use in riser applications. [After FAA]

off-axis optical system: An optical system in which the optical axis of the aperture is not coincident with the mechanical center of the aperture. *Note:* The principal applications of off-axis optical systems are to avoid obstruction of the primary aperture by secondary optical elements, instrument packages, or sensors, and to provide ready access to instrument packages or sensors at the focus. The engineering tradeoff of an off-axis optical system is an increase in image aberrations.

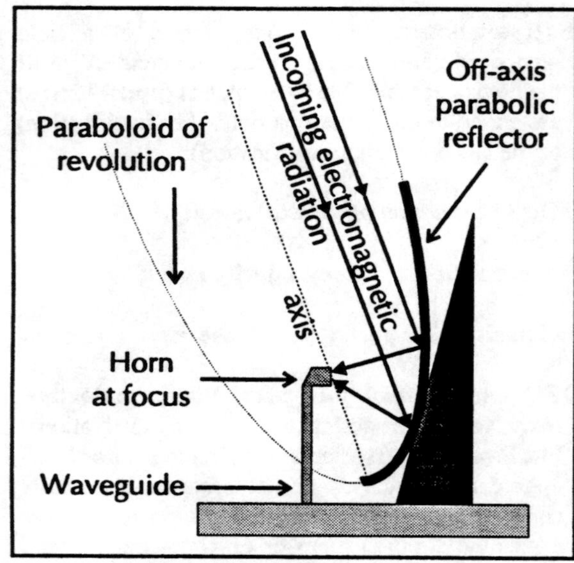

off-axis optical system used as a microwave antenna system

off-hook: **1.** In telephony, the condition that exists when an operational telephone instrument or other user instrument is in use, *i.e.,* during dialing or communicating. (188) *Note:* Off-hook originally referred to the condition that prevailed when the separate earpiece, *i.e.,* receiver, was removed from its switchhook, which extended from a vertical post that also supported the microphone, and which connected the instrument to the line when not depressed by the weight of the receiver. **2.** One of two possible signaling states, such as tone or no tone and ground connection versus battery connection. (188) *Note:* If off-hook pertains to one state, on-hook pertains to the other. **3.** The active state, *i.e.,* closed loop, of a subscriber or PBX user loop. (188) **4.** An operating state of a communications link in which data transmission is enabled either for (a) voice or data communications or (b) network signaling.

off-hook service: *Synonym* **hotline.**

off-hook signal: In telephony, of a circuit, a signal indicating seizure, request for service, or a busy condition. (188)

office classification: Prior to divestiture, numbers that were assigned to offices according to their hierarchical function in the U.S. public switched telephone network. *Note 1:* The following class numbers are used:
Class 1: Regional Center (RC)
Class 2: Sectional Center (SC)
Class 3: Primary Center (PC)
Class 4: Toll Center (TC) [Only if operators are present; otherwise Toll Point (TP)]
Class 5: End Office (EO) [Local central office]
Note 2: Any one center handles traffic from one center to two or more centers lower in the hierarchy. Since divestiture, these designations have become less firm.

off line: **1.** In computer technology, the state or condition of a device or equipment that is not under the direct control of another device. **2.** In computer technology, the status of a device that is disconnected from service.

off-line: **1.** Pertaining to the operation of a functional unit when not under the direct control of the system with which it is associated. (188) *Note 1:* Off-line units are not available for immediate use on demand by the system. *Note 2:* Off-line units may be independently operated. **2.** Pertaining to equipment that is disconnected from a system, is not in operation, and usually has its main power source disconnected or turned off.

off-line recovery: The process of recovering nonprotected message traffic by use of an off-line processor or central processing unit. (188)

off-line storage: Storage that is not under the control of a processing unit.

off-net calling: The process by which telephone calls that originate or pass through private switching systems in transmission networks are extended to stations in a public switched telephone system. (188)

off-premises extension (OPX): An extension telephone, PBX station, or key system station located on property that is not contiguous with that on which the main telephone, PBX, or key system is located.

off-the-air: **1.** In radio communications systems, pertaining to a station that is completely shut down, *i.e.*, that is not transmitting any signal, not even an unmodulated carrier. **2.** In a radio station, pertaining to a particular source of modulation, such as a specific microphone, that is disconnected, *i.e.*, is no longer capable of modulating the carrier. *Note:* The carrier may continue unmodulated or it may be modulated by another signal source.

off-the-air monitoring: **1.** In radio net operations, the listening, by the net-control station, to the transmissions of stations in the net, particularly to check the quality of their transmissions. *Note:* Off-the-air monitoring is usually performed during periods when the net-control station is not transmitting. **2.** The listening, by a radio station, to its own transmissions by receiving the signal that has been transmitted by the transmitting antenna, to discover the quality of the signal being transmitted to other stations or being broadcast. *Note:* In off-the-air monitoring, the received signal must have traveled through the air a reasonable distance from the transmitting antenna and not be a signal that is tapped on its way to the transmitting antenna internal to the station or in the antenna transmission line, *i.e.*, the feeder. The monitoring distance should be such that direct inductive or capacitive coupling between the transmitting antenna and monitor antenna does not occur. [From Weik '89]

off-the-shelf: Pertaining to equipment already manufactured and available for delivery from stock. (188)

OFN: *Abbreviation for* **optical fiber, nonconductive**. *Note:* OFN is the designation given by the National Fire Protection Association (NFPA) to interior fiber-optic cables which contain no electrically conductive component, and which are not certified for use in plenum or riser applications. [After FAA]

OFNP: *Abbreviation for* **optical fiber, nonconductive, plenum**. *Note:* OFNP is the designation given by the National Fire Protection Association (NFPA) to interior fiber-optic cables which contain no electrically conductive component, and which are certified for use in plenum applications. [After FAA]

OFNR: *Abbreviation for* **optical fiber, nonconductive, riser**. *Note:* OFNR is the designation given by the National Fire Protection Association (NFPA) to interior fiber-optic cables which contain no electrically conductive component, and which are certified for use in riser applications. [After FAA]

oligarchically synchronized network: A synchronized network in which the timing of all clocks is controlled by a selected few clocks.

Omega: A global radionavigation system that enables user with special receivers to obtain position information by measuring phase difference between precisely timed signals radiated by a network of eight transmitting stations deployed worldwide. (188) *Note:* The transmitted signals time-share transmission on frequencies of 10.2, 11.05, 11.33, and 13.6 kHz. Since the transmissions are coordinated with UTC (USNO), they also provide time reference.

omnidirectional antenna: An antenna that has a radiation pattern that is nondirectional in azimuth. (188) *Note:* The vertical radiation pattern may be of any shape.

omnidirectional range station: A radionavigation land station in the aeronautical radionavigation service providing direct indication of the bearing (omnibearing) of that station from an aircraft. [NTIA]

ONA: *Abbreviation for* **open network architecture.**

on-board communication station: A low-powered mobile station in the maritime mobile service intended for use for internal communications on board a ship, or between a ship and its lifeboats and liferafts during lifeboat drills or operations, or for communication within a group of vessels being towed or pushed, as well as for line handling and mooring instructions. [NTIA] [RR]

144-line weighting: In telephony, a noise weighting used in a noise measuring set to measure line noise as it would be perceived if the line were terminated with a No. 144-receiver, or a similar instrument. (188) *Note:* The meter scale readings are in dBrn (144-line).

144-receiver weighting: In telephony, a noise weighting used in a noise measuring set to measure noise across the receiver of an instrument equipped with a No. 144-receiver. (188) *Note:* The meter scale readings are in dBrn (144-receiver).

one-way communication: Communication in which information is always transferred in only one preassigned direction. *Note 1:* One-way communication is not necessarily constrained to one transmission path. *Note 2:* Examples of one-way communications systems include broadcast stations, one-way intercom systems, and wireline news services.

one-way-only channel: A channel capable of transmission in only one direction, which cannot be reversed. (188) *Synonym* **unidirectional channel.**

one-way operation: *Synonym* **simplex operation (def. #1).**

one-way reversible operation: *Synonym* **half-duplex operation.**

one-way trunk: A trunk between two switching centers, over which traffic may be originated from one preassigned location only. (188) *Note 1:* The traffic may consist of two-way communications; the expression *"one way"* refers only to the origin of the demand for a connection. *Note 2:* At the originating end, the one-way trunk is known as an *"outgoing trunk"*; at the other end, it is known as an *"incoming trunk"*.

on-hook: 1. In telephony, the condition that exists when an operational telephone, or other user instrument, is not in use. (188) *Note:* On-hook originally referred to the storage of an idle telephone receiver, *i.e.,* separate earpiece, on a hook that extended from a vertical post that supported the microphone also. The hook was mechanically connected to a switch that automatically disconnected the idle telephone from the network. **2.** One of two possible signaling states, such as tone or no tone, or ground connection versus battery connection. (188) *Note:* If on-hook pertains to one state, off-hook pertains to the other. **3.** The idle state, *i.e.,* open loop, of a subscriber or PBX user loop. (188) **4.** An operating state of a communications link in which data transmission is disabled and a high-impedance, *i.e.,* open circuit, is presented to the link by the end instrument(s). *Note:* During the on-hook condition, the link is responsive to ringing signals.

on-hook signal: In telephony, of a circuit, a signal indicating a disconnect, unanswered call, or an idle condition. (188)

on line: 1. In computer technology, the state or condition of a device or equipment that is under the direct control of another device. **2.** In computer technology, the status of a device that is functional and ready for service.

on-line: 1. Pertaining to the operation of a functional unit when under the direct control of the system with which it is associated. (188) *Note 1:* On-line units are available for immediate use on demand by the system without human intervention. *Note 2:* On-line units may not be independently operated. **2.** Pertaining to equipment that is connected to a system, and is in operation.

online computer system: A computer system that is a part of, or is embedded in, a larger entity, such as a communications system, and that interacts in real or near-real time with the entity and its users.

on-premises extension: An extension telephone, PBX station, or key system station located on property that is contiguous with that on which the main telephone, PBX, or key system is located.

on-premises wiring: Customer-owned metallic or optical-fiber communications transmission lines, installed within or between buildings. *Note:* On-premises wiring may consist of horizontal wiring, vertical wiring, and backbone wiring, and may extend from the external network interface to the user work station areas. It includes the total communications wiring to transport current or future data, voice, LAN, and image information.

on-the-air: **1.** In radio communications systems, pertaining to a station that is transmitting a carrier, whether or not the carrier is modulated. **2.** In a radio station, pertaining to a particular source of modulation, such as a specific microphone, that is connected, *i.e.,* is capable of modulating the carrier.

open circuit: **1.** In communications, a circuit available for use. (188) **2.** In electrical engineering, a circuit that contains an essentially infinite impedance. *Note:* An open circuit may be intentional, as in a switch, or may constitute a fault, as in a severed cable.

open dual bus: A dual bus in which the head-of-bus functions for both buses are at different locations.

open network architecture (ONA): In the context of the FCC's Computer Inquiry III, the overall design of a communication carrier's basic network facilities and services to permit all users of the basic network to interconnect to specific basic network functions and interfaces on an unbundled, equal-access basis. *Note:* The ONA concept consists of three integral components: (a) basic serving arrangements (BSAs), (b) basic service elements (BSEs), and (c) complementary network services.

open system: A system with characteristics that comply with specified, publicly maintained, readily available standards and that therefore can be connected to other systems that comply with these same standards.

open systems architecture: **1.** The layered hierarchical structure, configuration, or model of a communications or distributed data processing system that (a) enables system description, design, development, installation, operation, improvement, and maintenance to be performed at a given layer or layers in the hierarchical structure, (b) allows each layer to provide a set of accessible functions that can be controlled and used by the functions in the layer above it, (c) enables each layer to be implemented without affecting the implementation of other layers, and (d) allows the alteration of system performance by the modification of one or more layers without altering the existing equipment, procedures, and protocols at the remaining layers. *Note 1:* Examples of independent alterations include (a) converting from wire to optical fibers at a physical layer without affecting the data-link layer or the network layer except to provide more traffic capacity, and (b) altering the operational protocols at the network level without altering the physical layer. *Note 2:* Open systems architecture may be implemented using the Open Systems Interconnection—Reference Model (OSI—RM) as a guide while designing the system to meet performance requirements. **2.** Nonproprietary systems architecture.

Open Systems Interconnection (OSI): Pertaining to the logical structure for communications networks standardized by the International Organization for Standardization (ISO). *Note:* Adherence to the standard enables any OSI-compliant system to communicate with any other OSI-compliant system for a meaningful exchange of information.

Open Systems Interconnection (OSI)—Architecture: Communications system architecture that adheres to the set of ISO standards relating to open systems architecture.

Open Systems Interconnection (OSI)—Protocol Specification: The lowest level of abstraction within the OSI standards scheme. *Note:* Each OSI—Protocol Specification operates at a single layer. Each defines the primitive operations and permissible responses required to exchange information between peer processes in communicating systems to carry out all or a subset of the services defined within the OSI—Service Definitions for that layer.

Open Systems Interconnection—Reference Model (OSI—RM): An abstract description of the digital communications between application processes running in distinct systems. The model employs a hierarchical structure of seven layers. Each layer performs value-added service at the request of the adjacent higher layer and, in turn, requests more basic services from the adjacent lower layer:

➢ **Physical Layer:** Layer 1, the lowest of seven hierarchical layers. The Physical layer performs services requested by the Data Link Layer. The major functions and services performed by the physical layer are: (a) establishment and termination of a connection to a communications medium; (b) participation in the process whereby the communication resources are effectively shared among multiple users, *e.g.*, contention resolution and flow control; and, (c) conversion between the representation of digital data in user equipment and the corresponding signals transmitted over a communications channel.
➢**Data Link Layer:** Layer 2. This layer responds to service requests from the Network Layer and issues service requests to the Physical Layer. The Data Link Layer provides the functional and procedural means to transfer data between network entities and to detect and possibly correct errors that may occur in the Physical Layer. *Note:* Examples of data link protocols are HDLC and ADCCP for point-to-point or packet-switched networks and LLC for local area networks.
➢ **Network Layer:** Layer 3. This layer responds to service requests from the Transport Layer and issues service requests to the Data Link Layer. The Network Layer provides the functional and procedural means of transferring variable length data sequences from a source to a destination via one or more networks while maintaining the quality of service requested by the Transport Layer. The Network Layer performs network routing, flow control, segmentation/desegmentation, and error control functions.
➢**Transport Layer:** Layer 4. This layer responds to service requests from the Session Layer and issues service requests to the Network Layer. The purpose of the Transport Layer is to provide transparent transfer of data between end users, thus relieving the upper layers from any concern with providing reliable and cost-effective data transfer.
➢**Session Layer:** Layer 5. This layer responds to service requests from the Presentation Layer and issues service requests to the Transport Layer. The Session Layer provides the mechanism for managing the dialogue between end-user application processes. It provides for either duplex or half-duplex operation and establishes checkpointing, adjournment, termination, and restart procedures.
➢ **Presentation Layer:** Layer 6. This layer responds to service requests from the Application Layer and issues service requests to the Session Layer. The Presentation Layer relieves the Application Layer of concern regarding syntactical differences in data representation within the end-user systems. *Note:* An example of a presentation service would be the conversion of an EBCDIC-coded text file to an ASCII-coded file.
➢ **Application Layer:** Layer 7, the highest layer. This layer interfaces directly to and performs common application services for the application processes; it also issues requests to the Presentation Layer. The common application services provide semantic conversion between associated application processes. *Note:* Examples of common application services of general interest include the virtual file, virtual terminal, and job transfer and manipulation protocols.

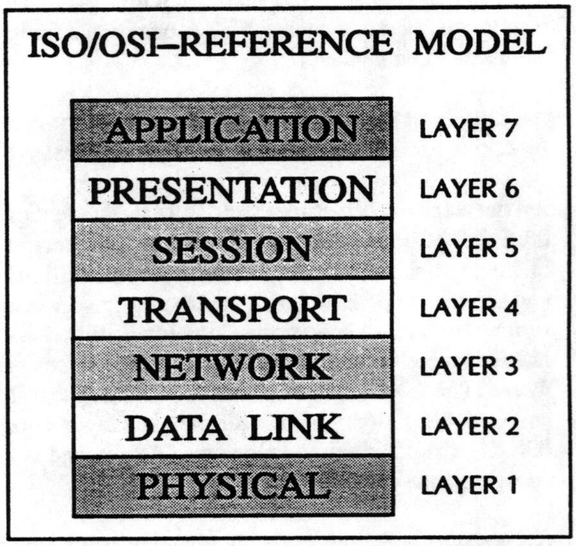

ISO/OSI–REFERENCE MODEL

APPLICATION	LAYER 7
PRESENTATION	LAYER 6
SESSION	LAYER 5
TRANSPORT	LAYER 4
NETWORK	LAYER 3
DATA LINK	LAYER 2
PHYSICAL	LAYER 1

Open Systems Interconnection—Reference Model

Open Systems Interconnection (OSI)—Service Definitions: The next lower level of abstraction below that of the OSI—Reference Model. The OSI—Service Definitions for each layer define the layer's abstract interface and the facilities provided to the user of the service independent of the mechanism used to accomplish the service.

Open Systems Interconnection (OSI)—Systems Management: In the Application Layer of the OSI—Reference Model (OSI—RM), the set of functions related to the management and status of various resources identified in all layers of the OSI—RM.

open waveguide: An all-dielectric waveguide in which electromagnetic waves are guided by a refractive index gradient so that the waves are confined to the guide by refraction or reflection from the outer surface of the guide or from surfaces within the guide. *Note 1:* In an open waveguide, the electromagnetic waves propagate, without radiation, within the waveguide, although evanescent waves coupled to internal waves may travel in the space immediately outside the waveguide. *Note 2:* Examples of open waveguides are (a) optical fibers and (b) planar waveguides in integrated optical circuits. [From Weik '89]

open wire: Conductors that are separately supported with insulators on poles or towers above the surface of the Earth. (188) *Note 1:* Open wire conductors may be insulated or uninsulated. *Note 2:* Open wire may be used in both communication applications and power applications.

operand: An entity on which an operation is performed.

operating system: An integrated collection of routines that service the sequencing and processing of programs by a computer. *Note:* An operating system may provide many services, such as resource allocation, scheduling, input/output control, and data management. Although operating systems are predominantly software, partial or complete hardware implementations may be made in the form of firmware. (188)

operating time: **1.** The time interval between the instant of occurrence of a specified input condition to a system and the instant of completion of a specified operation. **2.** In communications, computer, and information processing systems, the time interval between the instant a request for service is received from a user and the instant of final release of all facilities by the user or either of two users. (188) **3.** In communications systems conference calls, the time interval between the instant a request for service is received from one of a group of concurrent users and the instant all but one of the users have released all facilities.

operation: **1.** The method, act, process, or effect of using a device or system. (188) **2.** A well-defined action that, when applied to any permissible combination of known entities, produces a new entity, *e.g.*, the process of addition in arithmetic—in adding 5 and 3 to obtain 8, the numbers 5 and 3 are the operands, the number 8 is the result, and the plus sign is the operator indicating that the operation performed is addition. **3.** A program step, usually specified by a part of an instruction word, that is undertaken or executed by a computer. *Note:* Examples of operations include addition, multiplication, extraction, comparison, shift, transfer.

operational load: The total power requirements for communications facilities. (188)

operational service period: **1.** A period during which a telecommunications service remains in an operational state. *Note:* The operational state must be defined in accordance with specified criteria. **2.** A performance measurement period, or succession of performance measurement periods, during which a telecommunications service remains in an operational service state. (188) *Note:* An operational service period begins at the beginning of the performance measurement period in which the telecommunications service enters the operational service state, and ends at the beginning of the performance measurement period in which the telecommunications service leaves the operational service state.

operational service state: During any performance measurement period, a telecommunications service condition that existed when the calculated values of specified performance parameters were equal to or better than their associated outage thresholds. (188)

operations security: [The] process denying to potential adversaries information about capabilities and/or intentions by identifying, controlling and protecting generally unclassified evidence of the planning and execution of sensitive activities. [NIS]

operations system: In network management, a system that processes telecommunications management information and that supports and controls the performance of various telecommunications management functions. (188)

Note: An operations system performs surveillance and testing functions to support customer access maintenance.

OPSEC: *Acronym for* **operations security.**

optical amplifier: *See* **fiber amplifier, optical repeater.**

optical attenuator: In optical communications, a device used to reduce the power level of an optical signal. *Note 1:* Optical attenuators used in fiber optic communications systems may use a variety of principles for their functioning. Those using the gap-loss principle are sensitive to the modal distribution ahead of the attenuator, and should be used at or near the transmitting end, or they may introduce less loss than intended. Optical attenuators using absorptive or reflective techniques avoid this problem. *Note 2:* The basic types of optical attenuators are fixed, step-wise variable, and continuously variable.

optical axis: **1.** Of a refractive or reflective optical element, the straight line that is coincident with the axis of symmetry of the surfaces. *Note:* The optical axis of a system is often coincident with its mechanical axis, but it need not be, *e.g.*, in the case of an off-axis parabolic reflector used to transmit signals to, or receive signals from, a geosynchronous satellite. *Contrast with* **off-axis optical system.** **2.** In a lens element, the straight line which passes through the centers of curvature of the lens surfaces. [JP1] **3.** In an optical system, the line formed by the coinciding principal axes of the series of optical elements. [JP1] **4.** In an optical fiber, *synonym* **fiber axis,** which is the preferred term.

optical beamsplitter: *See* **beamsplitter.**

optical cable: *See* **fiber optic cable.**

optical cable assembly: *See* **cable assembly.**

optical cavity: A region bounded by two or more mirrors that are aligned to provide multiple reflections of lightwaves. *Note:* The resonator in a laser is an optical cavity. *In this sense, synonym* **resonant cavity.**

optical character reader (OCR): A device used for optical character recognition.

optical character recognition (OCR): The machine identification of printed characters through use of light-sensitive devices. (188)

optical conductor: *Deprecated synonym for* **optical fiber.**

optical connector: A demountable device for attaching a cabled or uncabled optical fiber to another, or to an active device such as a transmitter. *Note 1:* A connector is distinguished by the fact that it may be disconnected and reconnected, as opposed to a splice, which permanently joins two fibers. *Note 2:* Optical connectors are sometimes erroneously referred to as *"couplers."* Such usage is incorrect and is to be avoided. [After FAA]

optical coupler: *See* **directional coupler, star coupler (def.# 1), T-coupler.**

optical density (OD): For a given wavelength, an expression of the transmittance of an optical element. *Note 1:* Optical density is expressed by $\log_{10}(1/T)$ where T is transmittance. (188) *Note 2:* The higher the optical density, the lower the transmittance. *Note 3:* Optical density times 10 is equal to transmission loss expressed in decibels, *e.g.*, an optical density of 0.3 corresponds to a transmission loss of 3 dB.

optical detector: A transducer that generates an output signal when irradiated with optical energy. (188)

optical disk: A flat, circular, plastic disk coated with material on which bits may be stored in the form of highly reflective areas and significantly less reflective areas, from which the stored data may be read when illuminated with a narrow-beam source, such as a laser diode. *Note:* The bits are stored sequentially on a continuous spiral track.

optical dispersion: *See* **dispersion.**

optical fiber: A filament of transparent dielectric material, usually glass or plastic, and usually circular in cross section, that guides light. (188) *Note 1:* An optical fiber usually has a cylindrical core surrounded by, and in intimate contact with, a cladding of similar geometry. *Note 2:* The refractive index of the core must be slightly higher than that of the cladding for the light to be guided by the fiber. *Synonym* **lightguide.**

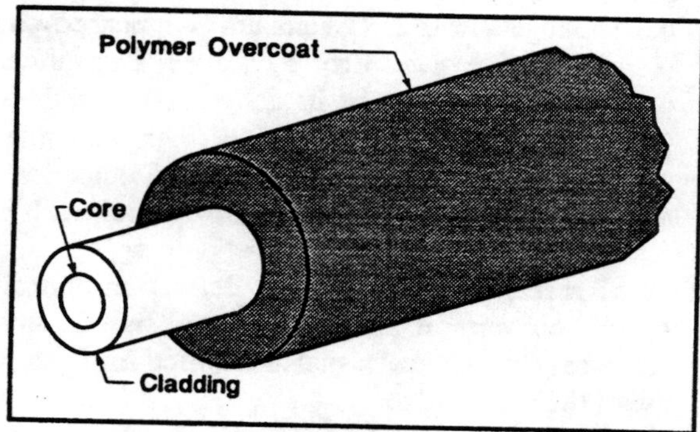

optical fiber

optical fiber cable: *See* **fiber optic cable.**

optical fiber coating: *See* **primary coating.**

optical fiber, conductive: *See* **OFC.**

optical fiber, conductive, plenum: *See* **OFCP.**

optical fiber, conductive, riser: *See* **OFCR.**

optical fiber jacket: *See* **sheath.**

optical fiber link: *See* **fiber optic link.**

optical fiber, nonconductive: *See* **OFN.**

optical fiber, nonconductive, plenum: *See* **OFNP.**

optical fiber, nonconductive, riser: *See* **OFNR.**

optical fiber nuclear hardening: Design allowances made to prevent or ameliorate the effects of gamma or high-energy neutron radiation or bombardment, that causes some optical fibers to darken, increase attenuation, or depart from normal operating parameters. *Note:* Light sources, such as LEDs and lasers, and photodetectors, also need to be hardened to prevent similar malfunctions. [From Weik '89]

optical fiber transfer function: *See* **transfer function.**

optical fiber waveguide: *See* **optical fiber.**

optical filter: In the optical regime, an element that selectively transmits or blocks a range of wavelengths.

optical heterodyning: *See* **optical mixing.**

optical interface: In a fiber optic communications link, a point at which an optical signal is passed from one equipment or medium to another without conversion to an electrical signal.

optical isolator: A device that uses a short optical transmission path to accomplish electrical isolation between elements of a circuit. *Note 1:* The optical path may be air or a dielectric waveguide. *Note 2:* The transmitting and receiving elements of an optical isolator may be contained within a single compact module, for mounting, *e.g.*, on a circuit board. *Synonym* **optoisolator.**

optical junction: Any physical interface in a fiber optic system. (188) *Note:* Source to fiber, fiber to fiber, fiber to detector, beam to prism (or lens), fiber to lens, lens to fiber, are examples of optical junctions.

optical line code: Sequences of optical pulses suitably structured to permit information transfer over an optical link.

optical link: An optical transmission channel, including any repeaters or regenerative repeaters, designed to connect two electronic or optoelectronic communications terminals. *Note:* An optical link is sometimes held to include the terminal optical transmitters and receivers, especially in the case of a communications link utilizing separate electronic terminals originally designed for metallic transmission, and retrofitted for optical transmission. [After FAA]

optically active material: A material that rotates the plane of polarization of light that passes through it. (188)

optical mixing: Optical beating, *i.e.*, the mixing, *i.e.*, heterodyning, of two lightwaves (incoming signal and local oscillator) in a nonlinear device to produce a beat frequency low enough to be further processed by conventional electronic circuitry. *Note:* Optical mixing is the optical analog of heterodyne reception of radio signals. [After FAA] *Synonym* **optical heterodyning.**

optical multiplexing: *See* **wavelength-division multiplexing.**

optical path length: **1.** In a medium of constant refractive index, *n*, the product of the geometric distance and the refractive index. **2.** In a medium of varying refractive index, the integral of $n\delta s$, where δs is an element of length along the path, and *n* is the local refractive index. (188) *Note:* Optical path length is proportional to the phase shift that a lightwave undergoes along a path.

optical power: *See* **radiant power.**

optical power budget: In a fiber-optic communication link, the allocation of available optical power (launched into a given fiber by a given source) among various loss-producing mechanisms such as launch coupling loss, fiber attenuation, splice losses, and connector losses, in order to ensure that adequate signal strength (optical power) is available at the receiver. *Note 1:* The optical power budget is usually specified or expressed in dB. *Note 2:* The amount of optical power launched into a given fiber by a given transmitter depends on the nature of its active optical source (LED or laser diode) and the type of fiber, including such parameters as core diameter and numerical aperture. Manufacturers sometimes specify an optical power budget only for a fiber that is optimum for their equipment—or specify only that their equipment will operate over a given distance, without mentioning the fiber characteristics. The user must first ascertain, from the manufacturer or by testing, (a) the transmission losses for the type of fiber to be used, (b) the required signal strength for a given level of performance. *Note 3:* In addition to transmission loss, including those of any splices and connectors, allowance should be made for at least several dB of optical power margin losses, to compensate for component aging and to allow for future splices in the event of a severed cable. *Contrast with* **optical power margin, bandwidth-limited operation.**

optical power margin: In an optical communications link, the difference between (a) the optical power that is launched by a given transmitter into the fiber, less transmission losses from all causes, and (b) the minimum optical power that is required by the receiver for a specified level of performance. *Note 1:* The optical power margin is usually expressed in dB. At least several dB of optical power margin should be included in the optical power budget. *Note 2:* The amount of optical power launched into a given fiber by a given transmitter depends on the nature of its active optical source (LED or laser diode) and the type of fiber, including such parameters as core diameter and numerical aperture. *Contrast with* **optical power budget.**

optical receiver: A device that detects an optical signal, converts it to an electrical signal, and processes the electrical signal as required for further use. (188)

optical regenerator: *See* **optical repeater.**

optical repeater: In an optical communication system, an optoelectronic device or module that receives an optical signal, amplifies it (or, in the case of a digital signal, reshapes, retimes, or otherwise reconstructs it), and retransmits it as an optical signal. (188)

optical source: **1.** In optical communications, a device that converts an electrical signal into an optical signal. *Note:* The two most commonly used optical sources are light-emitting diodes (LEDs) and laser diodes. **2.** Test equipment that generates a stable optical signal for the purpose of making optical transmission loss measurements. [After FAA]

optical spectrum: By custom and practice, the electromagnetic spectrum between the wavelengths of the vacuum ultraviolet at 0.001 μm and the far infrared at 100 μm. *Note:* The term "*optical spectrum*" originally applied only to that region of the electromagnetic spectrum visible to the normal human eye, but is now considered to include all wavelengths between the shortest wavelengths of radio and the longest of x-rays. At this writing, no formal spectral limits are recognized nationally or internationally.

optical splitter: *See* **directional coupler.**

optical switch: A switch that enables signals in optical fibers or integrated optical circuits (IOCs) to be selectively switched from one circuit to another. *Note 1:* An optical switch may operate by (a) mechanical means such as physically shifting an optical fiber to drive one or more alternative fibers, or (b) electro-optic effects, magneto-optic effects, or other methods. *Note 2:* Slow optical switches, such as those using moving fibers, may be used for alternate routing of an optical transmission path, *e.g.*, routing around a fault. Fast optical switches, such as those using electro-optic or magneto-optic effects, may be used to perform logic operations.

optical system power margin: *See* **power margin.**

optical thickness: **1.** The product of the physical thickness of an isotropic optical element and its refractive index. **2.** Of an optical system, the total optical path length through all elements. (188)

optical time domain reflectometer (OTDR): An optoelectronic instrument used to characterize an optical fiber. *Note 1:* An OTDR injects a series of optical pulses into the fiber under test. It also extracts, from the same end of the fiber, light that is scattered back and reflected back. The intensity of the return pulses is measured and integrated as a function of time, and is plotted as a function of fiber length. *Note 2:* An OTDR may be used for estimating the fiber's length and overall attenuation, including splice and mated-connector losses. It may also be used to locate faults, such as breaks.

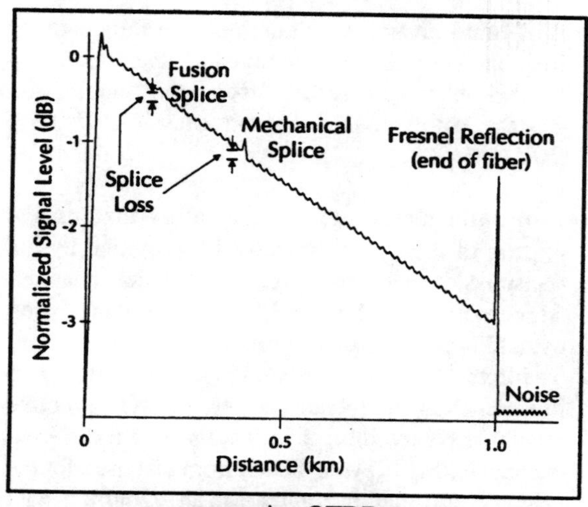

representative OTDR trace

optical transmittance: *See* **transmittance.**

optical transmitter: A device that accepts an electrical signal as its input, processes this signal, and uses it to modulate an opto-electronic device, such as an LED or an injection laser diode, to produce an optical signal capable of being transmitted via an optical transmission medium. (188)

optical waveguide: Any structure having the ability to guide optical energy. (188) *Note:* Optical waveguides may be (a) thin-film deposits used in integrated optical circuits (IOCs) or (b) optical fibers.

optimum traffic frequency: *Synonym* **FOT.**

optimum transmission frequency: *Synonym* **FOT.**

optimum working frequency: *Synonym* **FOT.**

optoelectronic: Pertaining to any device that functions as an electrical-to-optical or optical-to-electrical transducer, or an instrument that uses such a device in its operation. (188) *Note 1:* Photodiodes, LEDs, injection laser diodes, and integrated optical circuit (IOC) elements are examples of optoelectronic devices commonly used in optical fiber communications. *Note 2:* "*Electro-optical*" is often erroneously used as a synonym.

optoisolator: *Synonym* **optical isolator.**

OPX: *Abbreviation for* **off-premises extension.**

orbit: The path, relative to a specified frame of reference, described by the center of mass of a satellite or other object in space subjected primarily to natural forces, mainly the force of gravity. [NTIA] [RR]

orbit determination: The process of describing the past, present, or predicted position of a satellite in terms of orbital parameters. [JP1]

order of diversity: The number of independently fading propagation paths or frequencies, or both, used in diversity reception. (188)

orderwire circuit: A voice or data circuit used by technical control and maintenance personnel for coordination and control actions relative to activation, deactivation, change, rerouting, reporting, and maintenance of communication systems and services. (188) *Synonyms* **engineering channel, engineering orderwire, service channel.**

orderwire multiplex: A multiplex carrier set specifically designed for the purpose of carrying orderwire traffic, as opposed to one designed for carrying mission traffic. (188)

ordinary ray: *See* **birefringence.**

organizer: *See* **splice organizer**.

originating user: The user that initiates a particular information transfer transaction. *Note:* The originating user may be either the source user or the destination user.

originator: *See* **access originator, disengagement originator.**

originator-to-recipient speed of service: *Synonym* **speed of service (def. #1).**

orthogonal multiplex: A method of combining two or more digital signals that have mutually independent pulses, thus avoiding intersymbol interference. (188)

orthomode transducer: A device forming part of an antenna feed and serving to combine or separate orthogonally polarized signals.

oscillator: An electronic circuit designed to produce an ideally stable alternating voltage or current.

OSI: *Abbreviation for* **Open Systems Interconnection.**

OSI—RM: *Abbreviation for* **Open Systems Interconnection—Reference Model.**

OTAR: *Abbreviation for* **over-the-air rekeying.**

OTDR: *Abbreviation for* **optical time domain reflectometer.**

other common carrier (OCC): A communications common carrier—usually an interexchange carrier—that offers communications services in competition with AT&T and/or the established U.S. telephone local exchange carriers.

outage: A telecommunications system service condition in which a user is completely deprived of service by the system. (188) *Note:* For a particular system or a given situation, an outage may be a service condition that is below a defined system operational threshold, *i.e.,* below a threshold of acceptable performance. *See* **outage threshold.**

outage duration: That period of time between the onset of an outage and the restoration of service. (188)

outage probability: The probability that an outage will occur within a specified time period.

outage ratio: The sum of all the outage durations divided by the time period of measurement.

outage state: *See* **outage.**

outage threshold: For a supported performance parameter of a system, the value that establishes the minimum performance level at which the system is considered to remain in an operational state. (188) *Note:* A measured parameter value better than the outage threshold indicates that the system is in a system operational state.

out-of-band emission: Emission on a frequency or frequencies immediately outside the necessary bandwidth which results from the modulation process, but excluding spurious emission. [NTIA] [RR]

out-of-band signaling: **1.** Signaling that uses a portion of the channel bandwidth provided by the transmission medium, *e.g.*, the carrier channel, which portion is above the highest frequency used by, and is denied to, the speech or intelligence path by filters. *Note:* Out-of-band signaling results in a lowered high-frequency cutoff of the effective available bandwidth. **2.** Signaling via a different channel (either FDM or TDM) from that used for the primary information transfer. (188) *Contrast with*

common-channel signaling, in-band signaling, out-slot signaling.

out-of-frame-alignment time: The time during which frame alignment is effectively lost. (188) *Note:* The out-of-frame-alignment time includes the time to detect loss of frame alignment and the alignment recovery time.

outpulsing: The process of transmitting address information over a trunk from one switching center or switchboard to another. (188)

output: **1.** Information retrieved from a functional unit or from a network, usually after some processing. **2.** An output state, or sequence of states. **3.** Pertaining to a device, process, or channel involved in the production of data by a computer or by any of its components.

output angle: *Synonym* **radiation angle.**

output rating: **1.** The expression of the stated power available at the output terminals of a transmitter when connected to the normal load or its equivalent. (188) **2.** Under specified ambient conditions, the expression of the power that can be delivered by a device over a long period of time without overheating. (188)

outside plant: **1.** In telephony, all cables, conduits, ducts, poles, towers, repeaters, repeater huts, and other equipment located between a demarcation point in a switching facility and a demarcation point in another switching facility or customer premises. *Note:* The demarcation point may be at a distribution frame, cable head, or microwave transmitter. **2.** In DOD communications, the portion of intrabase communications equipment between the main distribution frame (MDF) and a user end instrument or the terminal connection for a user instrument. (188)

out-slot signaling: Signaling performed in digital time slots that are not within the channel time slot. *Contrast with* **out-of-band signaling.**

outward dialing: *See* **direct outward dialing.**

ovality: **1.** The attribute of an optical fiber, the cross section of the core or cladding of which deviates from a perfect circle. **2.** In an optical fiber, the degree of deviation, from perfect circularity, of the cross section of the core or cladding. *Note 1:* The cross sections of the core and cladding are assumed to first approximation to be elliptical. Quantitatively, the ovality of either the core or cladding is expressed as $2(a-b)/(a+b)$, where a is the length of the major axis and b is the length of the minor axis. The dimensionless quantity so obtained may be multiplied by 100 to express ovality as a percentage. *Note 2:* Alternatively, ovality of the core or cladding may be expressed or specified by a tolerance field consisting of two concentric circles, within which the cross section boundaries must lie. *Synonym* **noncircularity.**

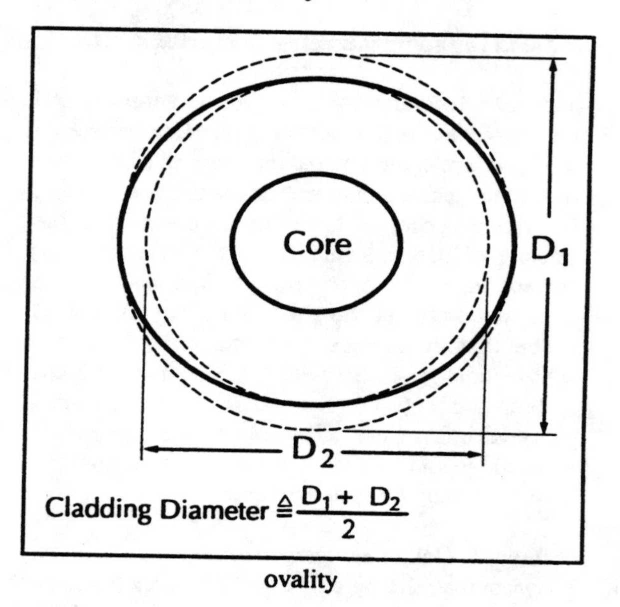

ovality

overfill: **1.** The condition that prevails when the numerical aperture of an optical source, such as a laser, light-emitting diode, or optical fiber, exceeds that of the driven element, *e.g.*, optical fiber core. **2.** The condition that prevails when the beam diameter of an optical source, such as a laser, light-emitting diode, or optical fiber, exceeds that of the driven element, *e.g.*, optical fiber core. *Note:* In optical communications testing, overfill in both numerical aperture and mean diameter (core diameter or spot size) is usually required.

overflow: **1.** In telephony, the generation of potential traffic that exceeds the capacity of a communications system or subsystem. (188) **2.** In telephony, a count of telephone call attempts made on groups of busy trunks or access lines. **3.** In telephony, traffic handled by overflow equipment. **4.** In telephony, traffic that exceeds the capacity of the switching equipment and is therefore lost. (188) **5.** In telephony, on a particular route, excess traffic that is offered to another route, *i.e.*, an alternate route. (188) **6.** In digital computing, *synonym for* **arithmetic overflow. 7.** In digital communications, the condition that exists when the incoming data rate exceeds that which can be accommodated by a buffer, resulting in the loss of information.

overhead bit: Any bit other than a user information bit. (188)

overhead communications: *See* **overhead bit.**

overhead information: Digital information transferred across the functional interface between a user and a telecommunications system, or between functional units within a telecommunications system, for the purpose of directing or controlling the transfer of user information or the detection and correction of errors. (188) *Note:* Overhead information originated by the user is not considered to be system overhead information. Overhead information generated within the communications system and not delivered to the user is system overhead information. Thus, the user throughput is reduced by both overheads while system throughput is reduced only by system overhead.

overlay: 1. One of several segments of a computer program that, during execution, occupies the same area of main storage, one segment at a time. **2.** To use repeatedly the same areas of internal storage during different stages of the execution of a program. (188)

overload: A load, placed on a device or facility, that is greater than the device or facility is capable of handling, *i.e.*, capable of performing the functions for which it was designed. *Note:* Examples of overloads are (a) traffic on a communications system greater than the traffic capacity of the system, (b) for analog inputs, voltage levels above which an analog-to-digital converter cannot distinguish a change, and (c) in electrical circuits, an electrical current that will result in damage from overheating. [From Weik '89]

overload point: *Synonym* **load capacity.**

overmodulation: 1. The condition that prevails when the instantaneous level of the modulating signal exceeds the value necessary to produce 100% modulation of the carrier. *Note 1:* Overmodulation results in spurious emissions by the modulated carrier, and distortion of the recovered modulating signal. *Note 2:* Overmodulation in the sense of this definition is almost always considered a fault condition. **2.** The condition that prevails when the mean level of the modulating signal is such that peaks in the modulating signal exceed the value necessary to produce 100% modulation of the carrier. *Note:* Overmodulation in the sense of this definition, if not excessive, is sometimes considered permissible. (188)

override: 1. To preempt, manually or automatically, a prescribed procedure. *Note:* For example, one might manually override a prescribed course of action programmed to occur in the event of a fault. **2.** In telephony, the entering of or seizure of, a busy circuit, *i.e.*, an occupied circuit, by a party other than those using the circuit. (188) *Note:* For example, an attendant might override a circuit after a busy verification, or a user with a higher precedence level might override a circuit.

overshoot: 1. In the transition of any parameter from one value to another, the transitory value of the parameter that exceeds the final value. *Note:* Overshoot occurs when the transition is from a lower value to a higher value. When the transition is from a higher value to a lower value, and the parameter takes a transitory value that is lower than the final value, the phenomenon is called *"undershoot."* **2.** The increased amplitude of a portion of a nonsinusoidal waveform, *i.e.*, signal, at the output of a nonlinear circuit, *e.g.*, a realizable amplifier, caused by the characteristics of the circuit. (188). *Note 1:* Overshoot causes distortion of the signal. *Note 2:* Overshoot may result from circuit design parameters that are intended to decrease the response time of the circuit. *Note 3:* The amount of overshoot in a given circuit is designed to minimize response time while maintaining distortion of the signal within acceptable limits. The absence or presence of overshoot, and if present, its magnitude, is a function of a circuit design parameter called *"damping." See illustration under* **waveform.** **3.** The result of an unusual atmospheric, *e.g.*, ionospheric, condition that causes microwave signals to be received where they are not intended.

over-the-air rekeying (OTAR): Changing traffic encryption key or transmission security key in remote crypto-equipment by sending new key directly to the remote crypto-equipment over the communication path it secures. [NIS]

over-the-horizon radar: A radar system that makes use of the atmospheric reflection and refraction phenomena to extend its range of detection beyond line of sight. Over-the-horizon radars may be either forward scatter or backscatter systems. [JP1]

overtone: Of a sinusoidal wave, an integral multiple of the frequency, *i.e.*, the fundamental, of the wave, other than the fundamental itself. *Note 1:* The first overtone is twice the frequency of the fundamental, and thus corresponds to the second harmonic; the second overtone is three times the frequency of the fundamental, and thus corresponds to the third harmonic, *etc.* *Note 2:* Use of the term *overtone* is generally confined to acoustic waves, especially in applications related to music. *Contrast with* **fundamental, harmonic.**

(this page intentionally left blank)

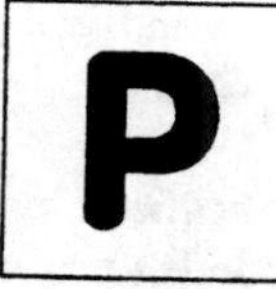

PABX: *Abbreviation for* **private automatic branch exchange.** *See* **PBX.** *Note:* Use of the term *"PBX"* is more common than *"PABX,"* regardless of automation.

packet: In data communication, a sequence of binary digits, including data and control signals, that is transmitted and switched as a composite whole. *Note:* The data, control signals, and possibly error control information, are arranged in a specific format. (188)

packet assembler/disassembler (PAD): A functional unit that enables data terminal equipment (DTE) not equipped for packet switching to access a packet-switched network.

packet format: The structure of data, address, and control information in a packet. (188) *Note:* The size and content of the various fields in a packet are defined by a set of rules that are used to assemble the packet.

packet Internet groper: *See* **ping.**

packet mode: A mode of operating a communications network in which packet switching is used rather than message switching. [From Weik '89]

packet-mode terminal: Data terminal equipment (DTE) that can control, format, transmit, and receive packets. (188)

packet-switched data transmission service: A service that (a) provides for the transmission of data in the form of packets, (b) switches data at the packet level, and (c) may provide for the assembly and disassembly of data packets. (188)

packet switching: The process of routing and transferring data by means of addressed packets so that a channel is occupied during the transmission of the packet only, and upon completion of the transmission the channel is made available for the transfer of other traffic. (188)

packet-switching network: A switched network that transmits data in the form of packets. (188)

packet-switching node: In a packet-switching network, a node that contains data switches and equipment for controlling, formatting, transmitting, routing, and receiving data packets. (188) *Note:* In the Defense Data Network (DDN), a packet-switching node is usually configured to support up to thirty-two X.25 56-kb/s host connections, as many as six 56-kb/s interswitch trunk (IST) lines to other packet-switching nodes, and at least one Terminal Access Controller (TAC).

packet transfer mode: A method of information transfer, by means of packet transmission and packet switching, that permits dynamic sharing of network resources among many connections.

packing density: The number of storage cells per unit length, area, or volume of storage media. *Note:* Examples of packing density are the number of bits or characters stored per unit length of magnetic tape and the number of bits stored per unit length of track on an optical disk.

pad: A network, of fixed resistors, that attenuates signals by a fixed amount with negligible distortion. (188) *Note:* The resistive network is called an *attenuator* if the resistance is adjustable.

PAD: *Acronym for* **packet assembler/disassembler.**

pager: A mobile receiver for paging communications, also known as a *"beeper."* [47CFR]

paging: A one-way communications service from a base station to mobile or fixed receivers that provide signaling or information transfer by such means as tone, tone-voice, tactile, optical readout, *etc.* [47CFR]

paired cable: A cable made up of one or more separately insulated twisted-wire pairs, none of which is arranged with another to form quads. (188)

paired disparity code: A code in which some or all of the characters are represented by two sets of digits of opposite disparity that are used in sequence so as to minimize the total disparity of a longer sequence of digits. (188) *Note 1:* An alternate mark inversion signal is an implementation of a paired disparity code. *Note 2:* The digits may be represented by disparate physical quantities, such as two different frequencies, phases, voltage levels, magnetic polarities, or electrical polarities, each one of the pair representing a 0 or a 1.

pair-gain system: A transmission system that uses concentrators or multiplexers so that fewer wire pairs may be used than would otherwise be required to provide service to a given number of subscribers. (188)

PAL: *Acronym for* **phase alternation by line.** A television signal standard (625 lines, 50 Hz, 220 V primary power) used in the United Kingdom, much of the rest of western Europe, several South American countries, some Middle East and Asian countries, several African countries, Australia, New Zealand, and other Pacific island countries.

PAL-M: A modified version of the phase-alternation-by-line (PAL) television signal standard (525 lines, 50 Hz, 220 V primary power), used in Brazil.

PAM: *Abbreviation for* **pulse-amplitude modulation.**

PAMA: *Abbreviation for* **pulse-address multiple access.**

panning: **1.** On the viewing screen of a display device, *e.g.*, a computer monitor, horizontal shifting of the entire displayed image. *Note:* The panning direction is at a right angle with respect to the scrolling direction. **2.** In video technology, the use of a camera to scan a subject horizontally. **3.** In antenna systems, successively changing the azimuth of a beam of radio-frequency energy over the elements of a given horizontal region, or the corresponding process in reception.

***p/a* r:** *Abbreviation for* **peak-to-average ratio.**

parabolic antenna: An antenna consisting of a parabolic reflector and a radiating or receiving element at or near its focus. (188) *Note:* If the reflector is in the shape of a paraboloid of revolution, it is called a paraboloidal reflector; cylindrical paraboloids and off-axis paraboloids of revolution are also used.

parabolic profile: In an optical fiber, a power-law index profile with the profile parameter, g, equal to 2. *Synonym* **quadratic profile.**

parallel computer: A computer that has multiple arithmetic units or logic units that are used to accomplish parallel operations or parallel processing.

parallel port: A port through which two or more data bits are passed simultaneously, such as all the bits of an 8-bit byte, and that requires as many input channels as the number of bits that are to be handled simultaneously. *Contrast with* **serial port.** [From Weik '89]

parallel processing: Pertaining to the concurrent or simultaneous execution of two or more processes in a single unit.

parallel-to-serial conversion: Conversion of a stream of multiple data elements, received simultaneously, into a stream of data elements transmitted in time sequence, *i.e.*, one at a time. *Contrast with* **serial-to-parallel conversion.**

parallel transmission: **1.** The simultaneous transmission of the signal elements of a character or other data item. **2.** In digital communications, the simultaneous transmission of related signal elements over two or more separate paths. (188) *Note:* Protocols for parallel transmission, such as those used for computer ports, have been standardized by ANSI.

parametric amplifier (paramp): An amplifier that (a) has a very low noise level, (b) has a main oscillator that is tuned to the received frequency, (c) has another pumping oscillator of a different frequency that periodically varies the parameters, *i.e.*, the capacitance or inductance, of the main oscillator circuit, and (d) enables amplification of the applied signal by making use of the energy from the pumping action. *Note:* Paramps with a variable-capacitance main-oscillator semiconductor diode are used in radar tracking and communications Earth stations, Earth satellite stations, and deep-space stations. The noise temperature of paramps cooled to the temperature of liquid helium, about 5 K, is in the range of 20 to 30 K. Paramp gains are about 40 dB. [From Weik '89]

parasitic element: Of an antenna, a directive element that is not connected to a radio transmitter or receiver either directly or via a feeder, but is coupled to the driven element only by the fields. (188) *Synonym* **passive element.**

parasitic emission: In a communications system in which one or more electromagnetic sources are used, electromagnetic radiation—such as lightwaves, radio waves, microwaves, X-rays, or gamma rays from one or more of the sources—that is not harmonically related, *i.e.*, is not coherent, with the transmitted carrier. *Note:* Parasitic emissions are usually caused by undesired oscillations or energy-level transitions in the sources. [From Weik '89]

paraxial ray: In optical systems, a ray that is close to and nearly parallel with the optical axis. (188)

parity: In binary-coded data, a condition that is maintained such that, in any permissible coded expression, the total number of 1s, or 0s, is always odd or always even. *Note 1:* Parity is used in error-detecting and error-correcting codes. *Note 2:* For example, in the ASCII code or in the International Telegraph Alphabet 5 (ITA-5) code as usually implemented, 7 bits are used to represent each character and 1 bit is used as a parity check bit.

parity check: A test that determines whether the number of ones or zeros in an array of binary digits is odd or even. (188) *Note:* Odd parity is standard for synchronous transmission and even parity for asynchronous transmission. *Synonym* **odd-even check.**

par meter: *Abbreviation for* **peak-to-average ratio meter**. A meter used to measure, calculate, and display the ratio of the peak power level to the time-averaged power level in a circuit, *i.e.*, the peak-to-average ratio (*p/a* r). (188) *Note 1:* A par meter is used as a quick means to identify degraded telephone channels. *Note 2:* A par meter is very sensitive to envelope delay distortion. The par meter may also be used for idle channel noise, nonlinear distortion, and amplitude-distortion measurements. *Note 3:* The peak-to-average ratio can be determined for many signal parameters, such as voltage, current, power, frequency, and phase.

part 68: The section of Title 47 of the *Code of Federal Regulations* governing (a) the direct connection of telecommunications equipment and customer premises wiring with the public switched telephone network and certain private line services, such as (1) foreign exchange lines at the customer premises end, (2) the station end of off-premises stations associated with PBX and Centrex® services, (3) trunk-to-station tie lines at the trunk end only, and (4) switched service network station lines, *i.e.*, common control switching arrangements; and (b) the direct connection of (1) all PBX and similar systems to private line services for tie trunk type interfaces, (2) off-premises station lines, and (3) automatic identified outward dialing and message registration. *Note:* Part 68 rules provide the technical and procedural standards under which direct electrical connection of customer-provided telephone equipment, systems, and protective apparatus may be made to the nationwide network without causing harm and without a requirement for protective circuit arrangements in the service-provider networks.

party line: In telephone systems, an arrangement in which two or more user end instruments, usually telephones, are connected to the same loop. *Note:* If selective ringing is not used, individual users may be alerted by different ringing signals, such as a different number of rings or a different combination of long and short rings. Party lines remain primarily in rural areas where loops are long. Privacy is limited and congestion often occurs. *Synonym* **multiparty line.**

passband: The portion of spectrum, between limiting frequencies (or, in the optical regime, limiting wavelengths), that is transmitted with minimum relative loss or maximum relative gain. (188) *Note 1:* The limiting frequencies are defined as those at which the relative intensity or power decreases to a specified fraction of the maximum intensity or power. This decrease in power is often specified to be the half-power points, *i.e.*, 3 dB below the maximum power. *Note 2:* The difference between the limiting frequencies is called the bandwidth, and is expressed in hertz (in the optical regime, in nanometers or micrometers).

passive device: A device that does not require a source of energy for its operation. *Note:* Examples of passive devices are electrical resistors, electrical capacitors, diodes, optical fibers, cables, wires, glass lenses, and filters.

passive element: *Synonym* **parasitic element.**

passive network: A network that does not require a power source for its operation. (188) [From Weik '89]

passive repeater: An unpowered device used to route a microwave beam over or around an obstruction. (188) *Note:* Examples of passive repeaters are (a) two parabolic antennas connected back-to-back, and (b) a flat reflector used as a mirror.

passive satellite: In a satellite communications system, a satellite that only reflects signals from one Earth station to another, or from several Earth stations to several others. *Note:* Although the satellite acts passively by reflecting signals, it may contain active devices for station keeping. (188)

passive sensor: A measuring instrument in the Earth exploration-satellite service or in the space research service by means of which information is obtained by reception of radio waves of natural origin. [NTIA] [RR]

passive star: *See* **star coupler.**

passive station: On a multipoint connection or a point-to-point connection using basic mode link control, any tributary station waiting to be polled or selected.

password: **1.** [A] protected/private character string used to authenticate an identity or to authorize access to data. [NIS] **2.** In data communications, a word, character, or combination thereof, that permits access to otherwise inaccessible data, information, or facilities. (188)

password length equation: An equation that determines an appropriate password length, M, which provides an acceptable probability, P, that a password will be guessed in its lifetime. *Note:* The password length is given by $M = (\log S)/(\log N)$ where S is the size of the password space and N is the number of characters available. The password space is given by $S = LR/P$, where L is the maximum lifetime of a password and R is the number of guesses per unit of time.

password length parameter: A basic parameter affecting the password length needed to provide a given degree of security. *Note 1:* Password length parameters are related by the expression $P = LR/S$, where P is the probability that a password can be guessed in its lifetime, L is the maximum lifetime a password can be used to log in to a system, R is the number of guesses per unit of time, and S is the number of unique algorithm-generated passwords (the password space). *Note 2:* The degree of password security is determined by the probability that a password can be guessed in its lifetime.

patch: **1.** To connect circuits together temporarily. *Note:* In communications, patches may be made by means of a cord, *i.e.,* a cable, known as a "patch cord." In automated systems, patches may be made electronically. (188) **2.** In a computer program, one or more statements inserted to circumvent a problem or to alter temporarily or permanently a usually limited aspect or characteristic of the functioning of the program, *e.g.,* to customize the program for a particular application or environment.

patch and test facility (PTF): A facility in which supporting functions, such as (a) quality control checking and testing of equipment, links, and circuits, (b) troubleshooting, (c) activating, changing, and deactivating of circuits, and (d) technical coordinating and reporting, are performed.

patch bay: An assembly of hardware so arranged that a number of circuits, usually of the same or similar type, appear on jacks for monitoring, interconnecting, and testing purposes. (188) *Note 1:* Patch bays are used at many locations, such as technical control facilities, patch and test facilities, and at telephone exchanges. *Note 2:* Patch bays facilitate flexibility in the use, routing or restoration of a variety of circuit types, such as dc, VF, group, coaxial, equal-level, and digital data circuits.

patch panel: One segment of a patch bay. (188)

path: **1.** In communications systems and network topologies, a route between any two points. [From Weik '89] **2.** In radio communications, the route that (a) lies between a transmitter and a receiver and (b) may consist of two or more concatenated links. *Note:* Examples of paths are line-of-sight paths and ionospheric paths. **3.** In a computer program, the logical sequence of instructions executed by a computer. **4.** In database management systems, a series of physical or logical connections between records or segments, usually requiring the use of pointers.

path attenuation: *Synonym* **path loss.**

path clearance: In microwave line-of-sight communications, the perpendicular distance from the radio-beam axis to obstructions such as trees, buildings, or terrain. (188) *Note:* The required path clearance is usually expressed, for a particular *k*-factor, as some fraction of the first Fresnel zone radius.

path intermodulation noise: *See* **intermodulation noise.**

path loss: In a communication system, the attenuation undergone by an electromagnetic wave in transit between a transmitter and a receiver. (188) *Note 1:* Path loss may be due to many effects such as free-space loss, refraction, reflection, aperture-medium coupling loss, and absorption. *Note 2:* Path loss is usually expressed in dB. *Synonym* **path attenuation.**

path profile: A graphic representation of the physical features of a propagation path in the vertical plane containing both endpoints of the path, showing the surface of the Earth and including trees, buildings, and other features that may obstruct the radio signal. (188) *Note:* Profiles are drawn either with an effective Earth radius simulated by a parabolic arc—in which case the ray paths are drawn as straight lines—or with a *"flat Earth"*—in which case the ray paths are drawn as parabolic arcs.

path quality analysis: In a communications path, an analysis that (a) includes the overall evaluation of the component quality measures, the individual link quality measures, and the aggregate path quality measures, and (b) is performed by evaluating communications parameters, such as bit error ratio, signal-plus-noise-plus-distortion to noise-plus-distortion ratio, and spectral distortion.

path quality matrix: A data bank that contains path-quality analyses used to support path selection and routing determination. *Note:* In adaptive radio automatic link establishment, path quality matrices contain path quality data for single-link and multilink paths. (188)

path survey: The assembling of pertinent geographical and environmental data required to design a radio communication system. (188)

pattern recognition: The identification of objects and images by their shapes, forms, outlines, color, surface texture, temperature, or other attribute, usually by automatic means. [From Weik '89]

Pawsey stub: A device for connecting an unbalanced coaxial feeder to a balanced antenna.

PAX: *Abbreviation for* **private automatic exchange.** *See* **PBX.**

payload: In a set of data, such as a data field, block, or stream, being processed or transported, the part that represents user information and user overhead information, and may include user-requested additional information, such as network management and accounting information. *Note:* The payload does not include system overhead information for the processing or transportation system. *Synonym* **mission bit stream.**

payload module: The portion of a payload that completely occupies one or more channels.

PBER: *Abbreviation for* **pseudo bit-error ratio.** In adaptive high-frequency (HF) radio, a bit error ratio derived by a majority decoder that processes redundant transmissions. *Note:* In adaptive HF radio automatic link establishment, PBER is determined by the extent of error correction, such as by using the fraction of non-unanimous votes in the 2-of-3 majority decoder. (188)

PBX: *Abbreviation for* **private branch exchange.** **1.** A subscriber-owned telecommunications exchange that usually includes access to the public switched network. **2.** A switch that serves a selected group of users and that is subordinate to a switch at a higher level military establishment. (188) **3.** A private telephone switchboard that provides on-premises dial service and may provide connections to local and trunked communications networks. (188) *Note 1:* A PBX operates with only a manual switchboard; a private automatic exchange (PAX) does not have a switchboard, a private automatic branch exchange (PABX) may or may not have a switchboard. *Note 2:* Use of the term *"PBX"* is far more common than *"PABX,"* regardless of automation.

PBX tie trunk: *See* **tie trunk.**

PBX trunk: *See* **trunk.**

PC: *Abbreviation for* **carrier power (of a radio transmitter).**

PCB: *Abbreviation for* **power circuit breaker.**

PCM: *Abbreviation for* **pulse-code modulation.**

PCM multiplex equipment: *See* **multiplexer.**

PCS: *Abbreviation for* **Personal Communications Service.** A set of capabilities that allows some combination of terminal mobility, personal mobility, and service profile management. *Note 1:* The flexibility offered by PCS can supplement existing telecommunications services, such as cellular radio, used for NS/EP missions. *Note 2:* PCS and UPT are sometimes mistakenly assumed to be the same service concept. UPT allows complete personal mobility across multiple networks and service providers. PCS may use UPT concepts to improve subscriber mobility in allowing roaming to different service providers, but UPT and PCS are not the same service concept. *Contrast with* **Universal Personal Telecommunications service.**

PCS switching center: In personal communications service, a facility that (a) supports access-independent call control/service control, and connection control (switching) functions, and (b) is responsible for interconnection of access and network systems to support end-to-end services. *Note 1:* The PCS switching center represents a collection of one or more network elements. *Note 2:* The term *"center"* does not imply a physical location.

PCS System: In personal communications service, a collection of facilities that provides some combination of personal mobility, terminal mobility, and service profile management. *Note:* As used here, "facilities" includes hardware, software, and network components such as transmission facilities, switching facilities, signaling facilities, and databases.

PDM: *Abbreviation for* **pulse-duration modulation.**

PDN: *Abbreviation for* **public data network.**

PDS: *Abbreviation for* **protected distribution system.**

PDU: *Abbreviation for* **protocol data unit.**

PE: *Abbreviation for* **phase-encoded.** *See* **phase-encoded recording.**

peak busy hour: *Synonym* **busy hour.**

peak emission wavelength: Of an optical emitter, the spectral line having the greatest power. *Synonym* **peak wavelength.**

peak envelope power (of a radio transmitter) [PEP, pX, PX]: The average power supplied to the antenna transmission line by a transmitter during one radio frequency cycle at the crest of the modulation envelope taken under normal operating conditions. [NTIA] [RR] (188)

peak limiting: A process by which the absolute instantaneous value of a signal parameter is prevented from exceeding a specified value. (188)

peak power output: The output power averaged over that cycle of an electromagnetic wave having the maximum peak value that can occur during transmission. (188)

peak signal level: **1.** In a transmission path, the maximum instantaneous signal power, voltage, or current at any point. (188) **2.** At a given point in a transmission path, the maximum instantaneous signal power, voltage, or current that occurs during a specified period.

peak spectral emission: *See* **peak emission wavelength.**

peak-to-average ratio (*p/a* r): The ratio of the instantaneous peak value, *i.e.*, maximum magnitude, of a signal parameter to its time-averaged value. (188) *Note:* The peak-to-average ratio can be determined for many signal parameters, such as voltage, current, power, frequency, and phase.

peak-to-peak value: The absolute value of the difference between the maximum and the minimum magnitudes of a varying quantity.

peak wavelength: **1.** *Synonym* **peak emission wavelength.** **2.** Of an optical bandpass filter, the wavelength that suffers the lowest loss. [After FAA]

peer entity: In layered systems, one of a set of entities that are in the same layer or the equivalent layer of another system.

peer group: In Open Systems Interconnection (OSI)—Architecture, a group of functional units in a given layer of a network in which all the functions performed by the functional units extend throughout the system at the same layer. [From Weik '89]

peg count: **1.** In communication systems, a count that is made of the number of times that an event or condition occurs. [From Weik '89] **2.** In telephone systems, the process that provides counts of the calls of different service classes that occur during intervals of such frequency as to reliability indicate the traffic load. [From Weik '89] **3.** A count of the attempts to seize, or a count of the actual seizures that occur, of various types of telephone trunks, access lines, switches, or other equipment. [From Weik '89]

pel: In a facsimile system, the smallest discrete scanning line sample containing only monochrome information, *i.e.*, not containing gray-scale information.

penetration: **1.** The passage through a partition or wall of an equipment or enclosure by a wire, cable, or other conductive object. (188) **2.** [The] unauthorized act of bypassing the security mechanisms of a cryptographic system or AIS. [NIS] **3.** The passage of a radio frequency through a physical barrier, such as a partition, a wall, a building, or earth.

PEP: *Deprecated abbreviation for* **peak envelope power.** Either *"PX"* or *"pX"* is now preferred. *See* **peak envelope power, power.**

percentage modulation: **1.** In angle modulation, the fraction of a specified reference modulation, expressed in percent. (188) **2.** In amplitude modulation, the modulation factor expressed in percent. (188) *Note:* Percentage modulation may also be expressed in dB below 100% modulation.

percent break: In telephone dialing, the ratio, expressed in percent, of the open circuit time to the sum of the open and closed circuit times allotted to a single dial pulse cycle. (188)

performance management: In network management, (a) a set of functions that evaluate and report the behavior of telecommunications equipment and the effectiveness of the network or network element and (b) a set of various subfunctions, such as gathering statistical information, maintaining and examining historical logs, determining system performance under natural and artificial conditions, and altering system modes of operation. (188)

performance measurement period: The period during which performance parameters are measured. (188) *Note:* A performance measurement period is determined by required confidence limits and may vary as a function of the observed parameter values. User time is divided into consecutive performance measurement periods to enable measurement of user information transfer reliability.

performance parameter: A quality, usually quantified by a numerical value, which quality characterizes a particular aspect, capability, or attribute of a system. *Note:* Examples of performance parameters are peg count and mean time between failures.

periapsis: In a satellite orbit, the point that is closest to the gravitational center of the system consisting of the primary body and the satellite. (188) *Note:* In an orbit about the Earth, periapsis is called *perigee.* In an orbit about the Moon, periapsis is called *perilune,* and in an orbit about the Sun, it is called *perihelion.*

perigee: Of a satellite orbiting the Earth, the point in the orbit at which the gravitational centers of the satellite and Earth are closest to one another.

perigee altitude: *See* **altitude of the apogee or of the perigee.**

periodic antenna: An antenna that has an approximately constant input impedance over a narrow range of frequencies. *Note:* An example of a periodic antenna is a dipole array antenna. *Synonym* **resonant antenna.**

period (of a satellite): The time elapsing between two consecutive passages of a satellite through a characteristic point on its orbit. [NTIA] [RR]

periods processing: The processing of various levels of classified or unclassified information at distinctly different times. *Note:* Under periods processing, the system must be purged of all information from one processing period before transitioning to the next when there are different users with differing authorizations. [NIS] (188)

peripheral device: *See* **peripheral equipment.**

peripheral equipment: In a data processing system, any equipment, distinct from the central processing unit, that may provide the system with additional capabilities. *Note:* Such equipment is often offline until needed for a specific purpose and may, in some cases, be shared among several users.

peripheral node: *Synonym* **endpoint node.**

periscope antenna: An antenna configuration in which the transmitting antenna is oriented to produce a vertical radiation pattern, and a flat or off-axis parabolic reflector, mounted above the transmitting antenna, is used to direct the beam in a horizontal path toward the receiving antenna. (188) *Note:* A periscope antenna facilitates increased terrain clearance without long transmission lines, while permitting the active equipment to be located at or near ground level for ease of maintenance.

permanent bond: A bond not expected to require disassembly for operational or maintenance purposes. (188)

permanent signal (PS): An extended off-hook condition not followed by dialing.

permanent storage: A storage device in which stored data are nonerasable.

permanent virtual circuit (PVC): A virtual circuit used to establish a long-term connection between data terminal equipments (DTE). *Note 1:* In a PVC, the long-term association is identical to the data transfer phase of a virtual call. *Note 2:* Permanent virtual circuits eliminate the need for repeated call set-up and clearing. *Deprecated synonym* **nailed-up circuit.**

permissible interference: Observed or predicted interference which complies with quantitative interference and sharing criteria contained in these *[Radio] Regulations* or in CCIR Recommendations or in special agreements as provided for in these *Regulations.* [NTIA] [RR]

Personal Communications Service: *See* **PCS.**

personal mobility: In universal personal telecommunications, (a) the ability of a user to access telecommunication services at any UPT terminal on the basis of a personal identifier, and (b) the capability of the network to provide those services in accord with the user's service profile. *Note 1:* Personal mobility involves the network's capability to locate the terminal associated with the user for the purposes of addressing, routing, and charging the user for calls. *Note 2:* "Access" is intended to convey the concepts of both originating and terminating services. *Note 3:* Management of the service profile by the user is not part of personal mobility. The personal mobility aspects of personal communications are based on the UPT number.

personal registration: In universal personal telecommunications, the process of associating a UPT user with a specific terminal.

personal terminal: In personal communications service, a lightweight, small, portable terminal that provides the capability for the user to be either stationary or in motion while accessing and using telecommunication services.

phantom circuit: A third circuit derived from two suitably arranged pairs of wires, called side circuits, with each pair of wires being a circuit in itself and at the same time acting as one conductor of the third circuit. *Note:* The side circuits are coupled to their respective drops by center-tapped transformers, usually called "repeat coils." The center taps are on the line side of the side circuits. Current from the phantom circuit is split evenly by the center taps. This cancels crosstalk from the phantom circuit to the side circuits. (188)

phantom group: Three circuits that are derived from simplexing two physical circuits to form a phantom circuit. (188)

phase: **1.** Of a periodic, varying phenomenon, *e.g.*, an electrical signal or electromagnetic wave, any distinguishable instantaneous state of the phenomenon, referred to a fixed reference or another periodic varying phenomenon. (188) *Note 1:* Phase, *i.e.*, *phase time* (frequently abbreviated simply to "phase" in colloquial usage), can be specified or expressed by time of occurrence relative to a specified reference. *Note 2:* The phase of a periodic phenomenon can also be expressed or specified by angular measure, with one period usually encompassing 360° (2π radians). *Note 3:* Phase may be represented (a) in polar coordinates by $M\angle\theta$, where M is the magnitude and θ is the phase angle, and (b) in Cartesian coordinates, *i.e.*, an Argand diagram, as $(a + jb)$, where a is a real component and b is an imaginary component such that $\tan \theta = (b/a)$, where θ is the phase angle, and the magnitude, M, is $(a^2 + b^2)^{1/2}$ **2.** A distinguishable state of a phenomenon. (188) **3.** That period of time during which a specified function occurs in a sequential list of functions. (188)

phase angle: Of a periodic wave, the number of suitable units of angular measure between a point on the wave and a reference point. *Note 1:* The reference point may be a point on another periodic wave. The waves may be plotted on a suitable coordinate system, such as a Cartesian plot, with degrees or other angular measure usually plotted on the abscissa and amplitude on the ordinate. Usually, at least one full cycle of each wave is plotted, with 360° (2π radians) encompassing one full cycle. The reference points may be any significant instants on the waves, such as where they cross the abscissa axis. *Note 2:* The use of angular measure to define the relationship between a periodic wave and a reference point is derived from the projection of a rotating vector onto the real axis of an Argand diagram. *Note 3:* The value of the phase angle of a point on the wave is the point on the abscissa that corresponds to the point on the wave. *Note 4:* The phase angle of a vector may be written as $M\angle\theta$, where M is the magnitude of the vector and θ is the phase angle relative to the specified reference.

phase bandwidth: Of a network or device, the width of the continuous frequency range over which the phase-vs.-frequency characteristic does not depart from linearity by more than a stated amount. (188)

phase coherence: The state in which two signals maintain a fixed phase relationship with each other or with a third signal that can serve as a reference for each. (188)

phase coherent: *See* **phase coherence.**

phase constant: The imaginary part of the axial propagation constant for a particular mode, usually expressed in radians per unit length. (188)

phased array: A group of antennas in which the relative phases of the respective signals feeding the antennas are varied in such a way that the effective radiation pattern of the array is reinforced in a desired direction and suppressed in undesired directions. (188) *Note 1:* The relative amplitudes of—and constructive and destructive interference effects among—the signals radiated by the individual antennas determine the effective radiation pattern of the array. *Note 2:* A phased array may be used to point a fixed radiation pattern, or to scan rapidly in azimuth or elevation.

phase delay: In the transmission of a single-frequency wave from one point to another, the delay of an arbitrary point in the wave that identifies its phase. (188) *Note:* Phase delay may be expressed in any convenient unit, such as seconds, degrees, radians, or wavelengths.

phase departure: **1.** A phase deviation from a specified value. **2.** An unintentional deviation from the nominal phase value.

phase detector: A circuit or instrument that detects the difference in phase between corresponding points on two signals. (188)

phase deviation: In phase modulation, the maximum difference between the instantaneous phase angle of the modulated wave and the phase angle of the unmodulated carrier. (188) *Note:* For a sinusoidal modulating wave, the phase deviation, expressed in radians, is equal to the modulation index.

phase diagram: A graphic representation of the phase relationships between two or more waveforms. *Note:* A phase diagram may be represented as a vector diagram or as an amplitude-vs.-time diagram.

phase difference: The time interval or phase angle by which one wave leads or lags another. (188) *Synonym* **phase offset.**

phase distortion: Distortion that occurs when (a) the phase-frequency characteristic is not linear over the frequency range of interest, *i.e.*, the phase shift introduced by a circuit or device is not directly proportional to frequency, or (b) the zero-frequency intercept of the phase-frequency characteristic is not 0 or an integral multiple of 2π radians. (188) *Synonym* **phase-frequency distortion.**

phase-encoded (PE) recording: Binary recording on magnetic media, such as magnetic disks, tapes, and cards, in which a "1" is represented by a magnetic flux reversal to the polarity of the interblock gap, and a "0" is represented by a magnetic flux reversal to the polarity opposite to that of the interblock gap when recording in the forward direction. (188)

phase equalizer: *See* **delay equalizer.**

phase flux reversal: In phase-encoded recording, a magnetic flux reversal written at the nominal midpoint between successive "1" bits, or between successive "0" bits, to establish proper polarity. (188)

phase-frequency characteristic: A Cartesian-coordinate plot of phase shift as the dependent variable, versus frequency as the independent variable. *Note:* The phase-frequency characteristic is linear if the phase shift introduced by a circuit or device is the same for all frequencies in the input signal.

phase-frequency distortion: *Synonym* **phase distortion.**

phase hit: *See* **hit.**

phase instability: The fluctuation of the phase of a wave, relative to a reference. *Note:* The fluctuation is often from unknown causes.

phase interference fading: The variation in signal amplitude produced by the interaction of two or more signal elements with different relative phases. (188)

phase inversion: Introduction of a phase difference of 180°. *Note:* Phase inversion may occur with a random or periodic, symmetrical or non-symmetrical waveform, although it is usually produced by the inversion of a symmetrical periodic signal, resulting in a change in sign. A symmetrical periodic signal represented by $f(t) = Ae^{j\omega t}$, after phase inversion, becomes $f_1(t) = Ae^{j(\omega t+\pi)}$, where t is time, A is the magnitude of the vector, ω is angular frequency ($\omega = 2\pi f$), where f is the frequency and $\pi \approx 3.1416$ and $e \approx 2.7183$. The algebraic sum of $f(t)$ and $f_1(t)$ will always be zero.

phase jitter: Rapid, repeated phase perturbations that result in the intermittent shortening or lengthening of signal elements. (188) *Note 1:* Phase jitter may be random or cyclic. *Note 2:* The phase departure in phase jitter usually is smaller, but more rapid, than that of phase perturbation. Phase jitter may be expressed in degrees, radians, or seconds. Phase jitter is usually random. However, if cyclic, phase jitter may be expressed in hertz as well as in degrees, radians, or seconds.

phase jump: A sudden phase change in a signal. (188)

phase linearity: Direct proportionality of phase shift to frequency over the frequency range of interest.

phase-locked loop (PLL): An electronic circuit that controls an oscillator so that it maintains a constant phase angle relative to a reference signal. (188) *Note:* Phase-locked loops are widely used in space communications for coherent carrier tracking and threshold extension, bit synchronization, and symbol synchronization.

phase measurement tolerance: The maximum allowable difference between a phase measurement and the actual phase value. (188)

phase modulation (PM): Angle modulation in which the phase angle of a carrier is caused to depart from its reference value by an amount proportional to the instantaneous value of the modulating signal. (188)

phase noise: In an oscillator, rapid, short-term, random fluctuations in the phase of a wave, caused by time-domain instabilities. *Note:* Phase noise, $\mathcal{L}(f)$ in decibels relative to carrier power (dBc) on a 1-Hz bandwidth, is given by $\mathcal{L}(f) = 10\log[0.5(S_{\phi}(f))]$ where $S_{\phi}(f)$ is the spectral density of phase fluctuations.

phase nonlinearity: Lack of direct proportionality of phase shift to frequency over the frequency range of interest.

phase offset: *Synonym* **phase difference.**

phase perturbation: Any shifting (often quite rapid), from whatever cause, in the phase of a signal. (188) *Note 1:* The shifting in phase may appear to be random, cyclic, or both. *Note 2:* The phase departure in phase perturbation usually is larger, but less rapid, than that of phase jitter. *Note 3:* Phase perturbation may be expressed in degrees, with any cyclic component expressed in hertz.

phase quadrature: *See* **quadrature.**

phase shift: The change in phase of a periodic signal with respect to a reference. (188)

phase-shift keying (PSK): **1.** In digital transmission, angle modulation in which the phase of the carrier is discretely varied in relation either to a reference phase or to the phase of the immediately preceding signal element, in accordance with data being transmitted. (188) **2.** In a communications system, the representing of characters, such as bits or quaternary digits, by a shift in the phase of an electromagnetic carrier wave with respect to a reference, by an amount corresponding to the symbol being encoded. *Note 1:* For example, when encoding bits, the phase shift could be 0° for encoding a "0," and 180° for encoding a "1," or the phase shift could be –90° for "0" and +90° for a "1," thus making the representations for "0" and "1" a total of 180° apart. *Note 2:* In PSK systems designed so that the carrier can assume only two different phase angles, each change of phase carries one bit of information, *i.e.,* the bit rate equals the modulation rate. If the number of recognizable phase angles is increased to 4, then 2 bits of information can be encoded into each signal element; likewise, 8 phase angles can encode 3 bits in each signal element. *Synonyms* **biphase modulation, phase-shift signaling.**

phase-shift signaling: *Synonym* **phase-shift keying.**

phase term: In the propagation of an electromagnetic wave in a uniform waveguide, such as an optical fiber or metal waveguide, the parameter that indicates the phase change per unit distance of the wave at any point along the waveguide. [From Weik '89]

phase velocity: The velocity of propagation of a uniform plane wave, given by (a) the product of the wavelength and the frequency divided by (b) the refractive index of the medium in which the wave is propagating. (188) *Note 1:* In free space, the refractive index may be considered as unity. *Note 2:* In free space, the group velocity and the phase velocity are equal.

phasing: In facsimile transmission and reception, the process by which the start of the scanning line or lines is made to correspond to one edge of the object being scanned. *Note:* If there is no correspondence between the object being scanned and the scanning line or lines, distortion, often in the form of a split image, will occur in the received image.

phon: In acoustics, a unit of subjective loudness level equal to the sound pressure level in dB compared to that of an equally loud standard sound. *Note:* The accepted standard is a 1-kHz pure sine-wave tone or narrowband noise centered at 1 kHz. [From Weik '89]

phone: **1.** *Abbreviation for* **telephone, telephony.** **2.** *Colloquially,* the voice-operation mode in radio communications.

phonetic alphabet: A list of standard words used to identify letters in a message transmitted by radio or telephone. The following are the currently authorized words for each letter in the alphabet: **Alpha, Bravo, Charlie, Delta, Echo, Foxtrot, Golf, Hotel, India, Juliet, Kilo, Lima, Mike, November, Oscar, Papa, Quebec, Romeo, Sierra, Tango, Uniform, Victor, Whiskey, X-ray, Yankee, Zulu.** [JP1]

phonon: A quantum of acoustic energy, the level of which is a function of the frequency of the acoustic wave. *Note:* Phonons in acoustics are analogous to photons in electromagnetics. The energy of a phonon is usually less than 0.1 eV (electron-volt) and thus is one or two orders of magnitude less than that of a photon. When photons and phonons interact in semiconductors used in communications systems, undesirable system behavior can occur. [From Weik '89]

phonon absorption: Absorption of light energy by its conversion to vibrational energy. *Note:* Phonon absorption determines the fundamental, *i.e.,* quantum limit of attenuation, *i.e.,* minimum attenuation, in silica-based glasses in the far infrared region. [After FAA]

photoconductive effect: In certain materials, the phenomenon that results in photoconductivity.

photoconductivity: In certain materials, the increase in electrical conductivity that results from increases in the number of free carriers generated when photons are absorbed. *Note:* The photons must have quantum energy sufficient to overcome the band-gap in the material in question.

photocurrent: The current that flows through a photosensitive device, such as a photodiode, as the result of exposure to radiant power. *Note 1:* The photocurrent may occur as a result of the photoelectric, photoemissive, or photovoltaic effect. *Note 2:* The photocurrent may be enhanced by internal gain caused by interaction among ions and photons under the influence of applied fields, such as occurs in an avalanche photodiode (APD).

photodetector (PD): A transducer capable of accepting an optical signal and producing an electrical signal containing the same information as in the optical signal. [2196] *Note:* The two main types of semiconductor photodetectors are the photodiode (PD) and the avalanche photodiode (APD).

photodiode: A semiconductor diode that produces, as a result of the absorption of photons, (a) a photovoltage or (b) free carriers that support the conduction of photocurrent. *Note:* Photodiodes are used for the detection of optical communication signals and for the conversion of optical power to electrical power.

photoelectric effect: In certain materials, the changes in the electrical characteristics caused by photon absorption.

photon: A discrete packet, *i.e.,* quantum, of electromagnetic energy. *Note:* The energy of a photon is $h\nu$, where h is Planck's constant and ν is the frequency of the electromagnetic wave.

photon noise: In an optical communication link, noise attributable to the statistical nature of optical quanta. [FAA] *See* **quantum noise.**

photosensitive recording: Facsimile recording by the exposure of a photosensitive surface to a signal-controlled light beam or spot. (188)

photovoltaic effect: The production, as a result of the absorption of photons, of a voltage difference across a pn junction. *Note:* The voltage difference is caused by the internal drift of holes and electrons.

physical frame: *See* **frame.**

Physical Layer: *See* **Open Systems Interconnection—Reference Model.**

physical optics: The branch of optics that treats light propagation as a wave phenomenon rather than a ray phenomenon, as in geometric optics.

physical security: *See* **communications security.**

physical signaling sublayer (PLS): In a local area network (LAN) or a metropolitan area network (MAN) using open systems interconnection (OSI) architecture, the portion of the physical layer that (a) interfaces with the medium access control sublayer, (b) performs character encoding, transmission, reception, and decoding, and (c) performs optional isolation functions.

physical topology: The physical configuration, *i.e.*, interconnection, of network elements, *e.g.*, cable paths, switches, concentrators. *Note:* Physical topology is in contrast to logical topology. For example, a logical loop may consist of a physical star configuration, or a physical loop.

picowatt: *See* **pW.**

picowatt, psophometrically weighted: *See* **noise weighting.**

picture: *Synonym* **image.**

picture black: In TV and facsimile, pertaining to the signal or signal level that corresponds to the darkest part, *i.e.*, the spot with the lowest luminance or reflectivity, of the object being scanned.

picture element: *See* **pel, pixel.**

picture frequency: In analog facsimile systems, a baseband frequency generated by scanning an object. (188) *Note:* Picture frequencies do not include frequencies that are present in a modulated carrier.

picture white: In TV and facsimile, pertaining to the signal level that corresponds to the brightest part, *i.e.*, the spot with the highest luminance or reflectivity, of the object being scanned.

piecewise linear encoding: *See* **segmented encoding law.**

piecewise linear encoding law: *Synonym* **segmented encoding law.**

pigtail: **1.** A short length of optical fiber that is permanently affixed to an active device, *e.g*, LED or laser diode, and is used to couple the device, using a splice or connector, to a longer fiber. [After FAA] (188) **2.** A short length of single-fiber cable, usually tight-buffered, that has an optical connector on one end and a length of exposed fiber at the other end. *Note:* The exposed fiber of the pigtail is then spliced to one fiber of a multifiber trunk, *i.e.*, arterial, cable, to enable the multifiber cable to be "broken out" into individual single-fiber cables that may be connected to a patch panel or an input or output port of an optical receiver or transmitter. [After FAA] **3.** A short length of electrical conductor permanently affixed to a component, used to connect the component to another conductor.

pilot: A signal, usually a single frequency, transmitted over a communications system for supervisory, control, equalization, continuity, synchronization, or reference purposes. (188) *Note:* Sometimes it is necessary to employ several independent pilot frequencies. Most radio relay systems use radio or continuity pilots of their own but transmit also the pilot frequencies belonging to the carrier frequency multiplex system.

pilot frequency: *See* **synchronizing pilot.**

pilot-make-busy (PMB) circuit: A circuit arrangement by which trunks provided over a carrier system are made busy to the switching equipment in the event of carrier system failure, or during a fade of the radio system. (188)

pilot tone: *See* **pilot.**

PIN diode: *Acronym for* **positive-intrinsic-negative diode.** A photodiode with a large, neutrally doped intrinsic region sandwiched between p-doped and n-doped semiconducting regions. *Note:* A PIN diode exhibits an increase in its electrical conductivity as a function of the intensity, wavelength, and modulation rate of the incident radiation. *Synonym* **PIN photodiode.**

ping: *Abbreviation for* **packet Internet groper.** In TCP/IP, a protocol function that tests the ability of a computer to communicate with a remote computer by sending a query and receiving a confirmation response.

pink noise: In acoustics, noise in which there is equal power per octave.

PIN photodiode: *Synonym* **PIN diode.**

piston: In a hollow metallic waveguide, a longitudinally movable metallic plane surface that reflects essentially all the incident energy. *Note:* A piston is used for tuning, *e.g.*, fine-tuning a resonant cavity. *Synonym* **plunger.**

pitch: *Synonym* **lay length.**

pixel: In a raster-scanned imaging system, the smallest discrete scanning line sample that can contain gray scale information. (188)

PLA: *Abbreviation for* **programmable logic array.**

plain text: Unencrypted information. [NIS] (188) *Note:* Plain text includes voice. *Synonym* **clear text.**

planar array: An antenna in which all of the elements, both active and parasitic, are in one plane. (188) *Note 1:* A planar array provides a large aperture and may be used for directional beam control by varying the relative phase of each element. *Note 2:* A planar array may be used with a reflecting screen behind the active plane.

planar waveguide: *Synonym* **slab-dielectric waveguide.**

Planck's constant: The constant of proportionality, represented by the symbol h, that relates the energy E of a photon with the frequency ν of the associated wave through the relation $E = h\nu$, where $h = 6.626 \times 10^{-34}$ joule•second.

Planck's law: The fundamental law of quantum theory that describes the essential concept of the quanta of electromagnetic energy. *Note 1:* Planck's law states that the quantum of energy, E, associated with an electromagnetic field is given by $E = h\nu$, where h is Planck's constant and ν is the frequency of the electromagnetic radiation. *Note 2:* Planck's constant is usually given in joule•seconds and the frequency in hertz. Thus, the quantum of energy is usually given in joules. *Note 3:* The product of energy and time is sometimes referred to as the elementary quantum of action. Hence, h is sometimes referred to as the elementary quantum of action.

plane polarization: *Synonym* **linear polarization.**

plane wave: **1.** A wave whose surfaces of constant phase are infinite parallel planes normal to the direction of propagation. (188) **2.** An electromagnetic wave that predominates in the far-field region of an antenna, and has a wavefront that is essentially in a plane. (188) *Note:* In free space, the characteristic impedance of a plane wave is 377 Ω.

plant: All the facilities and equipment used to provide telecommunications services. *Note:* Plant is usually characterized as *outside plant* or *inside plant*. Outside plant, for example, includes all poles, repeaters and unoccupied buildings housing them, ducts, and cables—including the "inside" portion of interfacility cables outward from the main distributing frame (MDF) in a central office or switching center. Inside plant includes the MDF and all equipment and facilities within a central office or switching center.

plastic-clad silica (PCS) fiber: An optical fiber that has a silica-based core and a plastic cladding. *Note 1:* The cladding of a PCS fiber should not be confused with the polymer overcoat of a conventional all-glass fiber. *Note 2:* PCS fibers in general have significantly lower performance characteristics, *i.e.*, higher transmission losses and lower bandwidths, than all-glass fibers. *Synonym* **polymer-clad silica fiber.**

plenum: In a building, an enclosure, created by building components such as a suspended ceiling or false floor, and used for the movement of environmental air. *Note 1:* A plenum may be used to contain communications and power cables, *e.g.*, to reach equipment installed in open office or laboratory space. *Note 2:* Cables installed in plenums must meet applicable environmental and fire protection regulations. This may mean enclosing them in suitable ducts or using cables having jackets and other components made of materials that are resistant to open flame and are non-toxic at high temperatures.

plesiochronous: That relationship between two signals such that their corresponding significant instants occur at nominally the same rate, any variations being constrained within a specified limit. (188) *Note:* There is no limit to the phase difference that can accumulate between corresponding significant instants over a long period of time.

PL/I: A programming language that is designed for use in a wide range of commercial and scientific computer applications.

PLL: *Abbreviation for* **phase-locked loop.**

plotter: An output unit that presents data in the form of a two-dimensional graphic representation.

PLS: *Abbreviation for* **physical signaling sublayer.**

plunger: *Synonym* **piston.**

PM: *Abbreviation for* **phase modulation, preventive maintenance.**

PMB: *Abbreviation for* **pilot-make-busy.** *See* **pilot-make-busy circuit.**

Pockels cell: An electro-optic device in which birefringence is modified under the influence of an applied voltage. *Note:* A Pockels cell may be used as an intensity modulator at optical wavelengths.

POI: *Abbreviation for* **point of interface.**

pointer: **1.** A function indicator that (a) is under the direct control of a computer operator, and (b) is used to indicate displayed information, to highlight data, to identify areas of interest, to serve as a graphic display cursor, and/or to select icons. **2.** In computer graphics, a manually operated functional unit used to specify an addressable point. **3.** In computer programming, an identifier that indicates the location of a data item.

point of interface (POI): In a telecommunications system, the physical interface between the local access and transport area (LATA) access and inter-LATA functions. *Note:* The POI is used to establish the technical interface, the test points, and the points of operational responsibility. *Synonym* **interface point.**

point of presence (POP): A physical layer within a local access and transport area (LATA) at which an inter-LATA carrier establishes itself for the purpose of obtaining LATA access and to which the local exchange carrier provides access services.

point of train: In infrared transmission systems, a steady infrared light that is used (a) to assist the transmitter in locating a receiving station and (b) for keeping the transmitted light pointed in the proper direction for satisfactory reception. [From Weik '89]

point source: A source of electromagnetic radiation such that (a) the source is so distant from a point of observation or measurement of the radiation that the wavefront of the radiation is a planar rather than a curved surface, regardless of the shape of the source, (b) the size or shape of the source has no influence on the shape of the wavefront at the point of observation or measurement, and (c) the source need not necessarily radiate with equal radiance in all directions. [From Weik '89]

point-to-point link: A dedicated data link that connects only two stations. (188)

point-to-point transmission: Communications between two designated stations only. (188)

Poisson distribution: A mathematical statement of the probability that exactly k discrete events will take place during an interval of length t, expressed by

$$P(k,t) = \frac{(\lambda t)^k e^{-\lambda t}}{k!} ,$$

where k is a non-negative integer, e is the base of the natural logarithms ($e \approx 2.71828$), λ is the constant rate that the events occur, and λt is the expected number of events occurring during an interval of length t.

polar direct-current telegraph transmission: A form of binary telegraph transmission in which positive and negative direct currents denote the significant conditions. (188) *Synonym* **double-current transmission.**

polarential telegraph system: A direct-current telegraph system employing polar transmission in one direction and a form of differential duplex transmission in the other. (188) *Note:* Two types of polarential systems, known as types A and B, are in use. In half-duplex operation of a type A polarential system, the direct-current balance is independent of line resistance. In half-duplex operation of a type B polarential system, the direct current is substantially independent of the line leakage. Type A is better for cable loops where leakage is negligible but resistance varies with temperature. Type B is better for open wire where variable line leakage is frequent.

polarization: Of an electromagnetic wave, the property that describes the orientation, *i.e.*, time-varying direction and amplitude, of the electric field vector. (188) *Note 1:* States of polarization are described in terms of the figures traced as a function of time by the projection of the extremity of a representation of the electric vector onto a fixed plane in space, which plane is perpendicular to the direction of propagation. In general, the figure, *i.e.*, polarization, is elliptical and is traced in a clockwise or counterclockwise sense, as viewed in the direction of propagation. If the major and minor axes of the ellipse are equal, the polarization is said to be *circular*. If the minor axis of the ellipse is zero, the polarization is said to be *linear*. Rotation of the electric vector in a clockwise sense is designated *right-hand polarization*, and rotation in a counterclockwise sense is designated *left-hand polarization*. *Note 2:* Mathematically, an elliptically polarized wave may be described as the vector sum of two waves of equal wavelength but unequal amplitude, and in quadrature (having their respective electric vectors at right angles and $\pi/2$ radians out of phase).

polarization diversity: Diversity transmission and reception wherein the same information signal is transmitted and received simultaneously on orthogonally polarized waves with fade-independent propagation characteristics (188)

polarization-maintaining (PM) optical fiber: An optical fiber in which the polarization planes of lightwaves launched into the fiber are maintained during propagation with little or no cross-coupling of optical power between the polarization modes. [2196] *Note 1:* Cross sections of polarization-maintaining optical fibers range from elliptical to rectangular. *Note 2:* Polarization-maintaining optical fibers are used in special applications, such as in fiber optic sensing and interferometry. *Synonym* **polarization-preserving (PP) optical fiber.**

polarization-preserving (PP) optical fiber: *Synonym* **polarization-maintaining optical fiber.**

polar operation: A telegraph system in which marking signals are formed by current or voltage pulses of one polarity and spacing signals by current or voltage pulses of equal magnitude but opposite polarity (bipolar signal). (188)

polar orbit: An orbit for which the angle of inclination is 90°. (188) *Note:* A satellite in polar orbit will pass over both the north and south geographic poles once per orbit.

polar relay: A dc relay in which the direction of movement of the armature depends on the direction of the current flow. (188)

polling: **1.** Network control in which the control station invites tributary stations to transmit in the sequence specified by the control station. **2.** In point-to-point or multipoint communication, the process whereby stations are invited one at a time to transmit. **3.** Sequential interrogation of devices for various purposes, such as avoiding contention, determining operational status, or determining readiness to send or receive data. **4.** In automated HF radio systems, a technique for measuring and reporting channel quality. (188)

polymer-clad silica fiber: *Synonym* **plastic-clad silica fiber.**

POP: *Acronym for* **point of presence.**

port: **1.** Of a device or network, a point of access where signals may be inserted or extracted, or where the device or network variables may be observed or measured. (188) **2.** In a communications network, a point at which signals can enter or leave the network en route to or from another network.

portability: 1. The ability to transfer data from one system to another without being required to recreate or reenter data descriptions or to modify significantly the application being transported. **2.** The ability of software or of a system to run on more than one type or size of computer under more than one operating system. *See* **POSIX. 3.** Of equipment, the quality of being able to function normally while being conveyed.

portable station: 1. A station capable of being carried by one or more persons. *Note:* A portable station usually has a self-contained power source and can be operated while being carried. **2.** A station designed to be carried by a person and capable of transmitting and/or receiving while in motion or during brief halts at unspecified locations. [NTIA] [RR]

port operations service: A maritime mobile service in or near a port, between coast stations and ship stations, or between ship stations, in which messages are restricted to those relating to the operational handling, the movement and the safety of ships and, in emergency, to the safety of persons. Messages which are of a public correspondence nature shall be excluded from this service. [NTIA] [RR]

portrait mode: 1. In facsimile, the mode of scanning lines across the shorter dimension of a rectangular original. (188) *Note:* CCITT Group 1, 2, and 3 facsimile machines use portrait mode. **2.** In computer graphics, the orientation of an image in which the shorter dimension of the image is horizontal. **3.** An orientation of printed text on a page such that the lines of text are perpendicular to the long dimension of the page.

port station: A coast station in the port operations service. [NTIA] [RR]

positioned channel: In integrated services digital networks (ISDN), a channel that occupies dedicated bit positions in the framed data stream. *Note:* Examples of positioned channels are the B, H, and D channels.

positioned interface structure: Within a framed interface, a structure in which positioned channels provide all services and signaling.

positioning time: *Synonym* **seek time.**

positive feedback: *Synonym* **regeneration (def. #1).**

positive justification: *Synonym* **bit stuffing.**

POSIX: *Acronym for* **portable operating system interface for computer environments.** A Federal Information Processing Standard Publication (FIPS PUB 151-1) for a vendor-independent interface between an operating system and an application program, including operating system interfaces and source code functions. *Note:* IEEE Standard 1003.1-1988 was adopted by reference and published as FIPS PUB 151-1.

postalize: In communications, to structure rates or prices so that they are not distance sensitive, but depend on other factors, such as call duration, type of service, and time of day.

post-detection combiner: *Synonym* **maximal-ratio combiner.**

post-development review: *Synonym* **system follow-up.**

post-implementation review: *Synonym* **system follow-up.**

post-production processing: In broadband ISDN (B-ISDN), applications, the processing of audio and video information after contribution and prior to final use.

power: 1. The rate of transfer or absorption of energy per unit time in a system. (188) **2.** Whenever the power of a radio transmitter *etc.* is referred to, it shall be expressed in one of the following forms, according to the class of emission, using the arbitrary symbols indicated:

- peak envelope power (PX or pX);
- mean power (PY or pY);

- carrier power (PZ or pZ).

For different classes of emission, the relationships between peak envelope power, mean power and carrier power, under the conditions of normal operation and of no modulation, are contained in CCIR Recommendations which may be used as a guide. For use in formulae, the symbol p denotes power expressed in watts and the symbol P denotes power expressed in decibels relative to a reference level. [NTIA] [RR]

power budget: The allocation, within a system, of available electrical power, among the various functions that need to be performed. *Note:* An example of a power budget in a communications satellite is the allocation of available power among various functions, such as maintaining satellite orientation, maintaining orbital control, performing signal reception, and performing signal retransmission. *Synonym* **system budget.**

power circuit breaker (PCB): **1.** The primary switch used to apply and remove power from equipment. (188) **2.** A circuit breaker used on ac circuits rated in excess of 1500V. (188)

power density: *Deprecated synonym for* **irradiance.**

power factor: In alternating-current power transmission and distribution, the cosine of the phase angle between the voltage and current. *Note 1:* When the load is inductive, *e.g.*, an induction motor, the current lags the applied voltage, and the power factor is said to be a *lagging* power factor. When the load is capacitive, *e.g.*, a synchronous motor or a capacitive network, the current leads the applied voltage, and the power factor is said to be a *leading* power factor. *Note 2:* Power factors other than unity have deleterious effects on power transmission systems, including excessive transmission losses and reduced system capacity. Power companies therefore require customers, especially those with large loads, to maintain, within specified limits, the power factors of their respective loads or be subject to additional charges.

power failure transfer: **1.** The switching of primary utilities to their secondary backup whenever the primary source operates outside its design parameters. **2.** In telephony, a function, which, when activated in the event of a commercial power failure or a low-voltage battery condition at a subscriber location, supplies power to predesigned subscriber equipment via the central office trunk. *Note:* Power-failure transfer is an emergency mode of operation in which one and only one instrument may be powered from each trunk line from the subscriber location to the central office.

power gain of an antenna: *Synonym* **antenna gain.**

power-law index profile: For optical fibers, a class of graded-index profiles characterized by

$$n(r) = \begin{cases} n_1\sqrt{1 - 2\Delta\left(\dfrac{r}{\alpha}\right)^g} & , r \le \alpha \\ n_1\sqrt{1 - 2\Delta} & , r \ge \alpha \end{cases}$$

$$\text{where } \Delta = \frac{n_1^2 - n_2^2}{2n_1^2},$$

where $n(r)$ is the nominal refractive index as a function of distance from the fiber axis, n_1 is the nominal refractive index on axis, n_2 is the refractive index of the homogeneous cladding ($n(r) = n_2$ when $r \ge \alpha$), α is the core radius, and g is a parameter that defines the shape of the profile. *Note 1:* α is often used in place of g. Hence, this is sometimes called an alpha profile. *Note 2:* For this class of profiles, multimode distortion is smallest when g takes a particular value depending on the material used. For most materials, this optimum value is approximately 2. When g increases without limit, the profile tends to a step-index profile.

power margin: The difference between available signal power and the minimum signal power needed to overcome system losses and still satisfy the minimum input requirements of the receiver for a given performance level. *Note:* System power margin reflects the excess signal level, present at the input of the receiver, that is available to compensate for (a) the effects of component aging in the transmitter, receiver, or physical transmission medium, and (b) a deterioration in propagation conditions. *Synonym* **system power margin.**

Poynting vector: The vector obtained in the direction of a right-hand screw from the cross-product (vector product) of the electric field vector rotated into the magnetic field vector of an electromagnetic wave. *Note:* The Poynting vector, with transmission media parameters and constants, gives the irradiance and direction of propagation of the electromagnetic wave. Mathematically: $P = E \times H$. [From Weik '89]

PPM: *Abbreviation for* **pulse-position modulation.**

pre-arbitrated slot: A slot dedicated by the head-of-bus function for transferring isochronous service octets.

preassignment access plan: In satellite communications system operations, a fixed communication channel access plan, as opposed to a demand assignment access plan in which allocation of accesses or the number of channels per access is varied in accordance with the demand. [From Weik '89]

precedence: In communications, a designation assigned to a message by the originator to indicate to communications personnel the relative order of handling and to the addressee the order in which the message is to be noted. [After JP1] *Note:* The descending order of precedence for military messages is FLASH, IMMEDIATE, PRIORITY, and ROUTINE.

precipitation attenuation: The loss of energy by an electromagnetic wave because of scattering, refraction, and/or absorption during its passage through a volume of the atmosphere containing precipitation such as rain, snow, hail, or sleet.

precipitation static (p-static): Radio interference caused by the impact of charged particles against an antenna. *Note:* Precipitation static may occur in a receiver during certain weather conditions, such as snowstorms, hailstorms, rainstorms, dust storms, or combinations thereof. (188)

precise frequency: A frequency that is maintained to the known accuracy of an accepted reference frequency standard. (188) *Note:* Current uncertainty among international standards is approximately 1 part in 10^{14} as of 1995.

precise time: A time mark that is accurately known with respect to an accepted reference time standard. (188) *Note:* Current uncertainty among international standards is approximately 1 part in 10^{14} as of 1995.

precise time and time interval (PTTI): The discipline that addresses precise timekeeping and time information transfer.

precision: **1.** The degree of mutual agreement among a series of individual measurements, values, or results; often, but not necessarily, expressed by the standard deviation. **2.** With respect to a set of independent devices of the same design, the ability of these devices to produce the same value or result, given the same input conditions and operating in the same environment. **3.** With respect to a single device, put into operation repeatedly without adjustments, the ability to produce the same value or result, given the same input conditions and operating in the same environment. *Synonym (for defs. 1, 2, and 3)* **reproducibility. 4.** In computer science, a measure of the ability to distinguish between nearly equal values. (188) **5.** The degree of discrimination with which a quantity is stated; for example, a three-digit numeral to the base 10 discriminates among 1000 possibilities.

precision-sleeve splicing: Optical fiber splicing that uses a capillary tube, of suitable material, to align the mating fibers. *Note:* The capillary tube has an inside diameter slightly larger than the cladding diameter of the two optical fibers to be spliced. The fibers are inserted, one from either end, to form a butt joint. The capillary tube may contain an index-matching gel, or the fibers may be secured with an adhesive having a refractive index that approximates that of the fibers. [From Weik '89]

precombining: The combining of multiplexed signals prior to the modulation of the carrier. [From Weik '89] *Synonym* **premodulation combining.**

predetection: Referring to that portion of the circuitry of a receiver which, with respect to the signal being processed, is chronologically prior to the detection. *Note:* Predetection signals contain the carrier signal and all modulation, and are basically at radio frequencies.

predetection combining: *Synonym* **maximal-ratio combiner.**

preemphasis: A system process designed to increase, within a band of frequencies, the magnitude of some (usually higher) frequencies with respect to the magnitude of other (usually lower) frequencies, in order to improve the overall signal-to-noise ratio by minimizing the adverse effects of such phenomena as attenuation differences, or saturation of recording media, in subsequent parts of the system. (188) *Note:* Preemphasis has applications, for example, in audio recording and FM transmission.

preemphasis improvement: In FM broadcasting, the improvement in the signal-to-noise ratio of the high-frequency portion of the baseband, *i.e.*, modulating, signal, which improvement results from passing the modulating signal through a preemphasis network. *Note:* Preemphasis increases the magnitude of the higher signal frequencies, thereby improving the signal-to-noise ratio. At the output of the discriminator in the FM receiver, a deemphasis network restores the original signal power distribution.

preemphasis network: A network inserted in a system in order to increase the magnitude of one range of frequencies with respect to another. (188) *Note:* Preemphasis is usually employed in FM or phase modulation transmitters to equalize the modulating signal drive power in terms of deviation ratio. The receiver demodulation process includes a reciprocal network, called a deemphasis network, to restore the original signal power distribution.

preempting call: *See* **multilevel precedence and preemption.**

preemption: The seizure, usually automatic, of military system facilities that are being used to serve a lower precedence call in order to serve immediately a higher precedence call. (188)

preemption tone: In military telephone systems, a distinctive tone that is used to indicate to connected users, *i.e.*, subscribers, that their call has been preempted by a call of higher precedence. *Note:* An example of preemption tone is a distinctive, steady, high-pitch tone transmitted for three seconds or until the preempted user hangs up. [From Weik '89]

prefix-free code: *Synonym* **comma-free code.**

pregroup combining: In communications systems, assembling a number of narrowband channels, such as 4-kHz-wide telephone channels, into a specified frequency band such that, after pregroup translation, they may be formed with other pregroups into a standard group, such as a CCITT basic group, by frequency-division multiplexing. [From Weik '89]

pregroup translation: In communications systems, the process of transposing, in frequency, a pregroup of channels, such as telephone or data channels, in such a manner that they may be formed into a standard group, such as a CCITT basic group, by frequency-division multiplexing. [From Weik '89]

preliminary call: In radio transmission, a call that (a) includes at least the identification of the calling station and the called station, (b) is designed to establish communications with a particular station, and (c) usually includes a request to the called station to reply, although the request may be implied by the recitation of the call signs. [From Weik '89] *Note:* A preliminary call may be made on a frequency dedicated to that purpose only, and the rest of the communications session take place on a different frequency or frequencies.

premises wiring: *See* **on-premises wiring.**

premodulation combining: *Synonym* **precombining.**

Presentation Layer: *See* **Open Systems Interconnection—Reference Model.**

preset conference: A service feature that permits the automatic connection of a fixed group of users, or a closed user group with outgoing access, by keying a single directory number. [From Weik '89]

preset jammer: A jammer in which the frequency of the jamming transmitter is fixed before the transmitter is placed in operation. *Note:* Preset jammers are most useful in airborne jamming operations where weight and space requirements may prohibit the use of operators or elaborate control equipment in flight. Preset jammers are usually used in barrage-jamming over a wide band, usually in overlapping series of frequency bands. [From Weik '89]

press-to-talk operation: *Synonym* **push-to-talk operation.**

press-to-type operation: *Synonym* **push-to-type operation.**

preventive maintenance (PM): **1.** The care and servicing by personnel for the purpose of maintaining equipment and facilities in satisfactory operating condition by providing for systematic inspection, detection, and correction of incipient failures either before they occur or before they develop into major defects. [JP1] **2.** Maintenance, including tests, measurements, adjustments, and parts replacement, performed specifically to prevent faults from occurring. (188)

PRF: *Abbreviation for* **pulse repetition frequency.**

PRI: *Abbreviation for* **primary rate interface.**

primary channel: **1.** The channel that is designated as a prime transmission channel and is used as the first choice in restoring priority circuits. (188) **2.** In a communications network, the channel that has the highest data rate of all the channels sharing a common interface. *Note:* A primary channel may support the transfer of information in one direction only, either direction alternately, or both directions simultaneously.

primary coating: The plastic overcoat in intimate contact with the cladding of an optical fiber, applied during the manufacturing process. *Note 1:* The primary coating typically has an outside diameter of approximately 250 to 750 µm, and serves to protect the fiber from mechanical damage and chemical attack. It also enhances optical fiber properties by stripping off cladding modes, and in the case where multiple fibers are used inside a single buffer tube, it suppresses cross-coupling of optical signals from one fiber to another. *Note 2:* The primary coating should not be confused with a tight buffer, or the plastic cladding of a plastic-clad-silica (PCS) fiber. [After FAA] *Note 3:* The primary coating, which typically consists of many layers, may be color-coded to distinguish fibers from one another, *e.g.*, in a buffer tube containing multiple fibers. *Synonyms* **primary polymer coating, primary polymer overcoat.**

primary distribution system: A system of alternating current distribution for supplying the primaries of distribution transformers from the generating station or substation distribution buses. (188)

primary frequency: **1.** A frequency that is assigned for usual use on a particular circuit. **2.** The first-choice frequency that is assigned to a fixed or mobile station for radiotelephone communications.

primary frequency standard: A frequency source that meets national standards for accuracy and operates without the need for calibration against an external standard. *Note:* Examples of primary frequency standards are hydrogen masers and cesium beam frequency standards.

primary group: *See* **group.**

primary polymer coating: *Synonym* **primary coating.**

primary polymer overcoat: *Synonym* **primary coating.**

primary power: The source of electrical power that usually supplies the station main bus. (188) *Note 1:* The primary power source may be a Government-owned generating plant or a public utility power system. *Note 2:* A Class A primary power source assures, to a high degree of reliability, a continuous supply of ac electrical power.

primary radar: A radiodetermination system based on the comparison of reference signals with radio signals reflected from the position to be determined. [NTIA] [RR]

primary radiation: Radiation that is incident upon a material and produces secondary emission from the material.

primary rate interface (PRI): An integrated services digital network (ISDN) interface standard (a) that is designated in North America as having a 23B+D channels, (b) in which all circuit-switched B channels operate at 64 kb/s, and (c) in which the D channel also operates at 64 kb/s. *Note:* The PRI combination of channels results in a digital signal 1 (T1) interface at the network boundary.

primary route: The predetermined path of a message from its source, *i.e.*, sending or originating station, to a message sink, *i.e.*, receiving, addressee, or destination station. *Note 1:* In telephone switchboard operations, the primary route is the route that is attempted first by the operators or equipment when completing a call. *Note 2:* Alternate routing is based on network traffic conditions and supervisory policy. [From Weik '89]

primary service area: The service area of a broadcast station in which the groundwave is not subject to objectionable interference or objectionable fading. [47CFR]

primary station: In a data communication network, the station responsible for unbalanced control of a data link. *Note:* The primary station generates commands and interprets responses, and is responsible for initialization of data and control information interchange, organization and control of data flow, retransmission control, and all recovery functions at the link level.

primary substation: Equipment that switches or modifies voltage, frequency, or other characteristics of primary power. (188)

primary time standard: A time standard that does not require calibration against another time standard. (188) *Note 1:* Examples of primary time, (*i.e.*, frequency standards) are cesium standards and hydrogen masers. *Note 2:* The international second is based on the microwave frequency (9,192,631,770 Hz) associated with the atomic resonance of the hyperfine ground-state levels of the cesium-133 atom in a magnetically neutral environment. Realizable cesium frequency standards use a strong electromagnet to deliberately introduce a magnetic field which overwhelms that of the Earth. The presence of this strong magnetic field introduces a slight, but known, increase in the atomic resonance frequency. However, very small variations in the calibration of the electric current in the electromagnet introduce minuscule frequency variations among different cesium oscillators.

principal clock: Of a set of redundant clocks, the clock that is selected for normal use. (188) *Note 1:* The principal clock may be selected because of a property, *e.g.*, superior accuracy, that makes it a unique member of the set. *Note 2:* The term *"principal clock"* should not be confused with, or used as a synonym for, the term *"primary frequency standard."*

print-through: A transfer of magnetically recorded data from one part of a data medium to another part of the data medium when these parts are brought into physical contact. *Note:* Print-through is most commonly observed in magnetic tape. Print-through may be avoided by rewinding at timely intervals. [From Weik '89]

priority: **1.** Priority, unless specifically qualified, is the right to occupy a specific frequency for authorized uses, free of harmful interference from stations of other agencies. [NTIA] **2.** *Synonym* **priority level.** **3.** In DOD record communications systems, one of the four levels of precedence used to establish the time frame for handling a given message. **4.** In DOD voice communications systems, one of the levels of precedence assigned to a subscriber telephone for the purpose of preemption of telephone services.

priority level: In the Telecommunications Service Priority system, the level that may be assigned to an NS/EP telecommunications service, which level specifies the order in which provisioning or restoration of the service is to occur relative to other NS/EP or non-NS/EP telecommunication services. *Note:* Priority levels authorized are designated (highest to lowest) "E," "1," "2," "3," "4," and "5" for provisioning and "1," "2," "3," "4," and "5" for restoration. *Synonym* **priority.**

priority level assignment: The priority level(s) designated for the provisioning or restoration of a particular NS/EP telecommunications service.

priority message: A category of precedence reserved for messages that require expeditious action by the addressee(s) and/or furnish essential information for the conduct of operations in progress when routine precedence will not suffice. [JP1]

privacy: **1.** In a communications system or network, the protection given to information to conceal it from persons having access to the system or network. *Synonym* **segregation.** **2.** In a communications system, protection given to unclassified information, such as radio transmissions of law enforcement personnel, that requires safeguarding from unauthorized persons. **3.** In a communications system, the protection given to prevent unauthorized disclosure of the information in the system. (188) *Note 1:* The required protection may be accomplished by various means, such as by communications security measures and by directives to operating personnel. *Note 2:* The limited protection given certain voice and data transmissions by commercial crypto-equipment is sufficient to deter a casual listener, but cannot withstand a competent cryptanalytic attack.

private automatic branch exchange (PABX): *See* **PBX.**

private automatic exchange (PAX): *See* **PBX.**

private branch exchange (PBX): *See* **PBX.**

private exchange (PX): A private telecommunication switch that usually includes access to the public switched network.

private line: In the telephone industry usage, a service that involves dedicated circuits, private switching arrangements, and/or predefined transmission paths, whether virtual or physical, which provide communications between specific locations. *Note:* Among subscribers to the public switched telephone network(s), the term *"private line"* is often used to mean a one-party switched access line.

private NS/EP telecommunications services: Non-common-carrier telecommunications services, including private line, virtual private line, and private switched network services.

procedure-oriented language: A problem-oriented computer programming language that facilitates expressing a procedure in the form of explicit algorithms. *Note:* Examples of procedure-oriented languages are Fortran, ALGOL, COBOL, and PL/I.

proceed-to-select: In communications systems operation, pertaining to a signal or event in the call-access phase of a data call, which signal or event confirms the reception of a call-request signal and advises the calling data terminal equipment to proceed with the transmission of the selection signals. *Note:* Examples of proceed-to-select pertain to a dial tone in a telephone system.

proceed-to-select signal: In a communications system, a signal that indicates that the system is ready to receive a selection signal. *Note:* An example of a proceed-to-select signal is a dial tone.

process computer system: A computer system, with a process interface system, that monitors or controls a technical process.

process control: Automatic control of a process, in which a computer system is used to regulate the usually continuous operations or processes.

process control equipment: Equipment that measures the variables of a technical process, directs the process according to control signals from the process computer system, and provides appropriate signal transformation. *Note:* Examples of process control equipment include actuators, sensors, and transducers.

process control system: A system consisting of a computer, process control equipment, and possibly a process interface system. *Note:* The process interface system may be part of a special-purpose computer.

process gain: In a spread-spectrum communications system, the signal gain, signal-to-noise ratio, signal shape, or other signal improvement obtained by coherent band spreading, remapping, and reconstitution of the desired signal.

processing unit: A functional unit that consists of one or more processors and their internal storage.

process interface system: A functional unit that adapts process control equipment to the computer system in a process computer system.

processor: In a computer, a functional unit that interprets and executes instructions. *Note:* A processor consists of at least an instruction control unit and an arithmetic unit.

procurement: In the Federal Government, the process of obtaining services, supplies, and equipment in conformance with applicable laws and regulations.

procurement lead time: The interval between the initiation of a procurement action and receipt of the products or services purchased as the result of such action.

profile dip: *Synonym* **index dip.**

profile dispersion: *See* **dispersion.**

profile parameter (*g*): In the power-law index profile of an optical fiber, the parameter, *g*, that defines the shape of the refractive-index profile. *Note:* The optimum value of *g* for minimum dispersion is approximately 2.

pro forma message: A standard form of message, that has elements that usually are understood by prearrangement among the originator, the addressee, and the communications system operators. [From Weik '89]

program: **1.** A plan or routine for solving a problem on a computer. *Note:* Processing may include the use of an assembler, a compiler, an interpreter, or a translator to prepare the program for execution, as well as the execution of the program. The sequence of instructions may include statements and necessary declarations. **2.** A sequence of instructions used by a computer to do a particular job or solve a given problem. **3.** To design, write, and test programs.

program architecture: For a computer program, (a) the structure, relationships, and arrangement of the components of the program, (b) the program interfaces, and (c) the interface requirements for the program operating environment. [From Weik '89]

programmable: Pertaining to a device that can accept instructions that alter its basic functions.

programmable logic array (PLA): An array of gates having interconnections that can be programmed to perform a specific logical function.

programmable read-only memory (PROM): A storage device that, after being written to once, becomes a read-only memory.

programmer: **1.** The part of digital equipment that controls the timing and sequencing of operations. (188) **2.** A person who prepares computer programs, *i.e.*, writes sequences of instructions for execution by a computer. (188)

programming language: An artificial language that is used to generate or to express computer programs. *Note:* The language may be a high-level language, an assembly language, or a machine language. (188) *See figure at* **assembly language.**

programming system: One or more programming languages and the software necessary for using these languages with particular automatic data processing equipment.

program origin: *See* **computer program origin.**

PROM: *Acronym for* **programmable read-only memory.**

prompt: **1.** In interactive display systems, a message on the display surface of a display device to help the user to plan and execute subsequent operations. *Note:* Examples of prompts include (a) a blinking message displayed on a screen to inform the system operator of the status, condition, or mode the system is in and requiring the operator to take some action, and (b) a message that the system is ready to accept a command. **2.** In a computer, communications, or data processing system, to inform a user that the system is ready for the next command, data element, or other input. [From Weik '89]

propagation: The motion of waves through or along a medium. *Note:* For electromagnetic waves, propagation may occur in a vacuum as well as in material media.

propagation constant: For an electromagnetic field mode varying sinusoidally with time at a given frequency, the logarithmic rate of change, with respect to distance in a given direction, of the complex amplitude of any field component. *Note:* The propagation constant, λ, is a complex quantity given by $\lambda = \alpha + i\beta$, where α, the real part, is the attenuation constant and β, the imaginary part, is the phase constant.

propagation mode: The manner in which radio signals travel from a transmitting antenna to a receiving antenna, such as ground wave, sky wave, direct wave, ground reflection, or scatter. (188)

propagation path obstruction: A man-made or natural physical feature that lies near enough to a radio path to cause a measurable effect on path loss, exclusive of reflection effects. (188) *Note:* An obstruction may lie to the side, above, or below the path. Ridges, bridges, cliffs, buildings, and trees are examples of obstructions. If the clearance from the nearest anticipated path position, over the expected range of Earth radius k-factor, exceeds 0.6 of the first Fresnel zone radius, the feature is not normally considered an obstruction.

propagation time delay: The time required for a signal to travel from one point to another. (188)

proprietary standard: Documentation by a commercial entity specifying equipment, practices, or operations unique to that commercial entity.

proration: **1.** The proportional distribution or allocation of parameters, such as noise power and transmission losses, among a number of tandem-connected items, such as equipment, cables, links, or trunks, in order to balance the performance of communications circuits. (188) *Synonym* **budgeting.** **2.** In a telephone switching center, the distribution or allocation of equipment or components proportionally among a number of functions, to provide a requisite grade of service. (188)

protected distribution system (PDS): [A] wireline or fiber-optics telecommunication system that includes terminals and adequate acoustical, electrical, electromagnetic, and physical safeguards to permit its use for the unencrypted transmission of classified information. [NIS] *Note:* A complete protected distribution system includes the subscriber and terminal equipment and the interconnecting lines. *Deprecated synonym* **approved circuit.**

protected frequency: A frequency that is not to be deliberately jammed by friendly forces, usually during a specified period. [From Weik '89]

protection: *Synonym* **lockout (def. #5).**

protection interval (PI): In high-frequency (HF) radio automatic link establishment, the period between changes in the time-of-day portion of the time-varying randomization data used for encrypting transmissions. (188)

protection ratio: The minimum value of the wanted-to-unwanted signal ratio, usually expressed in decibels, at the receiver input determined under specified conditions such that a specified reception quality of the wanted signal is achieved at the receiver output. [NTIA] [RR]

protector: In telecommunications systems, a device used to protect facilities and equipment from abnormally high voltages or currents. (188) *Note 1:* A protector may contain arresters. *Note 2:* Protectors may be designed to operate on short-duration phenomena, or long-duration phenomena. The duration should be specified.

protocol: **1.** A formal set of conventions governing the format and control of interaction among communicating functional units. (188) *Note:* Protocols may govern portions of a network, types of service, or administrative procedures. For example, a data link protocol is the specification of methods whereby data communications over a data link are performed in terms of the particular transmission mode, control procedures, and recovery procedures. **2.** In layered communications system architecture, a formal set of procedures that are adopted to facilitate functional interoperation within the layered hierarchy.

protocol-control information: **1.** The queries and replies among communications equipment to determine the respective capabilities of each end of the communications link. **2.** For layered systems, information exchanged between entities of a given layer, via the service provided by the next lower layer, to coordinate their joint operation.

protocol converter: A functional unit that uses a specified algorithm to translate a bit stream from one protocol to another, for interoperation.

protocol data unit (PDU): **1.** Information that is delivered as a unit among peer entities of a network and that may contain control information, address information, or data. **2.** In layered systems, a unit of data that is specified in a protocol of a given layer and that consists of protocol-control information of the given layer and possibly user data of that layer.

protocol hierarchy: In open systems architecture, the distribution of network protocol among the various layers of the network. [From Weik '89]

protocol translator: In a communications system, the collection of hardware, software, firmware, or any combination of these, that is required or used to convert the protocols used in one network to those used in another network. [From Weik '89]

prototype: **1.** A pre-production, functioning specimen(s) that is the first of its type, typically used for the evaluation of design, performance, and/or production potential. **2.** A model suitable for evaluation of design, performance, and production potential. [JP1]

provisioning: The act of acquiring telecommunications service from the submission of the requirement through the activation of service. *Note 1:* Provisioning includes all associated transmission, wiring, and equipment. *Note 2:* In NS/EP telecommunication services, *"provisioning"* and *"initiation"* are synonymous and include altering the state of an existing priority service or capability.

PS: *Abbreviation for* **permanent signal.**

pseudo bit-error ratio: *See* **PBER.**

pseudorandom noise: Noise that satisfies one or more of the standard tests for statistical randomness. (188) *Note 1:* Although it seems to lack any definite pattern, pseudorandom noise contains a sequence of pulses that repeat themselves, albeit after a long time or a long sequence of pulses. *Note 2:* For example, in spread-spectrum systems, modulated carrier transmissions appear as pseudorandom noise to a receiver (a) that is not locked on the transmitter frequencies or (b) that is incapable of correlating a locally generated pseudorandom code with the received signal.

pseudorandom number generator: **1.** A device that produces a stream of unpredictable, unbiased, and usually independent bits. **2.** In cryptosystems, a random bit generator used for key generation or to start all the crypto-equipment at the same point in the key stream.

pseudorandom number sequence: **1.** An ordered set of numbers that has been determined by some defined arithmetic process but is effectively a random number sequence for the purpose for which it is required. **2.** A sequence of numbers that satisfies one or more of the standard tests for statistical randomness. (188) *Note:* Although a pseudorandom number sequence appears to lack any definite pattern, it will repeat after a very long time interval or after a very long sequence of numbers.

PSK: *Abbreviation for* **phase-shift keying.**

PSN: *Abbreviation for* **public switched network.**

psophometer: An instrument that provides a visual indication of the audible effects of disturbing voltages of various frequencies. *Note:* A psophometer usually incorporates a weighting network. The characteristics of the weighting network depend on the type of circuit under investigation, such as whether the circuit is used for high-fidelity music or for normal speech.

psophometrically weighted dBm: *See* **dBm(psoph), dBm0p.**

psophometric voltage: Circuit noise voltage measured with a psophometer that includes a CCIF-1951 weighting network. *Note 1: "Psophometric voltage"* should not be confused with *"psophometric emf,"* *i.e.*, the emf in a generator or line with 600Ω internal resistance. For practical purposes, the psophometric emf is twice the corresponding psophometric voltage. *Note 2:* Psophometric voltage readings, *V*, in millivolts, are commonly converted to dBm(psoph) by dBm(psoph) = $20 \log_{10} V - 57.78$.

psophometric weighting: A noise weighting established by the International Consultative Committee for Telephony (CCIF, which became CCITT and, more recently, ITU-T), designated as CCIF-1951 weighting, for use in a noise measuring set or psophometer. (188) *Note:* The shape of this characteristic is virtually identical to that of F1A weighting. The psophometer is, however, calibrated with a tone of 800 Hz, 0 dBm, so that the corresponding voltage across 600 ohms produces a reading of 0.775 V. This introduces a 1-dBm adjustment in the formulas for conversion with dBa.

PSTN: *Abbreviation for* **public switched telephone network.**

PTF: *Abbreviation for* **patch and test facility.**

PTM: *Abbreviation for* **pulse-time modulation.**

PTT: *Abbreviation for* **postal, telegraph, and telephone (organization).** In countries having nationalized telephone and telegraph services, the organization, usually a governmental department, which acts as its nation's common carrier. (188)

PTTI: *Abbreviation for* **precise time and time interval.**

public correspondence: Any telecommunication which the offices and stations must, by reason of their being at the disposal of the public, accept for transmission. [NTIA] [RR]

public data network (PDN): A network established and operated by a telecommunications administration, or a recognized private operating agency, for the specific purpose of providing data transmission services for the public. (188)

public data transmission service: A data transmission service that is established and operated by a telecommunication administration, or a recognized private operating agency, and uses a public data network. *Note:* A public data transmission service may include circuit-switched, packet-switched, and leased-circuit data transmission.

public key cryptography: The type of cryptography in which the encryption process is publicly available and unprotected, but in which a part of the decryption key is protected so that only a party with knowledge of both parts of the decryption process can decrypt the cipher text. *Note:* Commonly called non-secret encryption in professional cryptologic circles. FIREFLY is an application of public key cryptography. [NIS]

public land mobile network (PLMN): A network that is established and operated by an administration or by a recognized operating agency (ROA) for the specific purpose of providing land mobile telecommunications services to the public. *Note:* A PLMN may be considered as an extension of a fixed network, *e.g.* the Public Switched Telephone Network (PSTN) or as an integral part of the PSTN.

public switched network (PSN): Any common carrier network that provides circuit switching among public users. (188) *Note:* The term is usually applied to public switched telephone networks, but it could be applied more generally to other switched networks, *e.g.*, packet-switched public data networks.

public switched NS/EP telecommunications services: Those NS/EP telecommunications services utilizing public switched networks. *Note:* Public switched NS/EP telecommunication services may include both interexchange and intraexchange network facilities (*e.g.*, switching systems, interoffice trunks, and subscriber loops).

public switched telephone network (PSTN): A domestic telecommunications network usually accessed by telephones, key telephone systems, private branch exchange trunks, and data arrangements. *Note:* Completion of the circuit between the call originator and call receiver in a PSTN requires network signaling in the form of dial pulses or multifrequency tones.

public utilities commission (PUC): In the United States, a state regulatory body charged with regulating intrastate utilities, including telecommunications systems. *Note:* In some states this regulatory function is performed by public service commissions or state corporation commissions.

PUC: *Abbreviation for* **public utilities commission.**

pull-in frequency range: The maximum frequency difference between the local oscillator or clock and the reference frequency of a phase-locked loop over which the local oscillator can be locked.

pulsating direct current: A direct current (dc) that changes in value at regular or irregular intervals. *Note:* A pulsating direct current may change in value, *i.e.*, be always present but at different levels, or it may be a current that is interrupted completely at regular or irregular intervals, but when present, is always in the same direction.

pulse: **1.** A rapid, transient change in the amplitude of a signal from a baseline value to a higher or lower value, followed by a rapid return to the baseline value. (188) **2.** A rapid change in some characteristic of a signal, *e.g.*, phase or frequency, from a baseline value to a higher or lower value, followed by a rapid return to the baseline value.

pulse-address multiple access (PAMA): The ability of a communication satellite to receive signals from several Earth terminals simultaneously and to amplify, translate, and relay the signals back to Earth, based on the addressing of each station by an assignment of a unique combination of time and frequency slots. (188) *Note:* This ability may be restricted by allowing only some of the terminals access to the satellite at any given time.

pulse amplitude: The magnitude of a pulse parameter, such as the field intensity, voltage level, current level, or power level. *Note 1:* Pulse amplitude is measured with respect to a specified reference and therefore should be modified by qualifiers, such as "average," "instantaneous," "peak," or "root-mean-square." *Note 2:* Pulse amplitude also applies to the amplitude of frequency- and phase-modulated waveform envelopes. *See illustration under* **pulse.**

pulse-amplitude modulation (PAM): Modulation in which the amplitude of individual, regularly spaced pulses in a pulse train is varied in accordance with some characteristic of the modulating signal. (188) *Note:* The amplitude of the amplitude-modulated pulses conveys the information.

pulse broadening: An increase in pulse duration. *Note:* Pulse broadening may be specified by the impulse response, the root-mean-square pulse broadening, or the full-duration-at-half-maximum pulse broadening.

pulse carrier: An electromagnetic wave that (a) consists of a series of pulses usually of constant length, amplitude, spacing, and repetition rate when not modulated and (b) usually is used as a subcarrier. [From Weik '89]

pulse-code modulation (PCM): Modulation in which a signal is sampled, and the magnitude (with respect to a fixed reference) of each sample is quantized and digitized for transmission over a common transmission medium. (188) *Note 1:* In conventional PCM, before being digitized, the analog data may be processed (*e.g.*, compressed), but once digitized, the PCM signal is not subjected to further processing (*e.g.*, digital compaction) before being multiplexed

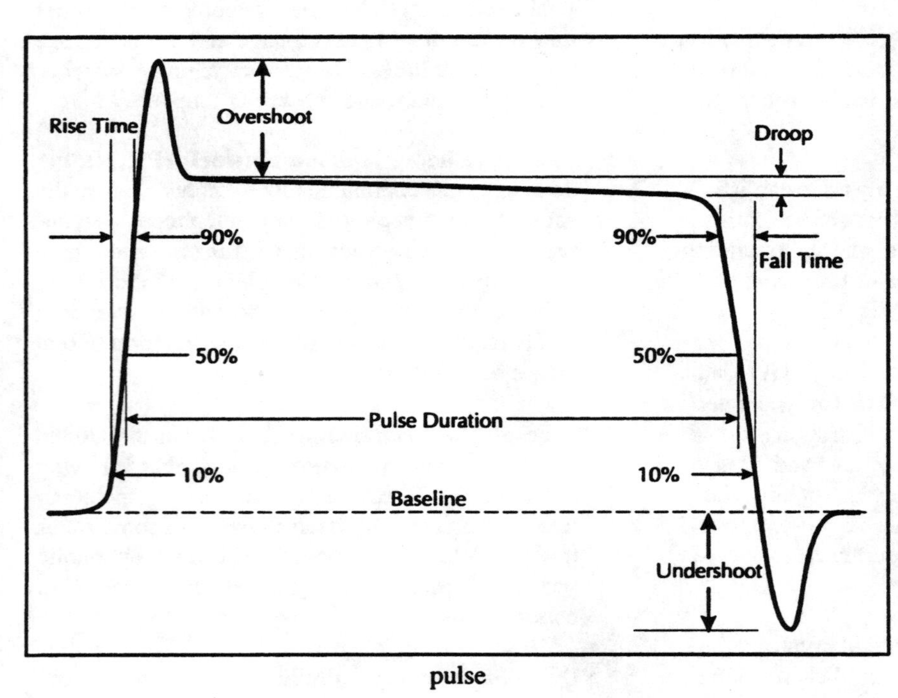

pulse

into the aggregate data stream. *Note 2:* PCM pulse trains may be interleaved with pulse trains from other channels.

pulse decay time: *Synonym* **fall time.**

pulse distortion: *See* **distortion.**

pulse duration: **1.** In a pulse waveform, the interval between (a) the time, during the first transition, that the pulse amplitude reaches a specified fraction (level) of its final amplitude, and (b) the time the pulse amplitude drops, on the last transition, to the same level. *Note:* The interval between the 50% points of the final amplitude is usually used to determine or define pulse duration, and this is understood to be the case unless otherwise specified. Other fractions of the final amplitude, *e.g.,* 90% or 1/e (where e = 2.71828. . .), may also be used, as may the root-mean-square (rms) value of the pulse amplitude. (188) *See illustration under* **pulse.** *Deprecated synonyms* **pulse length, pulse width.** **2.** In radar, measurement of pulse transmission time in microseconds, that is, the time the radar's transmitter is energized during each cycle.[JP1]

pulse-duration modulation (PDM): Modulation in which the duration of pulses is varied in accordance with some characteristic of the modulating signal. (188) *Deprecated synonyms* **pulse-length modulation, pulse-width modulation.**

pulse duty factor: In a periodic pulse train, the ratio of the pulse duration to the pulse period. (188)

pulse-frequency modulation (PFM): Modulation in which the pulse repetition rate is varied in accordance with some characteristic of the modulating signal. *Note:* Pulse-frequency modulation is analogous to frequency modulation of a carrier wave, in which the instantaneous frequency is a continuous function of the modulating signal.

pulse-interval modulation: *See* **pulse-position modulation.**

pulse length: *Deprecated synonym for* **pulse duration.**

pulse-length modulation *Deprecated synonym for* **pulse-duration modulation.**

pulse link repeater (PLR): A device that interfaces concatenated E&M signal paths. *Note 1:* A PLR converts a ground, received from the E lead of one signal path, to –48 Vdc, which is applied to the M lead of the concatenated signal path. *Note 2:* In many commercial carrier systems, the channel bank cards or modules have a "PLR" option that permits the direct connection, *i.e.,* concatenation, of E&M signaling paths, without the need for separate PLR equipment.

pulse period: The reciprocal of the pulse repetition rate.

pulse-position modulation (PPM): Modulation in which the temporal positions of the pulses are varied in accordance with some characteristic of the modulating signal. (188)

pulse-repetition frequency (PRF): In radar applications, *synonym* **pulse repetition rate.**

pulse repetition rate: The number of pulses per unit time.

pulse-repetition-rate modulation: *Synonym* **pulse-frequency modulation.**

pulse rise time: *See* **rise time.**

pulse string: *Synonym* **pulse train.**

pulse stuffing: *See* **bit stuffing.**

pulse-time modulation (PTM): The general class of pulse-code modulation in which the time of occurrence of some characteristic of the pulsed carrier is varied with respect to some characteristic of the modulating signal. (188) *Note:* PTM includes pulse-position modulation and pulse-duration modulation.

pulse train: A series of pulses having similar characteristics. *Synonym* **pulse string.**

pulse width: *Deprecated synonym for* **pulse duration.**

pulse-width modulation (PWM): *Deprecated synonym for* **pulse-duration modulation.**

pulsing: In telephony, the transmission of address information to a switching office by means of pulses, *i.e.,* signals, that originate from the subscriber, *i.e.,* user, equipment. *Note:* Examples of pulsing methods are dual-tone multifrequency (DTMF) signaling, in which a unique pair of audio frequencies represents each of the respective numerals or other characters on a keypad, and rotary dialing, in which dc pulses are generated by a rotary dial. *Synonyms* **key pulsing** (when using a keypad), **dial pulsing** (when using a rotary dial).

pump frequency: The frequency of an oscillator used to provide sustaining power to a device, such as a laser or parametric amplifier, that requires rf or optical power. (188)

pumping: The action of an oscillator that provides cyclic inputs to an oscillating reaction device. *Note:* Examples of pumping are the action that results in amplification of a signal by a parametric amplifier, and the action that provides a laser or maser with an input signal at the appropriate frequency to sustain stimulated emission.

pure binary numeration system: *See* **binary notation.**

pushbutton dialing: Dialing in which (a) pushbuttons or keys are used to actuate and connect audible tone oscillators to a line, (b) each button or key corresponds to a unique frequency or set of frequencies, and (c) each pushbutton or key represents a unique digit or symbol. [From Weik '89]

push-down file: *See* **last-in first-out.**

push-to-talk (PTT) operation: In telephone or two-way radio systems, that method of communication over a speech circuit in which the talker is required to keep a switch operated while talking. *Note:* In two-way radio, push-to-talk operation must be used when the same frequency is employed by both transmitters. For use in noisy environments, or for privacy, some telephone handsets have push-to-talk switches that allow the speaker to be heard only when the switch is activated. (188) *Synonym* **press-to-talk operation.**

push-to-type operation: In telegraph or data transmission systems, that method of communication in which the operator at a station must keep a switch operated in order to send messages. *Note 1:* Push-to-type operation is used in radio systems where the same frequency is employed for transmission and reception. (188) *Note 2:* Push-to-type operation is a derivative form of transmission and may be used in simplex, half-duplex, or duplex operation. *Synonym* **press-to-type operation.**

PVC: *Abbreviation for* **permanent virtual circuit.**

pW: *Abbreviation for* **picowatt.** A unit of power equal to 10^{-12} W, *i.e.,* –90 dBm. (188) *Note:* One picowatt is usually used as a reference level for both weighted and unweighted noise measurements. The type of measurement must be specified.

PWM: *Abbreviation for* **pulse-width modulation,** *which is a deprecated synonym for* **pulse-duration modulation.**

pWp: *Abbreviation for* **picowatt, psophometrically weighted.** *See* **noise weighting.**

pWp0: *Abbreviation for* **picowatts, psophometrically weighted,** measured at a zero-dBm transmission level point. *See* **dBm(psoph), psophometer.**

pX: *Abbreviation for* **peak envelope power (of a radio transmitter).**

PX: **1.** *Abbreviation for* **private exchange**; *See* **PBX.** **2.** *Abbreviation for* **peak envelope power.**

p x 64: In video teleconferencing, pertaining to a family of CCITT Recommendations, where p is a non-zero positive integer indicating the number of 64 kb/s channels. (188) *Note:* The p x 64 family includes CCITT Recommendations H.261, H.221, H.242, H.230, and H.320. These Recommendations form the basis for video telecommunications interoperability.

Q

QA: *Abbreviation for* **quality assurance.**

QAM: *Abbreviation for* **quadrature amplitude modulation.**

QC: *Abbreviation for* **quality control.**

QOS: *Abbreviation for* **quality of service.**

QPSK: *Abbreviation for* **quadrature phase-shift keying.**

quad: A group of four wires composed of two pairs twisted together. *Note:* The pairs have a fairly long length of twist and the quad a fairly short length of twist. (188)

quadded cable: A cable formed of multiples of quads, paired and separately insulated, and contained under a common jacket. (188)

quadratic profile: *Synonym* **parabolic profile.**

quadrature: **1.** The state of being separated in phase by 90° (π/2 radians). **2.** Pertaining to the phase relationship between two periodic quantities varying with the same period, that is, with the same frequency or repetition rate, when the phase difference between them is one-quarter of their period. (188)

quadrature amplitude modulation (QAM): Quadrature modulation in which the two carriers are amplitude modulated. (188)

quadrature modulation: Modulation using two carriers out of phase by 90° and modulated by separate signals. (188)

quadrature phase-shift keying (QPSK): Phase-shift keying in which four different phase angles are used. (188) *Note:* In QPSK, the four angles are usually out of phase by 90°. *Synonyms* **quadriphase, quaternary phase-shift keying.**

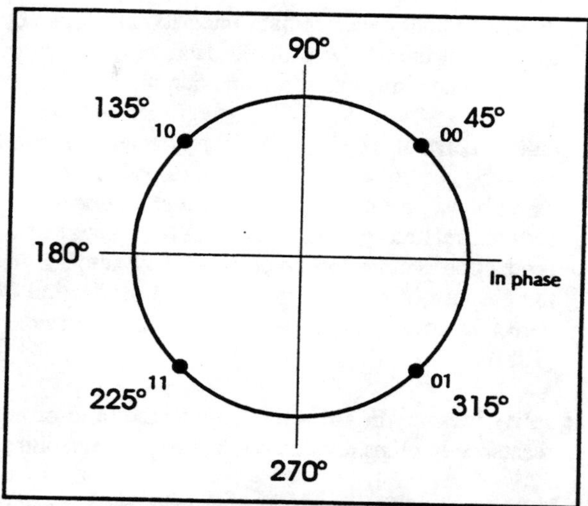

quadrature phase-shift keying

quadriphase: *Synonym* **quadrature phase-shift keying.**

quadruple diversity: In radio communication, diversity transmission and reception in which four independently fading signals are used. *Note:* Quadruple diversity may be accomplished through the use of space, frequency, angle, time, or polarization multiplexing, or combinations of these. (188)

quadruply clad fiber: A single-mode optical fiber that has four claddings. *Note 1:* Each cladding has a refractive index lower than that of the core. With respect to one another, their relative refractive indices are, in order of distance from the core, lowest, highest, lower, higher. *Note 2:* A quadruply clad fiber has the advantage of very low macrobending losses. It also has two zero-dispersion points, and moderately low dispersion over a wider wavelength range than a singly clad fiber or a doubly clad fiber. *See illustration under* **refractive index profile.**

quality assurance (QA): **1.** All actions taken to ensure that standards and procedures are adhered to and that delivered products or services meet performance requirements. (188) **2.** The planned systematic activities necessary to ensure that a component, module, or system conforms to established technical requirements. **3.** The policy, procedures, and systematic actions established in an enterprise for the purpose of providing and maintaining a specified degree of confidence in data integrity and accuracy throughout the lifecycle of the data, which includes input, update, manipulation, and output.

quality control (QC): A management function whereby control of the quality of (a) raw materials, assemblies, produced materiel, and components, (b) services related to production, and (c) management, production, and inspection processes is exercised for the purpose of preventing undetected production of defective materiel or the rendering of faulty services. (188)

quality factor: In a reactive circuit, the ratio of the reactance in ohms divided by the resistance in ohms.

quality of service (QOS): **1.** The performance specification of a communications channel or system. (188) *Note:* QOS may be quantitatively indicated by channel or system performance parameters, such as signal-to-noise ratio (S/N), bit error ratio (BER), message throughput rate, and call blocking probability. **2.** A subjective rating of telephone communications quality in which listeners judge transmissions by qualifiers, such as excellent, good, fair, poor, or unsatisfactory.

quantization: A process in which the continuous range of values of an analog signal is sampled and divided into nonoverlapping (but not necessarily equal) subranges, and a discrete, unique value is assigned to each subrange. *Note:* An application of quantization is its use in pulse-code modulation. If the sampled signal value falls within a given subrange, the sample is assigned the corresponding discrete value for purposes of modulation and transmission. (188)

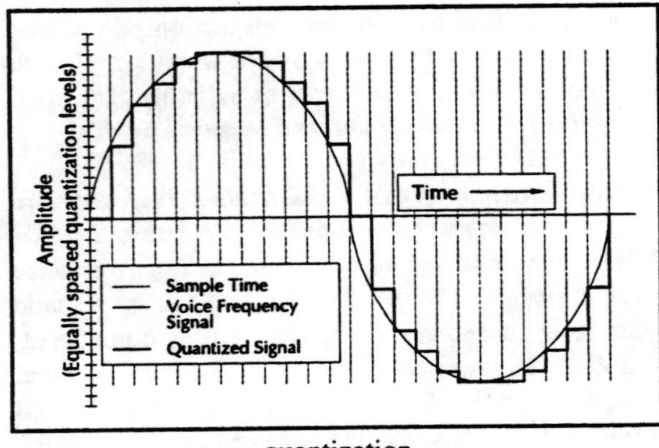

quantization

quantization error: *Synonym* **quantizing distortion.**

quantization level: In the quantization process, the discrete value assigned to a particular subrange of the analog signal being quantized. (188)

quantization noise: *Synonym* **quantizing noise.**

quantized feedback: In a digital feedback loop, the digital signal that is fed back. (188) *Note 1:* Several forms of analog-to-digital converters contain a quantized feedback loop following the basic A-D converter. *Note 2:* The feedback signal is often processed before introducing it to the loop.

quantizing distortion: Distortion that results from the quantization process. (188) *Synonym* **quantization error.**

quantizing levels: In digital transmission, the number of discrete signal levels transmitted as the result of signal digitization. (188)

quantizing noise: Noise caused by the error of approximation in quantization. (188) *Note:* Quantizing noise is dependent on the particular quantization process used and the statistical characteristics of the quantized signal. *Synonym* **quantization noise.**

quantum efficiency: In an optical source or detector, the ratio of the number of output quanta to the number of input quanta. *Note:* Input and output quanta need not both be photons.

quantum-limited operation: *Synonym* **quantum-noise-limited operation.**

quantum noise: Noise attributable to the discrete and probabilistic nature of physical phenomena and their interactions. *Note 1:* Quantum noise represents the fundamental limit of the achievable signal-to-noise ratio of an optical communication system. This limit is never achieved in practice. [After FAA] *Note 2:* Examples of quantum noise are photon noise in an optical signal and shot noise in an electrical conductor or semiconductor.

quantum-noise-limited operation: Operation wherein the minimum detectable signal is limited by quantum noise. (188) *Synonym* **quantum-limited operation.**

quarter common intermediate format (QCIF): A video format defined in CCITT Recommendation H.261 that is characterized by 176 luminance pixels on each of 144 lines, with half as many chrominance pixels in each direction. (188) *Note:* QCIF has one-fourth as many pixels as the full common intermediate format.

quartz clock: A clock containing a quartz oscillator that determines the accuracy and precision of the clock.

quartz oscillator: An oscillator in which a quartz crystal is used to stabilize the frequency. *Note:* The piezoelectric property of the quartz crystal results in a nearly constant output frequency, which is dependent upon the crystal size, shape, and excitation.

quasi-analog signal: A digital signal that has been converted to a form suitable for transmission over a specified analog channel. (188) *Note:* The specification of the analog channel should include frequency range, bandwidth, signal-to-noise ratio, and envelope delay distortion. When quasi-analog form of signaling is used to convey message traffic over dial-up telephone systems, it is often referred to as voice-data. A modem may be used for the conversion process.

quasi-analog transmission: Transmission in which a special-purpose modulator is used to convert digital signals into an analog form suitable for transmission over an analog voice-grade circuit. *Note:* A complementary demodulator is used to recover the digital signal at the other end of the circuit. (188) *See* **modem.**

quaternary phase-shift keying: *Synonym* **quadrature phase-shift keying.**

quaternary signal: A digital signal having four significant conditions.

query call: In adaptive high-frequency (HF) radio, an automatic-link-establishment call that requests responses from stations having connectivity to the destination specified in the call. (188)

queue: A set of items, such as telephone calls or packets, arranged in sequence. *Note:* Queues are used to store events occurring at random times and to service them according to a prescribed discipline that may be fixed or adaptive.

queue traffic: **1.** A series of outgoing or incoming calls waiting for service. (188) **2.** In a store-and-forward switching center, the outgoing messages awaiting transmission at the outgoing line position. (188)

queuing: The process of entering elements into or removing elements from a queue. (188)

queuing delay: **1.** In a switched network, the time between the completion of signaling by the call originator and the arrival of a ringing signal at the call receiver. (188) *Note:* Queues may be caused by delays at the originating switch, intermediate switches, or the call receiver servicing switch. **2.** In a data network, the sum of the delays between the request for service and the establishment of a circuit to the called data terminal equipment (DTE). **3.** In a packet-switched network, the sum of the delays encountered by a packet between the time of insertion into the network and the time of delivery to the addressee. (188)

queuing theory: The theoretical study of waiting lines, expressed in mathematical terms—including components such as number of waiting lines, number of servers, average wait time, number of queues or lines, and probabilities of queue times' either increasing or decreasing. *Note:* Queuing theory is directly applicable to network telecommunications, server queuing, mainframe computer queuing of telecommunications terminals, and advanced telecommunications systems.

quieting: In an FM receiver, the phenomenon that results in less noise when an unmodulated carrier is present than when there is no carrier present. *Note:* Quieting is expressed in dB.

(this page intentionally left blank)

raceway: Within a building, an enclosure, *i.e.*, channel, used to contain and protect wires, cables, or bus bars. (188)

rack: A frame upon which one or more units of equipment are mounted. (188) *Note:* DOD racks are always vertical.

racon: *See* **radar beacon.**

rad: *Acronym for* **radiation absorbed dose.** The basic unit of measure for expressing absorbed radiant energy per unit mass of material. *Note 1:* A rad corresponds to an absorption of 0.01 J/kg, *i.e.*, 100 ergs/g. *Note 2:* The absorbed radiant energy heats, ionizes, and/or destroys the material upon which it is incident.

rad.: *Abbreviation for* **radian(s).**

radar: *Acronym for* **radio detection and ranging. 1.** A radio detection system that transmits short bursts (pulses) of rf energy and detects their echos from objects (targets) such as aircraft or ships. *Note:* The round-trip propagation time for the echo return may be used to determine the target's range (distance from the radar's antenna). If the transmitting antenna has a narrow beam (the usual case), the azimuth or elevation of the target may also be determined. *Synonym* **primary radar.** *Contrast with* **secondary radar. 2.** A radio detection device that provides information on range, azimuth, and/or elevation of objects. [JP1] **3.** A radiodetermination system based on the comparison of reference signals with radio signals reflected, or retransmitted, from the position to be determined. [NTIA] [RR]

radar beacon (racon): 1. A transmitter-receiver associated with a fixed navigational mark which, when triggered by a radar, automatically returns a distinctive signal which can appear on the display of the triggering radar, providing range, bearing and identification information. [NTIA] [RR] **2.** A receiver-transmitter combination which sends out a coded signal when triggered by the proper type of pulse, enabling determination of range and bearing information by the interrogating station or aircraft. [JP1]

radar blind range: The range that corresponds to the situation in which a radar transmitter is on and hence the receiver must be off, so that the radar transmitted signal does not saturate, *i.e.*, does not blind, its own receiver. *Note:* Radar blind ranges occur because there is a time interval between transmitted pulses that corresponds to the time required for a pulse to propagate to the object, *i.e.*, to the target, and its reflection to travel back. This causes an attempt to measure the range just as the radar transmitter is transmitting the next pulse. However, the receiver is off, therefore this particular range cannot be measured. The width of the range value that cannot be measured depends on the duration of the time that the radar receiver is off, which depends on the duration of the transmitted pulse. The return-time interval could be coincident with the very next radar-transmitted pulse, *i.e.*, the first pulse following a transmitted pulse, or the second, or the third, and so on, giving rise to a succession of blind ranges. The blind ranges are given by $r_m = (mc)/(2fn)$, where r_m is the blind range for a given value of m, m is a positive integer that indicates which of the blind ranges is being determined, c is the velocity of electromagnetic wave propagation in vacuum (approximately 3×10^8 m/s), f is the radar pulse repetition rate, and n is the refractive index of the transmission medium (nearly 1 for air). The radar blind range is independent of the radar radio frequency (rf) of the radar pulse. [From Weik '89]

radar blind speed: The magnitude of the radial component of velocity of an object, *i.e.*, a target, relative to a radar site, that cannot be measured by the radar unit. *Note:* Radar blind speeds occur because of the relationship between the transmitted pulse repetition rate (PRR) and the received pulse-repetition rate. The Doppler pulse repetition rate is the difference between the transmitted and received pulse repetition rates. For example, when the object is stationary with respect to the radar site, the reflected PRR is the same as the transmitted PRR and therefore a net zero signal is indicated for the radial component of velocity. If it happens that the Doppler PRR is the same as the transmitted PRR, *i.e.*, the illuminating PRR, or it is a multiple of the transmitted PRR, a zero signal is also obtained and hence the radar is blind to these speeds, one for each multiple of the transmitted pulse repetition rate. It is not the absolute magnitude of the speed of the object that is measured, but only the radial component of the speed. The radial components of blind speeds, v_m, are given by $v_m = m\lambda f/102$, where v is the blind speed in knots, m is the multiple of the radar pulse repetition rate and the number of the blind speed, namely a positive integer, 1, 2, 3, 4, . . ., for the first, second, third, fourth, and so on, blind speed, λ is the wavelength of the illuminating radar in centimeters; f is the transmitter pulse repetition rate in pps (pulses per second); and the 102 is a units conversion factor. [From Weik '89]

radar cross section: An expression of the extent to which an object, *i.e.*, a target, reflects radar pulses, usually with respect to their point of origin. *Note:* The radar cross section of an aircraft can vary by a factor of over 100, depending on the aspect angle of the aircraft to the radar transmitter. Radar reflection off the nose of the aircraft usually represents the smallest radar cross section, while a broadside presentation to the signal produces the greatest cross section. Shape, surface roughness, and reflective material as well as orientation also affect the radar cross section. [From Weik '89]

radar intelligence (RADINT): Intelligence derived from data collected by radar. [JP1]

radar line-of-sight (LOS) equation: An equation that expresses the radar horizon range (RHR), given by

$$RHR_s = \sqrt{2h} + \sqrt{2a}$$

$$\approx 1.414\left(\sqrt{h} + \sqrt{a}\right) ,$$

where RHR_s is the radar horizon range in statute miles, h is the antenna height in feet, and a is the object critical altitude, *i.e.*, the target altitude in feet, below which the radar cannot illuminate the object. *Note:* The RHR is also given by

$$RHR_k \approx 4.12\left(\sqrt{h} + \sqrt{a}\right) ,$$

where RHR_k is the radar horizon range in kilometers when h and a are in meters. The effective Earth radius, namely 4/3 times the actual Earth radius, is used in deriving these formulas. The effective Earth radius for LOS varies with carrier frequency. Second-order differentials are neglected. They contribute less than 0.1%. [From Weik '89]

radar mile: The time required for a radar pulse to travel 1 mile (~1.6 km) to an object, *i.e.*, to a target; reflect; and return to the receiver. *Note:* A radar statute mile is approximately 10.8 µs (microseconds); a radar nautical mile is approximately 12.4 µs. The time for any other radar unit distance is readily determined, such as the radar meter or the radar kilometer. [From Weik '89]

radar resolution cell: The volume of space that is occupied by a radar pulse and that is determined by the pulse duration and the horizontal and vertical beamwidths of the transmitting radar. *Note:* The radar cannot distinguish between two separate objects that lie within the same resolution cell. The radar resolution cell depth (*RCD*) remains constant regardless of the distance from the transmitting antenna. It does not increase with range. The *RCD* is given by $RCD = 150d$, where the *RCD* is in meters and d is the pulse duration in microseconds. The height of the cell and the width of the cell do increase with range. These are given by $W = (HBW)(R/57)$ and $H = (VBW)(R/57)$, where W is the width of the cell, HBW is the horizontal beamwidth in degrees, R is the range, H is the height of the cell, and VBW is the vertical beamwidth in degrees. The range, R, is the distance from the radar antenna to the reflecting object, *i.e.*, the target. The width and height will come out in the same units in which the range is given. For example, if the range is given in meters, the width and height of the radar resolution cell will be in meters. The 57 merely converts degrees to radians. If the beamwidths are given in radian measure, the 57 is omitted. [From Weik '89]

radar signature: **1.** The detailed waveform of a detected radar echo. *Note:* Radar signatures may be used to identify or distinguish among objects, *i.e.*, targets, such as aircraft, decoys, missiles with warheads, and chaff. [From Weik '89] **2.** The detailed characteristics of a radar transmission. *Note:* Radar signatures based upon emission analysis may be used to identify or distinguish among specific radar types.

RADHAZ: *Acronym for* **electromagnetic radiation hazards.**

radian (rad.): A unit of plane angle measure equal to the angle subtended at the center of a circle by an arc equal in length to the radius of the circle. *Note:* One radian is equal to 360°/2π, which is approximately 57° 17' 44.6".

radiance: Radiant power, in a given direction, per unit solid angle per unit of projected area of the source, as viewed from the given direction. *Note:* Radiance is usually expressed in watts per steradian per square meter. (188)

radiant emittance: Radiant power emitted into a full sphere, *i.e.*, 4π sr (steradians), by a unit area of a source, expressed in watts per square meter. (188) *Synonym* **radiant exitance.**

radiant energy: Energy in the form of electromagnetic waves. *Note 1:* Radiant energy may be calculated by integrating radiant power with respect to time. *Note 2:* Radiant energy is usually expressed in joules. (188)

radiant exitance: *Synonym* **radiant emittance.**

radiant flux: *Deprecated synonym for* **radiant power.**

radiant intensity: Radiant power per unit solid angle, usually expressed in watts per steradian. (188)

radiant power: The rate of flow of electromagnetic energy, *i.e.*, radiant energy. *Note 1:* Radiant power is usually expressed in watts, *i.e.*, joules per second. *Note 2:* The modifier is often dropped and *"power"* is used to mean *"radiant power"*. *Deprecated synonyms* **flux, radiant flux.**

radiation: **1.** In communication, the emission of energy in the form of electromagnetic waves. (188) **2.** The outward flow of energy from any source in the form of radio waves. [NTIA] [RR]

radiation angle: In fiber optics, half the vertex angle of the cone of light emitted at the exit face of an optical fiber. (188) *Note:* The cone boundary is usually defined (a) by the angle at which the far-field irradiance has decreased to a specified fraction of its maximum value or (b) as the cone within which there is a specified fraction of the total radiated power at any point in the far field. *Synonym* **output angle.**

radiation efficiency: At a given frequency, the ratio of the power radiated to the total power supplied to the radiator. (188)

radiation field: *Synonym* **far-field region.**

radiation-hardened fiber: An optical fiber made with core and cladding materials that recover, within a specified period of time, a specified percentage of their intrinsic transparency after darkening from exposure to a radiation pulse.

radiation mode: For an optical fiber, an unbound mode. (188) *Note:* In an optical fiber, a radiation mode is one having fields that are transversely oscillatory everywhere external to the waveguide, and which exists even at the limit of zero wavelength. Specifically, a radiation mode is one for which

$$\beta = \sqrt{n^2(a)k^2 - (\ell/a)^2} \ ,$$

where β is the imaginary part (phase term) of the axial propagation constant, integer ℓ is the azimuthal index of the mode, $n(a)$ is the refractive index, where a is the core radius, and k is the free-space wave number, $k = 2\pi/\lambda$, where λ is the wavelength. Radiation modes correspond to refracted rays in the terminology of geometric optics. *Synonym* **unbound mode.**

radiation pattern: **1.** The variation of the field intensity of an antenna as an angular function with respect to the axis. (188) *Note:* A radiation pattern is usually represented graphically for the far-field conditions in either horizontal or vertical plane. **2.** In fiber optics, the relative power distribution at the output of a fiber or active device as a function of position or angle. (188) *Note 1:* The near-field radiation pattern describes the radiant emittance ($W \cdot m^{-2}$) as a function of position in the plane of the exit face of an optical fiber. *Note 2:* The far-field radiation pattern describes the irradiance as a function of angle in the far-field region of the exit face of an optical fiber. *Note 3:* The radiation pattern may be a function of the length of the fiber, the manner in which it is excited, and the wavelength. *Synonym* **directivity pattern.**

radiation resistance: The resistance that, if inserted in place of an antenna, would consume the same amount of power that is radiated by the antenna. (188)

radiation scattering: The diversion of radiation (thermal, electromagnetic, or nuclear) from its original path as a result of interaction or collisions with atoms, molecules, or larger particles in the atmosphere or other media between the source of radiation (*e.g.*, a nuclear explosion) and a point some distance away. As a result of scattering, radiation (especially gamma rays and neutrons) will be received at such a point from many directions instead of only from the direction of the source. [JP1]

RADINT: *Acronym for* **radar intelligence.**

radio: **1.** Telecommunication by modulation and radiation of electromagnetic waves. (188) **2.** A transmitter, receiver, or transceiver used for communication via electromagnetic waves. **3.** A general term applied to the use of radio waves. [NTIA] [RR]

radio altimeter: Radionavigation equipment, on board an aircraft or spacecraft, used to determine the height of the aircraft or the spacecraft above the Earth's surface or another surface. [NTIA] [RR]

radio and wire integration (RWI): The combining of wire circuits with radio facilities. [JP1]

radio baseband: *See* **baseband.**

radiobeacon station: A station in the radionavigation service the emissions of which are intended to enable a mobile station to determine its bearing or direction in relation to the radiobeacon station. [NTIA] [RR]

radio beam: A radiation pattern from a directional antenna, such that the energy of the transmitted electromagnetic wave is confined to a small angle in at least one dimension. (188)

radio channel: An assigned band of frequencies sufficient for radio communication. (188) *Note 1:* The bandwidth of a radio channel depends upon the type of transmission and the frequency tolerance. *Note 2:* A channel is usually assigned for a specified radio service to be provided by a specified transmitter.

radio common carrier (RCC): A common carrier engaged in the provision of Public Mobile Service, which is not also in the business of providing landline local exchange telephone service. These carriers were formerly called *"miscellaneous common carriers."* [47CFR]

radiocommunication: Telecommunication by means of radio waves. [NTIA] [RR]

radiocommunication service: A service as defined in this Section [of the *Radio Regulations*] involving the transmission, emission and/or reception of radio waves for specific telecommunication purposes. In these regulations, unless otherwise stated, any radiocommunication service relates to terrestrial radiocommunication. [NTIA] [RR]

radio control: The remote control of an apparatus by signals conveyed by electromagnetic waves. *Note:* Radio control may be used to control the movement of an aircraft, vehicle, missile, or other mobile unit, either manned or unmanned, from a radio station on the ground or in another mobile unit. [From Weik '89]

radio-coverage diagram: A diagram that shows the area within which a radio station is broadcasting an effective signal strength in relation to a given standard. *Note:* An example of a radio-coverage diagram is a polar plot, in each direction from the antenna, of the distance from the antenna at which the signal strength is equal to a specified value, *i.e.*, it is the locus of all points at which the signal strength is equal to a specified value. [From Weik '89]

radio coverage

radio detection: The detection of the presence of an object by radiolocation without precise determination of its position. [JP1]

radio detection and ranging: *See* **radar.**

radiodetermination: The determination of the position, velocity and/or other characteristics of an object, or the obtaining of information relating to these parameters, by means of the propagation properties of radio waves. [NTIA] [RR]

radiodetermination-satellite service: A radiocommunication service for the purpose of radiodetermination involving the use of one or more space stations. This service may also include feeder links necessary for its own operation. [NTIA] [RR]

radiodetermination station: A station in the radiodetermination service. [NTIA] [RR]

radio direction-finding [RDF]: Radiodetermination using the reception of radio waves for the purpose of determining the direction of a station or object. [NTIA] [RR]

radio direction-finding station: A radiodetermination station using radio direction-finding. [NTIA] [RR]

radio equipment: As defined in *Federal Information Management Regulations*, any equipment or interconnected system or subsystem of equipment (both transmission and reception) that is used to communicate over a distance by modulating and radiating electromagnetic waves in space without artificial guide. This does not include such items as microwave, satellite, or cellular telephone equipment.

radio fadeout: *See* **flutter.**

radio field intensity: *Synonym* **field strength.**

radio fix: **1.** The locating of a radio transmitter by bearings taken from two or more direction finding stations, the site of the transmitter being at the point of intersection. [JP1] **2.** The location of a ship or aircraft by determining the direction of radio signals coming to the ship or aircraft from two or more sending stations, the locations of which are known. [JP1]

radio frequency (rf): Any frequency within the electromagnetic spectrum normally associated with radio wave propagation. (188) *For designation of subdivisions, see* **electromagnetic spectrum** *and associated diagram.*

radio frequency assignment: *See* **frequency assignment.**

radio frequency channel assignment: *Synonym* **frequency assignment.**

radio frequency interference (RFI): *Synonym* **electromagnetic interference.**

radio horizon: The locus of points at which direct rays from an antenna are tangential to the surface of the Earth. (188) *Note:* If the Earth were a perfect sphere and there were no atmospheric anomalies, the radio horizon would be a circle. In practice, the distance to the radio horizon is affected by the height of the transmitting antenna, the height of the receiving antenna, atmospheric conditions, and the presence of obstructions, *e.g.*, mountains.

radio horizon range (RHR): The distance at which a direct radio wave can reach a receiving antenna of given height from a transmitting antenna of given height. *Note:* The radio horizon range in nautical miles, R, is given by the relation $R = 1.23(h_t^{1/2} + h_r^{1/2})$, where h_t and h_r are the heights of the transmitting and receiving antennas in feet. The radio horizon range, R, in nautical miles is also given by the relation $R = 2.23(h_t^{1/2} + h_r^{1/2})$, where h_t and h_r are the heights of the transmitting and receiving antennas in meters. The effective Earth radius, 4/3 times the actual Earth radius, is used in deriving the formulae. Second-order differentials are neglected. They are of the order of 0.1%. [From Weik '89]

radio interface: The common boundary between a mobile station and the radio equipment in the network, which is the boundary defined by functional characteristics, physical interconnection characteristics, signal characteristics, and other characteristics as appropriate.

radiolocation: Radiodetermination used for purposes other than those of radionavigation. [NTIA] [RR]

radiolocation land station: A station in the radiolocation service not intended to be used while in motion. [NTIA] [RR]

radiolocation mobile station: A station in the radiolocation service intended to be used while in motion or during halts at unspecified points. [NTIA] [RR]

radiolocation service: A radiodetermination service for the purpose of radiolocation. [NTIA] [RR]

radiological monitoring: *Synonym* **monitoring (def. #4).**

radiometry: The science of radiation measurement. *See Table of Radiometric Terms.*

TABLE OF RADIOMETRIC TERMS

Term	Symbol	Quantity	Unit
radiant energy	Q	energy	joule (J)
radiant power *Synonym* **optical power**	ϕ	power	watt (W)
irradiance	E	power incident per unit area (irrespective of angle)	$W \cdot m^{-2}$
spectral irradiance	E_λ	irradiance per unit wavelength interval at a given wavelength	$W \cdot m^{-2} \cdot nm^{-1}$
radiant emittance *Synonym* **radiant exitance**	W	power emitted (into a full sphere) per unit area	$W \cdot m^{-2}$
radiant intensity	I	power per unit solid angle	$W \cdot sr^{-1}$
radiance	L	power per unit angle per unit projected area	$W \cdot sr^{-1} \cdot m^{-2}$
spectral radiance	L_λ	radiance per unit wavelength interval at a given wavelength	$W \cdot sr^{-1} \cdot m^{-2} \cdot nm^{-1}$

radionavigation: **1.** Radio-location intended for the determination of position or direction or for obstruction warning in navigation. [JP1] **2.** Radiodetermination used for the purposes of navigation, including obstruction warning. [NTIA] [RR]

radionavigation land station: A station in the radionavigation service not intended to be used while in motion. [NTIA] [RR]

radionavigation mobile station: A station in the radionavigation service intended to be used while in motion or during halts at unspecified points. [NTIA] [RR]

radionavigation-satellite service: A radiodetermination-satellite service used for the purpose of radionavigation. This service may also include feeder links necessary for its operation. [NTIA] [RR]

radio net: **1.** An organization of radio stations that is capable of direct communication on a common frequency. (188) **2.** An organization of radio stations that broadcast common programming, not necessarily simultaneously, at different frequencies from different locations.

radio paging: The use of a pocket-size radio receiver capable of alerting its wearer that there is a phone call, either from a displayed phone number or to a predesignated number. *Note:* Radio paging may be considered a subset of paging. *Synonym* **beeping.**

radio personal terminal: *See* **personal terminal.**

radio range: **1.** The distance from a transmitter at which the signal strength remains above the minimum usable level for a particular antenna and receiver combination. **2.** A radio aid to air navigation that creates an infinite number of paths in space throughout a given sector or azimuth angle by various methods of transmission and reception of electromagnetic waves. [From Weik '89]

radio range station: A radionavigation land station in the aeronautical radionavigation service providing radial equisignal zones. (In certain instances a radio range station may be placed on board a ship.) [NTIA]

radio recognition and identification: *See* **identification, friend or foe.**

Radio Regulations Board: A permanent organization of the International Telecommunication Union (ITU) that implements frequency assignment policy and maintains the Master International Frequency Register (MIFR). *Note: Formerly* **International Frequency Registration Board (IFRB).**

radio relay: **1.** The reception and retransmission by a radio station of signals that are received either from another radio station or from a wire, fiber optic, microwave, coaxial cable, or other link of an integrated land line and radio communications system component. **2.** A terrestrial point-to-point communications system, such as a microwave-relay communications system or a satellite communications system. *Note:* The siting of radio-relay stations and the radio coverage diagrams of the antenna patterns are arranged for minimum interference with satellite Earth stations. The analog and digital baseband arrangements are similar to satellite systems. Radio-relay links may form part of the connection between an Earth station and a switching center. [From Weik '89]

radio relay system: A point-to-point radio transmission system in which signals are received, conditioned, and retransmitted by one or more intermediate radio stations. (188)

radiosonde: An automatic radio transmitter in the meteorological aids service usually carried on an aircraft, free balloon, kite, or parachute, and which transmits meteorological data. [NTIA] [RR]

radiotelegram: A telegram, originating in or intended for a mobile station or a mobile Earth station transmitted on all or part of its route over the radiocommunication channels of the mobile service or of the mobile-satellite service. [NTIA] [RR]

radio telegraphy: The transmission of telegraphic codes by means of radio. [JP1]

radiotelemetry: Telemetry by means of radio waves. [NTIA] [RR]

radiotelephone call: A telephone call, originating in or intended for a mobile station or a mobile Earth station, transmitted on all or part of its route over the radiocommunication channels of the mobile service or of the mobile-satellite service. [NTIA] [RR]

radiotelephone distress frequency: An international distress and calling frequency for mobile radiotelephone stations, survival craft, and emergency position-indicating radio beacons. *Note:* An example of a radiotelephone distress frequency is 2180 kHz. [From Weik '89]

radio telephony: The transmission of speech by means of modulated radio waves. [JP1]

radio teletypewriter (RTTY): A teletypewriter employed in a communication system using radio circuits. *Note:* Such systems are spoken of as RATT systems. (188)

radiotelex call: A telex call, originating in or intended for a mobile station or a mobile Earth station, transmitted on all or part of its route over the radiocommunication channels of the mobile service or the mobile-satellite service. [NTIA] [RR]

radio wave: An electromagnetic wave of a frequency arbitrarily lower than 3000 GHz. *Synonym* **Hertzian wave.**

radio watch shift: *Synonym* **area broadcast shift.**

radio-wire integration: *See* **radio and wire integration.**

RAM: *Acronym for* **random access memory.**

Raman amplifier: *Synonym* **fiber amplifier.**

Raman scattering: The generation of many different wavelengths of light from a nominally single-wavelength source (a) by means of lasing action and interaction with molecules, thereby creating many different excited molecular energy levels that will produce photons of various energy levels, *i.e.*, various wavelengths, when transitions to lower excited states occur and (b) by the beating together of two frequencies, thus inducing dipole moments in molecules at the difference frequencies and thereby causing modulation of laser-molecule interaction, which, in turn, produces light at side frequencies, *i.e.*, side wavelengths relative to the nominal wavelength. [From Weik '89]

random access discrete address (RADA): A communications technique in which radio users share one wide frequency band instead of each user's being assigned a narrow band.

random access memory (RAM): A read/write, non-sequential-access memory used for the storage of instructions and data. *Note 1:* RAM access time is essentially the same for all storage locations. *Note 2:* RAM is characterized by a shorter access time than disk or tape storage. *Note 3:* RAM is usually volatile.

randomizer: **1.** A device used to invert the sense of pseudorandomly selected bits of a bit stream to avoid long sequences of bits of the same sense. (188) *Note:* The same selection pattern must be used on the receive terminal in order to restore the original bit stream. **2.** [An] analog or digital source of unpredictable, unbiased, and usually independent bits. *Note:* Randomizers can be used for several different functions, including key generation or to provide a starting state for a key generator. [NIS]

random noise: Noise consisting of a large number of transient disturbances with a statistically random time distribution. (188) *Note:* Thermal noise is an example of random noise.

random number: **1.** A number selected from a known set of numbers in such a way that each number in the set has the same probability of occurrence. **2.** A number obtained by chance. **3.** One of a sequence of numbers considered appropriate for satisfying certain statistical tests or believed to be free from conditions that might bias the result of a calculation.

range: *See* **radio range.**

ranging: The measurement of the distance to a remote object (target), from a known observation or reference point. *Note:* Ranging may be accomplished by geometric means, *e.g.*, triangulation, or by the measurement of the transit time of an electromagnetic or acoustic signal. Ranging has application to navigation and cartography.

raster: A predetermined pattern of scanning lines within a display space. *Note:* An example of a raster is the pattern followed by an electron beam scanning the screen of a television camera or receiver. (188)

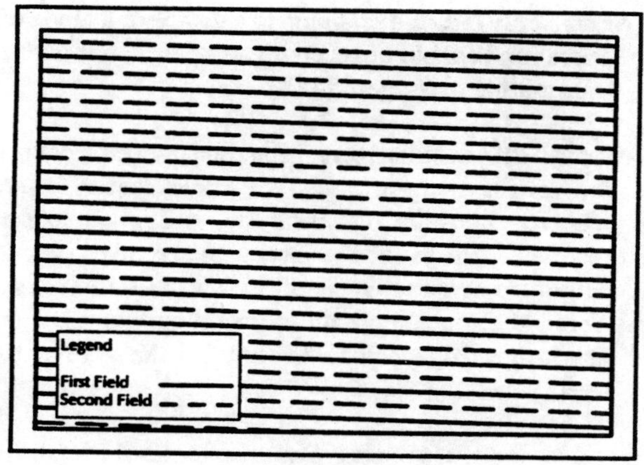

raster lines

raster count: The total number of raster scanning lines within a display space. [From Weik '89]

raster density: In display systems, the number of scanning lines per unit distance perpendicular to the scanning direction. [From Weik '89]

raster scanning: Scanning in which the motion of the scanning spot follows a raster. (188)

rated output power: That power available at a specified output of a device under specified conditions of operation. (188) *Note:* Rated output power may be further described; *e.g.*, maximum rated output power, average rated output power.

ratio-squared combiner: *Synonym* **maximal-ratio combiner.**

ray: A geometric representation of a lightwave by a line normal to the electromagnetic wavefront; *i.e.*, in the direction of propagation of the wave. [FAA]

Rayleigh distribution: A mathematical statement, usually applied to frequency distributions of random variables, for the case in which two orthogonal variables are independent and normally distributed with unit variance. (188)

Rayleigh fading: In electromagnetic wave propagation, phase-interference fading caused by multipath, and which may be approximated by the Rayleigh distribution. (188)

Rayleigh scattering: Of an electromagnetic wave propagating in a material medium, scattering caused by refractive-index inhomogeneities that are small compared to the wavelength. (188) *Note 1:* Rayleigh scattering losses vary as the reciprocal of the fourth power of the wavelength. *Note 2:* Ionospheric scattering is caused partly by Rayleigh scattering.

ray optics: *Synonym* **geometric optics.**

RBOC: *Acronym for* **Regional Bell Operating Company.**

RC: *Abbreviation for* **reflection coefficient.**

RCC: *Abbreviation for* **radio common carrier.**

RDF: *Abbreviation for* **radio direction finding.**

read head: A magnetic head capable of reading only.

reading: The acquisition or interpretation of data from a storage device, from a data medium, or from another source.

read-only memory (ROM): A memory in which data, under normal conditions, can only be read. *Synonym* **nonerasable storage.**

read-only storage: A storage device in which the contents cannot be modified, except by a particular user, or when operating under particular conditions, *e.g.*, a storage device in which writing is prevented by a lockout. (188) *Synonym* **fixed storage.**

read/write opening: *Synonym* **read/write slot.**

read/write slot: An opening in the jacket of a diskette to allow access to the read/write heads. *Synonym* **read/write opening.**

ready-for-data signal: **1.** A call-control signal that is transmitted by the data circuit-terminating equipment (DCE) to the data terminal equipment (DTE) to indicate that the connection is available for data transfer between both DTEs. **2.** A signal that (a) is sent in the backward direction in the interexchange data channel, to indicate that all the succeeding exchanges involved in the connection have been through-connected, or (b) is sent in the forward direction in the interexchange data channel to indicate that all the preceding exchanges involved in the connection have been through-connected. *Note:* The ready-for-data signal is sent by the user terminal. It corresponds to the ready-for-data state at the user interface. [From Weik '89]

real power: *See* **effective power.**

real time: **1.** The actual time during which a physical process occurs. (188) **2.** Pertaining to the performance of a computation during the actual time that the related physical process occurs, in order that results of the computation can be used in guiding the physical process.

reasonableness check: A test to determine whether a value conforms to specified criteria. *Note:* A reasonableness check can be used to eliminate questionable data points from subsequent processing. *Synonym* **wild-point detection.**

receive-after-transmit time delay: The time interval between (a) the instant of keying off the local transmitter to stop transmitting and (b) the instant the local receiver output has increased to 90% of its steady-state value in response to an rf signal from a distant transmitter. (188) *Note 1:* The rf signal from the distant transmitter must exist at the local receiver input prior to, or at the time of, keying off the local transmitter. *Note 2:* Receive-after-transmit time delay applies only to half-duplex operation.

received noise power: **1.** The calculated or measured noise power, within the bandwidth being used, at the receive end of a circuit, channel, link, or system. (188) **2.** The absolute power of the noise measured or calculated at a receive point. (188) *Note:* The related bandwidth and the noise weighting must also be specified. **3.** The value of noise power, from all sources, measured at the line terminals of telephone set's receiver. (188) *Note:* Either flat weighting or some other specific amplitude-frequency characteristic or noise weighting characteristic must be associated with the measurement.

received signal level (RSL): The signal level at a receiver input terminal. *Note 1:* The signal bandwidth and the established reference level must be specified. (188) *Note 2:* The RSL is usually expressed in dB with respect to 1 mW, *i.e.,* 0 dBm.

receive only (RO): Pertaining to a device or a mode of operation capable of receiving messages, but not of transmitting messages. (188)

receiver attack-time delay: The time interval from (a) the instant a step rf signal, at a level equal to the receiver threshold of sensitivity, is applied to the receiver input to (b) the instant the receiver output amplitude reaches 90% of its steady-state value. *Note:* If a squelch circuit is operating, the receiver attack-time delay includes the time for the receiver to break squelch. (188)

receiver lockout system: *Synonym* **lockout (def. #4).**

receiver release-time delay: The time interval from removal of rf energy at the receiver input until the receiver output is squelched. (188)

recognized operating agency (ROA): Any operating agency, as defined in the ITU Convention (Geneva, 1992), which operates a public correspondence or broadcasting service and upon which the obligations provided for in Article 6 of the ITU Constitution are imposed by the Member in whose territory the head office of the agency is situated, or by the Member which has authorized this operating agency to establish and operate a telecommunication service on its territory. *Formerly* **recognized private operating agency (RPOA).**

reconditioned carrier reception: *Synonym* **exalted-carrier reception.**

reconstructed sample: An analog sample generated at the output of a decoder when a specified character signal is applied at its input. (188) *Note:* The amplitude of the reconstructed sample is proportional to the value of the corresponding encoded sample.

record: **1.** A set of data treated as a unit. (188) **2.** To write data on a medium, such as magnetic tape, magnetic disk, or optical disk.

record communication: **1.** A telecommunications process that produces an electronic message that is transmitted, received, stored or archived, and may be retrieved. **2.** A telecommunications process, that produces a hard copy record of the transmission, such as a teletypewriter printout or a facsimile printout. (188)

recorder warning tone: A half-second burst of 1400 Hz applied to a telephone line every 15 seconds to indicate to the called party that the calling party is recording the conversation. *Note:* The recorder warning tone is required by law to be generated as an integral part of any recording device used for the purpose and is required to be not under the control of the calling party. The tone is recorded together with the conversation.

record information: All forms (*e.g.*, narrative, graphic, data, computer memory) of information registered in either temporary or permanent form so that it can be retrieved, reproduced, or preserved. [JP1]

recording density: *Synonym* **bit density.**

recording spot: In a facsimile recorder, the spot that is used to generate the recorded copy on the record medium. (188)

record medium: **1.** The physical medium on which information is stored in recoverable form. (188) **2.** In facsimile transmission, the physical medium on which the recorder forms an image of the object, *i.e.*, creates the recorded copy. (188) *Note:* The record medium and the record sheet may be identical. *Synonym* **record sheet.**

record sheet: *Synonym* **record medium.**

record traffic: **1.** Traffic that is recorded, in permanent or quasipermanent form, by the originator, the addressee, or both. (188) **2.** Traffic that is permanently or semipermanently recorded in response to administrative procedures or public law.

recovery: In a database management system, the procedures and capabilities available for reconstruction of the contents of a database to a state that prevailed before the detection of processing errors and before the occurrence of a hardware or software failure that resulted in the destruction of some or all of the stored data.

recovery procedure: **1.** The actions necessary to restore an automated information system's data files and computational capability after a system failure. **2.** In data communications, a process whereby a data station attempts to resolve conflicting or erroneous conditions arising during the transfer of data.

RED/BLACK concept: [The] separation of electrical and electronic circuits, components, equipment, and systems that handle classified plain text (RED) information, in electrical signal form, from those which handle unclassified (BLACK) information in the same form. [NIS] (188)

RED signal: **1.** [A] Telecommunications or automated information system signal that would divulge classified information if recovered and analyzed. [NIS] **2.** In cryptographic systems, a signal containing classified information that has NOT been encrypted.

reduced carrier single-sideband emission: A single-sideband emission in which the degree of carrier suppression enables the carrier to be reconstituted and to be used for demodulation. [NTIA] [RR]

reduced carrier transmission: A form of amplitude-modulation in which the carrier is transmitted at a controlled level below that which is required for demodulation, but at a level sufficient to serve as a frequency reference. (188)

redundancy: **1.** In the transmission of data, the excess of transmitted message symbols over that required to convey the essential information in a noise-free circuit. *Note:* Redundancy may be introduced intentionally (as in the case of error detection or correction codes) or inadvertently (such as by oversampling a band-limited signal, inefficient formats, *etc.*). **2.** In a communication system, surplus capability usually provided to improve the reliability and quality of service. (188)

redundancy check: **1.** A method of verifying that any redundant hardware or software in a communication system is in an operational condition. (188) **2.** A check that uses one or more extra binary digits or characters attached to data for the detection of errors. (188)

redundant code: A code using more signal elements than necessary to represent the intrinsic information. (188) *Note:* The redundancy may be used for error-control purposes.

reference antenna: An antenna that may be real, virtual, or theoretical, and has a radiation pattern that can be used as a basis of comparison with other antenna radiation patterns. *Note:* Examples of reference antennas are unit dipoles, half-wave dipoles, and isotropic, *i.e.,* omnidirectional antennas. [From Weik '89]

reference black level: [In television,] The level corresponding to the specified maximum excursion of the luminance signal in the black direction. [47CFR]

reference circuit: A hypothetical circuit of specified equivalent length and configuration, and having a defined transmission characteristic or characteristics, used primarily as a reference for measuring the performance of other, *i.e.,* real, circuits or as a guide for planning and engineering of circuits and networks. (188) *Note:* Normally, several types of reference circuits are defined, with different configurations, because communications are required over a wide range of distances. A group of related reference circuits is also called a *reference system.*

reference clock: **1.** A clock with which another clock is compared. (188) **2.** A clock, usually of high stability and accuracy, used to govern a network of mutually synchronized clocks of lower stability. *Note:* The failure of a reference clock does not necessarily cause loss of synchronism.

reference configuration: In ISDN, a combination and arrangement of functional groups and reference points that reflect possible network topology.

reference frequency: **1.** A standard fixed frequency from which operational frequencies may be derived or with which they may be compared. (188) *Note:* The reference frequency may be used to specify an assigned frequency or fix a characteristic or carrier frequency. **2.** A frequency having a fixed and specific position with respect to the assigned frequency. The displacement of this frequency with respect to the assigned frequency has the same absolute value and sign that the displacement of the characteristic frequency has with respect to the center of the frequency band occupied by the emission. [NTIA] [RR]

reference monitor: [An] access control concept that refers to an abstract machine that mediates all accesses to objects by subjects. [NIS]

reference noise: The magnitude of circuit noise chosen as a reference for measurement. (188) *Note:* Many different levels with a number of different weightings are in current use, and care must be taken to ensure that the proper parameters are stated. *See* **dBa, dBa(F1A), dBa(HA1), dBa0, dBm, dBm(psoph), dBm0, dBrn, dBrnC, dBrnC0, dBrn(f_1-f_2), dBrn(144-line), dBx.**

reference point: In ISDN, a logical point between two, nonoverlapping functional groups. *Note:* When equipment is placed at a reference point, that reference point is designated an interface.

reference surface: In optical-fiber technology, that surface of an optical fiber that is used to contact the transverse-alignment elements of a component such as a connector or mechanical splice. *Note:* For telecommunications-grade fibers, the reference surface is the outer surface of the cladding. For plastic-clad silica (PCS) fibers, which have a strippable polymer cladding (not to be confused with the polymer overcoat of an all-glass fiber), the reference surface may be the core.

reference system: A group of related reference circuits.

reference transmission level point: *See* **relative transmission level, transmission level point.**

reference white level: [In television,] The level corresponding to the specified maximum excursion of the luminance signal in the white direction. [47CFR]

reflectance: The ratio of reflected power to incident power, generally expressed in dB or percent. (188)

reflected code: *Synonym* **Gray code.**

reflecting layer: In the ionosphere, a layer that has a free-electron density sufficient to reflect radio waves. *Note:* The principal reflecting layers are the E, F_1, and F_2 layers in the daylight hemisphere. *Note 2:* A critical frequency is associated with the reflection by each layer.

reflecting loss: *See* **reflection loss.**

reflection: The abrupt change in direction of a wave front at an interface between two dissimilar media so that the wave front returns into the medium from which it originated. *Note 1:* Reflection may be specular (*i.e.*, mirror-like) or diffuse (*i.e.*, not retaining the image, only the energy) according to the nature of the interface. *Note 2:* Depending on the nature of the interface, *i.e.*, dielectric-conductor or dielectric-dielectric, the phase of the reflected wave may or may not be inverted.

reflection coefficient (RC): **1.** The ratio of the amplitude of the reflected wave and the amplitude of the incident wave. (188) **2.** At a discontinuity in a transmission line, the complex ratio of the electric field strength of the reflected wave to that of the incident wave. (188) *Note 1:* The reflection coefficient may also be established using other field or circuit quantities. *Note 2:* The reflection coefficient is given by

$$RC = \left|\frac{Z_1 - Z_2}{Z_1 + Z_2}\right| = \frac{SWR - 1}{SWR + 1} ,$$

where Z_1 is the impedance toward the source, Z_2 is the impedance toward the load, the vertical bars designate absolute magnitude, and *SWR* is the standing wave ratio.

reflection loss: **1.** At a discontinuity or impedance mismatch, *e.g.*, in a transmission line, the ratio of the incident power to the reflected power. (188) *Note 1:* Reflection loss is usually expressed in dB. *Note 2:* The reflection loss, L_r, is given by

$$L_r = 20 \log_{10}\left|\frac{Z_1 - Z_2}{Z_1 + Z_2}\right| = 10\log_{10}\frac{(Z_1 - Z_2)^2}{(Z_1 + Z_2)^2} ,$$

where Z_1 and Z_2 are the respective impedances, and the vertical bars designate absolute magnitude. (188) **2.** In an optical fiber, the loss that takes place at any discontinuity of refractive index, especially at an air-glass interface such as a fiber endface, at which a fraction of the optical signal is reflected back toward the source. *Note:* This reflection phenomenon is also called "*Fresnel reflection loss,*" or simply, "*Fresnel loss.*" At normal incidence, the fraction of reflected power is expressed by the formula

$$L_f = 10 \log_{10}\frac{(n_1 - n_2)^2}{(n_1 + n_2)^2} ,$$

where n_1 and n_2 are the respective indices of refraction.

reflective array antenna: An antenna, such as a billboard antenna, in which the driven elements are situated at a predetermined distance from a surface designed to reflect the signal in a desired direction. (188) *Note:* Reflective array antennas (a) usually have many driven elements working in conjunction with an electrically large reflecting surface to produce a unidirectional beam, (b) may be used to increase antenna gain, *i.e.*, reduce radiation in unwanted directions, and (c) may contain parasitic elements as well as driven elements.

reflectivity: The reflectance at the surface of a material so thick that the reflectance does not change with increasing thickness, *i.e.*, the intrinsic reflectance of the surface, irrespective of other parameters such as the reflectance of the rear surface. *Note:* The term "*reflectivity*" is no longer in common use. *See* **reflectance.**

reflector: In an antenna, one or more conducting elements or surfaces that reflect incident radiant energy. (188)

refracted ray: **1.** A ray that undergoes a change of velocity, or in the general case, both velocity and direction, as a result of interaction with the material medium in which it travels. **2.** In an optical fiber, a ray that is refracted from the core into the cladding. Specifically a ray having direction such that

$$\frac{n^2(r) - n^2(a)}{1 - (r/a)^2\cos^2\phi(r)} \le \sin^2\theta(r) ,$$

where r is the radial distance from the fiber axis, $\phi(r)$ is the azimuthal angle of projection of the ray at r on the transverse plane, $\theta(r)$ is the angle the ray makes with the fiber axis, $n(r)$ is the refractive index at r, $n(a)$ is the refractive index at the core radius, a. Refracted rays correspond to radiation modes in the terminology of mode descriptors.

refraction: Retardation, and—in the general case—redirection, of a wavefront passing through (a) a boundary between two dissimilar media or (b) a medium having a refractive index that is a continuous function of position, *e.g.*, a graded-index optical fiber. *Note:* For two media of different refractive indices, the angle of refraction is closely approximated by Snell's Law.

refraction profile: *Synonym* **refractive index profile.**

refractive index (n, η): Of a medium, the ratio of the velocity of propagation of an electromagnetic wave in vacuum to its velocity in the medium. (188) *Synonym* **index of refraction.**

refractive index contrast: In an optical fiber, a measure of the relative difference in refractive index of the core and cladding. (188) *Note:* Refractive index contrast, Δ, is given by $\Delta = (n_1^2 - n_2^2)/(2n_1^2)$, where n_1 is the maximum refractive index in the core and n_2 is the refractive index of the homogeneous cladding.

refractive index profile: Of the cross section of an optical fiber, the description, *i.e.*, plot, of the value of the refractive index as a function of distance from the fiber axis along a diameter. (188) *Synonyms* **index profile, refraction profile.**

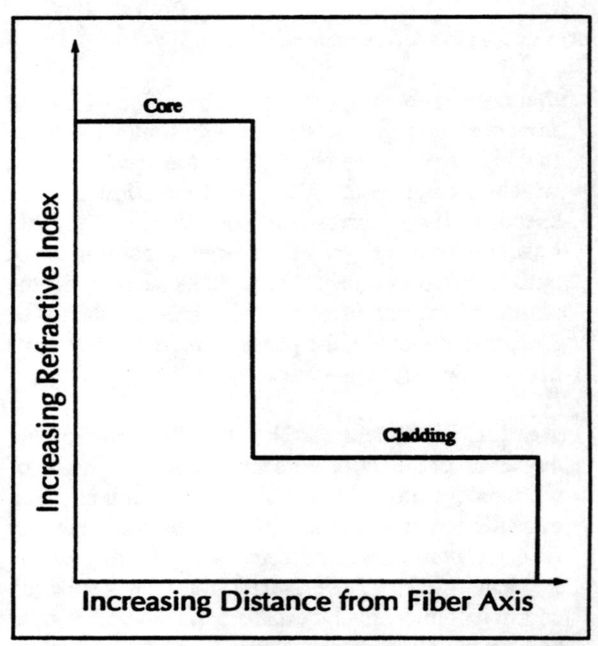

refractive index profile, multimode step-index fiber

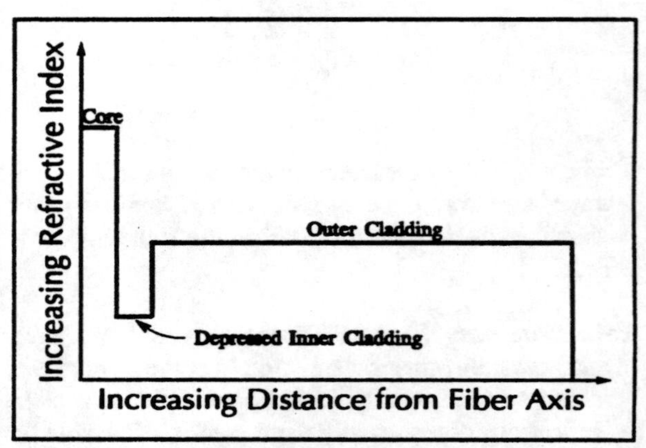

refractive index profile, doubly clad single-mode fiber

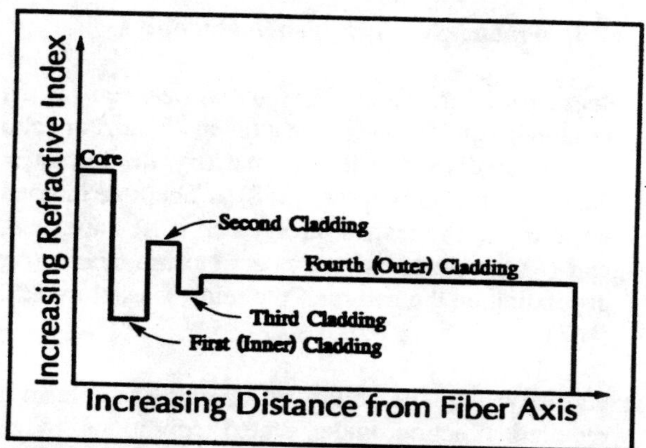

refractive index profile, quadruply clad single-mode fiber

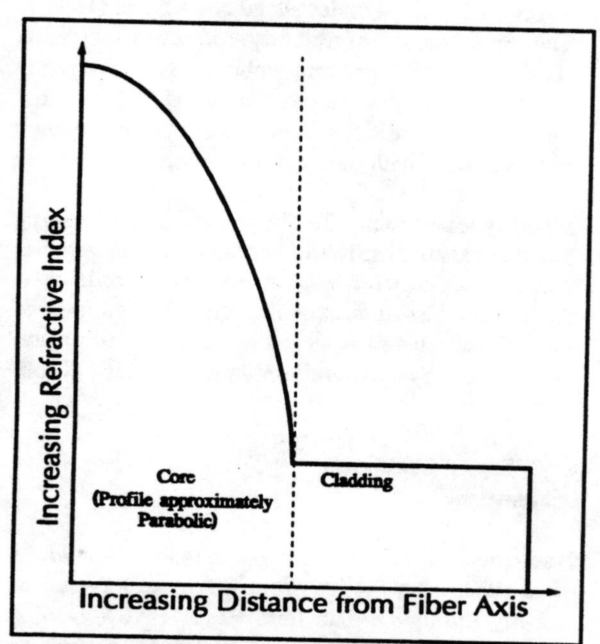

refractive index profile, graded-index multimode fiber

reframing time: The time interval between the instant at which a valid frame-alignment signal is available at the receiving data terminal equipment and the instant at which frame alignment is established. *Note:* The reframing time includes the time required for replicated verification of the validity of the frame-alignment signal. *Synonym* **frame-alignment recovery time.**

refresh: To reproduce, repeatedly, a display image on a display surface, so that the image remains visible.

regeneration: 1. In a regenerative repeater, the process by which digital signals are amplified, reshaped, retimed, and retransmitted. (188) *Synonym* **positive feedback.** 2. In a storage or display device, the restoration of stored or displayed data that have deteriorated. (188) *Note:* For example, conventional cathode-ray tube displays must be continually regenerated for the data to remain displayed. 3. In computer graphics, the sequence of events needed to generate a display image from its representation in storage.

regenerative feedback: Feedback in which the portion of the output signal that is returned to the input has a component that is in phase with the input signal.

regenerative repeater: A repeater, designed for digital transmission, in which digital signals are amplified, reshaped, retimed, and retransmitted. (188) *Synonym* **regenerator.**

regenerator: *Synonym* **regenerative repeater.**

Regional Bell Operating Company (RBOC): One of the seven holding companies formed by divestiture by the American Telephone and Telegraph Company of its local Bell System operating companies, and to which one or more of the Bell System local telephone companies were assigned.

regional center: *See* **office classification.**

register: 1. A device, accessible to one or more input circuits, that accepts and stores data. (188) *Note:* A register is usually used only as a device for temporary storage of data. 2. A temporary-memory device used to receive, hold, and transfer data (usually a computer word) to be operated upon by a processing unit. *Note:* Computers typically contain a variety of registers. General purpose registers may perform many functions, such as holding constants or accumulating arithmetic results. Special purpose registers perform special functions, such as holding the instruction being executed, the address of a storage location, or data being retrieved from or sent to storage.

registered jack (RJ): Any of the series of jacks, described in the *Code of Federal Regulations,* Title 47, part 68, used to provide interface to the public telephone network.

registration: **1.** The accurate positioning of an entity relative to a reference. (188) **2.** *See* **FCC registration program.**

registration program: *See* **FCC Registration Program.**

rekeying: The changing of one or more keys that are used for either COMSEC or TRANSEC functions. (188)

relative error: The ratio of an absolute error to the true, specified, or theoretically correct value of the quantity that is in error.

relative spectral width: *See* **spectral width.**

relative transmission level: The ratio of the signal power, at a given point in a transmission system, to a reference signal power. (188) *Note:* The ratio is usually determined by applying a standard test tone at zero transmission level point (or applying adjusted test tone power at any other point) and measuring the gain or loss to the location of interest. A distinction should be made between the standard test tone power and the expected median power of the actual signal required as the basis for the design of transmission systems.

relay: **1.** To retransmit a received message from one station to another station. **2.** An electromechanical or semiconductor switch (*i.e.,* solid-state relay) in which a current or voltage applied across one port or terminal controls electrical currents or voltages that appear across another terminal or terminals. (188)

relay configuration: An operating configuration in which a circuit is established between two stations via an intermediate relay station. Two links are used simultaneously and the channel connections at the relay station are accomplished completely within the station. (188)

relay station: An intermediate station that passes information between terminals or other relay stations.

released loop: *Synonym* **switched loop.**

release time: **1.** The time interval between (a) the instant that an enabling signal (as in a vogad or echo suppressor) is discontinued, and (b) the instant at which suppression ceases. (188) **2.** The time interval between (a) the instant a relay coil is de-energized, and (b) the instant that contact closure ceases (or, depending on the nature of the relay, is established). (188)

reliability: **1.** The ability of an item to perform a required function under stated conditions for a specified period of time. **2.** The probability that a functional unit will perform its required function for a specified interval under stated conditions. (188) **3.** The continuous availability of communication services to the general public, and emergency response activities in particular, during normal operating conditions and under emergency circumstances with minimal disruption.

reliability assessment: **1.** The process of determining whether existing hardware, firmware, or software has achieved a specified level of operational reliability. **2.** The process of determining the achieved level of operational reliability of existing hardware, firmware, or software. *Synonym* **reliability evaluation.** [From Weik '89]

reliability evaluation: *Synonym* **reliability assessment.**

remote access: **1.** Pertaining to communication with a data processing facility from a remote location or facility through a data link. **2.** A PABX service feature that allows a user at a remote location to access by telephone PABX features, such as access to wide area telephone service (WATS) lines. *Note:* For remote access, individual authorization codes are usually required.

remote access data processing: Data processing in which some input/output functions are performed by devices that are connected to a computer system by means of data communication.

remote batch entry: Submission of batches of data through an input unit that has access to a computer through a data link.

remote batch processing: Batch processing in which input-output units have access to a computer through a data link.

remote call forwarding: A service feature that allows calls coming to a remote call-forwarding number to be automatically forwarded to any answering location designated by the call receiver. *Note:* Customers may have a remote-forwarding telephone number in a central switching office without having any other local telephone service in that office.

remote clock: **1.** A clock that is remote from a particular facility, such as a communications station or node, with which it is associated. **2.** A clock that is remote from another clock to which it is to be compared.

remote control equipment: Devices used to perform monitoring, controlling, and/or supervisory functions, at a distance. (188)

remote job entry (RJE): In computer operations, a mode of operation that allows execution of job instructions received from a remote site and return of the output to the same or a different remote site via a communications link.

remote login: A login that allows a user terminal to connect to a host computer via a network or direct telecommunications link, and to interact with that host computer as if the user terminal were directly connected to that host computer. *Synonym* **remote logon.**

remote logon: *Synonym* **remote login.**

remote operations service element (ROSE) protocol: A protocol that (a) provides remote operation capabilities, (b) allows interaction between entities in a distributed application, and (c) upon receiving a remote operations service request, allows the receiving entity to attempt the operation and report the results of the attempt to the requesting entity. (188)

remote orderwire: An extension of a local orderwire to a point convenient for personnel to perform required operational and maintenance functions. (188)

remote rekeying: The encrypted transmission of keys from a remote source. (188)

remote trunk arrangement (RTA): Arrangement that permits the extension of TSPS functions to remote locations. [47CFR]

REN: *Acronym for* **ringer equivalency number.**

reorder tone: *See* **busy signal.**

repeater: **1.** An analog device that amplifies an input signal regardless of its nature, *i.e.,* analog or digital. **2.** A digital device that amplifies, reshapes, retimes, or performs a combination of any of these functions on a digital input signal for retransmission. (188) *Note:* The term *"repeater"* originated with telegraphy and referred to an electromechanical device used to regenerate telegraph signals. Use of the term has continued in telephony and data communications.

repeating coil: A voice-frequency transformer characterized by a closed core, a pair of identical balanced primary (line) windings, a pair of identical but not necessarily balanced secondary (drop) windings, and low transmission loss at voice frequencies. (188) *Note:* It permits transfer of voice currents from one winding to another by magnetic induction, matches line and drop impedances, and prevents direct conduction between the line and the drop.

repeat-request (RQ) system: *Synonym* **ARQ.**

reperforator: In teletypewriter systems, a device used to punch a tape in accordance with arriving signals, permitting reproduction of the signals for retransmission. (188)

repertory dialer: A telephone set that stores a group of numbers frequently called by a customer and transmits the dialing information to the central office by a single action. (188) *Contrast with* **speed calling.**

reproducibility: *Synonym* **precision (def. #1, #2, #3).**

reproducible fault: In computer and telecommunications systems, a fault that will occur each time the set of conditions causing the fault occurs. *Note:* The conditions under which the fault occurs and the precision with which the fault occurs must be specified when determining the reproducibility of a fault.

reproduction speed: **1.** In facsimile systems, the rate at which recorded copy is produced. *Note:* The reproduction speed is usually expressed (a) as the area of recorded copy produced per unit time, such as square meters per second or (b) as the number of pages per minute. **2.** In duplicating equipment, the rate at which copies are made. *Note:* The reproduction speed is usually expressed in pages per minute. (188)

request data transfer: A signal sent by the DTE to the DCE to request the establishment of a data connection.

request-repeat (RQ) signal: A signal from a receiver to a transmitter asking that a message be transmitted again. [From Weik '89]

request-to-send signal: A signal that is generated by a receiver in order to condition a remote transmitter to commence transmission. [From Weik '89]

reradiation: **1.** Radiation, at the same or different wavelengths, *i.e.*, frequencies, of energy received from an incident wave. (188) **2.** Undesirable radiation of signals locally generated in a radio receiver. *Note:* Radiation might cause interference or reveal the location of the device.

rerouting: Recommencement of route selection from the first point of routing control, when congestion is encountered at some intermediate switching point in the connection that is to be established.

resale carrier: A company that redistributes the services of a commercial carrier and retails the services to the public.

resale service: In FCC deliberations and rulings, the right of a buyer of basic telecommunication services, such as private lines, foreign exchanges, or WATS, to resell and/or share with others the unused capacity.

reserve capacity: Installed capacity of a system which is not normally utilized but can be made available when required. [NATO]

reserved circuit service: In ISDN applications, a telecommunications service that establishes a communication path at a preset time (requested by the user) in response to a user-network signaling request.

reserved word: In programming languages, a keyword whose definition is fixed by the programming language and which cannot be changed by the user. *Note:* In Ada® and COBOL all keywords are reserved words, while Fortran has no reserved words.

reset mode: The parameters initially programmed for basic operation. (188)

resettability: A measure of the ability to duplicate controllable conditions. *Note:* An example of resettability is the ability to reset the frequency controls of radio equipment. (188)

resident: Pertaining to computer programs that remain on a particular storage device.

residual error rate: *Synonym* **undetected error ratio.**

residual error ratio: **1.** The ratio of (a) the number of bits, unit elements, characters, or blocks incorrectly received but undetected or uncorrected by the error-control equipment to (b) the total number of bits, unit elements, characters or blocks. [After CCITT] **2.** The error ratio that remains after attempts at correction are made.

residual modulation: *Synonym* **carrier noise level.**

resistance hybrid: A network of resistors to which four branches of a circuit may be connected to make them conjugate in pairs. *Note:* The primary use of a resistance hybrid is to convert between 2-wire and 4-wire communications circuits. Such conversion is necessary when repeaters are introduced in a 2-wire circuit.

resolution: **1.** The minimum difference between two discrete values that can be distinguished by a measuring device. (188) *Note:* High resolution does not necessarily imply high accuracy. **2.** The degree of precision to which a quantity can be measured or determined. (188) **3.** A measurement of the smallest detail that can be distinguished by a sensor system under specific conditions. [JP1] (188)

resolving power: A measure of the ability of a lens or optical system to form separate and distinct images of two objects with small angular separation. (188) *Note:* The resolving power of an optical system is ultimately limited by diffraction by the aperture. Thus, an optical system cannot form a perfect image of a point.

resonance: In an electrical circuit, the condition that exists when the inductive reactance and the capacitive reactance are of equal magnitude, causing electrical energy to oscillate between the magnetic field of the inductor and the electric field of the capacitor. *Note 1:* Resonance occurs because the collapsing magnetic field of the inductor generates an electric current in its windings that charges the capacitor and the discharging capacitor provides an electric current that builds the magnetic field in the inductor, and the process is repeated. *Note 2:* At resonance, the series impedance of the two elements is at a minimum and the parallel impedance is a maximum. Resonance is used for tuning and filtering, because resonance occurs at a particular frequency for given values of inductance and capacitance. Resonance can be detrimental to the operation of communications circuits by causing unwanted sustained and transient oscillations that may cause noise, signal distortion, and damage to circuit elements. *Note 3:* At resonance the inductive reactance and the capacitive reactance are of equal magnitude. Therefore, $|\omega L| = |1/\omega C|$, where $\omega = 2\pi f$, in which f is the resonant frequency in hertz, L is the inductance in henrys, and C is the capacity in farads when standard SI units are used. Thus,

$$f = \frac{\pi}{2\sqrt{LC}} .$$

resonant antenna: *Synonym* **periodic antenna.**

resonant cavity: *See* **optical cavity.**

resource controller (RC): The processor(s) that control access to satellite payload communications resources within an individual satellite program. (188)

respond opportunity: In data transmission, the link level logical control condition during which a given secondary station may transmit a response.

response: **1.** A reply to a query. (188) **2.** In data transmission, the content of the control field of a response frame advising the primary station concerning the processing by the secondary station of one or more command frames. **3.** The effect of an active or passive device upon an input signal.

response frame: In data transmission, all frames that may be transmitted by a secondary station.

response PDU: A protocol data unit (PDU) transmitted by a logical link control (LLC) sublayer in which the PDU command/response (C/R) bit is equal to "1".

response time: The time a system or functional unit takes to react to a given input. (188) *Note:* For example, in data processing, the response time perceived by the end user is the interval between (a) the instant at which an operator at a terminal enters a request for a response from a computer and (b) the instant at which the first character of the response is received at a terminal. In a data system, the system response time is the interval between the receipt of the end of transmission of an inquiry message and the beginning of the transmission of a response message to the station originating the inquiry.

response timer (T_K): In multilevel precedence and preemption, the device that controls the length of time that the call receiver of the precedence call has to accept the incoming precedence call. *Note:* The length of the time is usually set in the range of 4 s to 30 s.

responsiveness: Ability of an entity to provide service within the required time. *Note:* The term *timeliness* is sometimes used incorrectly to mean responsiveness. [NATO]

responsivity: In a photodetector, the ratio of the electrical output to the optical input. (188) *Note 1:* Responsivity is usually expressed in amperes per watt, or volts per watt, of incident radiant power. *Note 2:* Responsivity is a function of the wavelength of the incident radiation and the bandgap of the material of which the photodetector is made. *Deprecated synonym* **sensitivity.**

restart: The resumption of the execution of a computer program using the data recorded at a checkpoint.

restitution: A series of significant conditions determined by the decisions taken according to the products of the demodulation process. (188)

restoration: Action taken to repair and return to service one or more telecommunications services, including repair of a damaged or impaired telecommunications facility, that have a degraded quality of service or have a service outage. (188) *Note:* Restoration may be done by various means, such as patching, routing, substitution of component parts, or selecting other pathways.

restricted access: A class of service in which users may be denied access to one or more of the system features or operating levels. (188)

restricted area: *Synonym* **controlled space.**

restricted channel: In digital communications systems, a channel that has a useful capacity of only 56 kb/s (kilobits per second), instead of 64 kb/s. (188) *Note:* The restricted channel, currently common in North America, was originally developed to satisfy a ones-density limitation in T1 circuits.

retrieval service: In interactive telecommunications, a service allowing access to and retrieval of stored information, *e.g.*, the information within a database.

retrograde orbit: Of a satellite orbiting the Earth, an orbit in which the projection of the satellite's position on the (Earth's) equatorial plane revolves in the direction opposite that of the rotation of the Earth.

return loss: The ratio, at the junction of a transmission line and a terminating impedance or other discontinuity, of the amplitude of the reflected wave to the amplitude of the incident wave. *Note 1:* Return loss is usually expressed in dB. (188) *Note 2:* Return loss is a measure of the dissimilarity between impedances in metallic transmission lines and loads, or between refractive indices in dielectric media, *e.g.*, optical fibers. *Note 3:* In a metallic transmission line, return loss is given by

$$L_r = 10 \log_{10} \left| \frac{Z_1 + Z_2}{Z_1 - Z_2} \right| ,$$

where Z_1 is the impedance toward the source and Z_2 is the impedance toward the load, and the vertical bars indicate magnitude. *Note 4:* For dielectric media, *e.g,* optical fibers, *see* **reflection loss (def #2).**

return-to-zero (RZ): A digital code having two information states, *e.g.*, "0" and "1" or "mark" and "space", in which code the signal returns to a rest state during a portion of the bit period. (188)

reverse-battery signaling: Loop signaling in which battery and ground are reversed on the tip and ring of the loop to give an "off-hook" signal when the call receiver answers. (188) *Note:* Reverse-battery signaling may be used either for a short period, or for the duration of a call, to indicate that it is a toll call.

revertive pulsing: In telephone networks, a means of controlling distant switching selections by pulsing. *Note:* In revertive pulsing, the near end receives signals from the far end.

rf: *Abbreviation for* **radio frequency.**

rf bandwidth: *See* **occupied bandwidth, necessary bandwidth.**

RFI: *Abbreviation for* **radio frequency interference.** *See* **electromagnetic interference.**

rf power margin: **1.** The amount of transmitter power above that which is computed by the link designer as the minimum required to meet specified link performance. *Note:* The rf power margin allows for uncertainties in (a) empirical components of the signal level prediction method, (b) terrain characteristics, (c) atmospheric conditions, and (d) equipment performance parameters. (188) **2.** At any given time in an operational link, the reserve transmitter power over that which is required to maintain specified link performance.

rf tight: Offering a high degree of electromagnetic shielding effectiveness. (188)

RGB: *Abbreviation for* **red-green-blue.** Pertaining to the use of three separate signals to carry the red, green, and blue components, respectively, of a color video image. (188) *Note:* The image is not NTSC-encoded; RGB typically results in higher resolution than that specified by the National Television Standards Committee.

rhombic antenna: A directional antenna that is composed of long-wire radiators that form the sides of a rhombus, the two halves of which are fed equally in opposite phase at one apex. [From Weik '89] *Note:* A rhombic antenna is usually terminated at the apex opposite the driven apex, which termination makes its radiation pattern unidirectional. It is bidirectional if the opposite apex is unterminated.

RI: *Abbreviation for* **routing indicator.**

ribbon cable: **1.** Any cable constructed as a ribbon with parallel elements. **2.** A fiber optic cable in which the optical fibers are held in grooves and laminated within a flat semirigid strip of material, such as plastic, that positions, holds, and protects them. *Note:* Ribbon cables may be stacked to produce fiber optic cables with large numbers of fibers. Buffers, strength members, fillers, and jacketing are usually added to produce the final cable. [After 2196].

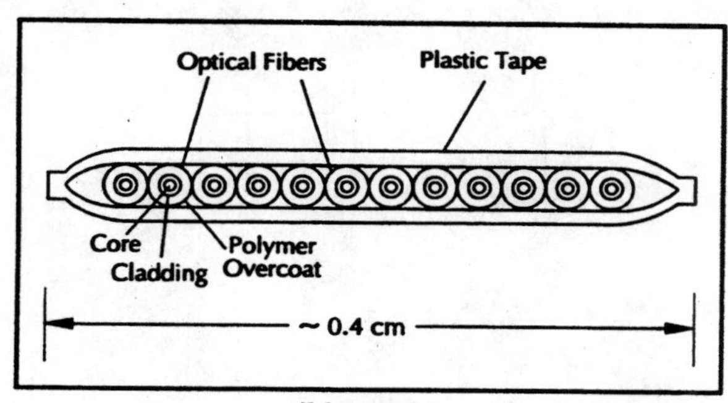

ribbon cable

right-hand (or clockwise) polarized wave: An elliptically or circularly polarized wave, in which the electric field vector, observed in any fixed plane, normal to the direction of propagation, whilst looking in the direction of propagation, rotates with time in a right-hand or clockwise direction. [NTIA] [RR] *Synonym* **clockwise polarized wave.**

right-hand rule: *Synonym* **Fleming's rule.**

ring: **1.** In telephony, a signal of specific duration and character that indicates to a user (customer, subscriber) that a calling party is engaged in an access attempt. (188) **2.** *Synonym* **ring network.** *See* **network topology.**

ringaround: **1.** The improper routing of a call back through a switching center already engaged in attempting to complete the same call. **2.** In secondary surveillance radar, the presence of false targets declared as a result of transponder interrogation by side lobes of the interrogating antenna.

ringback signal: **1.** In telephony, a signal, usually consisting of an audio tone interrupted at a slow rate, provided to a caller to indicate that the called-party instrument is receiving a ringing signal. (188) *Note:* This signal may be generated by the called-party servicing switch or by the calling-party switch. **2.** A ringing signal returned to a caller to indicate that one of the types of delayed automatic calling is now ringing the called party. (188)

ringback tone: *Synonym* **audible ringing tone.**

ringdown: In telephony, a method of signaling an operator in which telephone ringing current is sent over the line to operate a lamp and cause the drop of a self-locking relay. (188) *Note 1:* Ringdown (a) is used in manual operation, as distinguished from dialing, (b) uses a continuous or pulsed ac signal transmitted over the line, and (c) may be used with or without a switchboard. (188) *Note 2:* The term *"ringdown"* originated in magneto telephone signaling in which cranking the magneto in a telephone set would not only *"ring"* its bell but also cause a lever to fall *"down"* at the central office switchboard.

ringdown circuit: In telephony, a circuit in which manually generated signaling power is used to perform ringdown. (188)

ringdown signaling: In telephony, the application of a signal to a line (a) to operate a line signal lamp or a supervisory signal lamp at a switchboard or (b) to ring a called receiver instrument. (188)

ringer equivalency number (REN): A number determined in accordance with the *Code of Federal Regulations,* Title 47, part 68, which number represents the ringer loading effect on a line. *Note:* A ringer equivalency number of 1 represents the loading effect of a single traditional telephone set ringing circuit. Modern telephone instruments may have a REN lower than 1. The total REN expresses the total loading effect of the subscriber's equipment on the central office ringing current generator. The service provider usually sets a limit, *e.g.,* 3, 4, or 5 (representing "extension," *i.e.,* parallel-connected telephones), to the total REN on a subscriber's loop. The actual number of instruments across the loop may be greater than the service provider's REN limit, if their respective individual RENs are less than 1.

ring latency: In a ring network, such as a token ring network, the time required for a signal to propagate once around the ring. *Note 1:* Ring latency may be measured in seconds or in bits at the data transmission rate. *Note 2:* Ring latency includes signal propagation delays in (a) the ring medium, (b) the drop cables, and (c) the data stations connected to the ring network. (188)

ring network: *See* **network topology.**

ring topology: *See* **network topology.**

ring transit time: *See* **round-trip delay time.**

R interface: For a basic rate access in an ISDN environment, the interfacing specifications covering pre-ISDN standards (*e.g.,* EIA-232C).

rip cord: Of an optical cable, a parallel cord of strong yarn that is situated under the jacket(s) of the cable for the purpose of facilitating jacket removal preparatory to splicing or breaking out. *Note:* The rip cord is exposed by carefully removing or severing a portion of the jacket near the end of the cable. It is then grasped with the fingers, or usually, with a tool such as a pair of pliers, and pulled to sever the jacket for the remainder of the desired distance. [After FAA]

ripple voltage: **1.** In a dc voltage, the alternating component that is residually retained from rectification of ac power, or from generation and commutation. (188) **2.** In a dc voltage, the alternating component that is coupled into a circuit from a source of interference.

rise time: In the approximation of a step function, the time required for a signal to change from a specified low value to a specified high value. Typically, these values are 10% and 90% of the step height. (188)

RJ: *Abbreviation for* **registered jack.**

RJE: *Abbreviation for* **remote job entry.**

rms pulse duration: *See* **root-mean-square pulse duration.**

RO: *Abbreviation for* **receive only.**

ROM: *Acronym for* **read-only memory.**

roofing filter: A low-pass filter used to reduce unwanted higher frequencies. (188)

room noise level: *Synonym* **ambient noise level.**

root: In computer science, the highest level of a hierarchy.

root-mean-square (rms) deviation: A single quantity, σ_{rms}, characterizing a function, $f(x)$, given by

$$\sigma_{rms} = \sqrt{\frac{1}{M_0}\int_{-\infty}^{\infty}(x - M_1)^2 f(x)\,dx}\ , \text{ where}$$

$$M_0 = \int_{-\infty}^{\infty} f(x)\,dx\ , \text{ and}$$

$$M_1 = \frac{1}{M_0}\int_{-\infty}^{\infty} x\,f(x)\,dx\ .$$

Note: The term "*rms deviation*" is also used in probability and statistics, where the normalization,

M_0, is unity. Here, the term is used in a more general sense.

root-mean-square (rms) pulse broadening: The temporal rms deviation of the impulse response of a system.

root-mean-square (rms) pulse duration: A special case of root-mean-square deviation where the independent variable is time and *f*(*t*) describes the pulse waveform.

rotary dial: A signaling mechanism—usually incorporated within a telephone set—that when rotated and released, generates dc pulses required for establishing a connection in a telephone system.

rotary hunting: Hunting in which all the numbers in the hunt group are selected in a prescribed order. (188) *Note:* In modern electronic switching systems, the numbers in the hunt group are not necessarily selected in consecutive order.

rotary switching: In telephone systems, an electro-mechanical switching method whereby the selecting mechanism consists of a rotating element using several groups of wipers, brushes, and contacts. (188)

rotational position sensing: [In magnetic media,] A technique used to locate a given sector, a desired track, and a specific record by continuous comparison of the read/write head position with appropriate synchronization signals.

round-trip delay time: **1.** The elapsed time for transit of a signal over a closed circuit. (188) *Note:* Round-trip delay time is significant in systems that require two-way interactive communication such as voice telephony or ACK/NAK data systems where the round-trip time directly affects the throughput rate. It may range from a very few microseconds for a short line-of-sight (LOS) radio system to many seconds for a multiple-link circuit with one or more satellite links involved. This includes the node delays as well as the media transit time. **2.** In primary or secondary radar systems, the time required for a transmitted pulse to reach a target and for the echo or transponder reply to return to the receiver. (188)

route: **1.** In communications systems operations, the geographical path that is followed by a call or message over the circuits that are used in establishing a chain of connections. **2.** To determine the path that a message or call is to take in a communications network. *Note:* In a Transmission Control Protocol/Internet Protocol (TCP/IP) internet, each IP datagram is routed separately. The route a datagram follows may include many gateways and many physical networks. **3.** To construct the path that a call or message is to take in a communications network in going from one station to another or from a source user end instrument to a destination user end instrument. [From Weik '89]

route diversity: The allocation of circuits between two points over more than one geographic or physical route with no geographic points in common. (188)

route matrix: In communications network operations, a record that indicates the interconnections between pairs of nodes in the network, and is used to produce direct routes, alternate routes, and available route tables from point to point. [From Weik '89]

router: In data communications, a functional unit used to interconnect two or more networks. *Note 1:* Routers operate at the network layer (layer 3) of the ISO Open Systems Interconnection—Reference Model. *Note 2:* The router reads the network layer address of all packets transmitted by a network, and forwards only those addressed to another network.

routine: A computer program, called by another program, that may have some general or frequent use.

routine message: A category of precedence to be used for all types of messages that justify transmission by rapid means unless of sufficient urgency to require a higher precedence. [JP1]

routing: The process of determining and prescribing the path or method to be used for establishing telephone connections or forwarding messages.

routing diagram: In a communications system, a diagram that (a) shows all links between all switch-boards, exchanges, switching centers, and stations in the system, such as the links between primary relay, major relay, minor relay, and tributary stations as well as supplementary links, (b) is used to identify the stations and links, and (c) is used to indicate tape-relay routes, transfer circuits, refile circuits, radio links, operational status, line conditions, and other network information required for network operations and management. [From Weik '89]

routing directory: *See* **routing table.**

routing indicator (RI): **1.** A group of letters assigned to indicate: (a) the geographic location of a station; (b) a fixed headquarters of a command, activity, or unit at a geographic location; and (c) the general location of a tape relay or tributary station to facilitate the routing of traffic over the tape relay networks. [JP1] **2.** In a message header, an address, *i.e.*, group of characters, that specify routing instructions for the transmission of the message to its final destination. (188) *Note:* Routing indicators may also include addresses of intermediate points.

routing table: A matrix associated with a network control protocol, which gives the hierarchy of link routing at each node.

RQ: *Abbreviation for* **repeat-request.** *See* **ARQ.**

RSL: *Abbreviation for* **received signal level.**

RTA: *Abbreviation for* **remote trunk arrangement.**

RTTY: *Abbreviation for* **radio teletypewriter.**

rubidium clock: A clock containing a quartz oscillator stabilized by a rubidium standard.

rubidium standard: A frequency standard in which a specified hyperfine transition of electrons in rubidium-87 atoms is used to control the output frequency. *Note:* A rubidium standard consists of a gas cell, which has an inherent long-term instability. This instability relegates the rubidium standard to its status as a secondary standard.

run: The execution of one or more computer jobs or programs.

run-length encoding: A redundancy-reduction technique for facsimile in which a run of consecutive picture elements having the same state (gray scale or color) is encoded into a single code word. (188)

rural radio service: A public radio service rendered by fixed stations on frequencies below 1000 MHz used to provide (1) Basic Exchange Telecommunications Radio Service, which is public message communication service between a central office and subscribers located in rural areas, (2) public message communication service between landline central offices and different exchange areas which it is impracticable to interconnect by any other means, or (3) private line telephone, telegraph, or facsimile service between two or more points to which it is impracticable to extend service via landline. [47CFR]

rural subscriber station: A fixed station in the rural radio service used by a subscriber for communication within a central office station.

RWI: *Abbreviation for* **radio and wire integration.**

RX: *Abbreviation for* **receive, receiver.**

RZ: *Abbreviation for* **return-to-zero.**

s: *Abbreviation for* **second.**

safety service: Any radiocommunication service used permanently or temporarily for the safeguarding of human life and property. [NTIA] [RR]

sampling: *See* **signal sampling.**

sampling frequency: *See* **sampling rate.**

sampling interval: The reciprocal of the sampling rate, *i.e.*, the interval between corresponding points on two successive sampling pulses of the sampling signal.

sampling rate: The number of samples taken per unit time, *i.e.*, the rate at which signals are sampled for subsequent use, such as for modulation, coding, and quantization. *Deprecated synonym* **sampling frequency.** (188)

sampling theorem: *Synonym* **Nyquist's theorem.**

satellite: A body which revolves around another body of preponderant mass and which has a motion primarily and permanently determined by the force of attraction of that other body. [NTIA] [RR] (188) *Note:* A parent body and its satellite revolve about their common center of gravity.

satellite access: In satellite communications systems, the establishment of contact with a communications satellite space station. *Note:* An example of satellite access is access at the moment at which an Earth station commences to use a satellite space station as a signal repeater, *i.e.*, to use its transponder. Each radio frequency (rf) carrier that is relayed by a satellite space station at any time occupies an access channel. Accesses, *i.e.*, channels, are distinguishable by various system parameters, such as frequency, time, or code. [From Weik '89]

satellite communications: A telecommunications service provided via one or more satellite relays and their associated uplinks and downlinks. (188)

satellite Earth terminal: *Synonym* **Earth terminal.**

satellite emergency position-indicating radiobeacon: An Earth station in the mobile-satellite service the emissions of which are intended to facilitate search and rescue operations. [RR]

satellite link: A radio link between a transmitting Earth station and a receiving Earth station through one satellite. A satellite link comprises one uplink and one downlink. [NTIA] [RR]

satellite network: A satellite system or a part of a satellite system, consisting of only one satellite and the cooperating Earth stations. [RR]

satellite operation: *See* **satellite PBX.**

satellite PBX: A PBX system that is not equipped with attendant positions, and is associated with an attended main PBX system. *Note:* The main attendant provides attendant functions for the satellite system.

satellite period: *See* **period (of a satellite).**

satellite relay: An active or passive satellite repeater that relays signals between two Earth terminals. (188)

satellite system: A space system using one or more artificial Earth satellites. [NTIA] [RR]

saturation: **1.** In a communications system, the condition in which a component of the system has reached its maximum traffic handling capacity. *Note:* Saturation is equivalent to one erlang per circuit. **2.** The point at which the output of a linear device, such as a linear amplifier, deviates significantly from being a linear function of the input when the input signal is increased. (188) *Note:* Modulation often requires that amplifiers operate below saturation.

scan: **1.** To examine sequentially, part by part. **2.** To examine every reference in every entry in a file routinely as part of a retrieval scheme. **3.** In radar, one complete rotation of the interrogating antenna. **4.** In SONAR, to search 360° or a specific search sector by the use of phased array of transducers. **5.** To sweep, *i.e.*, rotate, a beam about a point or about an axis.

scan line: **1.** The line produced on a recording medium frame by a single sweep of a scanner. [JP1] **2.** *Synonym* **scanning line.**

scanner: A device that examines a spatial pattern, one part after another, and generates analog or digital signals corresponding to the pattern. *Note:* Scanners are often used in mark sensing, pattern recognition, and character recognition. (188)

scanning: **1.** In telecommunications systems, examination of traffic activity to determine whether further processing is required. *Note:* Scanning is usually performed periodically. **2.** In television, facsimile, and picture transmission, the process of successively analyzing the colors and densities of the object according to a predetermined pattern. (188) **3.** The process of tuning a device through a predetermined range of frequencies in prescribed increments and at prescribed times. *Note:* Scanning may be performed at regular or random increments and intervals. (188) **4.** In radar and radio direction finding, the slewing of an antenna or radiation pattern for the purpose of probing in a different direction. *Note 1:* In radar, scanning may be mechanical, using a rotary microwave joint to feed the antenna, or electronic, using a phased array of radiators, the radiated pattern (beam) of which depends on the relative phases of the signals fed to the individual radiators. *Note 2:* In civilian air traffic control radar, scanning usually implies continuous rotation of the antenna or beam about a vertical axis. In military radars, scanning may occur about other than a vertical axis, and may not encompass a full 360°.

scanning direction: In facsimile transmitting equipment, the scanning of an object, such as a message surface or the developed plane in the case of a drum, along parallel lines in a specified pattern. *Note 1:* The scanning direction is equivalent to scanning over a right-hand helix on a drum. *Note 2:* The orientation of the message on the scanning plane will depend upon its dimensions. *Note 3:* In facsimile receiving equipment, scanning from right to left and top to bottom, is called "positive" reception and from left to right and top to bottom, is called "negative" reception. (188) *Note 4:* Scanning direction conventions are included in CCITT Recommendations for phototelegraphic equipment.

scanning field: In facsimile systems, the total of the areas that are actually explored by the scanning spot during the scanning of the object by the transmitter or during scanning of the record medium by the receiver. [From Weik '89]

scanning line: In an imaging system, the path traversed by a scanning spot during a single line sweep.

scanning line frequency: In facsimile, the frequency at which a fixed line perpendicular to the direction of scanning is crossed by a scanning spot. (188) *Note:* The scanning line frequency is equivalent to drum speed in some mechanical systems. *Synonym* **scanning line rate.**

scanning line length: In facsimile systems, the total length of a scanning line, equal to the spot speed divided by the scanning line frequency. (188) *Note:* The scanning line length is usually greater than the length of the available line.

scanning line period: In facsimile systems, the time interval between (a) the instant at which the scanning spot probes or writes to a given spot on one scanning line, and (b) the instant at which the scanning spot probes or writes to the corresponding spot on the next scanning line.

scanning line rate: *Synonym* **scanning line frequency.**

scanning pitch: The distance between the centers of consecutive scanning lines.

scanning rate: In facsimile and television systems, the rate of displacement of the scanning spot along the scanning line. (188)

scanning spot: In facsimile systems, the area on the object, *i.e.*, the original, covered instantaneously by the pickup system of the scanner. (188)

scan-stop lockup: In automatic link establishment (ALE) radios, the undesired condition in which the normal process of (a) scanning radio channels, (b) stopping on the desired channel, or (c) returning to scan is terminated by the equipment.

scatter: *See* **scattering.**

scattering: Of a wave propagating in a material medium, a phenomenon in which the direction, frequency, or polarization of the wave is changed when the wave encounters discontinuities in the medium, or interacts with the material at the atomic or molecular level. (188) *Note:* Scattering results in a disordered or random change in the incident energy distribution.

scattering center: In the microstructure of a transmission medium, a site at which electromagnetic waves are scattered. *Note 1:* Examples of scattering centers are vacancy defects; interstitial defects; inclusions, such as a gas molecules, hydroxide ions, iron ions, and trapped water molecules; and microcracks or fractures in dielectric waveguides. *Note 2:* Scattering centers are frozen in the medium when it solidifies and may not necessarily cause Rayleigh scattering, which varies inversely as the fourth power of the wavelength. For example, in glass optical fibers, there is a high attenuation band at 0.95 μm, primarily caused by scattering and absorption by OH^- (hydroxyl) ions. [From Weik '89]

scattering coefficient: The factor that expresses the attenuation caused by scattering, *e.g.*, of radiant or acoustic energy, during its passage through a medium. *Note:* The scattering coefficient is usually expressed in units of reciprocal distance.

scattering cross section: The area of an incident wavefront, at a reflecting surface or medium, such as an object in space, through which will pass radiant energy, that, if isotropically scattered from that point, would produce the same power at a given receiver as is actually provided by the entire reflecting surface. [From Weik '89]

scattering loss: The part of the transmission (power) loss that results from scattering within a transmission medium or from roughness of a reflecting surface. (188)

SCC: *Abbreviation for* **specialized common carrier.**

scene cut: Video imagery in which consecutive frames are highly uncorrelated.

scene cut response: In video systems, the perceived impairments associated with a scene cut.

schematic: **1.** A diagram, drawing, or sketch that details the elements of a system, such as the elements of an electrical circuit or the elements of a logic diagram for a computer or communications system. **2.** Pertaining to a diagram, drawing, or sketch that details the elements of a system, such as the elements of an electrical circuit or the elements of a logic diagram for a computer or communications system.

scintillation: In electromagnetic wave propagation, a small random fluctuation of the received field strength about its mean value. (188) *Note:* Scintillation effects become more significant as the frequency of the propagating wave increases.

scrambler: A device that transposes or inverts signals or otherwise encodes a message at the transmitter to make the message unintelligible at a receiver not equipped with an appropriately set descrambling device. (188) *Note:* Scramblers usually use a fixed algorithm or mechanism. However, a scrambler provides communications privacy that is inadequate for classified traffic.

screen: **1.** In a telecommunications, computing, or data processing system, to examine entities that are being processed to determine their suitability for further processing. **2.** A nonferrous metallic mesh used to provide electromagnetic shielding. (188) **3.** To reduce undesired electromagnetic signals and noise by enclosing devices in electrostatic or electromagnetic shields. (188) **4.** A viewing surface, such as that of a cathode ray tube or liquid crystal display (LCD).

scroll: In a display device, to move the display window of the screen vertically to view the contents of a stored document. *Note:* Scrolling may be performed continuously or incrementally. (188)

SDLC: *Abbreviation for* **synchronous data link control.**

search time: In data processing systems, the time interval required to locate a particular data element, record, or file in a storage device.

SECAM: *Acronym for* **système electronique couleur avec memoire.** A television signal standard (625 lines, 50 Hz, 220 V primary power) used in France, eastern European countries, the former USSR, and some African countries.

second (s): In the International System of Units (SI), the time interval equal to 9,192,631,770 periods of the radiation corresponding to the transition between the two hyperfine levels of the ground state of the cesium-133 atom. (188)

secondary channel: In a system in which two channels share a common interface, a channel that has a lower data signaling rate (DSR) capacity than the primary channel.

secondary emission: Particles or radiation, such as photons, Compton recoil electrons, delta rays, secondary cosmic rays, and secondary electrons, that are produced by the action of primary radiation on matter. (188)

secondary frequency standard: A frequency standard that does not have inherent accuracy, and therefore must be calibrated against a primary frequency standard. *Note:* Secondary standards include crystal oscillators and rubidium standards. A crystal oscillator depends for its frequency on its physical dimensions, which vary with fabrication and environmental conditions. A rubidium standard is a secondary standard even though it uses atomic transitions, because it takes the form of a gas cell through which an optical signal is passed. The gas cell has inherent inaccuracies because of gas pressure variations, including those induced by temperature variations. There are also variations in the concentrations of the required buffer gases, which variations cause frequency deviations.

secondary radar: A radiodetermination system based on the comparison of reference signals with radio signals retransmitted from the position to be determined. [NTIA] [RR] *Note:* An example of secondary radar is the transponder-based surveillance of aircraft. *Synonym* **secondary surveillance radar.**

secondary radiation: *See* **secondary emission.**

secondary service area: [T]he service area of a broadcast station served by the skywave and not subject to objectionable interference and in which the signal is subject to intermittent variations in strength. [47CFR]

secondary station: In a communications network, a station that (a) is responsible for performing unbalanced link-level operations as instructed by the primary station and (b) interprets received commands and generates responses.

secondary surveillance radar: *Synonym* **secondary radar.**

secondary time standard: A time standard that requires periodic calibration against a primary time standard.

second dialtone: 1. Dialtone presented to the call originator after an access code has been dialed for access to a second, outside, telecommunications system or service. **2.** Dialtone returned to the call originator after she/he has dialed an access number and has reached a switch providing access to modem, to a fax machine, to another telephone, *etc.*

second window: Of silica-based optical fibers, the transmission window at approximately 1.3 µm. *Note:* The second window is the minimum-dispersion window in silica-based glasses. [After FAA]

SECORD: *Acronym for* **secure voice cord board.** A desk-mounted patch panel that provides the capability for controlling (a) sixteen 50-kb/s wideband or sixteen 2400-b/s narrowband user lines and (b) 5 narrowband trunks to DSN or other narrowband facilities. (188)

SECTEL: *Acronym for* **secure telephone.** *See* **STU.**

sector: A predetermined, addressable angular part of a track or band on a magnetic drum or magnetic disk.

sectoring: In magnetic or optical disk storage media, the division of tracks into a specified number of segments, for the purpose of organizing the data stored thereon.

secure communications: Telecommunications deriving security through use of type 1 products and/or protected distribution systems. [NIS]

secure telephone unit: *See* **STU.**

secure transmission: **1.** In transmission security, *see* **secure communications.** **2.** In spread-spectrum systems, the transmission of binary coded sequences that represent information that can be recovered only by persons or systems that have the proper key for the spread-spectrum code-sequence generator, *i.e.*, have a synchronized generator that is identical to that used for transmission. [From Weik '89]

secure voice cord board: *See* **SECORD.**

security: **1.** A condition that results from the establishment and maintenance of protective measures that ensure a state of inviolability from hostile acts or influences. [JP1] **2.** With respect to classified matter, the condition that prevents unauthorized persons from having access to official information that is safeguarded in the interests of national security. [After JP1] **3.** Measures taken by a military unit, an activity or installation to protect itself against all acts designed to, or which may, impair its effectiveness. [JP1]

security filter: **1.** In communications security, the hardware, firmware, or software used to prevent access to specified data by unauthorized persons or systems, such as by preventing transmission, preventing forwarding messages over unprotected lines or circuits, or requiring special codes for access to read-only files. [From Weik '89] **2.** [An] AIS trusted subsystem that enforces security policy on the data that passes through it. [NIS]

security kernel: **1.** In computer and communications security, the central part of a computer or communications system hardware, firmware, and software that implements the basic security procedures for controlling access to system resources. **2.** A self-contained usually small collection of key security-related statements that (a) works as a part of an operating system to prevent unauthorized access to, or use of, the system and (b) contains criteria that must be met before specified programs can be accessed. **3.** Hardware, firmware, and software elements of a trusted computing base that implement the reference monitor concept. [NIS]

security management: In network management, the set of functions (a) that protects telecommunications networks and systems from unauthorized access by persons, acts, or influences and (b) that includes many subfunctions, such as creating, deleting, and controlling security services and mechanisms; distributing security-relevant information; reporting security-relevant events; controlling the distribution of cryptographic keying material; and authorizing subscriber access, rights, and privileges. (188)

seek: To position selectively the access mechanism of a direct access [storage] device.

seek time: The time required for the access arm of a direct-access storage device to be positioned on the appropriate track. *Synonym* **positioning time.**

segment: In a distributed queue dual bus (DQDB) network, a protocol data unit (PDU) that (a) consists of 52 octets transferred between DQDB-layer peer entities as the information payload of a slot, (b) contains a header of 4 octets and a payload of 48 octets, and (c) is either a pre-arbitrated segment or a queued arbitrated segment.

segmented encoding law: An encoding law in which an approximation to a curve defined by a smooth encoding law is obtained by a number of linear segments. *Synonym* **piecewise linear encoding law.**

segregation: *Synonym* **privacy (def. #1).**

seizing: The temporary dedication of various parts of a communications system to a specific use, usually in response to a user request for service. (188) *Note:* The parts seized may be automatically connected, such as by direct distance dialing (DDD), or may require operator intervention.

seizure signal: In telephone systems, a signal used by the calling end of a trunk or line to indicate a request for service. (188) *Note:* A seizure signal also locks out the trunk or line to other demands for service.

selcall: *Acronym for* **selective calling.** Calling from one station in which call identification is sent to signal automatically one or more remote stations and to establish links among them. (188) *Note 1:* Selective calling may be used to un-mute the speakers at designated stations or to initiate a handshake for link establishment. *Note 2:* Selective calling is specified in CCIR Recommendations for HF and VHF/UHF radio, generally for ship-to-shore, ship-to-ship, aircraft-to-aircraft, and aircraft-to-ground communications.

selection position: *Synonym* **decision instant.**

selective calling: *See* **selcall.**

selective combiner: *Synonym* **maximal-ratio combiner.**

selective fading: Fading in which the components of the received radio signal fluctuate independently. (188)

selective jamming: *See* **electronic warfare.**

selective ringing: In a party line, ringing only the desired user instrument. (188) *Note:* Without selective ringing, all the instruments on the party line will ring at the same time, selection being made by the number of rings.

selectivity: A measure of the ability of a receiver to discriminate between a wanted signal on one frequency and unwanted signals on other frequencies. (188)

self-authentication: **1.** A procedure in which a transmitting station, *i.e.*, a calling station, establishes its own validity without the participation of the receiving station, *i.e.*, the called station. *Note:* The calling station establishes its own authenticity and the called station is not required to challenge the calling station. Self-authentication is usually used only when one-time authentication systems are used to derive the authentication. [From Weik '89] **2.** Implicit authentication, to a predetermined level, of all transmissions on a secure communications system. [NIS]

self-delineating block: A block in which a bit pattern or a flag identifies the beginning or end of a block.

self-synchronizing code: A code in which the symbol stream formed by a portion of one code word, or by the overlapped portion of any two adjacent code words, is not a valid code word. (188) *Note 1:* A self-synchronizing code permits the proper framing of transmitted code words provided that no uncorrected errors occur in the symbol stream. *Note 2:* External synchronization is not required. *Note 3:* High-level data link control (HDLC) and Advanced Data Communication Control Procedures (ADCCP) frames represent self-synchronizing code words.

semiautomated tactical command and control system: A machine-aided command and control system in which human intervention is required in varying degrees to operate the system.

semiautomatic switching system: **1.** In telephone systems, a switching system in which telephone operators receive call instructions orally from users and complete them by automatic equipment. (188) **2.** At tape-relay intermediate stations, the manual routing or rerouting of taped messages without rekeying them. (188)

semiconductor laser: *Synonym* **injection laser diode.**

semiduplex operation: **1.** A method which is simplex operation at one end of the circuit and duplex operation at the other. *RR Footnote:* In general, duplex operation and semiduplex operation require two frequencies in radiocommunication; simplex operation may use either one or two. [NTIA] [RR] **2.** Operation of a communications network in which a base station operates in a duplex mode with a group of remote stations operating in a half-duplex mode. (188) *Note:* The terms "*half-duplex*" and "*simplex*" are used differently in wire and radio communications.

sender: A device that accepts address information from a register or routing information from a translator, and then transmits the proper routing information to a trunk or to local equipment. *Note:* Sender and register functions are often combined in a single unit. (188)

sending-end crossfire: In teletypewriter (TTY) systems, interference, in a given channel, caused by transmissions from one or more adjacent TTY channels transmitting from the end at which the crossfire, *i.e.*, interference, is measured. (188)

sensitive information: Information, the loss, or misuse, or unauthorized access to or modification of which could adversely affect the national interest or the conduct of federal programs, or the privacy to which individuals are entitled to under 5 U.S.C. Section 552a (the Privacy Act), but that has not been specifically authorized under criteria established by an Executive Order or an Act of Congress to be kept secret in the interest of national defense or foreign policy. [NIS]

sensitivity: In an electronic device, *e.g.*, a communications system receiver, or detection device, *e.g.*, PIN diode, the minimum input signal required to produce a specified output signal having a specified signal-to-noise ratio, or other specified criteria. (188) *Note 1:* The signal input may be expressed as power in dBm or as field strength in microvolts per meter, with input network impedance stipulated. *Note 2: "Sensitivity"* is sometimes improperly used as a synonym for *"responsivity."*

sensor: A device that responds to a physical stimulus, such as thermal energy, electromagnetic energy, acoustic energy, pressure, magnetism, or motion, by producing a signal, usually electrical.

sentinel: *See* **flag.**

separate channel signaling: Signaling in which the whole or a part of one or more channels in a multichannel system is used to provide for supervisory and control signals for the message traffic channels. (188) *Note:* The same channels, such as frequency bands or time slots, that are used for signaling are not used for message traffic. *Contrast with* **common-channel signaling.**

septet: A byte composed of seven binary elements. *Synonym* **seven-bit byte.**

sequence: An arrangement of items according to a specified set of rules, for example, items arranged alphabetically, numerically, or chronologically.

sequential access: *Synonym* **serial access.**

sequential logic element: A device that has at least one output channel and one or more input channels, all characterized by discrete states, such that the state of each output channel is determined by the previous states of the input channels.

sequential transmission: *Synonym* **serial transmission.**

serial: 1. Pertaining to a process in which all events occur one after the other; for example, the serial transmission of the bits of a character according to the CCITT V.25 protocol. 2. Pertaining to the sequential or consecutive occurrence of two or more related activities in a single device or channel. 3. Pertaining to the sequential processing of the individual parts of a whole, such as the bits of a character or the characters of a word, using the same facilities for successive parts.

serial access: 1. Pertaining to the sequential or consecutive transmission of data into or out of a device, such as a computer, transmission line, or storage device. (188) 2. A process by which data are obtained from a storage device or entered into a storage device in such a way that the process depends on the location of those data and on a reference to data previously accessed. *Synonym* **sequential access.**

serial computer: 1. A computer that has a single arithmetic and logic unit. 2. A computer, some specified characteristic of which is serial; for example, a computer than manipulates all bits of a word serially.

serializer: *See* **parallel-to-serial conversion.**

serial port: A port through which data are passed serially, *i.e.*, one bit at a time, and that requires only one input channel to handle a set of bits, *e.g.*, all the bits of a byte. *Contrast with* **parallel port.**

serial-to-parallel conversion: Conversion of a stream of data elements received in time sequence, *i.e.*, one at a time, into a data stream consisting of multiple data elements transmitted simultaneously. *Contrast with* **parallel-to-serial conversion.**

serial transmission: The sequential transmission of the signal elements of a group representing a character or other entity of data. *Note:* The characters are transmitted in a sequence over a single line, rather than simultaneously over two or more lines, as in parallel transmission. The sequential elements may be transmitted with or without interruption. (188) *Synonym* **sequential transmission.**

series T junction: A three-port waveguide junction that has an equivalent circuit in which the impedance of the branch waveguide is predominantly in series with the impedance of the main waveguide at the junction.

server: A network device that provides service to the network users by managing shared resources. *Note 1:* The term is often used in the context of a client-server architecture for a local area network (LAN). *Note 2:* Examples are a printer server and a file server.

service: In the Open Systems Interconnection—Reference Model (OSI—RM), a capability of a given layer, and the layers below it, that (a) is provided to the entities of the next higher layer and (b) for a given layer, is provided at the interface between the given layer and the next higher layer.

service access: In personal communications service (PCS), the ability for the network to provide user access to features and to accept user service requests specifying the type of bearer services or supplementary service that the users want to receive from the PCS network.

service access point (SAP): **1.** A physical point at which a circuit may be accessed. (188) **2.** In an Open Systems Interconnection (OSI) layer, a point at which a designated service may be obtained.

service bit: A system overhead bit used for providing a network service, such as a request for a repetition or for a numbering sequence. (188) *Note:* A service bit is not a check bit.

service channel: *Synonym* **orderwire circuit.**

service class: *See* **class of service.**

service data unit (SDU): In layered systems, a set of data that is sent by a user of the services of a given layer, and is transmitted to a peer service user semantically unchanged.

service feature: In telephony, any of a number of special functions that may be specified initially, or added to, the user's basic service. (188) *Note:* Modern telephone switches are capable of providing a wide variety of service features, such as call forwarding and call waiting.

service identification: The information that uniquely identifies an NS/EP telecommunications service to the service vendor and the service user.

service integrity: The degree to which a service is provided without excessive impairment, once obtained. [NATO]

service probability: The probability of obtaining a specified (or higher) grade of service during a given period of time. (188)

service profile: *Synonym* **UPT service profile.**

service profile management: *Synonym* **UPT service profile management.**

service program: *Synonym* **utility program.**

service routine: *Synonym* **utility program.**

service signals: Signals that enable data systems equipment to function correctly, and possibly to provide ancillary facilities. *Synonym* **housekeeping signals.**

service termination point: The last point of service rendered by a commercial carrier under applicable tariffs. *Note 1:* The service termination point is usually on the customer premises. *Note 2:* The customer is responsible for equipment and operation from the service termination point to user end instruments. *Note 3:* The service termination point usually corresponds to the demarcation point.

service user: An individual or organization, including a service vendor, that is provided a telecommunications service for which a priority level has been requested or assigned.

Session Layer: *See* **Open Systems Interconnection—Reference Model.**

set: **1.** A finite or infinite number of objects, entities, or concepts, that have a given property or properties in common. **2.** To configure all or part of a device into a specified state.

seven-bit byte: *Synonym* **septet.**

seven-hundred (700) service: A personal telephone service that allows individuals to receive, via a single number, telephone calls in various locations (*e.g.*, home, office, or car) from call originators using the same common carrier.

sexadecimal: *Synonym* **hexadecimal.**

sextet: A byte composed of six binary elements. *Synonym* **six-bit byte.**

S-F: *Abbreviation for* **store-and-forward.**

SF: *Abbreviation for* **single-frequency.** *See* **single-frequency signaling.**

SFTS: *Abbreviation for* **standard frequency and time signal.** *See* **standard time and frequency signal service.**

SGML: *Abbreviation for* **Standard Generalized Mark-up Language.** A file format for storage of text and graphics files.

shadow loss: **1.** The attenuation caused to a radio signal by obstructions in the propagation path. (188) **2.** In a reflector antenna, the relative reduction in the effective aperture of the antenna caused by the masking effect of other antenna parts, such as a feed horn or a secondary reflector, which parts obstruct the radiation path. (188)

shannon (Sh): The unit of information derived from the occurrence of one of two equiprobable, mutually exclusive, and exhaustive events. *Note:* A bit may, with perfect formatting and source coding, contain 1 Sh of information. However, the information content of a bit is usually be less than 1 Sh.

Shannon's law: A statement defining the theoretical maximum rate at which error-free digits can be transmitted over a bandwidth-limited channel in the presence of noise, usually expressed in the form $C = W \log_2(1 + S/N)$, where C is the channel capacity in bits per second, W is the bandwidth in hertz, and S/N is the signal-to-noise ratio. (188) *Note:* Error-correction codes can improve the communications performance relative to uncoded transmission, but no practical error correction coding system exists that can closely approach the theoretical performance limit given by Shannon's law.

shaping network: A network inserted in a circuit for the purpose of improving or modifying the waveform of signals. (188)

sheath: Of a communications or power cable, the outer covering or coverings of tough material, often plastic, that is resistant to environmental hazards such as abrasion, liquid intrusion, solar radiation, *etc.*, and is used to protect cable component(s) such as optical fibers or metallic conductors that transport the signal or power. *Note:* There may be more than one sheath surrounding a given cable. For example, some cable designs use an inner sheath surrounded by metallic armor, over which is an outer sheath. *Synonym* **jacket.**

sheath miles: The actual length of cable in route miles. [47CFR]

shell: In a computer environment, an operating system command interpreter, *i.e.*, a software utility that reads an input specifying an operation, and that may perform, direct, or control the specified operation. *Note 1:* For example, a shell may permit a user to switch among application programs without terminating any of them. *Note 2:* A shell may take its input from either a user terminal or from a file.

SHF: *Abbreviation for* **super high frequency.** *See* **electromagnetic spectrum.**

shield: **1.** A housing, screen, sheath, or cover that substantially reduces the coupling of electric, magnetic, or electromagnetic fields into or out of circuits or transmission lines. (188) **2.** A protective cover that prevents the accidental contact of objects or persons with parts or components operating at hazardous voltage levels. (188)

shielded pair: A 2-wire transmission line surrounded by a sheath of conductive material that protects it from the effects of external fields and confines fields produced within the line. (188)

shielded twisted pair: A transmission line composed of a twisted 2-wire metallic transmission line surrounded by a sheath of conductive material that protects it from the effects of external fields and confines fields produced within the line. (188)

shielding: *See* **shield.**

shielding effectiveness: The factor that expresses the attenuation caused by scattering, *e.g.*, of radiant or acoustic energy, during its passage through a medium. *Note:* The scattering coefficient is usually expressed in units of reciprocal distance.

shift: **1.** The movement of some or all of the characters or bits of a word by the same number of character or bit positions in the direction of a specified end of a word. **2.** In radar, the ability to move the origin of a radial display away from the center of the cathode ray tube.

shift register: A storage device, usually in a central processing unit (CPU), in which device a serially ordered set of data may be moved, as a unit, into a discrete number of storage locations. (188) *Note 1:* Shift registers may be configured so that the stored data may be moved in more than one direction. *Note 2:* Shift registers may be configured so that data may be entered and stored from multiple inputs. *Note 3:* Shift registers may be grouped into arrays of two or more dimensions in order to perform more complex data operations.

ship Earth station: A mobile Earth station in the maritime mobile-satellite service located on board ship. [NTIA] [RR]

ship's emergency transmitter: A ship's transmitter to be used exclusively on a distress frequency for distress, urgency or safety purposes. [NTIA] [RR]

ship station: A mobile station in the maritime mobile service located on board a vessel which is not permanently moored, other than a survival craft station. [NTIA] [RR]

shock excitation: *Synonym* **impulse excitation.**

short haul toll traffic: A general term applied to message toll traffic between nearby points. In common usage, this term is ordinarily applied to message toll traffic between points less than 20 to 50 miles apart. [47CFR]

shortwave: In radio communications, pertaining to the band of frequencies approximately between 3 MHz and 30 MHz. *Note: "Shortwave"* is not a term officially recognized by the international community.

short wavelength: In optical communication, optical radiation having a wavelength less than approximately 1 µm.

shot noise: The noise caused by random fluctuations in the motion of charge carriers in a conductor. (188) *Note:* There is often a minor inconsistency in referring to shot noise in an optical system: many authors refer to shot noise loosely when speaking of the mean square shot noise current (amperes^2) rather than noise power (watts).

SI: *Abbreviation for* **International System of Units.**

SID: *Abbreviation for* **sudden ionospheric disturbance.**

sideband: In amplitude modulation (AM), a band of frequencies higher than or lower than the carrier frequency, containing energy as a result of the modulation process. (188) *Note:* Amplitude modulation results in two sidebands. The frequencies above the carrier frequency constitute what is referred to as the *"upper sideband"*; those below the carrier frequency, constitute the *"lower sideband."* In conventional AM transmission, both sidebands are present. Transmission in which one sideband is removed is called *"single-sideband transmission."*

sideband transmission: *See* **single-sideband transmission.**

side circuit: Either of the two circuits used to derive a phantom circuit. (188)

side lobe: In a directional antenna radiation pattern, a lobe in any direction other than that of the main lobe. (188)

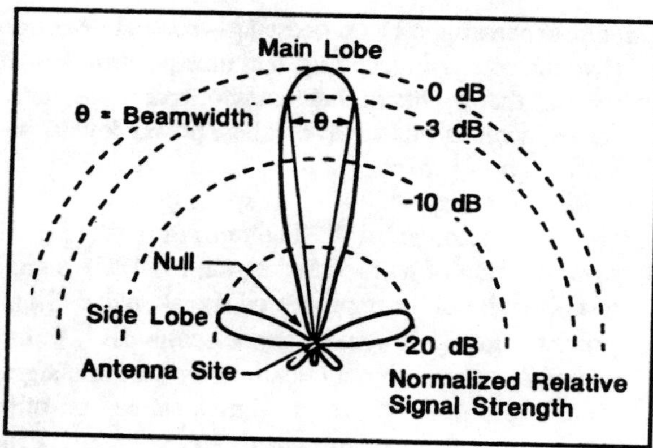

side lobe

sidetone: The sound of the speaker's own voice (and background noise) as heard in the speaker's telephone receiver. (188) *Note:* Sidetone volume is usually suppressed relative to the transmitted volume. *Synonym* **telephone sidetone.**

SIGINT: *Acronym for* **signals intelligence.**

signal: **1.** Detectable transmitted energy that can be used to carry information. **2.** A time-dependent variation of a characteristic of a physical phenomenon, used to convey information. **3.** As applied to electronics, any transmitted electrical impulse. [JP1] **4.** Operationally, a type of message, the text of which consists of one or more letters, words, characters, signal flags, visual displays, or special sounds, with prearranged meaning and which is conveyed or transmitted by visual, acoustical, or electrical means. [JP1]

signal center: A combination of signal communication facilities operated by the Army in the field and consisting of a communications center, telephone switching central and appropriate means of signal communications. [JP1]

signal compression: **1.** In analog (usually audio) systems, reduction of the dynamic range of a signal by controlling it as a function of the inverse relationship of its instantaneous value relative to a specified reference level. (188) *Note 1:* Signal compression is usually expressed in dB. *Note 2:* Instantaneous values of the input signal that are low, relative to the reference level, are increased, and those that are high are decreased. *Note 3:* Signal compression is usually accomplished by separate devices called *"compressors."* It is used for many purposes, such as (a) improving signal-to-noise ratios prior to digitizing an analog signal for transmission over a digital carrier system, (b) preventing overload of succeeding elements of a system, or (c) matching the dynamic ranges of two devices. *Note 4:* Signal compression (in dB) may be a linear or nonlinear function of the signal level across the frequency band of interest and may be essentially instantaneous or have fixed or variable delay times. *Note 5:* Signal compression always introduces distortion, which is usually not objectionable, if the compression is limited to a few dB. *Note 6:* The original dynamic range of a compressed signal may be restored by a circuit called an *"expander."* (188) **2.** In facsimile systems, a process in which the number of pels scanned on the original is larger than the number of encoded bits of picture information transmitted. (188)

signal contrast: In facsimile, the ratio of the level of the white signal to the level of the black signal. (188) *Note:* Signal contrast is usually expressed in dB.

signal conversion equipment: *Synonym* **modem.**

signal distance: **1.** A measure of the difference between a given signal and a reference signal. *Note:* For analog signals, the signal distance is the root mean square difference between the given signal and a reference signal over a symbol period. **2.** *Synonym* **Hamming distance.**

signal distortion: *See* **distortion.**

signal droop: In an otherwise essentially flat-topped rectangular pulse, distortion characterized by a decline of the pulse top. *See illustration under* **waveform.**

signal element: A part of a signal, distinguished by its nature, magnitude, duration, transition, or relative position. *Note:* Examples of signal elements include signal transitions, significant conditions, significant instants, and binary digits (bits). (188)

signal expansion: Restoration of the dynamic range of a compressed signal. *Contrast with* **signal compression.**

signal frequency shift: *See* **frequency shift.**

signaling: **1.** The use of signals for controlling communications. **2.** In a telecommunications network, the information exchange concerning the establishment and control of a connection and the management of the network, in contrast to user information transfer. (188) **3.** The sending of a signal from the transmitting end of a circuit to inform a user at the receiving end that a message is to be sent. (188)

signaling path: In a transmission system, a path used for system control, synchronization, checking, signaling, and service signals used in system management and operations rather than for the data, messages, or calls of the users. (188)

signaling rate: *See* **data signaling rate.**

Signaling System No. 7 (SS7): A common-channel signaling system defined by the CCITT in the 1988 Blue Book, in Recommendations Q.771 through Q.774. *Note:* SS7 is a prerequisite for implementation of an Integrated Services Digital Network (ISDN).

signaling time slot: In TDM carrier systems, a time slot starting at a particular phase or instant in each frame and allocated to the transmission of signaling (supervisory and control) data. (188)

signal intelligence: *See* **signals intelligence.**

signal level: In a communications system, the signal power or intensity at a specified point and with respect to a specified reference level, *e.g.*, 1 mW.

signal message: In communications systems, a message, *i.e.*, an assembly of signaling information, that (a) includes associated message alignment and service indications, (b) pertains to a call, and (c) is transferred via the message transfer part.

signal-plus-noise-plus-distortion to noise-plus-distortion ratio: *See* **SINAD.**

signal-plus-noise-to-noise ratio ($(S+N)/N$): At a given point in a communications system, the ratio of (a) the power of the desired signal plus the noise to (b) the power of the noise. *Note:* The $(S+N)/N$ ratio is usually expressed in dB. (188)

signal processing: The processing—such as detection, shaping, converting, coding, and time positioning—of signals, that results in their transformation into other forms, such as other waveshapes, power levels, and coding arrangements.

signal processing gain: **1.** The ratio of (a) the signal-to-noise ratio of a processed signal to (b) the signal-to-noise ratio of the unprocessed signal. *Note:* Signal processing gain is usually expressed in dB. **2.** In a spread-spectrum communications system, the signal gain, signal-to-noise ratio, signal shape, or other signal improvement obtained by coherent band spreading, remapping, and reconstitution of the desired signal.

signal reference subsystem: The portion of a facility grounding system that (a) provides reference planes, such as ground-return circuits, for all of the signal paths in the facility and (b) is isolated from other circuits, especially isolated from circuits that carry fault, lightning discharge, and power distribution currents.

signal regeneration: Signal processing that restores a signal so that it conforms to its original characteristics. (188)

signal-return circuit: A current-carrying return path from a load back to the signal source, *i.e.*, the low side of the closed loop energy transfer circuit between a source-load pair. (188)

signal sample: The value of a particular characteristic of a signal at a chosen instant. (188)

signal sampling: The process of obtaining a sequence of instantaneous values of a particular signal characteristic, usually at regular time intervals. (188)

signal security: A generic term that includes both communications security and electronics security. [JP1]

signals intelligence (SIGINT): **1.** A category of intelligence comprising, either individually or in combination, all communications intelligence, electronics intelligence, and foreign instrumentation signals intelligence, however transmitted. [JP1] **2.** Intelligence derived from communications, electronics, and foreign instrumentation signals. [JP1]

signals security: [A] generic term encompassing communications security and electronic security. [NIS]

signal-to-crosstalk ratio: At a specified point in a circuit, the ratio of the power of the wanted signal to the power of the unwanted signal from another channel. *Note 1:* The signals are adjusted in each channel so that they are of equal power at the zero transmission level point in their respective channels. *Note 2:* The signal-to-crosstalk ratio is usually expressed in dB. (188)

signal-to-noise ratio (SNR): The ratio of the amplitude of the desired signal to the amplitude of noise signals at a given point in time. [JP1] *Note 1:* SNR is expressed as 20 times the logarithm of the amplitude ratio, or 10 times the logarithm of the power ratio. *Note 2:* SNR is usually expressed in dB and in terms of peak values for impulse noise and root-mean-square values for random noise. In defining or specifying the SNR, both the signal and noise should be characterized, *e.g.,* peak-signal-to-peak-noise ratio, in order to avoid ambiguity.

signal-to-noise ratio per bit: The ratio given by E_b/N_0, where E_b is the signal energy per bit and N_0 is the noise energy per hertz of noise bandwidth. (188)

signal transfer point (STP): In a common-channel signaling network, a switching center that provides for the transfer from one signaling link to another. *Note:* In nonassociated common-channel signaling, the signal transfer point need not be the point through which the call, which is associated with the signaling being switched, passes.

signal transition: In the modulation of a carrier, a change from one significant condition to another. *Note 1:* Examples of signal transitions are a change from one electrical current, voltage, or power level to another; a change from one optical power level to another; a phase shift; or a change from one frequency or wavelength to another. *Note 2:* Signal transitions are used to create signals that represent information, such as "0" and "1" or "mark" and "space."

signature: **1.** The complete set of electromagnetic and/or acoustic signals received, *e.g.,* from an infrared source, a radio or radar transmitter, an aircraft, or a ship. *Note:* Signatures may consist of analog or digital signals, or both, and may be analyzed to indicate the nature of their source and assist in its recognition. **2.** The attributes of an electromagnetic or acoustic wave that has been reflected from or transmitted through an object and contains information indicating the attributes of the object.

significant condition: In the modulation of a carrier, one of the values of the signal parameter chosen to represent information. (188) *Note 1:* Examples of significant conditions are an electrical current, voltage, or power level; an optical power level; a phase value; or a frequency or wavelength chosen to represent a "0" or a "1"; or a "mark" or a "space." *Note 2:* The duration of a significant condition is the time interval between successive significant instants. *Note 3:* A change from one significant condition to another is called a "*signal transition.*" *Note 4:* Signal transitions are used to create signals that represent information, such as "0" and "1" or "mark" and "space." *Note 5:* Significant conditions are recognized by an appropriate device. Each significant instant is determined when the appropriate device assumes a condition or state usable for performing a specific function, such as recording, processing, or gating.

significant digit: In a representation of a number, a digit that is needed for a given purpose; in particular, a digit that must be kept to preserve a given accuracy or a given precision.

significant instant: In a signal, any instant at which a significant condition of a signal begins or ends. (188) *Note:* Examples of significant instants include the instant at which a signal crosses the baseline or reaches 10% or 90% of its maximum value.

significant interval: The time interval between two consecutive significant instants. (188)

silent zone: *Synonym* **skip zone.**

silica: Silicon dioxide (SiO_2). *Note 1:* Silica may occur in crystalline or amorphous form, and occurs naturally in impure forms such as quartz and sand. *Note 2:* Silica is the basic material of which the most common communication-grade optical fibers are presently made. [After FAA]

silicon dioxide (SiO_2): *See* **silica.**

silicon photodiode: A silicon-based PN- or PIN-junction photodiode. *Note 1:* Such photodiodes are useful for direct detection of optical wavelengths shorter than approximately 1 µm. *Note 2:* Because of their greater bandgap, silicon-based photodiodes are quieter than germanium-based photodiodes, but germanium photodiodes must be used for wavelengths longer than approximately 1 µm. [FAA]

simple buffering: The assigning of buffer storage for the duration of the execution of a computer program.

Simple Mail Transfer Protocol (SMTP): The Transmission Control Protocol/Internet Protocol (TCP/IP) standard protocol that facilitates transfer of electronic-mail messages, specifies how two systems are to interact, and specifies the format of messages used to control the transfer of electronic mail.

Simple Network Management Protocol (SNMP): The Transmission Control Protocol/Internet Protocol (TCP/IP) standard protocol that (a) is used to manage and control IP gateways and the networks to which they are attached, (b) uses IP directly, bypassing the masking effects of TCP error correction, (c) has direct access to IP datagrams on a network that may be operating abnormally, thus requiring management, (d) defines a set of variables that the gateway must store, and (e) specifies that all control operations on the gateway are a side-effect of fetching or storing those data variables, *i.e.,* operations that are analogous to writing commands and reading status.

simple scanning: In facsimile transmission, scanning using only one spot at a time. (188)

simplex circuit: **1.** A circuit that provides transmission in one direction only. (188) **2.** *Deprecated definition:* A circuit using ground return and affording communication in either direction, but in only one direction at a time. *Note:* The above two definitions are contradictory; however, both are in common use. The user is cautioned to verify the nature of the service specified by this term.

simplex operation: **1.** Operation in which transmission occurs in one and only one preassigned direction. *Synonym* **one-way operation.** (188) *Note:* Duplex operation may be achieved by simplex operation of two or more simplex circuits or channels. **2.** Operating method in which transmission is made possible alternately in each direction of a telecommunication channel, for example by means of manual control. *Note:* In general, duplex operation and semiduplex operation require two frequencies in radiocommunication; simplex operation may use either one or two. [NTIA] [RR] *Note 2:* These two definitions are contradictory, however, both are in common use. The first one is used in telephony and the second one is used in radio. The user is cautioned to verify the nature of the service specified by this term.

simplex (SX) signaling: Signaling in which two conductors are used for a single channel, and a center-tapped coil, or its equivalent, is used to split the signaling current equally between the two conductors. (188) *Note:* SX signaling may be one-way, for intra-central-office use, or the simplex legs may be connected to form full duplex signaling circuits that function like composite (CX) signaling circuits with E & M lead control.

simulate: To represent certain features of the behavior of a physical or abstract system by the behavior of another system. *Note 1:* For example, delay lines may be used to simulate propagation delay and phase shift caused by an actual transmission line. *Note 2:* A simulator may imitate only a few of the operations and functions of the unit it simulates. *Contrast with* **emulate.**

SINAD: *Abbreviation for* **signal-plus-noise-plus-distortion to noise-plus-distortion ratio.** **1.** The ratio of (a) total received power, *i.e.,* the received signal-plus-noise-plus-distortion power to (b) the received noise-plus-distortion power. (188) **2.** The ratio of (a) the recovered audio power, *i.e.,* the original modulating audio signal plus noise plus distortion powers from a modulated radio frequency carrier to (b) the residual audio power, *i.e.,* noise-plus-distortion powers remaining after the original modulating audio signal is removed. (188) *Note:* The SINAD is usually expressed in dB.

singing: An undesired self-sustained audio oscillation in a circuit. *Note:* Singing is usually caused by positive feedback, excessive gain, or unbalance of a hybrid termination, or by some combination of these. (188)

singing margin: The difference in power levels between the singing point and the operating gain of a system or component. (188)

singing point: The threshold point at which additional gain in the system will cause self-oscillation. (188)

single-current system: *Synonym* **neutral direct-current telegraph system.**

single-current transmission system: *Synonym* **neutral direct-current telegraph system.**

single-ended control: *Synonym* **single-ended synchronization.**

single-ended synchronization: Synchronization between two locations, in which phase error signals used to control the clock at one location are derived by comparing the phase of the incoming signals to the phase of the internal clock at that location. *Synonym* **single-ended control.**

single-frequency interference: Interference caused by a single-frequency source. *Note 1:* An example of single-frequency interference is interference in a transmission channel induced by a 60-Hz source. (188) *Note 2:* The interference caused by the single-frequency source may have other frequencies and may also appear in many channels.

single-frequency (SF) signaling: In telephony, signaling in which dial pulses or supervisory signals are conveyed by a single voice-frequency tone in each direction. (188) *Note 1:* An SF signaling unit converts E & M signaling to a format (characterized by the presence or absence of a single voice-frequency tone), which is suitable for transmission over an ac path, *e.g.*, a carrier system. The SF tone is present in the idle state and absent during the seized state. In the seized state, dial pulses are conveyed by bursts of SF tone, corresponding to the interruptions in dc continuity created by a rotary dial or other dc dialing mechanism. *Note 2:* The SF tone may occupy a small portion of the user data channel spectrum, *e.g.*, 1600 Hz or 2600 Hz ("in-band" SF signaling), usually with a notch filter at the precise SF frequency, to prevent the user from inadvertently disconnecting a call if user data has a sufficiently strong spectral content at the SF frequency. The SF tone may also be just outside the user voice band, *e.g.*, 3600 Hz. *Note 3:* The Defense Data Network (DDN) transmits dc signaling pulses or supervisory signals, or both, over carrier channels or cable pairs on a 4-wire basis using a 2600-Hz signal tone. The conversion into tones, or vice versa, is done by SF signal units.

single-harmonic distortion: Of a fundamental frequency, the ratio of the power of a specified harmonic to the power of the fundamental frequency. *Note:* Single-harmonic distortion is measured at the output of a device under specified conditions and is expressed in dB. (188)

single-mode fiber: *Synonym* **single-mode optical fiber.**

single-mode optical fiber: An optical fiber in which only the lowest order bound mode can propagate at the wavelength of interest. *Note 1:* The lowest order bound mode is ascertained for the wavelength of interest by solving Maxwell's equations for the boundary conditions imposed by the fiber, *e.g.*, core (spot) size and the refractive indices of the core and cladding. *Note 2:* The solution of Maxwell's equations for the lowest order bound mode will permit a pair of orthogonally polarized fields in the fiber, and this is the usual case in a communication fiber. *Note 3:* In step-index guides, single-mode operation occurs when the normalized frequency, V, is less than 2.405. For power-law profiles, single-mode operation occurs for a normalized frequency, V, less than approximately

$$2.405\sqrt{\frac{g+2}{g}} ,$$

where g is the profile parameter. *Note 4:* In practice, the orthogonal polarizations may not be associated with degenerate modes. *Synonyms* **monomode optical fiber, single-mode fiber, single-mode optical waveguide, unimode fiber.**

single-mode optical waveguide: *Synonym* **single-mode optical fiber.**

single-Morse system: *Synonym* **neutral direct-current telegraph system.**

single-polarized antenna: An antenna that radiates or receives radio waves with a specific polarization. *Note:* For a singly polarized antenna, the desired sense of polarization is usually maintained only for certain directions or within the major portion of the radiation pattern. (188)

single-sideband (SSB) emission: An amplitude modulated emission with one sideband only. [NTIA] [RR] (188)

single-sideband (SSB) equipment reference level: The power of one of two equal tones that, when used together to modulate a transmitter, cause it to develop its full rated peak power output. (188)

single-sideband suppressed carrier (SSB-SC) transmission: Single-sideband transmission in which the carrier is suppressed. (188) *Note:* In SSB-SC the carrier power level is suppressed to the point where it is insufficient to demodulate the signal.

single-sideband (SSB) transmission: Sideband transmission in which only one sideband is transmitted. (188) *Note:* The carrier may be suppressed.

single-tone interference: An undesired discrete frequency appearing in a transmission channel. (188) *Note:* The single-tone interference frequency is the frequency that appears in the channel regardless of the nature of the source.

sink: **1.** An absorber of energy. **2.** In communications, a device that receives information, control, or other signals from a source. (188)

S interface: For basic rate access in an Integrated Services Digital Network (ISDN) environment, a user-to-network interface reference point that (a) is characterized by a 4-wire, 144-kb/s (2B+D) user rate, (b) serves as a universal interface between ISDN terminals or terminal adapters and the network channel termination, (c) allows a variety of terminal types and subscriber networks, such as PBXs, local area networks (LANs), and controllers, to be connected to the network, and (d) operates at 4000 48-bit frames per second, *i.e.*, 192 kb/s, with a user portion of 36 bits per frame, *i.e.*, 144 kb/s.

six-bit byte: *Synonym* **sextet.**

skew: **1.** In parallel transmission, the difference in arrival time of bits transmitted at the same time. (188) **2.** For data recorded on multichannel magnetic tape, the difference between reading times of bits recorded in a single transverse line. (188) *Note:* Skew is usually interpreted to mean the difference in reading times between bits recorded on the tracks at the extremities, *i.e.*, edges, of the tape. **3.** In facsimile systems, the angular deviation of the received frame from rectangularity caused by asynchronism between the scanner and the recorder. *Note:* Skew is expressed numerically as the tangent of the deviation angle. (188) **4.** In facsimile, the angle between the scanning line, or recording line, and the perpendicular to the paper path.

skew ray: In a multimode optical fiber, a bound ray that travels in a helical path along the fiber and thus (a) is not parallel to the fiber axis, (b) does not lie in a meridional plane, and (c) does not intersect the fiber axis.

skin effect: The tendency of alternating current to flow near the surface of a conductor, thereby restricting the current to a small part of the total cross-sectional area and increasing the resistance to the flow of current. *Note:* The skin effect is caused by the self-inductance of the conductor, which causes an increase in the inductive reactance at high frequencies, thus forcing the carriers, *i.e.*, electrons, toward the surface of the conductor. At high frequencies, the circumference is the preferred criterion for predicting resistance than is the cross-sectional area. The depth of penetration of current can be very small compared to the diameter. [From Weik '89]

skip distance: At a given azimuth, the minimum distance between the transmitting station and the closest point of return to the Earth of a transmitted wave reflected from the ionosphere. (188)

skip zone: An annular region within the transmission range of an antenna, within which signals from the transmitter are not received. *Note:* The skip zone is bounded by the locus of the farthest points at which the ground wave can be received and the nearest points at which reflected sky waves can be received. (188) *Synonyms* **silent zone, zone of silence.**

sky wave: A radio wave that travels upward from the antenna. (188) *Note:* A sky wave may be reflected to Earth by the ionosphere.

slab-dielectric waveguide: An electromagnetic waveguide (a) that consists solely of dielectric materials, (b) in which the dielectric propagation medium has a rectangular cross section, (c) that has a width, thickness, and refractive indices that determine the operating wavelength and the modes the guide will support beyond the equilibrium length, (d) that may be cladded, protected, distributed, and electronically controllable, and (e) that may be used in various applications, such as in integrated optical circuits (IOCs) in which their shape is geometrically more convenient than the optical fibers that are circular in cross section, that are used in fiber optic cables for long-distance transmission. *Note:* Their principle of operation is the same as that for optical fibers that are circular in cross section. [After 2196]. *Synonym* **planar waveguide.**

slant range: The line-of-sight distance between two points, not at the same level relative to a specific datum. [JP1] (188) *Note:* An example of slant range is the distance to an airborne radar target, *e.g.,* an airplane flying at high altitude with respect to that of the radar antenna. The slant range is the hypotenuse of the triangle represented by the altitude of the airplane and the distance between the radar antenna and the airplane's ground track (the point on the Earth at which it is directly overhead). In the absence of altitude information, the aircraft location would be plotted farther from the antenna than its actual ground track.

slave clock: A clock that is coordinated with a master clock. *Note 1:* Slave clock coordination is usually achieved by phase-locking the slave clock signal to a signal received from the master clock. *Note 2:* To adjust for the transit time of the signal from the master clock to the slave clock, the phase of the slave clock may be adjusted with respect to the signal from the master clock so that both clocks are in phase. Thus, the time markers of both clocks, at the output of the clocks, occur simultaneously.

slave station: **1.** In a data network, a station that is selected and controlled by a master station. *Note:* Usually a slave station can only call, or be called by, a master station. **2.** In navigation systems using precise time dissemination, a station having a clock is synchronized by a remote master station. *Synonym* **subordinate station.**

slewing: **1.** Rotating a directional antenna or transducer rapidly about one or more axes. **2.** Changing the frequency or pulse repetition rate of a signal source. **3.** Changing the tuning of a receiver, usually by sweeping through many or all frequencies. [From Weik '89] **4.** Redirecting the beam of a fixed antenna array by changing the relative phases of the signals feeding the antenna elements.

sliding window: A variable-duration window that allows a sender to transmit a specified number of data units before an acknowledgement is received or before a specified event occurs. *Note:* An example of a sliding window in packet transmission is one in which, after the sender fails to receive an acknowledgement for the first transmitted packet, the sender "slides" the window, *i.e.,* resets the window, and sends a second packet. This process is repeated for the specified number of times before the sender interrupts transmission. *Synonym (loosely)* **acknowledgement delay period.**

slip: In a sequence of transmitted symbols, *e.g.,* digital bits, a signal phase shift, *i.e.,* a signal positional displacement, that causes the loss of one or more symbols or the insertion of one or more extraneous symbols. *Note:* Slips are usually caused by inadequate synchronization of the two clocks controlling the transmission and reception of the signals that represent the symbols.

SLIP: *Acronym for* **serial line Internet protocol.** A protocol that allows a computer to use the Internet protocol (IP) with a standard telephone line and a high-speed modem.

slip-free operation: Operation of a communications system with sufficient phase-locking to avoid overflowing or emptying buffers. (188)

slit source: *Synonym* **line source.**

slope: In a transmission line, the rate of change of attenuation with respect to frequency over the frequency spectrum. (188) *Note 1:* The slope is usually expressed in dB per hertz or dB per octave. *Note 2:* In metallic lines, the slope is usually greater at high frequencies than at low frequencies.

slope equalizer: A device or circuit used to achieve a specified slope in a metallic transmission line. (188)

slope-keypoint compaction: Data compaction accomplished by stating (a) a specific keypoint of departure, (b) a direction or slope of departure, (c) the maximum deviation from a prescribed specific value, and (d) a new keypoint and a new slope. *Note:* An example of slope-keypoint compaction is the storage or transmission of a slope and one point on a straight line instead of storing and transmitting a large number of values, *i.e.*, of points, on the line. [From Weik '89]

slot: In a distributed-queue dual-bus (DQDB) network, a protocol data unit (PDU) that (a) consists of 53 octets used to transfer segments of user information, (b) has the capacity to contain a segment of 52 octets and a 1-octet access control field, and (c) may be either a pre-arbitrated (PA) slot or a queued arbitrated (QA) slot.

slot antenna: A radiating element formed by a slot in a conducting surface or in the wall of a waveguide. (188)

slotted-ring network: A ring network that allows unidirectional data transmission between data stations by transferring data in predefined slots in the transmission stream over one transmission medium such that the data return to the originating station.

slot time: In networks using carrier sense multiple access with collision detection (CSMA/CD), the length of time that a transmitting station waits before attempting to retransmit following a collision. *Note:* Slot time varies from station to station.

SMDR: *Abbreviation for* **station message-detail recording.**

smearing: In video displays, a localized distortion over a sub-region of the image, characterized by reduced sharpness of edges and spatial detail.

smooth Earth: Idealized surfaces, such as water surfaces or very level terrain, having radio horizons that are not formed by prominent ridges or mountains but are determined solely as a function of antenna height above ground and the effective Earth radius. (188)

SNA: *Abbreviation for* **systems network architecture.** A proprietary communications architecture.

sneak current: In a communications circuit, an anomalous current that presents no immediate danger, but may cause improper operation or damage. (188)

Snell's law: A law of geometric optics that defines the amount of bending that takes place when a light ray strikes a refractive boundary, *e.g.*, an air-glass interface, at a non-normal angle. *Note 1:* Snell's law states that

$$n_1 \sin\theta_1 = n_2 \sin\theta_2 \quad ,$$

where n_1 is the index of refraction of the medium in which the incident ray travels, θ_1 is the angle, with respect to the normal at the refractive boundary, at which the incident ray strikes the boundary, n_2 is the index of refraction of the medium in which the refracted ray travels, and θ_2 is the angle, with respect to the normal at the refractive boundary, at which the refracted ray travels. The incident ray and refracted ray travel in the same plane, on opposite sides of the normal at the point of incidence. *Note 2:* If a ray travels from a medium of lower refractive index into a medium of higher refractive index, it is bent toward the normal; if it travels from a medium of higher refractive index to a medium of lower index, it is bent away from the normal. *Note 3:* If the incident ray travels in a medium of higher refractive index toward a medium of lower refractive index at such an angle that Snell's law would call for the sine of the refracted ray to be greater than unity (a mathematical impossibility); *i.e.*,

$$\sin\theta_2 = \frac{n_1}{n_2}\sin\theta_1 > 1 \quad ,$$

then the "refracted" ray in actuality becomes a reflected ray and is totally reflected back into the medium of higher refractive index, at an angle equal to the incident angle (and thus still "obeys" Snell's Law). This reflection occurs even in the absence of a metallic reflective coating (*e.g.*, aluminum or silver). This phenomenon is called *"total internal reflection."* The smallest angle of incidence, with respect to the normal at the refractive boundary, which angle will support total internal reflection, is called the *"critical angle."* [After FAA]

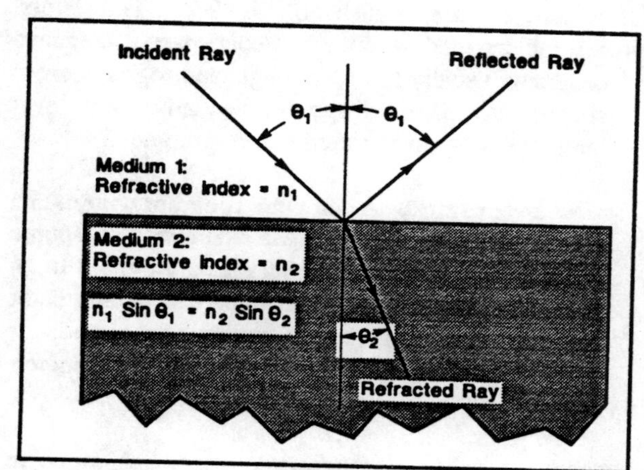

Snell's law

SNMP: *Abbreviation for* **Simple Network Management Protocol.**

snow: In video display systems, noise that (a) is uniformly distributed on the display surface, such as that of a television or radar screen, (b) has the appearance of a uniform distribution of fixed or moving spots, mottling, or speckling, and (c) is usually caused by random noise on an intensity-modulated signal in a display device, such as a cathode-ray tube.

SNR: *Abbreviation for* **signal-to-noise ratio.**

soft copy: A nonpermanent display image, for example, a cathode ray tube display.

soft limiting: *See* **limiting.**

soft sectoring: On magnetic disks, magnetic drums, and optical disks, the identification of sector boundaries by using recorded information.

software: **1.** A set of computer programs, procedures, and associated documentation concerned with the operation of a data processing system; *e.g.*, compilers, library routines, manuals, and circuit diagrams. [JP1] **2.** Information (generally copyrightable) that may provide instructions for computers; data for documentation; and voice, video, and music for entertainment or education.

software engineering: The discipline devoted to the design, development, and use of computer software. *Note:* Software engineering must address various aspects of data processing, including compatibility with the computer system which is to execute the software, and tradeoffs among maintainability, flexibility, efficiency, processing time, and costs. [From Weik '89]

software package: A package that consists of (a) one or more computer programs and possibly related material such as utility programs or tutorial programs, recorded on a medium suitable for delivery to the user, and from which the user can transfer the program(s) to a data-processing device, and (b) instructional materials such as handbooks and manuals, update information, and possibly support services information. *Note 1:* The computer programs may consist, for example, of application programs or operating systems, and are usually written in a high-level or low-level language, respectively. *Note 2:* The recording medium is usually a magnetic diskette or an optical compact disk.

software tool: Software, such as a computer program, routine, subroutine, program block, or program module, that can be used to develop, test, analyze, or maintain a computer program or its documentation. *Note:* Examples of software tools are automated software verification routines, compilers, program maintenance routines, bootstraps, program analyzers, and software monitors. [From Weik '89]

SOH: *Abbreviation for* **start-of-heading character.**

solid-state scanning: In facsimile, scanning in which all or a part of the scanning process is performed by electronic commutation of an array of solid-state photosensitive elements. (188)

soliton: An optical pulse having a shape, spectral content, and power level designed to take advantage of nonlinear effects in an optical fiber waveguide, for the purpose of essentially negating dispersion over long distances.

sonar: *Acronym for* **sound navigation and ranging.** A device that is used primarily for the detection and location of underwater objects by reflecting acoustic waves from them, or by the interception of acoustic waves from an underwater, surface, or above-surface acoustic source. *Note:* Sonar operates with acoustic waves in the same way that radar and radio direction-finding equipment operate with electromagnetic waves, including use of the Doppler effect, radial component of velocity measurement, and triangulation. [From Weik '89]

SONET: *Acronym for* **synchronous optical network.** An interface standard for synchronous 2.46-Gb/s optical-fiber transmission, applicable to the Physical Layer of the OSI Reference Model. *Note 1:* SONET uses a basic data rate of 51.840 Mb/s, called OC1 (optical carrier 1). The SONET hierarchy is defined in multiples of OC1, up to and including OC48, for a maximum data rate of 2.48832 Gb/s. *Note 2:* SONET was developed by the Exchange Carriers Standards Association (ECSA).

sonobuoy: In sonar systems, a device (a) that is used to detect acoustic waves, such as those produced by ships and submarines, (b) that, when activated, relays information by radio, (c) that may be active or passive, and (d) that may be directional or nondirectional. [From Weik '89]

sounder prediction station: A station equipped with an ionosphere sounder for realtime monitoring of upper atmosphere phenomena or to obtain data for the prediction of propagation conditions. [NTIA]

sounding: In automated HF radio systems, the broadcasting of a very brief signal, containing the station address, station identifier, or call sign, to permit receiving stations to measure link quality. (188)

sound navigation and ranging: *See* **sonar.**

sound-powered telephone: A telephone in which the operating power is derived from the speech input only. (188)

sound wave: *See* **acoustic wave.**

source: In communications, that part of a system from which messages are considered to originate. (188)

source efficiency: In optical systems, the ratio of emitted optical power of a source to the input electrical power. (188)

source language: In computing, data processing, and communications systems, a language from which statements are translated. *Note:* Translators, assemblers, and compilers prepare target language programs, usually machine-language programs, from source language programs, usually high-level language programs written by programmers.

source program: **1.** A computer program written in a source language. *Note:* An example of a source program is a program that serves as the input to an assembler, compiler, or translator. **2.** A computer program that must be assembled, compiled, or translated before it can be executed by a computer. [From Weik '89]

source quench: A congestion-control technique in which a computer experiencing data traffic congestion sends a message back to the source of the messages or packets causing the congestion, requesting that the source stop transmitting.

source user: The user providing the information to be transferred to a destination user during a particular information transfer transaction. *Synonym* **information source.**

space: In telegraphy, one of the two significant conditions of encoding. (188) *Note 1:* The complementary significant condition is called a *"mark."* *Note 2:* In modern digital communications, the two corresponding significant conditions of encoding are called *"zero"* and *"one."* *Synonyms* **spacing pulse, spacing signal.**

spacecraft: A man-made vehicle which is intended to go beyond the major portion of the Earth's atmosphere. [NTIA] [RR]

space diversity: A method of transmission or reception, or both, in which the effects of fading are minimized by the simultaneous use of two or more physically separated antennas, ideally separated by one or more wavelengths. (188)

space-division multiplexing: *A misnomer. Note: Space-division multiplexing* has been improperly applied to the use of multiple physical transmission channels, *e.g.*, twisted pairs or optical fibers, under one sheath.

space-division switching: In telephony, switching in which single transmission-path routing determination is accomplished in a switch by using a physically separated set of matrix contacts or cross-points. (188)

space operation service: A radiocommunication service concerned exclusively with the operation of spacecraft, in particular space tracking, space telemetry and space telecommand. These functions will normally be provided within the service in which the space station is operating. [NTIA] [RR]

space radiocommunication: Any radiocommunication involving the use of one or more space stations or the use of one or more reflecting satellites or other objects in space. [NTIA] [RR]

space research service: A radiocommunication service in which spacecraft or other objects in space are used for scientific or technological research purposes. [NTIA] [RR]

space station: A station located on an object which is beyond, is intended to go beyond, or has been beyond, the major portion of the Earth's atmosphere. [NTIA] [RR]

space subsystem: In satellite communications, that portion of the satellite link that is in orbit. (188)

space system: Any group of cooperating Earth stations and/or space stations employing space radiocommunication for specific purposes. [NTIA] [RR]

space telecommand: The use of radiocommunication for the transmission of signals to a space station to initiate, modify or terminate functions of equipment on an associated space object, including the space station. [NTIA] [RR]

space telemetry: The use of telemetry for the transmission from a space station of results of measurements made in a spacecraft, including those relating to the functioning of spacecraft. [NTIA] [RR]

space tracking: Determination of the orbit, velocity or instantaneous position of an object in space by means of radiodetermination, excluding primary radar, for the purpose of following the movement of the object. [NTIA] [RR]

spacing bias: The uniform lengthening of all spacing signal pulses at the expense of the pulse width of all marking signal pulses. (188)

spacing end distortion: *See* **end distortion.**

spacing pulse: *Synonym* **space.**

spacing signal: *Synonym* **space.**

spare: An individual part, subassembly, or assembly supplied for the maintenance or repair of systems or equipment.

spatial application: An application requiring high spatial resolution, possibly at the expense of reduced temporal positioning accuracy, *i.e.*, increased jerkiness. *Note:* Examples of spatial applications include the requirement to display small characters and to resolve fine detail in still video, or in motion video that contains very limited motion.

spatial coherence: *See* **coherent.**

spatial edge noise: In a video display, that form of edge busyness that is characterized by spatially varying distortion that occurs in close proximity to the edges of objects.

spatially coherent radiation: *See* **coherent.**

special grade access line: In the Defense Switched Network, an access line specially conditioned, usually by providing amplitude and delay equalization, to give it characteristics suitable for handling special services, such as reducing data signaling rates (DSR) to a rate between 600 b/s and 2400 b/s. (188)

special grade of service: In the Defense Switched Network, a network-provided service in which specially conditioned interswitch trunks and access lines are used to provide secure voice, data, and facsimile transmission. (188)

special interest group: *Synonym* **community of interest.**

specialized common carrier (SCC): A common carrier offering a limited type of service or serving a limited market.

special purpose computer: A computer that is designed to operate on a restricted class of problems.

special service: A radiocommunication service, not otherwise defined in this Section [of the *Radio Regulations*], carried on exclusively for specific needs of general utility, and not open to public correspondence. [RR with editor's note in brackets]

specification: **1.** An essential technical requirement for items, materials, or services, including the procedures to be used to determine whether the requirement has been met. (188) *Note:* Specifications may also include requirements for preservation, packaging, packing, and marking. **2.** An official document intended primarily for supporting procurement, which document clearly and accurately describes the essential technical requirements for items, materials, or services, including the procedures by which it will be determined that the requirements have been met. (188) *Note:* An example of a Federal specification is FIPS-PUB 159, *Detail Specification for 62.5-µm Core Diameter/125-µm Cladding Diameter Class Ia Multimode Optical Fibers.*

specific detectivity: For a photodetector, a figure of merit used to characterize performance, equal to the reciprocal of noise equivalent power (*NEP*), normalized to unit area and unit bandwidth. *Note:* Specific detectivity, D^*, is given by

$$D^* = \frac{\sqrt{A \Delta f}}{NEP},$$

where A is the area of the photosensitive region of the detector and Δf is the effective noise bandwidth. *Synonym* **D-Star.**

speckle noise: *Synonym* **modal noise.**

speckle pattern: In optical systems, a field-intensity pattern produced by the mutual interference of partially coherent beams that are subject to minute temporal and spatial fluctuations. (188) *Note:* In a multimode fiber, a speckle pattern results from a superposition of mode field patterns. If the relative modal group velocities change with time, the speckle pattern will also change with time. If differential mode attenuation occurs, modal noise results.

spectral bandwidth: *See* **spectral width.**

spectral density: For a specified bandwidth of radiation consisting of a continuous frequency spectrum, the total power in the specified bandwidth divided by the specified bandwidth. *Note:* Spectral density is usually expressed in watts per hertz.

spectral irradiance: Irradiance per unit wavelength interval at a given wavelength, usually expressed in watts per unit area per unit wavelength interval. (188)

spectral line: A narrow range of emitted or absorbed wavelengths.

spectral loss curve: Of an optical fiber, a plot of attenuation as a function of wavelength. (188) *Note:* Spectral loss curves must be normalized with respect to distance before meaningful comparison among fibers can be made.

spectral purity: The degree to which a signal is monochromatic.

spectral radiance: Radiance per unit wavelength interval at a given wavelength, expressed in watts per steradian per unit area per wavelength interval. (188)

spectral responsivity: The ratio of an optical detector's electrical output to its optical input, as a function of optical wavelength.

spectral width: The wavelength interval over which the magnitude of all spectral components is equal to or greater than a specified fraction of the magnitude of the component having the maximum value. (188) *Note 1:* In optical communications applications, the usual method of specifying spectral width is the full width at half maximum. This method may be difficult to apply when the spectrum has a complex shape. Another method of specifying spectral width is a special case of root-mean-square deviation where the independent variable is wavelength, λ, and $f(\lambda)$ is a suitable radiometric quantity. *Note 2:* The *relative spectral width*, $\Delta\lambda/\lambda$, is frequently used where $\Delta\lambda$ is obtained according to note 1, and λ is the center wavelength.

spectral window: *See* **window.**

spectrum: *See* **electromagnetic spectrum, optical spectrum.**

spectrum designation of frequency: *See* **electromagnetic spectrum.**

spectrum signature: The pattern of radio signal frequencies, amplitudes, and phases, which pattern characterizes the output of a particular device and tends to distinguish it from other devices. (188)

specular reflection: Reflection from a smooth surface, such as a mirror, which maintains the integrity of the incident wavefront.

speech digit signaling: *Synonym* **bit robbing.**

speech-plus: Pertaining to a circuit that was designed and used for speech transmission, but to which other uses, such as digital data transmission, facsimile transmission, telegraph, or signaling superimposed on the speech signals, have been added by means of multiplexing. [From Weik '89]

speech-plus-duplex operation: Operation in which speech and telegraphy (duplex or simplex) are transmitted simultaneously over the same circuit, and mutual interference is eliminated by the use of filters. (188)

speech-plus-signaling: Pertaining to equipment that permits the use of part of a voice-frequency band for signaling. (188)

speech power: *See* **volume unit.**

speech synthesizer: A device that is capable of accepting digital or analog data and developing intelligible speech sounds that correspond to the input data, without resorting to recorded sounds or without simply being a speech scrambler operating in reverse. [From Weik '89]

speed calling: A service feature that enables a switch or station to store certain telephone numbers and dial them automatically when a short (1-, 2-, or 3-digit) code is entered. (188) *Contrast with* **repertory dialer, speed dialing.**

speed dialing: 1. *Synonym* **abbreviated dialing. 2.** Dialing at a speed greater than the normal ten pulses per second. (188)

speed of light (*c*): The speed of an electromagnetic wave in free space, precisely 299,792,458 m/s. *Note 1:* The preceding figure is precise because by international agreement the meter is now defined in terms of the speed of light. *Note 2:* The speed of an electromagnetic wave, *e.g.*, light, is equal to the product of the wavelength and the frequency. *Note 3:* In any physical medium, the speed of light is lower than in free space. Since the frequency is not changed, the wavelength is also decreased. [After FAA]

speed of service: 1. The time between release of a message by the originator to receipt of the message by the addressee, as perceived by the end user. (188) *Synonym* **originator-to-recipient speed of service. 2.** The time between entry of a message into a communications system and receipt of the message at the terminating communications facility, *i.e.*, the communications facility serving the addressee, as measured by the system. (188)

speed-up tone: *Synonym* **camp-on busy signal.**

spike: An extremely short pulse of relatively high amplitude.

spike file: *See* **last-in first-out.**

spill forward: In automatic switching, the transfer of full control on a call to the succeeding office by sending forward the complete telephone address of the called party. (188)

spill-forward feature: A service feature, in the operation of an intermediate office, that, acting on incoming trunk service treatment indications, assumes routing control of the call from the originating office. (188) *Note:* This increases the chances of completion by offering the call to more trunk groups than are available in the originating office.

spillover: In an antenna, the part of the radiated energy from the feed that does not impinge on the reflectors.

spiral-four cable: A quadded cable with four conductors. (188) *Synonym* **star quadded cable.**

splice: **1.** To join, permanently, physical media that conduct or transmit power or a communication signal. **2.** A device that so joins conducting or transmitting media. **3.** The completed joint.

splice closure: A usually weatherproof encasement, commonly made of tough plastic, that envelops the exposed area between spliced cables, *i.e.,* where the jackets have been removed to expose the individual transmission media, optical or metallic, to be joined. *Note 1:* The closure usually contains some device or means to maintain continuity of the tensile strength members of the cables involved, and also may maintain electrical continuity of metallic armor, and/or provide external connectivity to such armor for electrical grounding. *Note 2:* In the case of fiber optic cables, it also contains a splice organizer to facilitate the splicing process and protect the exposed fibers from mechanical damage. *Note 3:* In addition to the seals at its seams and points of cable entry, the splice closure may be filled with an encapsulant to further retard the entry of water. [After FAA] *Synonym* **closure.**

splice loss: In fiber optic systems, any loss of optical power at a splice. *Note:* A practical splice, of physically realizable fibers, has losses attributable to a number of mechanisms, some of which are intrinsic to the fibers, and some of which are intrinsic to the method or device being used to join them. [After FAA]

splice organizer: In optical communication, a device that facilitates the splicing or breaking out of fiber optic cables. *Note:* The organizer provides means to separate and secure individual buffer tubes, fibers, and/or pigtails. It also provides means to secure mechanical splices or protective sleeves used in connection with fusion splices, and has means to contain the slack fiber that remains after the splicing process is completed. [After FAA]

split homing: The connection of a terminal facility to more than one switching center by separate access lines, each of which has a separate directory number. (188)

split screen: On a display device, display space that has been divided into two or more areas, so that each area can display different portions of the same file or portions of different files. *Note 1:* The split screen excludes the data lying between the portions of the file or files being displayed and includes the desired data in the two or more windows afforded by the split screen. *Note 2:* Examples of split screens are screens in which different portions of a spreadsheet, database, graph, or picture that are too far apart in storage to be viewed or displayed simultaneously as a single image, are viewed adjacently on a single screen. [From Weik '89]

splitter: *See* **directional coupler.**

(S+N)/N: *Abbreviation for* **signal-plus-noise-to-noise ratio.**

spontaneous emission: Radiation emitted when the internal energy of a quantum mechanical system drops from an excited level to a lower level without regard to the simultaneous presence of similar radiation. *Note:* Examples of spontaneous emission include radiation from an LED, and radiation from an injection laser below the lasing threshold.

spoofing: **1.** (COMSEC) [The] interception, alteration, and retransmission of a cipher signal or data in such a way as to mislead the recipient. [NIS] **2.** (AIS)

[An] attempt to gain access to an AIS by posing as an authorized user. [NIS]

spooling: The use of auxiliary storage as buffer storage to reduce processing delays when transferring data between peripheral equipment and the processors of a computer. *Note:* The term is derived from the expression *"simultaneous peripheral operation on line."*

sporadic E: Irregular scattered patches of relatively dense ionization that develop seasonally within the E region and that reflect and scatter frequencies up to 150 MHz. *Note 1:* The sporadic E is a regular day-time occurrence over the equatorial regions and is common in the temperate latitudes in late spring, early summer and, to a lesser degree, in early winter. *Note 2:* At high, *i.e.*, polar, latitudes, sporadic E can accompany auroras and associated disturbed magnetic conditions. *Note 3:* The sporadic E can sometimes support reflections for distances up to 2,400 km at frequencies up to 150 MHz. *Synonym* **sporadic E propagation.**

sporadic E propagation: *Synonym* **sporadic E.**

spot beam: In satellite communications systems, a narrow beam from a satellite station antenna that illuminates, with high irradiance, a limited area of the Earth by using beam (directive) antennas rather than Earth-coverage antennas.

spot jamming: The jamming of a specific channel or frequency. [JP1]

spot projection: In facsimile systems, optical scanning in which a scanning spot is moved across the object and the scanning spot size is determined by the illuminated area of the spot. (188)

spot size: **1.** The size of the electron spot on the face of a cathode ray tube. *Note:* The spot size is larger than the diameter of the electron beam because of the spillover of electrons into adjacent areas of the screen near the spot. The spot size is a function of the ability of the tube to focus the electron beam, as well as of the electron gun aperture. [From Weik '89] **2.** In facsimile systems, the diameter of the scanning spot or the recording spot. [From Weik '89] **3.** In single-mode optical fibers, the effective core diameter.

spot speed: In facsimile systems, the speed of the scanning or recording spot along the available line. (188) *Note:* The spot speed is usually measured on the object or on the recorded copy.

spread spectrum: **1.** Telecommunications techniques in which a signal is transmitted in a bandwidth considerably greater than the frequency content of the original information. (188) *Note:* Frequency hopping, direct sequence spreading, time scrambling, and combinations of these techniques are forms of spread spectrum. [NIS] **2.** A signal structuring technique that employs direct sequence, frequency hopping or a hybrid of these, which can be used for multiple access and/or multiple functions. This technique decreases the potential interference to other receivers while achieving privacy and increasing the immunity of spread spectrum receivers to noise and interference. Spread spectrum generally makes use of a sequential noise-like signal structure to spread the normally narrowband information signal over a relatively wide band of frequencies. The receiver correlates the signals to retrieve the original information signal. [NTIA] (188)

spur: A secondary route having a junction to the primary route in a network.

spurious emission: Emission on a frequency or frequencies which are outside the necessary bandwidth and the level of which may be reduced without affecting the corresponding transmission of information. Spurious emissions include harmonic emissions, parasitic emissions, intermodulation products and frequency conversion products, but exclude out-of-band emissions. [NTIA] [RR] (188)

spurious radiation: Any unintentional emission. (188)

spurious response: In radio reception, a response in the receiver intermediate frequency (IF) stage produced by an undesired emission in which the fundamental frequency (or harmonics above the fundamental frequency) of the undesired emission mixes with the fundamental or harmonic of the receiver local oscillator. (188)

square wave: A wave that has two significant conditions, *i.e.*, two levels of amplitude, that change from one condition to the other in a relatively short time compared to the wavelength. *Note:* When the instantaneous amplitude is plotted versus time or distance, the waveform has a rectangular shape. [From Weik '89]

squelch: A circuit function that acts to suppress the audio output of a receiver. [NTIA] (188) *Note:* The squelch function is activated in the absence of a sufficiently strong desired input signal, in order to exclude undesired lower-power input signals that may be present at or near the frequency of the desired signal. *Contrast with* **noise suppression.**

sr: *Abbreviation for* **steradian.**

SSB: *Abbreviation for* **single sideband.** *See* **single-sideband emission.**

SSB-SC: *Abbreviation for* **single-sideband suppressed carrier.** *See* **single-sideband suppressed carrier transmission.**

SS7: *Abbreviation for* **Signaling System No. 7.**

stability: The invariability of a specified property of a substance, device, or apparatus with time, or under the influence of typically extrinsic factors.

stagger: In facsimile systems, periodic error in the position of the recorded spot along the recorded line. (188)

standard: **1.** Guideline documentation that reflects agreements on products, practices, or operations by nationally or internationally recognized industrial, professional, trade associations or governmental bodies. *Note:* This concept applies to formal, approved standards, as contrasted to de facto standards and proprietary standards, which are exceptions to this concept. **2.** An exact value, a physical entity, or an abstract concept, established and defined by authority, custom, or common consent to serve as a reference, model, or rule in measuring quantities or qualities, establishing practices or procedures, or evaluating results. A fixed quantity or quality. [JP1]

standard frequency and time signal-satellite service: A radiocommunication service using space stations on Earth satellites for the same purpose as those of the standard frequency and time signal service. This service may also include feeder links necessary for its operation. [NTIA] [RR]

standard frequency and time signal service: A radiocommunication service for scientific, technical and other purposes, providing the transmission of specified frequencies, time signals, or both, of stated high precision, intended for general reception. [NTIA] [RR]

standard frequency and time signal station: A station in the standard frequency and time signal service. [NTIA] [RR]

Standard Generalized Mark-up Language: *See* **SGML.**

standardized profile: A profile that specifies one or more interoperable open systems interconnection stacks that are intended to cover one or more specific functional areas. (188) *Note:* Examples of standardized profiles are the ISO standardized profiles and the NATO standardized profiles.

standard optical source: A reference optical source to which emitting and detecting devices are compared for calibration purposes. (188) *Note:* In the United States, recognized standard optical sources must be traceable to the National Institute of Standards and Technology (NIST), formerly the National Bureau of Standards (NBS).

standard telegraph level (*STL*): The power per individual telegraph channel required to yield the standard composite data level. *Note:* For example, for a composite data level of –13 dBm at 0-dBm transmission level point (0TLP), the *STL* would be approximately –25 dBm for a 16-channel VFCT terminal computed from $STL = -(13+10\log_{10} n)$, where n is the number of telegraph channels and the *STL* is in dBm. (188)

standard test signal: A single-frequency signal with standardized level used for testing the peak power transmission capability and for measuring the total harmonic distortion of circuits or parts of a circuit. (188) *Note:* Standardized test signal levels and frequencies are listed in MIL-STD-188-100 and in the *Code of Federal Regulations,* Title 47, part 68.

standard test tone: A single-frequency signal with a standardized level generally used for level alignment of single links and of links in tandem. (188) *Note:* For standardized test signal levels and frequencies, see MIL-STD-188-100 for DOD use, and the *Code of Federal Regulations*, Title 47, part 68 for other Government agencies.

standard time and frequency signal (STFS) service: In the United States, standard time and frequency signals, broadcast on very precise carrier frequencies by the U.S. Naval Observatory and the National Institute of Standards and Technology (NIST), formerly the National Bureau of Standards (NBS). *Note:* The *Radio Regulations* (RR) define an identical international service as **standard frequency and time signal service.**

standby: **1.** In computer and communications systems operations, pertaining to a power-saving condition or status of operation of equipment that is ready for use but not in use. *Note:* An example of a standby condition is a radio station operating condition in which the operator can receive but is not transmitting. **2.** Pertaining to a dormant operating condition or state of a system or equipment that permits complete resumption of operation in a stable state within a short time. **3.** Pertaining to spare equipment that is placed in operation only when other, in-use equipment becomes inoperative. *Note:* Standby equipment is usually classified as (a) *hot* standby equipment, which is warmed up, *i.e.*, powered and ready for immediate service, and which may be switched into service automatically upon detection of a failure in the regular equipment, or (b) *cold* standby equipment, which is turned off or not connected to a primary power source, and which must be placed into service manually.

standing wave: In a transmission line, a wave in which the distribution of current, voltage, or field strength is formed by the superposition of two waves propagating in opposite directions, and which wave is characterized by a series of nodes (maxima) and anti-nodes (minima) at fixed points along the transmission line. *Note:* A standing wave may be formed when a wave is transmitted into one end of a transmission line and is reflected from the other end by an impedance mismatch, *i.e.*, discontinuity, such as an open or a short. *Synonym* **stationary wave.**

standing wave ratio (*SWR*): The ratio of the amplitude of a standing wave at an anti-node (minimum) to the amplitude at an adjacent node (maximum). (188) *Note 1:* The standing wave ratio (*SWR*) in a uniform transmission line is given by

$$SWR = \frac{1 + \rho}{1 - \rho} ,$$

where ρ is the reflection coefficient. *Note 2:* Reflections occur as a result of discontinuities, such as an imperfection in an otherwise uniform transmission line, or when a transmission line is terminated with other than its characteristic impedance.

star coupler: A passive optical coupler having a number of input and output ports, used in network applications. *Note:* An optical signal introduced into any input port is distributed to all output ports. Because of the nature of the construction of a passive star coupler, the number of ports is usually a power of 2; *i.e.*, two input ports and two output ports (a "two-port" coupler, customarily called a *"directional coupler,"* or *"splitter"*); four input ports and four output ports (a "four-port" coupler); eight input ports and eight output ports (an "eight-port" coupler); *etc.* [FAA]

star network: *See* **network topology.**

star quadded cable: *Synonym* **spiral-four cable.**

starting frame delimiter: A specified bit pattern that indicates the start of a transmission frame.

start message: *Synonym* **go-ahead notice.**

start notice: *Synonym* **go-ahead notice.**

start-of-heading character (SOH): A transmission control character used as the first character of a message heading.

start-of-text character (STX): A transmission control character that precedes a text and may be used to terminate the message heading.

star topology: *See* **network topology.**

start pulse: *See* **A-condition, start signal.**

start-record signal: In facsimile systems, a signal used for starting the process of converting the electrical signal to an image on the record medium. (188)

start signal: **1.** A signal that prepares a device to receive data or to perform a function. *Contrast with* **A-condition. 2.** In start-stop transmission, a signal at the beginning of a character that prepares the receiving device for the reception of the code elements. *Note:* A start signal is limited to one signal element usually having the duration of a unit interval. (188)

start-stop character: A character that includes one start signal at the beginning and one or two stop signals at the end.

start-stop distortion: In start-stop modulation, the ratio of (a) the maximum absolute difference between the actual and the theoretical intervals that separate any significant instant of modulation or demodulation from the significant instant of the start signal element immediately preceding it to (b) the unit interval. (188)

start-stop margin: In start-stop modulation, the maximum amount of overall start-stop distortion that is compatible with correct translation by the start-stop equipment of all the character signals that appear singly, that appear at the maximum allowable speed, or that appear at the standard modulation rate. (188)

start-stop modulation: A method of modulation in which the time of occurrence of the bits within each character, or block of characters, relates to a fixed time frame, but the start of each character, or block of characters, is not related to this fixed time frame. (188)

start-stop system: *Synonym* **asynchronous communications system.**

start-stop transmission: **1.** Asynchronous transmission in which a start pulse and a stop pulse are used for each symbol. (188) **2.** Signaling in which each group of code elements corresponding to an alphanumeric character is (a) preceded by a start signal that serves to prepare the receiving mechanism for the reception and registration of a character and (b) followed by a stop signal that serves to bring the receiving mechanism to rest in preparation for the reception of the next character. (188)

start-stop TTY distortion: *Synonym* **teletypewriter signal distortion.**

statement: **1.** In programming languages, a language construct that represents a set of declarations or a step in a sequence of actions. **2.** In computer programming, a symbol string or other arrangement of symbols. **3.** In computer programming, a meaningful expression or generalized instruction, represented in a source language.

staticizer: *See* **serial-to-parallel conversion.**

station: One or more transmitters or receivers or a combination of transmitters and receivers, including the accessory equipment, necessary at one location for carrying on a radiocommunication service, or the radio astronomy service. Each station shall be classified by the service in which it operates permanently or temporarily. [NTIA] [RR]

stationary satellite: *See* **geostationary orbit.**

stationary wave: *Synonym* **standing wave.**

station battery: Within a facility, a separate battery power source that satisfies all significant requirements for dc input power associated with the facility. (188) *Note:* Station batteries are usually centrally located. The batteries may power radio and telephone equipment as well as provide emergency lighting and controls for the equipment.

station clock: In a station, the principal clock, or alternate clock, that provides the timing reference at the station. (188)

station equipment: *See* **customer premises equipment.**

station load: The total power requirements of the integrated station facilities. (188)

station message-detail recording (SMDR): A record of all calls originated or received by a switching system. *Note:* SMDRs are usually generated by a computer.

statistical multiplexing: Multiplexing in which channels are established on a statistical basis; *i.e.*, connections are made according to probability of need.

statistical time-division multiplexing: Time-division multiplexing in which connections to communication circuits are made on a statistical basis.

statute mile: A unit of distance equal to 1.609 km (0.869 nmi, 5280 ft.). (188)

STDM: *Abbreviation for* **statistical time-division multiplexing.**

steady-state condition: **1.** In a communications circuit, a condition in which some specified characteristic of a condition, such as a value, rate, periodicity, or amplitude, exhibits only negligible change over an arbitrarily long period. **2.** In an electrical circuit, the condition that exists after all initial transients or fluctuating conditions have damped out, and all currents, voltages, or fields remain essentially constant, or oscillate uniformly. (188) **3.** In fiber optics, *synonym for* **equilibrium mode distribution.**

step-by-step (SXS) switching system: An automatic dial telephone system in which calls are switched by a succession of switches that move a step at a time, from stage to stage, each step being made in response to the dialing of a number. (188)

step-index fiber: An optical fiber with a core having a uniform refractive index. (188)

step-index profile: For an optical fiber, a refractive index profile characterized by a uniform refractive index within the core and a sharp decrease in refractive index at the core-cladding interface. (188) *Note 1:* The step-index profile corresponds to a power-law index profile with the profile parameter approaching infinity. *Note 2:* The step-index profile is used in most single-mode fibers and some multimode fibers.

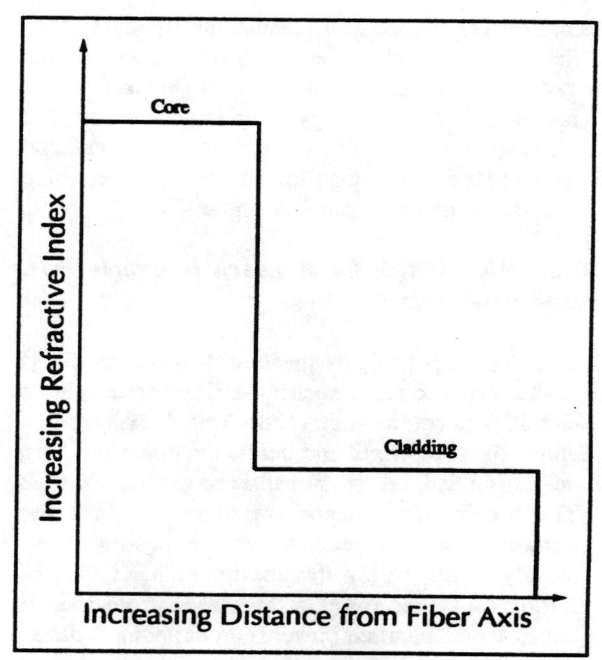

step-index profile

steradian (sr): The metric unit of solid angle. *See* **metric system.**

stereophonic crosstalk: An undesired signal occurring in the main channel from modulation of the stereophonic channel or that occurring in the stereophonic channel from modulation of the main channel. [47CFR]

stereophonic sound subcarrier: A subcarrier within the FM broadcast baseband used for transmitting signals for stereophonic sound reception of the main broadcast program service. [47CFR]

STFS: *Abbreviation for* **standard time and frequency signal.** *See* **standard time and frequency signal service.**

still image: Nonmoving visual information, *i.e.*, fixed images, such as graphs, drawings, and pictures. (188)

still video: Video imagery that is not intended to convey the appearance of movement. *Contrast with* **freeze frame, freeze frame television.**

stimulated emission: In a quantum mechanical system, the radiation emitted when the internal energy of the system drops from an excited level (induced by the presence of radiant energy at the same frequency) to a lower level. *Note:* An example of stimulated emission is the radiation from an injection laser diode operated above the lasing threshold.

STL: *Abbreviation for* **standard telegraph level, studio-to-transmitter link.**

stopband: A band of frequencies, between specified limits, that a circuit, such as a filter or telephone circuit, does not transmit. (188) *Note 1:* Frequencies above the lower limit and below the upper limit are not transmitted, *i.e.*, are not allowed to pass. *Note 2:* The limiting frequencies are those at which the transmitted power level increases to a specified level, usually 3 dB below the maximum level, as the frequency is decreased or increased from that at which the transmitted power is a minimum. *Note 3:* The difference between the limits is the stopband bandwidth, usually expressed in hertz.

stop element: *See* **stop signal.**

stop-record signal: In facsimile systems, a signal used for stopping the process of converting the electrical signal to an image on the record medium. (188)

stop signal: **1.** In start-stop transmission, a signal at the end of a character that prepares the receiving device for the reception of a subsequent character. A stop signal is usually limited to one signal element having any duration equal to or greater than a specified minimum value. (188) **2.** A signal to a receiving mechanism to wait for the next signal.

storage: **1.** The retention of data in any form, usually for the purpose of orderly retrieval and documentation. [JP1] **2.** A device consisting of electronic, electrostatic, electrical, hardware or other elements into which data may be entered, and from which data may be obtained, as desired. [JP1]

storage cell: **1.** An addressable storage unit. **2.** The smallest subdivision of storage into which a unit of data can be entered, stored, and retrieved. *Synonym* **storage element.**

storage element: *Synonym* **storage cell.**

storage register: *See* **register.**

store-and-forward (S-F): Pertaining to communications systems in which messages are received at intermediate routing points and recorded *i.e.*, stored, and then transmitted, *i.e.*, forwarded, to the next routing point or to the ultimate recipient. (188)

store-and-forward switching center: A message switching center in which a message is accepted from the originating user, *i.e.*, sender, when it is offered, held in a physical storage, and forwarded to the destination user, *i.e.*, receiver, in accordance with the priority placed upon the message by the originating user and the availability of an outgoing channel. (188)

stored-program computer: A computer that (a) is controlled by internally stored instructions, (b) can synthesize and store instructions, and (c) can subsequently execute those instructions.

STP: *Abbreviation for* **signal transfer point.**

strap: *See* **cross-connection.**

stray current: Electrical current through a path other than the intended path. (188)

streamer: *Synonym* **streaming tape drive.**

streaming tape drive: A magnetic tape unit capable of recording from, and dumping to, another storage medium without stopping at interblock gaps. *Note:* Streaming tape drives are often used for bulk transfer of data between tape and disk storage. *Synonym* **streamer.**

streaming tape recording: A method of recording on magnetic tape, which method maintains continuous tape motion without the requirement to start and stop within the interrecord gap.

strength member: Any component of a communication cable, metallic or optical, the function of which is to protect the transport medium, *i.e.*, conductor or fiber, from excessive tensile and bending stresses during installation and while in service. [After FAA]

stressed environment: In radiocommunications, an environment that is under the influence of extrinsic factors that degrade communications integrity, such as when (a) the benign communications medium is disturbed by natural or man-made events (such as an intentional nuclear burst), (b) the received signal is degraded by natural or man-made interference (such as jamming signals or co-channel interference), (c) an interfering signal can reconfigure the network, and/or (d) an adversary threatens successful communications, in which case radio signals may be encrypted in order to deny the adversary an intelligible message, traffic flow information, network information, or automatic link establishment (ALE) control information.

string: A sequence of data elements, such as bits or characters, considered as a whole.

stroke: A straight line or arc that is used as a segment of a graphic character.

stroke edge: In character recognition, the line of discontinuity between a side of a stroke and the background, obtained by averaging, over the length of the stroke, the irregularities resulting from the printing and detecting processes.

stroke speed: In facsimile systems, the rate at which a fixed line perpendicular to the direction of scanning is crossed in one direction by a scanning or recording spot. (188) *Note 1:* Stroke speed is usually expressed as a number of strokes per minute. When the system scans in both directions, the stroke speed is twice this number. *Note 2:* In most conventional mechanical systems, the stroke speed is equivalent to drum speed.

stroke width: In character recognition, the distance between the two edges of a stroke, measured perpendicular to the stroke centerline.

structured programming: A technique for organizing and coding computer programs in which a hierarchy of modules is used, each having a single entry and a single exit point, and in which control is passed downward through the structure without unconditional branches to higher levels of the structure. Three types of control flow are used: sequential, test, and iteration.

STU: *Acronym for* **secure telephone unit**. A U.S. Government-approved telecommunications terminal that protects the transmission of sensitive or classified information in voice, data, and facsimile systems.

studio-to-transmitter link (STL): A communications link used for the transmission of broadcast material from a studio to the transmitter. *Note:* The STL may be a microwave, radio, or landline link.

stuffing: *See* **bit stuffing, de-stuffing.**

stunt box: A device that controls the nonprinting functions of a printer at a terminal.

STX: *Abbreviation for* **start-of-text character.**

SUB: *Acronym for* **substitute character.**

sub-band adaptive differential pulse code modulation (SB-ADPCM): Modulation in which (a) an audio frequency band is split into two sub-bands, *i.e.*, a higher and a lower band, and (b) the signals in each sub-band are encoded using ADPCM. (188)

subcarrier: A carrier used to modulate another carrier. *Note:* The modulated carrier can be used to modulate another carrier, and so on, so that there can be several levels of subcarriers, *i.e.*, several intermediate carriers. (188)

sublayer: **1.** In a layered open communications system, a specified subset of the services, functions, and protocols included in a given layer. **2.** In the Open Systems Interconnection—Reference Model, a subdivision of a given layer, *e.g.*, a conceptually complete group of the services, functions, and protocols included in the given layer.

subnet address: In an Internet Protocol (IP) address, an extension that allows users in a network to use a single IP network address for multiple physical subnetworks. *Note:* The IP address contains three parts: the network, the subnet, and host addresses. Inside the subnetwork, gateways and hosts divide the local portion of the IP address into a subnet address and a host address. Outside of the subnetwork, routing continues as usual by dividing the destination address into a network portion and a local portion.

subnetwork: A collection of equipment and physical transmission media that forms an autonomous whole and that can be used to interconnect systems for purposes of communication.

subordinate station: *Synonym* **slave station.**

subroutine: A set of computer instructions to carry out a predefined function or computation. *Note: "Open"* subroutines are integrated into the main program. *"Closed"* subroutines are arranged so that program control is shifted to them for execution of their task(s) and then returned to the main program.

subscriber: In a public switched telecommunications network, the ultimate user, *i.e.,* customer, of a communications service. *Note 1:* Subscribers include individuals, activities, organizations, etc. *Note 2:* Subscribers use end instruments, such as telephones, modems, facsimile machines, computers, and remote terminals, that are connected to a central office. *Note 3:* Subscribers are usually subject to tariff. *Note 4:* Subscribers do not include communications systems operating personnel except for their personal terminals.

subscriber line: *Synonym* **loop (def. #1).**

substitute character (SUB): A control character that is used in the place of a character that is recognized to be invalid or in error or that cannot be represented on a given device.

substitution method: In optical fiber technology, a method of measuring the transmission loss by (a) using a stable optical source, at the wavelength of interest, to drive a mode scrambler, the output of which overfills (drives) a 1-meter to 2-meter reference fiber having physical and optical characteristics matching those of the fiber under test, (b) measuring the power level at the output of the reference fiber, (c) repeating the procedure, substituting the fiber under test for the reference fiber, and (d) subtracting the power level obtained at the output of the fiber under test from the power level obtained at the output of the reference fiber, to get the transmission loss of the fiber under test. *Note 1:* The substitution method has certain shortcomings with regard to its accuracy, but its simplicity makes it a popular field test method. It is conservative, in that if it were used to measure the individual losses of several long fibers, and the long fibers were concatenated, the total loss obtained (excluding splice losses) would be expected to be lower than the sum of the individual fiber losses. *Note 2:* Some modern optical power meters have the capability to set to zero the reference level measured at the output of the reference fiber, so that the transmission loss of the fiber under test may be read out directly.

subvoice-grade channel: A channel with a bandwidth narrower than that of a voice-grade channel. *Note:* A subvoice-grade channel is usually a subchannel of a voice-grade line.

successful block delivery: The transfer of a nonduplicate user information block between the source user and intended destination user. *Note:* Successful block delivery includes the delivery of correct and incorrect blocks. *Contrast with* **successful block transfer.**

successful block transfer: The transfer of a correct, nonduplicate, user information block between the source user and intended destination user. *Note:* Successful block transfer occurs when the last bit of the transferred block crosses the functional interface between the telecommunications system and the intended destination user. Successful block transfer can only occur within a defined maximum block transfer time after initiation of a block transfer attempt. *Contrast with* **successful block delivery.**

successful disengagement: The termination of user information transfer between a source user and a destination user in response to a disengagement request. *Note:* Successful disengagement occurs at the earliest moment at which either user is able to initiate a new information transfer transaction.

sudden ionospheric disturbance (SID): An abnormally high ionization density in the D region caused by an occasional sudden solar flare, *i.e.,* outburst of ultraviolet light from the Sun. *Note:* The SID results in a sudden increase in radio-wave absorption that is most severe in the upper medium-frequency (MF) and lower high-frequency (HF) ranges. (188)

sum check: *Synonym* **summation check.**

summation check: **1.** A check based on the formation of the sum of the digits of a numeral. *Note:* The sum of the individual digits is usually compared with a previously computed value. **2.** A comparison of checksums on the same data on different occasions or on different representations of the data in order to verify data integrity. *Synonym* **sum check.**

sunspot: In the photosphere, *i.e.*, visible disk of the Sun, a dark marking that manifests a magnetic anomaly that is associated with interference with radio communications on Earth. *Note:* Sunspot activity, *i.e.*, the number of sunspots occurring at a given time or on a given day, is cyclic. The period of a cycle, from maximum through minimum and back to maximum sunspot count, is approximately 11 years.

superencryption: [The] process of encrypting encrypted information. *Note:* [This process] occurs when a message, encrypted off-line, is transmitted over a secured, on-line circuit, or when information encrypted by the originator is multiplexed into a communications trunk, which is then bulk encrypted. [NIS]

supergroup: *See* **group, multiplex hierarchy.**

supergroup distribution frame (SGDF): In frequency-division multiplexing (FDM), the distribution frame that provides terminating and interconnecting facilities for group modulator output, group demodulator input, supergroup modulator input, and supergroup demodulator output circuits of the basic supergroup spectrum of 312 kHz to 552 kHz. (188)

super high frequency (SHF): *See* **electromagnetic spectrum.**

superluminescent LED: A light-emitting diode in which there is stimulated emission with amplification but insufficient feedback for oscillations to build up to achieve lasing action.

superradiance: In a gain medium, amplification of spontaneously emitted radiation characterized by moderate spectral line narrowing and moderate directionality. *Note:* Superradiance is usually distinguished from lasing action by the absence of positive feedback, and hence the absence of well-defined modes of oscillation.

supervisor: *Synonym* **supervisory program.**

supervisory control: The use of characters or signals for the automatic actuation of equipment or indicators.

supervisory program: **1.** A program, usually part of an operating system, that controls the execution of other routines and regulates work scheduling, input-output operations, error actions, and similar functions. (188) **2.** A program that allocates computer component space and schedules computer events by task queuing and system interrupts. *Note:* Control of the system is returned to the supervisory program frequently enough to ensure that demands on the system are met. *Synonym* **supervisory routine. 3.** A computer program, usually part of an operating system, that controls the execution of other computer programs and regulates the flow of work in a data processing system. *Synonyms* **executive program, supervisor.**

supervisory routine: *Synonym* **supervisory program.**

supervisory signals: Signals used to indicate, or to indicate and control, the various operating states of the circuits or circuit combinations involved in a particular connection. (188)

suppressed carrier single-sideband emission: A single-sideband emission in which the carrier is virtually suppressed and not intended to be used for demodulation. [NTIA] [RR]

suppressed carrier transmission: Amplitude modulation (AM) transmission in which the carrier level is reduced below that required for demodulation. *Note 1:* Reduction of the carrier level permits higher power levels in the sidebands than would be possible with conventional AM transmission. *Note 2:* Carrier power must be restored by the receiving station to permit demodulation. *Note 3:* Suppressed carrier transmission is a special case of reduced carrier transmission.

surface refractivity: The refractive index of the Earth's atmosphere, calculated from observations of pressure, temperature, and humidity at the surface of the Earth. (188) *Note:* The surface refractivity gradient is the difference in refractive index between the surface and a given altitude, such as between the surface and 1000 m.

surface wave: A wave that is guided along the interface between two different media or by a refractive index gradient. (188) *Note 1:* The field components of the wave diminish with distance from the interface. *Note 2:* Optical energy is not converted from the surface wave field to another form of energy and the wave does not have a component directed normal to the interface surface. *Note 3:* In optical fiber transmission, evanescent waves are surface waves. *Note 4:* In radio transmission, ground waves are surface waves that propagate close to the surface of the Earth, the Earth having one refractive index and the atmosphere another, thus constituting an interface surface.

surge: *Synonym* **impulse.**

surge suppressor: *Synonym* **arrester.**

survey: *See* **path survey.**

survivability: A property of a system, subsystem, equipment, process, or procedure that provides a defined degree of assurance that the named entity will continue to function during and after a natural or man-made disturbance; *e.g.,* nuclear burst. (188) *Note:* For a given application, survivability must be qualified by specifying the range of conditions over which the entity will survive, the minimum acceptable level or post-disturbance functionality, and the maximum acceptable outage duration.

survivable operation: *See* **survivability.**

survival craft station: A mobile station in the maritime mobile service or the aeronautical mobile service intended solely for survival purposes and located on any lifeboat, life-raft or other survival equipment. [NTIA] [RR]

susceptibility: In electronic warfare, the degree to which electronic equipment is affected by electromagnetic energy radiated by an enemy's equipment, such as jamming transmitters. (188)

susceptibility threshold: The amount of undesired signal power required at the input terminals of a receiver to cause barely perceptible interference at the receiver output terminals.

susceptiveness: In telephone systems, the extent to which circuits pick up noise and low-frequency energy by induction from power systems. *Note:* Susceptiveness depends on telephone circuit balance, wire and connection transpositions, wire spacing, and isolation from ground. (188)

sweep acquisition: A technique whereby the frequency of the local oscillator is slowly swept past the reference in order to assure that the pull-in range is reached.

sweep jamming: Jamming in which (a) a narrow frequency band of jamming energy is repeatedly swept over a relatively wide frequency band, (b) the sweep rate is such as to be on any given frequency only long enough to accomplish its jamming task, returning to that frequency again before the expiration of the jammed circuit recovery time. *Note 1:* Sweep jamming combines the advantages of both spot- and barrage-jamming by rapid electronic sweeping of a narrow band of jamming signals over a broad frequency spectrum. *Note 2:* The disadvantage of sweep-jamming is its high susceptibility to electronic counter-countermeasures. [From Weik '89]

swim: Slow, graceful, undesired movements of display elements, groups, or images about their mean position on a display surface, such as that of a monitor. *Note 1:* Swim can be followed by the human eye, whereas jitter usually appears as a blur. *Note 2:* Jitter, swim, wander, and drift have increasing periods of variation in that order.

switch: **1.** In communications systems, a mechanical, electro-mechanical, or electronic device for making, breaking, or changing the connections in or among circuits. (188) **2.** *Deprecated synonym for* **central office, switching center.** **3.** In communications systems, to transfer a connection from one circuit to another. **4.** In a computer program, a conditional instruction and a flag that is interrogated by the instruction. **5.** In a computer program, a parameter that controls branching and that is bound, prior to the branch point being reached. *Synonym* **switchpoint.** **6.** In computer programming, a programming technique or statement for making a selection, such as a conditional jump. **7.** In computer software applications, a functional unit, such as a toggle button, used to make selections.

switchboard: Equipment used for manual switching operations. (188)

switch busy hour: In telephony, the busy hour for a single switch. (188)

switched circuit: In a communications network, a circuit that may be temporarily established at the request of one or more of the connected stations. (188)

switched loop: In telephony, a circuit that automatically releases a connection from a console or switchboard, once the connection has been made to the appropriate terminal. *Note:* Loop buttons or jacks are used to answer incoming listed directory number calls, dial "0" internal calls, transfer requests, and intercepted calls. The attendant can handle only one call at a time. *Synonym* **released loop.**

switched multimegabit data services (SMDS): A connectionless, broadband, packet-switched data service that provides LAN-like performance and features in metropolitan or wide areas. *Note:* Currently SMDS operates at 1.544 Mb/s (megabits per second) or 44.736 Mb/s. These are the T1 and T3 rates, respectively, over switched fiber optic networks.

switched network: **1.** A communications network, such as the public switched telephone network, in which any user may be connected to any other user through the use of message, circuit, or packet switching and control devices. **2.** Any network providing switched communications service. (188)

switching: The controlling or routing of signals in circuits to execute logical or arithmetic operations or to transmit data between specific points in a network. *Note:* Switching may be performed by electronic, optical, or electromechanical devices. [From Weik '89]

switching center: In communications systems, a facility in which switches are used to interconnect communications circuits on a circuit-, message-, or packet-switching basis. (188) *Synonyms, in telephony,* **central office, switching exchange, switching facility.** *Deprecated synonym* **switch.**

switching exchange: *Synonym* **switching center.**

switching facility: *Synonym* **switching center.**

switching system: **1.** A communications system consisting of switching centers and their interconnecting media. (188) **2.** Part of a communication system organized to temporarily associate functional units, transmission channels or telecommunication circuits for the purpose of providing a desired telecommunication facility. *Note:* Examples of NATO-owned switching system are IVSN and TARE. [NATO]

switchpoint: *Synonym* **switch (def. #5).**

SWR: *Abbreviation for* **standing wave ratio.**

SX: *Abbreviation for* **simplex signaling.**

SXS: *Abbreviation for* **step-by-step switching system.**

syllable: A character string or a bit string in a word.

symbolic language: A computer programming language used to express addresses and instructions with symbols convenient to humans rather than to machines.

symbolic logic: The discipline in which valid arguments and operations are dealt with using an artificial language designed to avoid the ambiguities and logical inadequacies of natural languages.

symmetrical channel: A channel in which the send and receive circuits have the same data signaling rate.

symmetrical pair: A balanced transmission line, in a multipair cable, having equal conductor resistances per unit length, equal impedances from each conductor to earth, and equal impedances to other lines. (188)

SYN: *Acronym for* **synchronous idle character.**

sync pulse: *Synonym* **synchronization pulse.**

synchronism: **1.** The state of being synchronous. **2.** For repetitive events with the same, multiple, or submultiple repetition rates, a relationship among the events such that a significant instant of one event bears a fixed time relationship to a corresponding instant in another event. *Note:* Synchronism is maintained when there is a fixed, *i.e.*, constant, phase relationship among the group of repetitive events. **3.** The simultaneous occurrence of two or more events at the same instant on the same coordinated time scale. (188)

synchronization: **1.** The attaining of synchronism. **2.** The obtaining of a desired fixed relationship among corresponding significant instants of two or more signals. (188) **3.** A state of simultaneous occurrences of significant instants among two or more signals.

synchronization bit: A bit used to achieve or maintain synchronism. (188) *Note:* The term *"synchronization bit"* is usually applied to digital data streams, whereas the term *"synchronization pulse"* is usually applied to analog signals.

synchronization code: In digital systems, a sequence of bits introduced into a transmitted signal to achieve or maintain synchronism.

synchronization pulse: A pulse used to achieve or maintain synchronism. *Note:* The term *"synchronization pulse"* is usually applied to analog signals, whereas the term *"synchronization bit"* is usually applied to digital data streams. *Synonym* **sync pulse.**

synchronizing: **1.** Achieving and maintaining synchronism. **2.** In facsimile, achieving and maintaining predetermined speed relations between the scanning spot and the recording spot within each scanning line. (188) *Note:* In the civilian community, the noun "*synchronization*" is preferred to "*synchronizing*."

synchronizing pilot: In FDM, a reference frequency used for achieving and maintaining syntonization of the oscillators of a carrier system or for comparing the frequencies or phases of the signals generated by those oscillators. (188)

synchronizing signal: In facsimile systems, the signal that maintains predetermined speed relations between the scanning spot and recording spot within each facsimile scanning line. (188)

synchronous: **1.** Pertaining to the relationship of two or more repetitive signals that have simultaneous occurrences of significant instants. (188) *Note:* "*Isochronous*" and "*anisochronous*" pertain to characteristics. "*Synchronous*" and "*asynchronous*" pertain to relationships. **2.** Pertaining to synchronism (def.#2).

synchronous crypto-operation: [A] method of on-line crypto-operation in which crypto-equipment and associated terminals have timing systems to keep them in step. [NIS]

synchronous data link control (SDLC): In a data network, a bit-oriented protocol for the control of synchronous transmission over data links.

synchronous data network: A data network in which synchronism is achieved and maintained between data circuit-terminating equipment (DCE) and the data switching exchange (DSE), and between DSEs. (188) *Note:* The data signaling rates are controlled by timing equipment within the network.

synchronous height: *See* **synchronous orbit.**

synchronous idle character (SYN): A transmission control character used in synchronous transmission systems to provide a signal from which synchronism or synchronous correction may be achieved between data terminal equipment, particularly when no other character is being transmitted.

synchronous network: A network in which clocks are controlled to run, ideally, at identical rates, or at the same mean rate with a fixed relative phase displacement, within a specified limited range. (188) *Note:* Ideally, the clocks are synchronous, but they may be mesochronous in practice. By common usage, such mesochronous networks are frequently described as *"synchronous."*

synchronous optical network: *See* **SONET**.

synchronous orbit: Any orbit in which an orbiting object has a period equal to the average rotational period of the body being orbited, and in the same direction of rotation as that body. *Note 1:* A synchronous orbit need not be equatorial, but it usually is, ideally. A body in a nonequatorial synchronous orbit will, when observed from a fixed point on the orbited body, appear to move up and down, *i.e.,* northward and southward. If the synchronous orbit is not perfectly circular, the orbiting body will appear to move back and forth, eastward and westward. The combination of these two motions will produce a figure-8 pattern as seen from the orbited body. *Note 2:* A synchronous orbit about the Earth that is circular and lies in the equatorial plane is called a geostationary orbit.

synchronous satellite: A satellite in a synchronous orbit. (188)

synchronous system: A system in which events, such as signals, occur in synchronism. *Note:* An example of a synchronous system is one in which a transmitter and receiver operate with a fixed time relationship. (188)

synchronous TDM: A multiplexing scheme in which timing is obtained from a clock that controls both the multiplexer and the channel source. (188)

synchronous transfer mode: In a Broadband Integrated Services Digital Network (B-ISDN), a proposed transport level technique in which time-division multiplexing and switching is to be used across the user's network interface.

synchronous transmission: Digital transmission in which the time interval between any two similar significant instants in the overall bit stream is always an integral number of unit intervals. (188) *Note:* "*Isochronous*" and "*anisochronous*" are characteristics, while "*synchronous*" and "*asynchronous*" pertain to relationships.

syntax: **1.** In a language, the relationships among characters or groups of characters, independent of their meanings or the manner of their interpretation and use. **2.** The structure of expressions in a language. **3.** The rules governing the structure of a language. **4.** In a language, the relationship among symbols. *Note:* In computer languages, as in all artificial languages, syntax is developed, and usually described, before their use begins. In natural languages, syntax is developed, and sometimes never described, after use has begun.

syntonization: The process of setting the frequency of one oscillator equal to that of another.

SYSGEN: *Acronym for* **system generation.**

system: **1.** Any organized assembly of resources and procedures united and regulated by interaction or interdependence to accomplish a set of specific functions. [JP1] **2.** A collection of personnel, equipment, and methods organized to accomplish a set of specific functions. (188)

system administration: In computer technology, a set of functions that provides support services, ensures reliable operations, promotes efficient use of the system, and ensures that prescribed service-quality objectives are met. *Synonym* **system management.**

system analysis: A systematic investigation of a real or planned system to determine the functions of the system and how they relate to each other and to any other system. *Synonym* **systems analysis.**

system blocking: *Synonym* **access denial.**

system blocking signal: A control message generated within a telecommunications system to indicate temporary unavailability of system resources required to complete a requested access. *Note:* The system blocking signal is part of system overhead information.

system budget: *See* **power budget.**

system documentation: The collection of documents that describes the requirements, capabilities, limitations, design, operation, and maintenance of a system, such as a communications, computing, or information processing system.

system failure transfer: In the event of a catastrophic failure, the ability to transfer central office trunks or interoffice trunking to predetermined stations to allow incoming and outgoing calls to be completed.

system follow-up: The study of the effects of a system after it has reached a stabilized state of operational use. *Synonyms* **post-development review, post-implementation review.**

system generation (SYSGEN): The process of selecting optional parts of an operating system and of creating a particular operating system tailored to the requirements of a data processing installation.

system integration: The progressive linking and testing of system components to merge their functional and technical characteristics into a comprehensive, interoperable system. *Note:* Integration of data systems allows data existing on disparate systems to be shared or accessed across functional or system boundaries.

system integrity: **1.** That condition of a system wherein its mandated operational and technical parameters are within the prescribed limits. (188) **2.** [The] quality of an AIS when it performs its intended function in an unimpaired manner, free from deliberate or inadvertent unauthorized manipulation of the system. [NIS]

system lifecycle: The course of developmental changes through which a system passes from its conception to the termination of its use and subsequent salvage. *Note:* For example, a system lifecycle might include the phases and activities associated with the analysis, acquisition, design, development, test, integration, operation, maintenance, and modification of the system.

system loading: In a frequency-division multiplexed (FDM) transmission system, the absolute power level of the composite signal transmitted in one direction. (188) *Note 1:* The absolute power level is referred to a zero transmission level point (0TLP). *Note 2:* The composite signal contains signaling, speech, and digital signals.

system management: **1.** Network management functions extended to include subscriber elements or user end instruments. (188) **2.** In computer systems, *synonym* **system administration.**

system operational threshold: For a supported performance parameter of a system, the value that establishes the minimum operational service performance level for the parameter. (188) *Note:* A measured parameter value worse than the system operational threshold indicates that the system is in an outage state.

system overhead information: *See* **overhead information.**

system power margin: *Synonym* **power margin.**

system reliability: The probability that a system, including all hardware, firmware, and software, will satisfactorily perform the task for which it was designed or intended, for a specified time and in a specified environment. [From Weik '89]

system robustness: The measure or extent of the ability of a system, such as a computer, communications, data processing, or weapons system, to continue to function despite the existence of faults in its component subsystems or parts. *Note:* System performance may be diminished or otherwise altered until the faults are corrected.

systems analysis: *Synonym* **system analysis.**

systems control: In a communications system, the control and implementation of a set of functions that (a) prevent or eliminate degradation of any part of the system, (b) initiate immediate response to demands that are placed on the system, (c) respond to changes in the system to meet long range requirements, and (d) may include various subfunctions, such as (i) immediate circuit utilization actions, (ii) continuous control of circuit quality, (iii) continuous control of equipment performance, (iv) development of procedures for immediate repair, restoration, or replacement of facilities and equipment, (v) continuous liaison with system users and with representatives of other systems, and (vi) the provision of advice and assistance in system use. (188)

systems design: **1.** A process of defining the hardware and software architecture, components, modules, interfaces, and data for a system to satisfy specified requirements. **2.** The preparation of an assembly of methods, procedures, or techniques united by regulated interaction to form an organized whole. [JP1]

systems engineering: *See* **systems design.**

system signaling and supervision: In transmission systems, any scheme used to provide such functions as system control, addressing, routing, error detection and correction, level control, priority, traffic control, message accountability, and/or other required overhead information. (188)

system software: Application-independent software that supports the running of application software.

system standard: In the military community, the system-specific characteristics, not dictated by the individual components' electrical performance characteristics, but necessary in order to permit internal and external interoperability. (188)

system supervision: In telephone systems, the use of signals and techniques to perform system management functions, such as system control, addressing, routing, error detection and correction, level control, priority, traffic control, message accountability, and other overhead functions that may be described in system overhead portions of messages. [From Weik '89]

system support: The continued provision of services and material necessary for the use and improvement of a system during its lifecycle.

system test time: The part of operating time during which a functional unit is tested for proper operation. *Note:* In a computer, the system test time may include the time for testing programs belonging to the operating system.

(this page intentionally left blank)

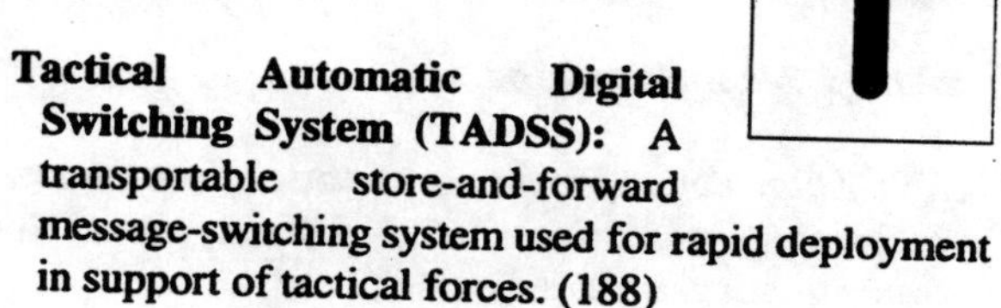

T: *Abbreviation for* **tera** (10^{12}). *See* **International System of Units.**

Tactical Automatic Digital Switching System (TADSS): A transportable store-and-forward message-switching system used for rapid deployment in support of tactical forces. (188)

tactical command and control (C^2) systems: The equipment, communications, procedures, and personnel essential to a commander for planning, directing, coordinating, and controlling tactical operations of assigned forces pursuant to assigned missions.

tactical communications: Communications in which information of any kind, especially orders and decisions, are conveyed from one command, person, or place to another within the tactical forces, usually by means of electronic equipment, including communications security equipment, organic to the tactical forces. (188) *Note:* Tactical communications do not include communications provided to tactical forces by the Defense Communications System (DCS), to nontactical military commands, and to tactical forces by civil organizations.

tactical communications system: A communications system that (a) is used within, or in direct support of, tactical forces, (b) is designed to meet the requirements of changing tactical situations and varying environmental conditions, (c) provides securable communications, such as voice, data, and video, among mobile users to facilitate command and control within, and in support of, tactical forces, and (d) usually requires extremely short installation times, usually on the order of hours, in order to meet the requirements of frequent relocation. (188)

tactical data information link (TADIL): A standardized communications link, approved by the Joint Staff, that is suitable for transmission of digital information, and is characterized by standardized message formats and transmission characteristics. (188)

tactical data information link—A (TADIL—A): A netted link in which one unit acts as a net control station and interrogates each unit by roll call. *Note:* Once interrogated, that unit transmits its data to the net. This means that each unit receives all the information transmitted. This is a direct transfer of data and no relaying is involved. (188)

tactical data information link—B (TADIL—B): A point-to-point data link between two units which provides for simultaneous transmission and reception of data (duplex). (188)

tactical load: For the host service tactical forces, the total power requirements for communications, including the requirements for weapons, detection, command and control systems, and related support functions. (188) *Note:* The tactical load is a part of the operational load.

TADIL: *Acronym for* **tactical data information link.**

TADSS: *Acronym for* **Tactical Automatic Digital Switching System.**

tag: *See* **flag, label.**

tag image file format (TIFF): A file format used to store an image using the particular data structure of the file. (188)

TAI: *Abbreviation for* **International Atomic Time.**

tail circuit: A communications line from the end of a major transmission link, such as a microwave link, satellite link, or LAN, to the end-user location. *Note:* A tail circuit is a part of a user-to-user connection.

tailing: In facsimile systems, the excessive prolongation of the decay of the signal. (188) *Synonym* **hangover.**

takeoff angle: *Synonym* **departure angle.**

tandem: Pertaining to an arrangement or sequencing of networks, circuits, or links, in which the output terminals of one network, circuit, or link are connected directly to the input terminals of another network, circuit, or link. *Note:* For example, concatenated microwave links constitute a tandem connection. (188)

tandem center: In a switched public telecommunications network, a facility that connects trunks to trunks and does not connect any local loops. (188)

tandem office: A central office that serves local subscriber loops, and also is used as an intermediate switching point for traffic between central offices.

tandem tie trunk network (TTTN): An arrangement that permits sequential connection of tie trunks between PBX and Centrex® locations by using tandem operation. *Note:* Tandem operation permits two or more dial tie trunks to be connected at a tandem center to form a through connection.

tap: **1.** To draw energy from a circuit. **2.** To monitor, with or without authorization, the information that is being transmitted via a communications circuit. **3.** To extract a portion of the signal from an optical fiber or communications link. *Note:* One method of tapping an optical fiber is to bend it to a relatively short radius, thus promoting radiation of a portion of the optical signal. [After FAA]

tapered fiber: An optical fiber in which the cross section, *i.e.*, cross-sectional diameter or area, varies, *i.e.*, increases or decreases, monotonically with length.

tape relay: A method of retransmitting TTY traffic from one channel to another, in which messages arriving on an incoming channel are recorded in the form of perforated tape, this tape then being either fed directly and automatically into an outgoing channel, or manually transferred to an automatic transmitter for transmission on an outgoing channel. (188)

target language: In computing, data processing, and communications systems, a language into which statements are translated. *Note:* Translators, assemblers, and compilers prepare target language programs, usually machine-language programs, from source language programs, usually high-level language programs written by programmers.

tariff: The published schedule of rates or charges for a specific unit of equipment, facility, or type of service such as might be provided by a telecommunications common carrier.

TASI: *Acronym for* **time-assignment speech interpolation.**

tasking: *See* **multitasking.**

TAT: *Abbreviation for* **trans-Atlantic telecommunications (cable).** *Note: TAT* formerly stood for *transatlantic telephone (cable).*

T-carrier: The generic designator for any of several digitally multiplexed telecommunications carrier systems. *Note 1:* The designators for T-carrier in the North American digital hierarchy correspond to the designators for the digital signal (DS) level hierarchy. *See the table on the following page.* *Note 2:* T-carrier systems were originally designed to transmit digitized voice signals. Current applications also include digital data transmission. (188) *Note 3:* If an "F" precedes the "T", a fiber optic cable system is indicated at the same rates. *Note 4:* The table below lists the designators and rates for current T-Carrier systems. *Note 5:* The North American and Japanese hierarchies are based on multiplexing 24 voice-frequency channels and multiples thereof, whereas the European hierarchy is based on multiplexing 30 voice-frequency channels and multiples thereof. *See table on following page.*

TCB: *Abbreviation for* **trusted computing base.**

TCF: *Abbreviation for* **technical control facility.**

T-coupler: A passive optical coupler having three ports (three fibers). *Note 1:* Two isolated inputs may be combined into one output; or one input, into two isolated outputs. *Note 2:* The amount of coupling loss, usually expressed in dB, between ports is determined by the design and construction of the coupler. [After FAA] *Synonym* **splitter.**

TCP: *Abbreviation for* **Transmission Control Protocol.** In the Internet Protocol suite, a standard, connection-oriented, full-duplex, host-to-host protocol used over packet-switched computer communications networks. *Note 1:* TCP corresponds closely to the ISO Open Systems Interconnection—Reference Model (OSI—RM) Layer 4 (Transport Layer). *Note 2:* The OSI—RM uses TP-0 or TP-4 protocols for transmission control.

T-Carrier Systems	North American	Japanese	European (CEPT)
Level zero (Channel data rate)	64 kb/s (DS0)	64 kb/s	64 kb/s
First level	1.544 Mb/s (DS1) (24 user channels)	1.544 Mb/s (24 user channels)	2.048 Mb/s (30 user channels)
(Intermediate level, North American Hierarchy only)	3.152 Mb/s (DS1C) (48 Ch.)	-	-
Second level	6.312 Mb/s (DS2) (96 Ch.)	6.312 Mb/s (96 Ch.), or 7.786 Mb/s (120 Ch.)	8.448 Mb/s (120 Ch.)
Third level	44.736 Mb/s (DS3) (672 Ch.)	32.064 Mb/s (480 Ch.)	34.368 Mb/s (480 Ch.)
Fourth level	274.176 Mb/s (DS4) (4032 Ch.)	97.728 Mb/s (1440 Ch.)	139.268 Mb/s (1920 Ch.)
Fifth level	400.352 Mb/s (5760 Ch.)	565.148 Mb/s (7680 Ch.)	565.148 Mb/s (7680 Ch.)

Note 1: The DS designations are used in connection with the North American hierarchy only.

Note 2: There are other data rates in use, *e.g.*, military systems that operate at six and eight times the DS1 rate. At least one manufacturer has a commercial system that operates at 90 Mb/s, twice the DS3 rate. New systems, which take advantage of the high data rates offered by optical communications links, are also deployed or are under development.

TCP/IP: *Abbreviation for* **Transmission Control Protocol/Internet Protocol.** Two interrelated protocols that are part of the Internet protocol suite. *Note 1:* TCP operates on the OSI Transport Layer and breaks data into packets. IP operates on the OSI Network Layer and routes packets. *Note 2:* TCP/IP was originally developed by the U.S. Department of Defense.

TCP/IP Suite: The suite of interrelated protocols associated with Transmission Control Protocol/Internet Protocol. *Note 1:* The TCP/IP Suite includes, but is not limited to, protocols such as TCP, IP, UDP, ICMP, FTP, and SMTP. *Note 2:* Additional application and management protocols are sometimes considered part of the TCP/IP Suite. This includes protocols such as SNMP.

TCS: *Abbreviation for* **trusted computer system.**

TCU: *Abbreviation for* **teletypewriter control unit.**

TDD: *Abbreviation for* **Telecommunications Device for the Deaf.**

TDM: *Abbreviation for* **time division multiplex.**

TDMA: *Abbreviation for* **time-division multiple access.**

TE: *Abbreviation for* **transverse electric.** *See* **transverse electric mode.**

technical area: In the military community, an area in which temperature, humidity, or access is controlled because it contains equipment, such as communications, computing, control, or support equipment, that requires such controls.

technical control facility (TCF): A physical plant, or a designated and specially configured part thereof, that (a) contains the equipment necessary for ensuring fast, reliable, and secure exchange of information, (b) typically includes distribution frames and associated panels, jacks, and switches and monitoring, test, conditioning, and orderwire equipment, and (c) allows telecommunications systems control personnel to exercise operational control of communications paths and facilities, make quality analyses of communications and communications channels, monitor operations and maintenance functions, recognize and correct deteriorating conditions, restore disrupted communications, provide requested on-call circuits, and take or direct such actions as may be required and practical to provide effective telecommunications services. (188)

technical control hubbing repeater: *Synonym* **data conferencing repeater.**

technical load: The portion of the operational load required for communications, tactical operations, and ancillary equipment including necessary lighting, air-conditioning, or ventilation required for full continuity of communications. (188)

technical vulnerability: In information handling, a hardware, software, or firmware weakness, or design deficiency, that leaves a system open to assault, harm, or unauthorized exploitation, either externally or internally, thereby resulting in unacceptable risk of information compromise, information alteration, or service denial.

TED: *Abbreviation for* **trunk encryption device.**

tee coupler: A passive coupler that has three ports.

TEK: *Abbreviation for* **traffic encryption key.**

teleaction service: In Integrated Services Digital Network (ISDN) applications, a telecommunications service that uses very short messages with very low data transmission rates between the user and the network.

telecommand: The use of telecommunication for the transmission of signals to initiate, modify or terminate functions of equipment at a distance. [NTIA] [RR]

telecommunication: **1.** Any transmission, emission, or reception of signs, signals, writing, images and sounds or intelligence of any nature by wire, radio, optical or other electromagnetic systems. [NTIA] [RR] **2.** Any transmission, emission, or reception of signs, signals, writings, images, sounds, or information of any nature by wire, radio, visual, or other electromagnetic systems. [JP1]

telecommunication architecture: *See* **network architecture.**

telecommunications center: *See* **communications center.**

Telecommunications Device for the Deaf (TDD): A machine that uses typed input and output, usually with a visual text display, to enable individuals with hearing or speech impairments to communicate over a telecommunications network.

telecommunications facilities: The aggregate of equipment, such as radios, telephones, teletypewriters, facsimile equipment, data equipment, cables, and switches, used for providing telecommunications services.

telecommunications management network (TMN): A network that interfaces with a telecommunications network at several points in order to receive information from, and to control the operation of, the telecommunications network. (188) *Note:* A TMN may use parts of the managed telecommunications network to provide for the TMN communications.

telecommunications security: *See* **communications security.**

telecommunications service: **1.** Any service provided by a telecommunication provider. **2.** A specified set of user-information transfer capabilities provided to a group of users by a telecommunications system. (188) *Note:* The telecommunications service user is responsible for the information content of the message. The telecommunications service provider has the responsibility for the acceptance, transmission, and delivery of the message.

Telecommunications Service Priority (TSP) service: A regulated service provided by a telecommunications provider, such as an operating telephone company or a carrier, for NS/EP telecommunications. *Note:* The TSP service replaced Restoration Priority (RP) service effective September 1990.

Telecommunications Service Priority (TSP) system: A system that provides a means for telecommunications users to obtain priority treatment from service providers for the NS/EP telecommunications requirements. *Note:* The TSP system replaced the Restoration Priority (RP) system effective September 1990.

Telecommunications Service Priority (TSP) system user: Any individual, organization, or activity that interacts with the NS/EP TSP System.

telecommunications system: *See* **communications system.**

telecommunications system operator: The organization responsible for providing telecommunications services to users.

teleconference: The live exchange of information among persons and machines remote from one another but linked by a telecommunications system. *Note:* The telecommunications system may support the teleconference by providing audio, video, and data services by one or more means, such as telephone, telegraph, teletype, radio, and television. (188)

telegram: Written matter intended to be transmitted by telegraphy for delivery to the addressee. This term also includes radiotelegrams unless otherwise specified. In this definition the term telegraphy has the same general meaning as defined in the [1979 General Worldwide Administrative Radio Conference] Convention. [RR with editor's note in brackets]

telegraph: *See* **telegraphy.**

telegraphy: A form of telecommunication which is concerned in any process providing transmission and reproduction at a distance of documentary matter, such as written or printed matter or fixed images, or the reproduction at a distance of any kind of information in such a form. For the purposes of the *Radio Regulations*, unless otherwise specified therein, telegraphy shall mean a form of telecommunication for the transmission of written matter by the use of a signal code. [NTIA] [RR]

telemetry: The use of telecommunication for automatically indicating or recording measurements at a distance from the measuring instrument. [RR]

telephone: A user end instrument that is used to transmit and receive voice-frequency signals.

telephone exchange: *Synonym* **central office.**

telephone frequency: *See* **audio frequency, voice frequency.**

telephone number: The unique network address that is assigned to a telephone user, *i.e.*, subscriber, for routing telephone calls.

telephone sidetone: *Synonym* **sidetone.**

telephony: **1.** The branch of science devoted to the transmission, reception, and reproduction of sounds, such as speech and tones that represent digits for signaling. *Note 1:* Transmission may be via various media, such as wire, optical fibers, or radio. *Note 2:* Analog representations of sounds may be digitized, transmitted, and, on reception, converted back to analog form. *Note 3:* *"Telephony"* originally entailed only the transmission of voice and voice-frequency data. Currently, it includes new services, such as the transmission of graphics information. **2.** A form of telecommunication set up for the transmission of speech or, in some cases, other sounds. [NTIA] [RR]

telephoto: Pertaining to pictures transmitted via a telecommunications system.

teleprinter: A teletypewriter that can only receive data and does not have a keyboard for transmission.

teleprocessing: The combining of telecommunications and computer operations interacting in the automatic processing, reception, and transmission of data and/or information. [JP1] *Note:* Teleprocessing includes human-machine interface equipment. (188)

teleseminar: *See* **teletraining.**

teleservice: *See* **telecommunications service.**

teletex: An international store-and-forward essentially error-free communications service that is defined by the CCITT, has a data signaling rate (DSR) of 2400 b/s over switched telephone networks, and has a communications protocol that supports the CCITT Group 4 facsimile service.

teletext: A type of one-way information service in which a subscriber can receive data on a video display. *Note:* The information is transmitted to the subscriber's video display over a common carrier channel. A proprietary video adapter unit is required for reception. *Contrast with* **viewdata.**

teletraining: Training that (a) in which usually live instruction is conveyed in real time via telecommunications facilities, (b) that may be accomplished on a point-to-point basis or on a point-to-multipoint basis, and (c) may assume many forms, such as a teleseminar, a teleconference, or an electronic classroom, usually including both audio and video. (188) *Synonyms* **distance learning, distance training, electronic classroom, virtual instruction.**

teletypewriter (TTY): A printing telegraph instrument that has a signal-actuated mechanism for automatically printing received messages. *Note 1:* A TTY may have a keyboard similar to that of a typewriter for sending messages. (188) *Note 2:* Radio circuits carrying TTY traffic are called "RTTY circuits" or "RATT circuits."

teletypewriter control unit (TCU): A device that controls and coordinates operations between teletypewriters and message switching centers. (188)

teletypewriter exchange service (TWX): A switched teletypewriter service in which suitably arranged teletypewriter stations are provided with lines to a central office for access to other such stations.

teletypewriter signal distortion: The shifting of signal pulse transitions from their proper positions relative to the beginning of the start pulse. *Note:* The magnitude of the distortion is expressed in percent of a perfect unit pulse length. (188) *Synonym* **start-stop TTY distortion.**

television (TV): A form of telecommunication for the transmission of transient images of fixed or moving objects. [NTIA] [RR] *Note 1:* The picture signal is usually accompanied by the sound signal. *Note 2:* In North America, TV signals are generated, transmitted, received, and displayed in accordance with the NTSC standard.

television broadcast translator: *See* **translator (def. #3).**

Telex®: A communication service involving teletypewriters connected through automatic exchanges.

Telnet: The TCP/IP standard network virtual terminal protocol that is used for remote terminal connection service and that allows a user at one site to interact with systems at other sites as if that user terminal were directly connected to computers at those sites.

TEM: *Abbreviation for* **transverse electric and magnetic.**

TEMPEST: **1.** [A] Short name referring to investigation, study, and control of compromising emanations from telecommunications and automated information systems equipment. [NIS] (188) **2.** To shield against compromising emanations.

temporal application: A video application requiring high temporal resolution, *i.e.,* reduced jerkiness, possibly at the expense of reduced spatial resolution. *Note:* An example of temporal applications is the ability to accurately discern moving image features such as facial expressions and lip movements.

temporal coherence: *See* **coherent.**

temporal edge noise: In a video display, that form of edge busyness that is characterized by time-varying sharpness at the edges of objects.

temporally coherent radiation: *See* **coherence time.**

terahertz (THz): A unit denoting one trillion (10^{12}) hertz. (188)

terminal: A device capable of sending, receiving, or sending and receiving information over a communications channel. (188)

Terminal Access Controller (TAC): A host computer that accepts terminal connections, usually from dial-up lines, and that allows the user to invoke Internet remote log-on procedures, such as Telnet.

terminal adapter: An interfacing device employed at the "R" reference point in an ISDN environment that allows connection of a non-ISDN terminal at the physical layer to communicate with an ISDN network. *Note:* Typically, a terminal adapter will support standard RJ-11 telephone connection plugs for voice and RS-232C, V.35 and RS-449 interfaces for data.

terminal endpoint (TE) functional group: A functional group that includes functions broadly belonging to Layer 1 and higher layers of the CCITT Recommendation X.200 Reference Model. (188) *Note 1:* The functions of a TE functional group are performed on various types of equipment, or combinations of equipment, such as digital telephones, data terminal equipment, and/or integrated work stations. *Note 2:* Examples of TE functions are protocol-handling, maintenance, interface, and connection functions.

terminal equipment: **1.** Communications equipment at either end of a communications link, used to permit the stations involved to accomplish the mission for which the link was established. **2.** In radio-relay systems, equipment used at points where data are inserted or derived, as distinct from equipment used only to relay a reconstituted signal. **3.** Telephone and telegraph switchboards and other centrally located equipment at which communications circuits are terminated.

terminal impedance: **1.** The impedance as measured at the unloaded output terminals of transmission equipment or a line that is otherwise in normal operating condition. (188) **2.** The ratio of voltage to current at the output terminals of a device, including the connected load.

terminal mobility: In commercial wireless networks, the ability of a terminal, while in motion, to access telecommunication services from different locations, and the capability of the network to identify and locate that terminal.

terminal mobility management: In personal communications service (PCS), (a) providing authentication of terminal information, (b) maintaining terminal location and capability information for each terminal, and (c) providing translation between terminal identification and location (routing address) for the completion of calls to terminals.

termination: **1.** The load connected to a transmission line, circuit, or device. *Note:* For a uniform transmission line, if the termination impedance is equal to the characteristic impedance of the line, wave reflections from the end of the line will be avoided. [From Weik '89] **2.** In hollow metallic waveguides, the point at which energy propagating in the waveguide continues in a nonwaveguide propagation mode into a load. [From Weik '89] **3.** An impedance, often resistive, that is connected to a transmission line or piece of equipment as a dummy load, for test purposes.

terminus: A device used to terminate, position, and hold an optical fiber within a connector.

ternary signal: A signal that can assume, at any given instant, one of three significant conditions, such as power level, phase position, pulse duration, or frequency. *Note:* Examples of ternary signals are (a) a pulse that can have a positive, zero, or negative voltage value at any given instant, (b) a sine wave that can assume phases of 0°, 120°, or 240° relative to a clock pulse, and (c) a carrier wave that can assume any one of three different frequencies depending on three different modulation signal significant conditions.

terrestrial radiocommunication: Any radiocommunication other than space radiocommunication or radio astronomy. [NTIA] [RR]

terrestrial station: A station effecting terrestrial radiocommunication. In these *[Radio] Regulations,* unless otherwise stated, any station is a terrestrial station. [NTIA] [RR]

test and validation: Physical measurements taken (a) to verify conclusions obtained from mathematical modeling and analysis or (b) for the purpose of developing mathematical models. (188)

test antenna: An antenna of known performance characteristics used in determining transmission characteristics of equipment and associated propagation paths. (188)

test center: *See* **patch and test facility, technical control facility.**

test point: A point within a piece of equipment or an equipment string that provides access to signals for the purpose of fault isolation. (188)

test tone: A tone sent at a predetermined level and frequency through a transmission system for test purposes, such as for facilitating measurements and for aligning gains and losses in the system. (188)

text processing: *Synonym* **word processing.**

T4 (carrier): *See* **T-carrier.**

T5 (carrier): *See* **T-carrier.**

TG: *Abbreviation for* **telegraph.** *See* **telegraphy.**

TGM: *Abbreviation for* **trunk group multiplexer.**

THD: *Abbreviation for* **total harmonic distortion.**

thermal noise: The noise generated by thermal agitation of electrons in a conductor. The noise power, P, in watts, is given by $P = kT\Delta f$, where k is Boltzmann's constant in joules per kelvin, T is the conductor temperature in kelvins, and Δf is the bandwidth in hertz. (188) *Note 1:* Thermal noise power, per hertz, is equal throughout the frequency spectrum, depending only on k and T. *Note 2:* For the general case, the above definition may be held to apply to charge carriers in any type of conducting medium. *Synonym* **Johnson noise.**

thermal radiation: **1.** Electromagnetic radiations emitted from a heat or light source as a consequence of its temperature; it consists essentially of ultraviolet, visible, and infrared radiations. [JP1] **2.** The heat and light produced by a nuclear explosion. [JP1]

thermodynamic temperature: A measure, in kelvins (K), proportional to the thermal energy of a given body at equilibrium. *Note 1:* A temperature of 0 K is called "absolute zero," and coincides with the minimum molecular activity (*i.e.*, thermal energy) of matter. *Note 2:* Thermodynamic temperature was formerly called "absolute temperature." *Note 3:* In practice, the International Temperature Scale of 1990 (ITS-90) serves as the basis for high-accuracy temperature measurements in science and technology.

THF: *Abbreviation for* **tremendously high frequency.** *See* **electromagnetic spectrum.**

thin-film laser: A laser that is constructed by thin-film deposition techniques on a substrate for use as a light source, is usually used to drive thin-film optical waveguides, and may be used in integrated optical circuits.

thin-film optical modulator: A modulator that consists of multilayered films of material of different optical characteristics, is capable of modulating transmitted light by using electro-optic, electro-acoustic, or magneto-optic effects to obtain signal modulation, and may be used as a component in integrated optical circuits. [From Weik '89]

thin-film optical multiplexer: A multiplexer that consists of multilayered films of material of different optical characteristics, is capable of multiplexing transmitted light by using electro-optic, electro-acoustic, or magneto-optic effects to obtain signal multiplexing, and may be used as a component in integrated optical circuits. [From Weik '89]

thin-film optical switch: A switch that consists of multilayered films of material of different optical characteristics, that is capable of switching transmitted light by using electro-optic, electro-acoustic, or magneto-optic effects to obtain signal switching, and is usually used as a component in integrated optical circuits. *Note:* Thin-film optical switches may support only one propagation mode. [From Weik '89]

thin-film optical waveguide: A slab-dielectric waveguide that consists of multilayered films of material of different optical characteristics, is capable of guiding an optical signal, and may be used as a component in integrated optical circuits. [From Weik '89]

third-order intercept point: A point (a) that is an extrapolated convergence—not directly measurable—of intermodulation distortion products in the desired output and (b) that indicates how well a receiver performs in the presence of strong nearby signals. (188) *Note:* Determination of a third-order intercept point is accomplished by using two test frequencies that fall within the first intermediate frequency mixer passband. Usually, the test frequencies are about 20 to 30 kHz apart.

third window: Of silica-based optical fibers, the transmission window at approximately 1.55 µm. *Note:* The third window is the minimum-loss window in silica-based fibers. [After FAA]

threat: Capabilities, intentions, and attack methods of adversaries to exploit, or any circumstance or event with the potential to cause harm to, information or an information system. [NIS]

three-bit byte: *Synonym* **triplet.**

three-way calling: A switching system service feature that permits users to add a third party at a different number during a call, without the assistance of an attendant.

threshold: **1.** The minimum value of a signal that can be detected by the system or sensor under consideration. (188) **2.** A value used to denote predetermined levels, such as those pertaining to volume of message storage, *i.e.*, in-transit storage or queue storage, used in a message switching center. (188) **3.** The minimum value of the parameter used to activate a device. (188) **4.** The minimum value a stimulus may have to create a desired effect.

threshold current: In a laser, the driving current corresponding to lasing threshold.

threshold extension: *See* **FM threshold extension.**

threshold frequency: In optoelectronics, the frequency of incident radiant energy below which there is no photoemissive effect. (188)

through group: A group of 12 voice-frequency channels transmitted as a unit through a carrier system. (188)

through-group equipment: In carrier telephone transmission, equipment that accepts the signal from a group receiver output and attenuates it to the proper signal level for insertion, without frequency translation, at the input of a group transmitter. (188)

throughput: **1.** The number of bits, characters, or blocks passing through a data communication system, or portion of that system. *Note 1:* Throughput may vary greatly from its theoretical maximum. (188) *Note 2:* Throughput is expressed in data units per period of time; *e.g.*, in the DDN, as blocks per second. **2.** The maximum capacity of a communications channel or system. **3.** A measure of the amount of work performed by a system over a period of time, *e.g.*, the number of jobs per day.

through supergroup: An aggregate of 60 voice-frequency channels, *i.e.*, five groups, transmitted as a unit through a carrier system. (188)

through-supergroup equipment: In carrier telephone transmission, equipment that accepts the multiplexed signal from a supergroup receiver output, amplifies it without frequency translation, and provides the proper signal level to the input of a supergroup transmitter equipment. (188)

THz: *Abbreviation for* **terahertz.** *See* **International System of Units.**

TIA: *Abbreviation for* **Telecommunications Industry Association.** *See* **EIA interface.**

ticketed call: A call for which a record is made of certain facts concerning the call, such as the time it was placed, the duration, the call originator, call destination numbers, and, where applicable, the attendant's name or initials. [From Weik '89]

TIE: *Acronym for* **time interval error.**

tie line: *See* **tie trunk.**

tie trunk: A telephone line that directly connects two private branch exchanges (PBXs).

TIFF: *Acronym for* **tagged image file format.**

tight buffer: *See* **buffer.**

tiling: *See* **block distortion.**

time: **1.** An epoch, *i.e.*, the designation of an instant on a selected time scale, astronomical or atomic. It is used in the sense of time of day [JP1] (188) **2.** On a time scale, the interval between two events, or the duration of an event. (188) **3.** An apparently irreversible continuum of ordered events.

time ambiguity: A situation in which more than one different time or time measurement can be obtained under the stated conditions.

time-assignment speech interpolation (TASI): An analog technique used on certain long transmission links to increase voice-transmission capacity. *Note:* TASI works by switching additional users onto any channel temporarily idled because an original user has stopped speaking. When the original user resumes speaking, that user will, in turn, be switched to any channel that happens to be idle. (188)

time availability: *Synonym* **circuit reliability.**

time block: An arbitrary grouping of several consecutive hours of a day, usually for a particular season, during which it is assumed that propagation data are statistically homogeneous. (188)

time code: A code used for the transmission and identification of time signals. (188) *Note:* In telecommunications systems, the format of the time code must be specified.

time code ambiguity: The shortest interval between successive repetitions of the same time code value. *Note:* For example, in a time code in which year-of-century is the most slowly changing field, the time code ambiguity would be 100 years; for a digital clock in which hours and minutes up to a maximum of 11:59 are displayed, the time code ambiguity would be 12 hours.

time code resolution: The interval between two successive time code states. *Note:* Time code resolution is determined by the most rapidly changing symbol position within the time code. For example, for a digital clock that displays hours and minutes, the time code resolution would be 1 minute.

time constant: The interval required for a system or circuit to change a specified fraction from one state or condition to another. *Note 1:* The time constant is used in the expression

$$A(t) = A(0)e^{-\frac{t}{a}} ,$$

where $A(t)$ is the value of the state at time t, $A(0)$ is the value of the state at time $t = 0$, a is the time constant, and t is the time that has elapsed from the start of the exponential decay. *Note 2:* When $t = a$, $A(t)/A(0) = 1/e$, or approximately 0.37, and the system has changed about 63% toward its new value in one time constant. A system is considered to have changed its state after the elapse of three time constants, which corresponds to a 95% change in state. For example, if an electrical capacitor, having a capacitance of C farads, is discharged through a resistor, having a resistance of R ohms, the capacitor will be approximately 95% discharged after the elapse of $3RC$ seconds. *Note 3:* Time constants are expressed in seconds, such as 3.5×10^{-6} seconds, *i.e.*, 3.5 µs. [From Weik '89]

time-delay distortion: *Synonym* **delay distortion.**

time-derived channel: *See* **time-division multiplexing.**

time diversity: Transmission in which signals representing the same information are sent over the same channel at different times. (188) *Note:* Time diversity is often used over systems subject to burst error conditions, and at intervals adjusted to be longer than an error burst.

time division: *See* **time-division multiplexing.**

time-division multiple access (TDMA): A communications technique that uses a common channel (multipoint or broadcast) for communications among multiple users by allocating unique time slots to different users. (188) *Note:* TDMA is used extensively in satellite systems, local area networks, physical security systems, and combat-net radio systems.

time-division multiplexing (TDM): Digital multiplexing in which two or more apparently simultaneous channels are derived from a given frequency spectrum, *i.e.*, bit stream, by interleaving pulses representing bits from different channels. (188) *Note:* Successive pulses represent bits from successive channels, *e.g.*, voice channels in a T1 system.

time-division switching: Switching of time-division multiplexed (TDM) channels by shifting bits between time slots in a TDM frame. (188)

time-domain reflectometer (TDR): An electronic instrument used to characterize and locate faults in metallic cables (*e.g.*, twisted pair, coax). *Note 1:* A TDR transmits a fast rise time pulse along the conductor. The resulting reflected pulse is measured at the input as a function of time and displayed on the instrument or plotted, as a function of cable length. *Note 2:* A TDR may be used to verify cable impedance characteristics, splice and connector location and associated losses, and estimate cable lengths.

time-gated direct-sequence spread spectrum: Direct-sequence spread spectrum where the transmitter is on only for a short fraction of a time interval. The on-time can be periodic or random within a time interval. [NTIA]

time guard band: A time interval left vacant on a channel to provide a margin of safety against intersymbol interference in the time domain between sequential operations, such as detection, integration, differentiation, transmission, encoding, decoding, or switching. (188)

time instability: The fluctuation of the time interval error caused by the instability of a real clock.

time interval error (TIE): The time difference between a real clock and an ideal uniform time scale, after a time interval following perfect synchronization between the clock and the scale.

time jitter: Short-term variation or instability in the duration of a specified time interval. (188)

timeliness: *See* **responsiveness.**

time marker: A reference signal, often repeated periodically, enabling the correlation of specific events with a time scale, such as for establishing synchronization.

time of occurrence: The date of an event, *i.e.*, the instant an event occurs, with reference to a specified time scale. (188)

time-out: **1.** A network parameter related to an enforced event designed to occur at the conclusion of a predetermined elapsed time. (188) **2.** A specified period of time that will be allowed to elapse in a system before a specified event is to take place, unless another specified event occurs first; in either case, the period is terminated when either event takes place. (188) *Note:* A time-out condition can be canceled by the receipt of an appropriate time-out cancellation signal. **3.** An event that occurs at the end of a predetermined period of time that began at the occurrence of another specified event. The time-out can be prevented by an appropriate signal.

time scale: **1.** A time measuring system defined to relate the passage of temporal events since a selected epoch. (188) *Note:* The internationally recognized time interval is the second. Time scales are graduated in intervals such as seconds, minutes, hours, days, and years, and in fractions of a second, such as milliseconds, nanoseconds, and picoseconds. **2.** Time coordinates placed on the abscissa (x-axis) of Cartesian-coordinate graphs used for depicting waveforms and similar phenomena.

time scale factor: A multiplier used to transform the real time of occurrence of an event or a problem into system time, such as that of a telecommunications system or a computer.

time-sharing: 1. The interleaving of two or more independent processes on one functional unit. (188) **2.** Pertaining to the interleaved use of computer time that enables two or more users to execute programs concurrently.

time slot: 1. Period of time during which certain activities are governed by specific regulations. [JP1] **2.** A time interval that can be recognized and uniquely defined. (188)

time standard: A stable device that emits signals at equal intervals such that their count may be used as a clock.

time tick: A time mark output of a clock system.

timing extraction: *Synonym* **timing recovery.**

timing recovery: The derivation of a timing signal from a received signal. (188) *Synonym* **timing extraction.**

timing signal: 1. The output of a clock. (188) **2.** A signal used to synchronize interconnected equipment. (188)

timing tracking accuracy: A measure of the ability of a timing synchronization system to minimize the clock difference between a master clock and any slaved clock.

T-interface: For basic rate access in an Integrated Services Digital Network (ISDN) environment, a user-to-network interface reference point that (a) is characterized by a four-wire, 144-kb/s (2B+D) user rate, (b) accommodates the link access and transport layer function in the ISDN architecture, (c) is located at the user premises, (d) is distance sensitive to the servicing network terminating equipment, and (e) functions in a manner analogous to that of the Channel Service Units (CSUs) and the Data Service Units (DSUs).

T junction: *See* **series T junction.**

TLM: *Abbreviation for* **telemetry.**

TLP: *Abbreviation for* **transmission level point.**

TM: *Abbreviation for* **transverse magnetic.** *See* **transverse magnetic mode.**

TOD: *Abbreviation for* **time of day.** *See* **time of occurrence.**

token: In certain local-area-network protocols, a group of bits that serves as a symbol of authority, is passed among data stations, and is used to indicate the station that is temporarily in control of the transmission medium.

token-bus network: A bus network in which a token passing procedure is used.

token passing: A network access procedure in which a token passes from station to station and the only station allowed to transmit information is the station with the token.

token ring adapter: A network interface card (NIC) designed to attach a client workstation to a token ring computer network and operate as a token-passing interface.

token-ring network: *See* **network topology.**

tolerance: The permissible range of variation of some characteristic from its nominal value.

tolerance field: 1. The region between two curves, such as circles or rectangles, used to specify the tolerance on component size and geometry. **2.** Pertaining to the cross section of an optical fiber, when used to specify the respective diameters and ovalities of, and concentricity error between, the core and cladding; two concentric annular regions which define the core-cladding boundary and the cladding outer boundary. *Note:* Dimensions are usually expressed in micrometers (μm). The larger annular region is defined by concentric circles of diameter $[D_C+\Delta D_C]$ and $[D_C-\Delta D_C]$, where D_C is the nominal diameter of the cladding and ΔD_C is the cladding diameter tolerance. The smaller annular region is defined by concentric circles of diameter $[D_c+\Delta D_c]$ and $[D_c-\Delta D_c]$, where D_c is the nominal diameter of the core and ΔD_c is the core diameter tolerance. When the core and cladding boundaries of the cross section of the fiber in question fall entirely within their respective defined areas, the fiber meets the specification. [After FAA] **3.** Of the cross section of a given optical fiber, when used to characterize the respective diameters and ovalities of the core and cladding, and the concentricity error between the core and cladding; two such pairs of concentric circles, the concentric pairs not necessarily being concentric with one another. *Note 1:* One pair of concentric circles characterizes the core, and the other pair, the cladding. The cladding *ovality* is characterized by the smallest circle that circumscribes its cross section, and the largest circle that fits within its cross section. (The cross section is assumed, to a first approximation, to be elliptical in shape, so these defining circles will be concentric.) The core cross section is characterized by an analogous pair of circles, also concentric with one another, but not necessarily with those defining the cladding cross section. *Note 2:* The distance between the centers of the two concentric pairs (core pair and cladding pair) defines the offset between the core and cladding (the *"core-cladding offset,"* also called the *"concentricity error"*). The width of the annulus defined by the cladding circles determines the ovality of the cladding, and the width of the annulus defined by the core determines the ovality of the core. [After FAA]

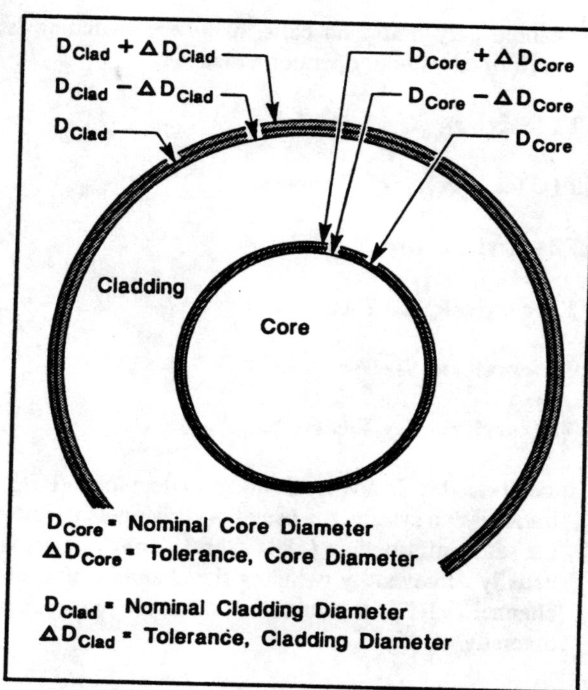

tolerance field

toll call: *See* **long-distance call.**

toll diversion: A system service feature by which users are denied the ability to place toll calls without the assistance of an attendant.

toll office: A central office used primarily for supervising and switching toll traffic.

toll quality: The voice quality resulting from the use of a nominal 4-kHz telephone channel. (188) *Note:* Toll quality may be quantized in terms of a specified bit error ratio.

toll restriction: *See* **classmark.**

toll switching trunk: A trunk connecting one or more end offices to a toll center as the first stage of concentration for intertoll traffic. (188) *Note:* Operator assistance or participation may be an optional function. In U.S. common carrier telephony service, a toll center designated *"Class 4C"* is an office where assistance in completing incoming calls is provided in addition to other traffic; a toll center designated *"Class 4P"* is an office where operators handle only outbound calls, or where switching is performed without operator assistance.

T1 (carrier): *See* **T-carrier.**

T1C (carrier): *See* **T-carrier.**

T2 (carrier): *See* **T-carrier.**

T3 (carrier): *See* **T-carrier.**

T4 (carrier): *See* **T-carrier.**

T5 (carrier): *See* **T-carrier.**

tone diversity: In a voice frequency telegraph (VFTG) transmission system, the use of two channels to carry the same information. (188) *Note:* Tone diversity is usually achieved by twinning the channels of a 16-channel VFTG to obtain 8 channels with dual diversity.

tone signaling: *See* **dual-tone multifrequency signaling.**

tool: *Synonym* **utility program.**

topography: The specification and arrangement in physical locations of actual communication and information system components which implement the topology. [NATO]

topology: *See* **network topology.**

torn-tape relay: An antiquated tape relay system in which the perforated tape is manually transferred by an operator to the appropriate outgoing transmitter. (188)

total channel noise: The sum of random noise, intermodulation noise, and crosstalk. (188) *Note:* Total channel noise does not include impulse noise because different techniques are required for its measurement.

total harmonic distortion (THD): Of a signal, the ratio of (a) the sum of the powers of all harmonic frequencies above the fundamental frequency to (b) the power of the fundamental frequency. *Note 1:* The THD is usually expressed in dB. *Note 2:* Measurements for calculating the THD are made at the output of a device under specified conditions. (188)

total internal reflection: The reflection that occurs when light, in a higher refractive-index medium, strikes an interface, with a medium with a lower refractive index, at an angle of incidence (with respect to the normal) greater than the critical angle. (188) *See* **Snell's law** (Note 3).

total line length: In facsimile, the spot speed divided by the scanning line frequency. *Note:* The total line length may be greater than the length of the available line.

touch panel: *See* **touch-sensitive.**

touch screen: *See* **touch-sensitive.**

touch-sensitive: Pertaining to a device that allows a user to interact with a computer system by touching an area on the surface of the device with a finger, pencil, or other object; for example, a touch-sensitive keypad or screen.

trace packet: In a packet-switching network, a unique packet that causes a report of each stage of its progress to be sent to the network control center from each visited system element.

trace program: A computer program that performs a check on another computer program by exhibiting the sequence in which the instructions are executed and usually the results of executing the instructions.

track: On a data medium, a path associated with a single read/write head position as data move past the head.

trackball: A ball that can be rotated about its center and that is used as an input device, *e.g.*, to position a cursor. *Synonym* **control ball.**

track density: The number of tracks per unit length, measured in a direction perpendicular to the direction in which the tracks are read.

tracking error: The deviation of a dependent variable with respect to a reference function.

tracking mode: An operational mode during which a system is operating within specified movement limits relative to a reference. (188)

tracking phase: *See* **tracking mode.**

traffic: 1. The information moved over a communication channel. (188) 2. A quantitative measurement of the total messages and their length, expressed in CCS or other units, during a specified period of time. (188)

traffic analysis: 1. In a communications system, the analysis of traffic rates, volumes, densities, capacities, and patterns specifically for system performance improvement. [From Weik '89] 2. [The] study of communications characteristics external to the text. [NIS] 3. The analysis of the communications-electronic environment for use in the design, development, and operation of new communications systems. [From Weik '89]

traffic capacity: The maximum traffic per unit of time that a given telecommunications system, subsystem, or device can carry under specified conditions. (188)

traffic encryption key (TEK): [A] key used to encrypt plain text or to superencrypt previously encrypted text and/or to decrypt cipher text. [NIS]

traffic engineering: The determination of the numbers and kinds of circuits and quantities of related terminating and switching equipment required to meet anticipated traffic loads throughout a communications system. (188)

traffic-flow security: 1. The protection resulting from features, inherent in some cryptoequipment, that conceal the presence of valid messages on a communications circuit; normally achieved by causing the circuit to appear busy at all times. [After JP1] 2. Measures used to conceal the presence of valid messages in an on-line cryptosystem or secure communications system. *Note:* Encryption of sending and receiving addresses and causing the circuit to appear busy at all times by sending dummy traffic are two methods of traffic-flow security. A more common method is to send a continuous encrypted signal, whether or not traffic is being transmitted.

traffic intensity: A measure of the average occupancy of a facility during a specified period of time, normally a busy hour, measured in traffic units (erlangs) and defined as the ratio of the time during which a facility is occupied (continuously or cumulatively) to the time this facility is available for occupancy. (188) *Note:* A traffic intensity of one traffic unit (one erlang) means continuous occupancy of a facility during the time period under consideration, regardless of whether or not information is transmitted. *Synonym* **call intensity.**

traffic load: The total traffic carried by a trunk or trunk group during a specified time interval. (188)

traffic monitor: In a communications network, a service feature that provides basic data on the amount and type of traffic handled by the network. (188)

traffic overflow: 1. That condition wherein the traffic offered to a portion of a communication system exceeds its capacity and the excess may be blocked or may be provided with alternate routing. (188) 2. The excess traffic itself. (188)

traffic register: *See* **register.**

traffic service position system (TSPS): A stored program electronic system associated with one or more toll switching systems which provides centralized traffic service position functions for several local offices at one location. [47CFR part 67, Appendix.]

traffic unit: *Synonym* **erlang.**

traffic usage recorder: A device for measuring and recording the amount of telephone traffic carried by a group, or several groups, of switches or trunks. (188)

transceiver: 1. A device that performs, within one chassis, both transmitting and receiving functions. 2. In military communications, the combination of transmitting and receiving equipment that (a) is in a common housing, (b) usually is designed for portable or mobile use, (c) uses common circuit components for both transmitting and receiving, and (d) provides half-duplex operation. (188)

transcoding: The direct digital-to-digital conversion from one encoding scheme, such as voice LPC-10, to a different encoding scheme without returning the signals to analog form. (188) *Note:* The transcoded signals, *i.e.*, the digital representations of analog signals may be any digital representation of any analog signal, such as voice, facsimile, or quasi-analog signals.

transducer: A device for converting energy from one form to another for the purpose of measurement of a physical quantity or for information transfer.

TRANSEC: *Abbreviation for* **transmission security.** *See* **communications security.**

transfer: To send information from one location and to receive it at another.

transfer characteristics: Those intrinsic parameters of a system, subsystem, or equipment which, when applied to the input of the system, subsystem, or equipment, will fully describe its output.

transfer function: **1.** A mathematical statement that describes the transfer characteristics of a system, subsystem, or equipment. **2.** The relationship between the input and the output of a system, subsystem, or equipment in terms of the transfer characteristics. *Note 1:* When the transfer function operates on the input, the output is obtained. Given any two of these three entities, the third can be obtained. *Note 2:* Examples of simple transfer functions are voltage gains, reflection coefficients, transmission coefficients, and efficiency ratios. An example of a complex transfer function is envelope delay distortion. *Note 3:* For a negative feedback circuit, the transfer function, *T*, is given by

$$T = \frac{e_0}{e_i} = \frac{G}{1 + GH} ,$$

where e_o is the output, e_i is the input, *G* is the forward gain, and *H* is the backward gain, *i.e.*, the fraction of the output that is fed back and combined with the input in a subtracter. **3.** Of an optical fiber, the complex mathematical function that expresses the ratio of the variation, as a function of modulation frequency, of the instantaneous power of the optical signal at the output of the fiber, to the instantaneous power of the optical signal that is launched into the fiber. *Note:* The optical detectors used in communication applications are square-law devices. Their output current is proportional to the input optical power. Because electrical power is proportional to current, when the optical power input drops by one-half (3 dB), the electrical power at the output of the detector drops by three-quarters (6 dB). [FAA]

transfer mode: In an integrated services digital network, (ISDN), a method of transmitting, multiplexing, and switching.

transfer rate: *See* **data transfer rate.**

transient: *See* **dynamic variation.**

transit delay: Between two given points in an integrated services digital network (ISDN), the time between the moment that the first bit of a data unit, such as a frame or block, passes the first given point and the moment that bit passes the second given point, plus the transmission time of the data unit.

transition: In a signal, the changing from one significant condition to another. *Note:* Examples of transitions are the changing from one voltage level to another in a data stream, the shifting from one phase position to another in phase-shift keying, and the translation from one frequency to another in frequency-shift keying. [From Weik '89]

transition frequency: The frequency associated with the difference between two discrete energy levels in an atomic system, given by

$$f_{2,1} = \frac{E_2 - E_1}{\hbar} ,$$

where $f_{2,1}$ is the frequency associated with the difference between two energy levels, E_2 and E_1 ($E_2 > E_1$), and $\hbar$ is Planck's constant. *Note:* If a transition from E_2 to E_1 occurs, a photon with frequency $f_{2,1}$ is likely to be emitted. If the atomic system is at energy level E_1, and a photon of frequency $f_{2,1}$ is absorbed, the energy level will be raised to E_2. [From Weik '89]

transition zone: *Synonym* **intermediate-field region.**

transit network identification: A network service feature that specifies the sequence of networks used to establish or partially establish a virtual circuit.

transit time: *Synonym* **phase delay.**

translating program: *Synonym* **translator (def. #2).**

translator: **1.** A device that converts information from one system of representation into equivalent information in another system of representation. (188) *Note:* An example of a translator in telephony is the device that converts dialed digits into call-routing information. **2.** A computer program that translates from one language into another language and in particular from one programming language into another programming language. *Synonym* **translating program.** **3.** In FM and TV broadcasting, a repeater station that receives a primary station's signal, amplifies it, shifts it in frequency, and rebroadcasts it. **4.** A device that converts one frequency to another.

transliterate: To convert the characters of one alphabet to the corresponding characters of another alphabet.

transmission: **1.** The dispatching, for reception elsewhere, of a signal, message, or other form of information. **2.** The propagation of a signal, message, or other form of information by any means, such as by telegraph, telephone, radio, television, or facsimile via any medium, such as wire, coaxial cable, microwave, optical fiber, or radio frequency. (188) **3.** In communications systems, a series of data units, such as blocks, messages, or frames. **4.** The transfer of electrical power from one location to another via conductors. (188)

transmission block: **1.** A group of bits or characters transmitted as a unit and usually containing an encoding procedure for error control purposes. **2.** In data transmission, a group of records sent, processed, or recorded as a unit. *Note:* A transmission block is usually terminated by an end-of-block character (EOB), end-of-transmission-block character (ETB), or end-of-text character (EOT or ETX).

transmission channel: *See* **channel.**

transmission coefficient: **1.** The ratio of the transmitted field strength to the incident field strength of an electromagnetic wave when it is incident upon an interface surface between media with two different refractive indices. **2.** In a transmission line, the ratio of the amplitude of the complex transmitted wave to that of the incident wave at a discontinuity in the line. **3.** The probability that a portion of a communications system, such as a line, circuit, channel or trunk, will meet specified performance criteria. *Note:* The value of the transmission coefficient is inversely related to the quality of the line, circuit, channel or trunk. (188)

transmission control character: *See* **control character.**

transmission control protocol: A network protocol that controls host-to-host transmissions over packet-switched communication networks.

transmission frame: A data structure, beginning and ending with delimiters, that consists of fields predetermined by a protocol for the transmission of user data and control data.

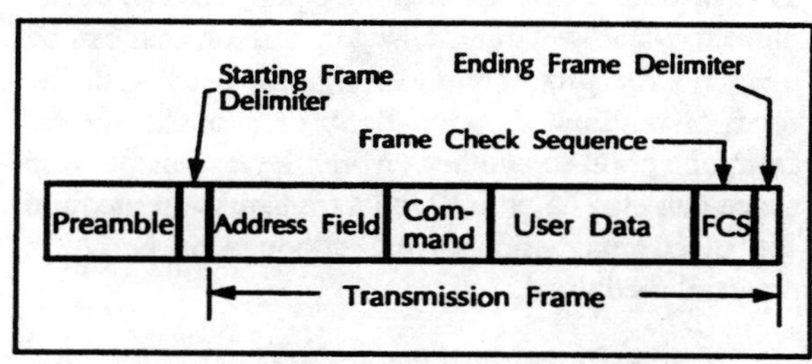

transmission frame

transmission level: At a specified point in a telecommunications system, the power that is measured when a standard test signal, *e.g.,* 0 dBm or −16 dBm at 1000 Hz, is transmitted from a corresponding reference point. (188) *Note:* The transmission level is usually expressed in dBm.

transmission level point (TLP): In a telecommunications system, a test point, *i.e.,* a point where a signal may be inserted or measured, and for which the nominal power of a test signal is specified. (188) *Note 1:* In practice, the abbreviation, TLP, is usually used, and it is modified by the nominal level for the point in question. For example, where the nominal level is 0 dBm, the expression 0 dBm TLP, or simply, 0TLP, is used. Where the nominal level is

–16 dBm, the expression –16 dBm TLP, or –16TLP, is used. *Note 2:* The nominal transmission level at a specified TLP is a function of system design and is an expression of the design gain or loss. *Note 3:* Voice-channel transmission levels, *i.e.*, TLPs, are usually specified for a frequency of approximately 1000 Hz. *Note 4:* The TLP at a point at which an end instrument, *e.g.*, a telephone set, is connected is usually specified as 0 dBm.

transmission line: The material medium or structure that forms all or part of a path from one place to another for directing the transmission of energy, such as electric currents, magnetic fields, acoustic waves, or electromagnetic waves. *Note:* Examples of transmission lines include wires, optical fibers, coaxial cables, rectangular closed waveguides, and dielectric slabs.

transmission loss: The decrease in power that occurs during transmission from one point to another. *Note:* Transmission loss is usually expressed in dB. (188)

transmission medium: Any material substance, such as fiber-optic cable, twisted-wire pair, coaxial cable, dielectric-slab waveguide, water, and air, that can be used for the propagation of signals, usually in the form of modulated radio, light, or acoustic waves, from one point to another. *Note:* By extension, free space can also be considered a transmission medium for electromagnetic waves, although it is not a material medium.

transmission security: *See* **communications security.**

transmission security key (TSK): [A] key that is used in the control of transmission security processes, such as frequency hopping and spread spectrum. [NIS]

transmission service channel: In video systems, the one-way transmission path between two designated points.

transmission system: Part of a communication system organized to accomplish the transfer of information from one point to one or more other points by means of signals. *Note:* Examples of NATO-owned transmission systems are SATCOM, ACE HIGH and CIP-67. [NATO]

transmission time: In facsimile, the interval between the start of picture signals and the detection of the end-of-message signal by the receiver for a single document.

transmission window: *Synonym* **spectral window.** *See* **window.**

transmissivity: *Obsolete. See* **transmittance.**

transmit-after-receive time delay: The time interval from removal of rf energy at the local receiver input until the local transmitter is automatically keyed on and the transmitted rf signal amplitude has increased to 90% of its steady-state value. (188) *An Exception:* High-frequency (HF) transceiver equipment is normally not designed with an interlock between receiver squelch and transmitter on-off key. The transmitter can be keyed on at any time, independent of whether or not a signal is being received at the receiver input.

transmit flow control: In data communications systems, control of the rate at which data are transmitted from a terminal so that the data can be received by another terminal. *Note 1:* Transmit flow control may occur between data terminal equipment (DTE) and a switching center, via data circuit-terminating equipment (DCE), or between two DTEs. The transmission rate may be controlled because of network or DTE requirements. *Note 2:* Transmit flow control can occur independently in the two directions of data transfer, thus permitting the transfer rates in one direction to be different from the transfer rates in the other direction.

transmittance: The ratio of the transmitted power to the incident power. (188) *Note 1:* In optics, transmittance is usually expressed as optical density or in percent. *Note 2:* Transmittance was formerly called *"transmission."*

transmitter attack-time delay: The interval from the instant a transmitter is keyed-on to the instant the transmitted radio frequency (rf) signal amplitude has increased to a specified level, usually 90% of its key-on steady-state value. *Note:* The transmitter attack-time delay excludes the time required for automatic antenna tuning. (188)

transmitter central wavelength range ($\lambda_{tmax} - \lambda_{tmin}$): In optical communication, the total allowed range of transmitter central wavelengths caused by the combined worst-case variations due to manufacturing, temperature, aging, and any other significant factors.

transmitter power output rating: The power output of a radio transmitter under stated conditions of operation and measurement. (188) *Note:* Power output ratings may be made against a number of criteria, *e.g.*, peak envelope power, peak power, mean power, carrier power, noise power, or stated intermodulation level.

transmitter release-time delay: The interval from the instant a transmitter is keyed-off to the instant the transmitted radio frequency (rf) signal amplitude has decreased to a specified level, usually 10% of its key-on steady-state value. (188)

transmultiplexer: Equipment that transforms signals derived from frequency-division multiplex equipment, such as group or supergroups, to time-division-multiplexed signals having the same structure as those derived from PCM multiplex equipment, such as primary or secondary PCM multiplex signals, and vice versa. (188)

transparency: **1.** The property of an entity that allows another entity to pass thorough it without altering either of the entities. **2.** In telecommunications, the property that allows a transmission system or channel to accept, at its input, unmodified user information, and deliver corresponding user information at its output, unchanged in form or information content. *Note:* The user information may be changed internally within the transmission system, but it is restored to its original form prior to the output without the involvement of the user. (188) **3.** The quality of a data communications system or device that uses a bit-oriented link protocol that does not depend on the bit sequence structure used by the data source. **4.** An image fixed on a clear base by means of a photographic printing, chemical, or other process, especially adaptable for viewing by transmitted light. [JP1]

transparent interface: An interface that allows the connection and operation of a system, subsystem, or equipment with another without modification of system characteristics or operational procedures on either side of the interface. (188)

transparent network: *See* **transparency (def. #2).**

transponder: **1.** An automatic device that receives, amplifies, and retransmits a signal on a different frequency. (188) **2.** An automatic device that transmits a predetermined message in response to a predefined received signal. (188) *Note:* An example of transponders is in identification-friend-or-foe systems and air-traffic-control secondary radar (beacon radar) systems. **3.** A receiver-transmitter that will generate a reply signal upon proper interrogation. [JP1]

transportability: **1.** In communications, the quality of equipment, devices, systems, and associated hardware that permits their being moved from one location to another to interconnect with locally available complementary equipment, devices, systems, associated hardware, or other complementary facilities. *Note:* Transportability implies the use of standardized components, such as standardized plugs and transmission media. **2.** The capability of material to be moved by towing, self-propulsion, or carrier via any means, such as railways, highways, waterways, pipelines, oceans, and airways. [JP1]

transportable station: A station which is transferred to various fixed locations but is not intended to be used while in motion. [NTIA]

Transport Layer: *See* **Open Systems Interconnection—Reference Model.**

transposition: **1.** In data transmission, a transmission defect in which, during one character period, one or more signal elements are changed from one significant condition to the other, and an equal number of elements are changed in the opposite sense. (188) **2.** In outside plant construction, an interchange of spatial positions of the several conductors of a cable between successive concatenated sections. *Note:* Transposition is usually used to minimize inductive coupling and thus reduce interference in communications circuits. (188)

transverse electric and magnetic (TEM) mode: A mode whose electric and magnetic field vectors are both normal to the direction of propagation. *Note:* The TEM mode is the most useful mode in a coaxial cable.

transverse electric (TE) mode: A mode whose electric field vector is normal to the direction of propagation. *Note:* TE modes may be useful modes in waveguides. In an optical fiber, TE and TM modes correspond to meridional rays.

transverse magnetic (TM) mode: A mode whose magnetic field vector is normal to the direction of propagation. *Note:* TM modes may be useful in waveguides. In an optical fiber, TE and TM modes correspond to meridional rays.

transverse offset loss: *Synonym* **lateral offset loss.**

transverse parity check: A parity check performed on a group of binary digits recorded on parallel tracks of a data medium, such as a magnetic disk, tape, drum, or card. [From Weik '89]

transverse redundancy check (TRC): In synchronized parallel bit streams, a redundancy check (a) that is based on the formation of a block check following preset rules, (b) in which the check-formation rule applied to blocks is also applied to characters, and (c) in which the check is made on parallel bit patterns. (188) *Note 1:* When the TRC is based on a parity bit applied to each character and block, the TRC can only detect, with limited certainty, whether or not there is an error. It cannot correct the error. Detection cannot be guaranteed because an even number of errors in the same character or block will escape detection, regardless of whether odd or even parity is used. *Note 2:* Two-dimensional arrays of bits may be used to represent characters or blocks in synchronized parallel data streams. When TRC is combined with longitudinal redundancy checking (LRC), individual erroneous bits can be corrected. *Synonym* **vertical redundancy check.**

transverse resolution: In a facsimile receiver, the dimension that (a) is perpendicular to a scanning line and (b) is the smallest recognizable detail of the image produced by the shortest signal capable of actuating the facsimile receiver under specified conditions. [From Weik '89]

trapped electromagnetic wave: An electromagnetic wave that enters a layer of material that is surrounded on both sides by a layer of material of a lesser refractive index such that, if the wave is traveling parallel or nearly parallel to the surfaces of the layers and hence the incident angles with the surfaces are greater than the critical angle, *i.e.*, the angles are grazing with the surface, total internal reflection will occur on both sides and hence trap the wave. *Note:* Dielectric slabs, optical fibers, and layers of air can serve as an electromagnetic wave trap, thus confining the wave to a given direction of propagation and to a given point. [From Weik '89]

trapped mode: *Synonym* **bound mode.**

trapped ray: *Synonym* **guided ray.**

traveling wave: A wave that (a) propagates in a transmission medium, (b) has a velocity determined by the launching conditions and the physical properties of the medium, and (c) may be a longitudinal or transverse wave. *Note 1:* For the purposes of this definition, free space may be considered a medium, although it is not a physical medium. *Note 2:* A traveling wave is not a wave that is reduced to a standing wave by reflections from a distant boundary. *Note 3:* Examples of traveling waves are radio waves propagating in free space, lightwaves propagating in optical fibers, water waves on the surface of the ocean, and seismic waves. [From Weik '89]

tree network: *See* **network topology.**

tree search: In a tree structure, a search in which it is possible to decide, at each step, which part of the tree may be rejected without a further search.

tree structure: A hierarchical organization in which a given node is considered to be an ancestor of all the lower level nodes to which the given node is connected. *Note 1:* The root node, *i.e.*, the base node, is an ancestor of all the other nodes. *Note 2:* In a tree structure there is one and only one path from any point to any other point.

tree topology: *See* **network topology.**

T reference point: In Integrated Services Digital Networks (ISDN), the conceptual point dividing NT2 and NT1 functional groupings in a particular ISDN arrangement. (188)

tremendously high frequency (THF): Frequencies from 300 GHz to 3000 GHz. (188)

tributary office: A local office, located outside the exchange in which a toll center is located, that has a different rate center from its toll center. [After 47CFR]

tributary station: **1.** In a data network, a station other than the control station. **2.** On a multipoint connection or a point-to-point connection using basic mode link control, any data station other than the control station.

trim effect: In a crystal oscillator, the degradation of frequency-vs.-temperature stability, and marked frequency offset, resulting from frequency adjustment which produces a rotation or distortion, or both, of the inherent frequency-vs.-temperature characteristic.

triple precision: Characterized by the use of three computer words to represent a number in accordance with required precision.

triplet: A byte composed of three bits. *Synonym* **three-bit byte.**

TRI-TAC: *Acronym for* **tri-services tactical.** *See* **tactical communications.**

TRI-TAC equipment: Equipment that (a) accommodates the transition from current manual and analog systems to fully automated digital systems and (b) provides for message switching, voice communications circuit switching, and the use of secure voice terminals, digital facsimile systems, and user digital voice terminals.

troposcatter: *Synonym* **tropospheric scatter.**

troposphere: **1.** The lower layers of atmosphere, in which the change of temperature with height is relatively large. It is the region where clouds form, convection is active, and mixing is continuous and more or less complete. [JP1] **2.** The layer of the Earth's atmosphere, between the surface and the stratosphere, in which temperature decreases with altitude and which contains approximately 80% of the total air mass. (188) *Note:* The thickness of the troposphere varies with season and latitude. It is usually 16 km to 18 km thick over tropical regions, and less than 10 km thick over the poles.

tropospheric duct: *See* **atmospheric duct.**

tropospheric scatter: **1.** The propagation of radio waves by scattering as a result of irregularities or discontinuities in the physical properties of the troposphere. [NTIA] [RR] [JP1] **2.** A method of transhorizon communications using frequencies from approximately 350 MHz to approximately 8400 MHz. (188) *Note:* The propagation mechanism is still not fully understood, though it includes several distinguishable but changeable mechanisms such as propagation by means of random reflections and scattering from irregularities in the dielectric gradient density of the troposphere, smooth-Earth diffraction, and diffraction over isolated obstacles (knife-edge diffraction). *Synonym* **troposcatter.**

tropospheric wave: A radio wave that is propagated by reflection from a place of abrupt change in the dielectric constant, or its gradient, in the troposphere. (188) *Note:* In some cases, a ground wave may be so altered that new components appear to arise from reflection in regions of rapidly changing dielectric constant. When these components are distinguishable from the other components, they are called "*tropospheric waves.*"

true power: *Synonym* **effective power.**

truncated binary exponential backoff: In carrier sense multiple access with collision avoidance (CSMA/CA) networks and in carrier sense multiple access with collision detection (CSMA/CD) networks, the algorithm used to schedule retransmission after a collision such that the retransmission is delayed by an amount of time derived from the slot time and the number of attempts to retransmit.

truncation: The deletion or omission of a leading or a trailing portion of a string in accordance with specified criteria.

truncation error: In the representation of a number, the error introduced when one or more digits are dropped.

trunk: **1.** In a communications network, a single transmission channel between two points that are switching centers or nodes, or both. (188) **2.** [A] circuit between switchboards or other switching equipment, as distinguished from circuits which extend between central office switching equipment and information origination/termination equipment. [47CFR] *Note:* Trunks may be used to interconnect switches, such as major, minor, public and private switches, to form networks.

trunk encryption device (TED): A bulk encryption device used to provide secure communications over a wideband digital transmission link. (188) *Note:* A TED is usually located between the output of a trunk group multiplexer and a wideband radio or cable facility.

trunk group: Two or more trunks of the same type between two given points. (188)

trunk group multiplexer (TGM): A time-division multiplexer that combines individual digital trunk groups into a higher rate bit stream for transmission over wideband digital communications links.

trunk hunting: *See* **hunting (def. #1).**

trusted computer system (TCS): [An] AIS that employs sufficient hardware and software assurance measures to allow simultaneous processing of a range of classified or sensitive information. [NIS] (188)

trusted computing base (TCB): [The] totality of protection mechanisms within a computer system, including hardware, firmware, and software, the combination of which is responsible for enforcing a security policy. *Note:* The ability of a trusted computing base to enforce correctly a unified security policy depends on the correctness of the mechanisms within the trusted computing base, the protection of those mechanisms to ensure their correctness, and the correct input of parameters related to the security policy. [NIS]

truth table: **1.** An operation table for a logic operation. **2.** A table that describes a logic function by listing all possible combinations of input values and indicating, for each combination, the output value.

TSK: *Abbreviation for* **transmission security key.**

TSP: *Abbreviation for* **Telecommunications Service Priority.**

TSPS: *Abbreviation for* **traffic service position system.**

TSP system: *See* **Telecommunications Service Priority system.**

TTTN: *Abbreviation for* **tandem tie trunk network.**

TTY: *Abbreviation for* **teletypewriter.**

TTY/TDD: A unique telecommunication device for the deaf, using TTY principles.

tuning: Adjusting the parameters and components of a circuit so that it resonates at a particular frequency or so that the current or voltage is either maximized or minimized at a specific point in the circuit. *Note:* Tuning is usually accomplished by adjusting the capacitance or the inductance, or both, of elements that are connected to or in the circuit. [From Weik '89]

tunneling mode: *Synonym* **leaky mode.**

tunneling ray: *Synonym* **leaky ray.**

Turing machine: A mathematical model of a device that changes its internal state and reads from, writes on, and moves a potentially infinite tape, all in accordance with its present state, thereby constituting a model for computer-like behavior.

turnaround time: In a half-duplex circuit, the time required to reverse the direction of transmission from transmit to receive or vice versa. (188)

turnkey: Pertaining to a procurement process that (a) includes contractual actions at least through the system, subsystem, or equipment installation phase and (b) may include follow-on contractual actions, such as testing, training, logistical, and operational support. (188) *Note:* Precise definition of the types of allowable contractual features are contained in the Federal Acquisition Regulations (FAR).

twin cable: A cable composed of two parallel conductors separated from each other by a ribbon-like insulator or encased by a foam insulator. (188) *Synonym* **twin-lead.**

twin-lead: *Synonym* **twin cable.**

twinplex: A frequency-shift-keyed (FSK) carrier telegraphy system in which four unique tones, *i.e.*, two pairs of tones, are transmitted over a single transmission channel, such as one twisted pair. *Note:* One tone of each pair represents a *"mark"* and the other a *"space."*

twin sideband transmission: *See* **independent-sideband transmission.**

twist: In telephony, a change, as a function of temperature, in the shape of the frequency-vs.-attenuation response curve, *i.e.*, characteristic, of a transmission line.

twisted pair cable: *See* **paired cable.**

two-out-of-five code: A binary-coded decimal notation in which (a) each decimal digit is represented by a binary numeral consisting of five binary digits of which two are of one kind, called *"ones,"* and three are of the other kind, called *"zeros"* and (b) the usual weights assigned to the digit positions are 0-1-2-3-6, except that *"zero"* is represented as 01100.

two-pilot regulation: In frequency-division multiplexed (FDM) systems, the use of two pilot frequencies within a band so that the differential change in attenuation with respect to temperature, *i.e.*, twist, can be detected and compensated by a regulator. (188)

two-sample deviation: The square root of the Allan variance.

two-sample variance: *Synonym* **Allan variance.**

two-source frequency keying: *Synonym* **frequency-exchange signaling.**

two-tone keying: In telegraphy systems, keying in which the modulating wave causes the carrier to be modulated with a single tone for the *"mark"* and modulated with a different single tone for the *"space."* (188)

two-tone telegraph: *See* **two-tone keying.**

two-way alternate operation: *Synonym* **half-duplex operation.**

two-way simultaneous operation: *Synonym* **duplex operation.**

two-wire circuit: A communications circuit formed by two metallic conductors insulated from each other. (188) *Contrast with* **four-wire circuit.**

TWX®: *Acronym for* **teletypewriter exchange service.**

TX: *Abbreviation for* **transmitter, transmit.**

type 1 product: [A] classified or controlled cryptographic item endorsed by the National Security Agency for securing classified and sensitive U.S. Government information, when appropriately keyed. *Note:* The term refers only to products, and not to information, key, services, or controls. Type 1 products contain classified National Security Agency algorithms. They are available to U.S. Government users, their contractors, and federally sponsored non-U.S. Government activities subject to export restrictions in accordance with International Traffic in Arms Regulation. [NIS]

type 2 product: Unclassified cryptographic equipment, assembly, or component, endorsed by the National Security Agency, for use in telecommunications and automated information systems for the protection of national security information. *Note:* The term refers only to products, and not to information, key, services, or controls. Type 2 products may not be used for classified information, but contain classified National Security Agency algorithms that distinguish them from products containing the unclassified data algorithm. Type 2 products are subject to export restrictions in accordance with the International Traffic in Arms Regulation. [NIS]

type 3 algorithm: [A] cryptographic algorithm that has been registered by the National Institute of Standards and Technology and has been published as a Federal Information Processing Standard for use in protecting unclassified sensitive information or commercial information. [NIS]

type 4 algorithm: [An] unclassified cryptographic algorithm that has been registered by the National Institute of Standards and Technology, but is not a Federal Information Processing Standard. [NIS]

UDP: *Abbreviation for* **user datagram protocol.** An Internet protocol for datagram service.

UHF: *Abbreviation for* **ultra high frequency.** *See* **electromagnetic spectrum.**

U interface: For basic-rate access in an Integrated Services Digital Network (ISDN) environment, a user-to-network interface reference point that is characterized by the use of a 2-wire-loop transmission system that (a) conveys information between the 4-wire user-to-network interface, *i.e.*, the S/T reference point, and the local exchange, (b) is located in the servicing central office, and (c) is not as distance sensitive as a service using a T interface.

ULF: *Abbreviation for* **ultra low frequency.** *See* **electromagnetic spectrum.**

ultra high frequency (UHF): Frequencies from 300 MHz to 3000 MHz. (188)

ultra low frequency (ULF): Frequencies from 300 Hz to 3000 Hz. (188)

ultraviolet (uv): The portion of the electromagnetic spectrum in which the longest wavelength is just below the visible spectrum, extending from approximately 4 nm to approximately 400 nm. *Note:* Some authorities place the lower limit of uv at values between 1 and 40 nm, 1 nm being the upper wavelength limit of x-rays. The 400-nm limit is the lowest visible wavelength, *i.e.*, the highest visible frequency, violet.

unallowable character: *Synonym* **illegal character.**

unavailability: A expression of the degree to which a system, subsystem, or equipment is not operable and not in a committable state at the start of a mission, when the mission is called for at an unknown, *i.e.*, random, time. *Note 1:* The conditions determining operability and committability must be specified. (188) *Note 2:* Expressed mathematically, unavailability is 1 minus the availability. *Note 3:* Unavailability may also be expressed mathematically as the ratio of the total time a functional unit is not capable of being used during a given interval to the length of the interval, *e.g.*, if the unit is not capable of being used for 68 hours a week, the unavailability is 68/168.

unbalanced line: A transmission line, such as a coaxial cable, in which the magnitudes of the voltages on the two conductors are not equal with respect to ground. (188)

unbalanced modulator: A modulator in which the modulation factor is different for the alternate half-cycles of the carrier. (188) *Synonym* **asymmetrical modulator.**

unbalanced wire circuit: A circuit in which the two sides are inherently electrically dissimilar. (188)

unbound mode: *Synonym* **radiation mode.**

unbundling: In the context of the FCC's Computer III Inquiry, the process of separating individual tariffed offerings and services that are associated with a specific element in the CEI or ONA tariff from other tariffed basic service offerings. [After para. 158, FCC *Report and Order*, June 16, 1986.]

underflow: In computing, a condition occurring when a machine calculation produces a non-zero result that is smaller than the smallest non-zero quantity that the machine's storage unit is capable of storing or representing.

underground cable: A communication cable designed to be placed under the surface of the Earth in a duct system that isolates it from direct contact with the soil. (188) *Contrast with* **direct-buried cable.**

underlap: In facsimile, a defect that occurs when the width of the scanning line is less than the scanning pitch.

undershoot: *See* **overshoot.**

undesired signal: Any signal that tends to produce degradation in the operation of equipment or systems. (188)

undetected error rate: *Deprecated synonym for* **undetected error ratio.**

undetected error ratio: The ratio of the number of bits, unit elements, characters, or blocks incorrectly received and undetected, to the total number of bits, unit elements, characters, or blocks sent. *Synonyms* **residual error rate, undetected error rate** *[deprecated].*

undisturbed day: A day during which neither sunspot activity nor ionospheric disturbance causes detectable interfere with radio communications.

UNICOM station: *Synonym* **aeronautical advisory station.**

unidirectional channel: *Synonym* **one-way-only channel.**

unidirectional operation: Operation in which data are transmitted from a transmitter to a receiver in only one direction. (188)

uniform encoding: An analog-to-digital conversion process in which, except for the highest and lowest quantization steps, all of the quantization subrange values are equal. *Synonym* **uniform quantizing.**

uniform linear array: An antenna composed of a relatively large number of usually identical elements arranged in a single line or in a plane with uniform spacing and usually with a uniform feed system. (188)

uniform quantizing: *Synonym* **uniform encoding.**

uniform-spectrum random noise: *See* **white noise.**

uniform time scale: A time scale made up of equal intervals.

uniform transmission line: A transmission line that has distributed electrical properties, *i.e.,* resistance, inductance, and capacitance per unit length, that are constant along the line, and in which the voltage-to-current ratio does not vary with distance along the line, if the line is terminated in its characteristic impedance. *Note 1:* Examples of uniform transmission lines are coaxial cables, twisted pairs, and single wires at constant height above ground, all of which have no changes in geometry, materials, or construction along their length. *Note 2:* In a uniform transmission line, signal attenuation is a function of the length of the line and the frequency of the signal. [From Weik '89]

unilateral control system: *Synonym* **unilateral synchronization system.**

unilateral synchronization system: A system of synchronization in which signals from a single location are used to synchronize clocks at one or more other locations. *Synonym* **unilateral control system.**

unimode fiber: *Synonym* **single-mode optical fiber**.

unintelligible crosstalk: Crosstalk that consists of unintelligible signals, hence from which information cannot be derived. (188)

unintentional interference: *See* **interference.**

uninterruptible power supply (UPS): A device that is inserted between a primary power source, such as a commercial utility, and the primary power input of equipment to be protected, *e.g.,* a computer system, for the purpose of eliminating the effects of transient anomalies or temporary outages. *Note 1:* An UPS consists of an inverter, usually electronic, that is powered by a battery that is kept trickle-charged by rectified ac from the incoming power line fed by the utility. In the event of an interruption, the battery takes over without the loss of even a fraction of a cycle in the ac output of the UPS. The battery also provides protection against transients. The duration of the longest outage for which protection is ensured depends on the battery capacity, and to a certain degree, on the rate at which the battery is drained. *Note 2:* An UPS should not be confused with a standby generator, which may not provide protection from a momentary power interruption, or which may result in a momentary power interruption when it is switched into service, whether manually or automatically.

unipolar signal: A two-state signal where one of the states is represented by voltage or current and the other state is represented by no voltage or current. (188) *Note:* The current flow can be in either direction.

unique key: Key held only by one crypto-equipment and its associated distribution center. (188)

unit-distance code: An unweighted code that changes at only one digit position when going from one number to the next in a consecutive sequence of numbers. *Note 1:* Use of one of the many unit-distance codes can minimize errors at symbol transition points when converting analog quantities into digital quantities. *Note 2:* An example of a unit-distance code is the Gray code. [From Weik '89]

unit element: In the representation of a character, a signal element that has a duration equal to the unit interval. (188)

unit impulse: A mathematical artifice consisting of an impulse of infinite amplitude and zero width, and having an area of unity. *Note:* The unit impulse is useful for the mathematical expression of the impulse response, *i.e.*, the transfer function, of a device. *Synonym* **Dirac delta function.**

unit interval: In isochronous transmission, the longest interval of which the theoretical durations of the significant intervals of a signal are all whole multiples. (188)

universal personal telecommunications number: *Synonym* **UPT number.**

Universal Personal Telecommunications (UPT) service: A telecommunications service that provides personal mobility and service profile management. *Note 1:* UPT service involves the network capability of identifying uniquely a UPT user by means of a UPT number. *Note 2:* The general principles of UPT are given in ITU-T Recommendation F.850. *Note 3:* UPT and PCS are sometimes mistakenly assumed to be the same service concept. UPT allows complete personal mobility across multiple networks and service providers. PCS may use UPT concepts to improve subscriber mobility in allowing roaming to different service providers, but UPT and PCS are not the same service concept.

universal service: The concept of making basic local telephone service (and, in some cases, certain other telecommunications and information services) available at an affordable price to all people within a country or specified jurisdictional area.

Universal Time (UT): **1.** The basis for coordinated dissemination of time signals, counted from 0000 at midnight. **2.** In celestial navigation applications, the time which gives the exact rotational orientation of the Earth obtained from UTC by applying increments determined by the U.S. Naval Observatory. **3.** A measure of time that conforms, within a close approximation, to the mean diurnal rotation of the Earth and serves as the basis of civil timekeeping. *Note:* Universal Time (UT1) is determined from observations of the stars, radio sources, and also from ranging observations of the Moon and artificial Earth satellites. The scale determined directly from such observations is designated Universal Time Observed (UTO); it is slightly dependent on the place of observation. When UTO is corrected for the shift in longitude of the observing station caused by polar motion, the time scale UT1 is obtained. When an accuracy better than one second is not required, Universal Time can be used to mean Coordinated Universal Time (UTC). [JP1] **4.** The official civil time of the United Kingdom. *Formerly called* **Greenwich Mean Time.** *Contrast with* **Coordinated Universal Time.**

UNIX™: A portable, multiuser, time-shared operating system that supports process scheduling, job control, and a programmable user interface. *Note 1:* There are many proprietary operating systems that are based on UNIX™ and are colloquially referred to as UNIX™, but are not necessarily interoperable. *Note 2:* Most UNIX™-based operating systems are POSIX compliant.

unnumbered command: In a data transmission, a command that does not contain sequence numbers in the control field.

unnumbered response: In data transmission, a response that does not contain sequence numbers in the control field.

unsuccessful call: *Synonym* **unsuccessful call attempt.**

unsuccessful call attempt: A call attempt that does not result in the establishment of a connection. *Synonym* **unsuccessful call.**

unused character: *Synonym* **illegal character.**

up-converter: A device that translates frequencies from lower to higher frequencies. (188)

update: The regeneration of a display to show current status, based on changes to the previously displayed data. *Note:* An update can be accomplished upon user request or by automatic means.

uplink (U/L): **1.** The portion of a communications link used for the transmission of signals from an Earth terminal to a satellite or to an airborne platform. *Note:* An uplink is the converse of a downlink. (188) **2.** Pertaining to data transmission from a data station to the headend. (188)

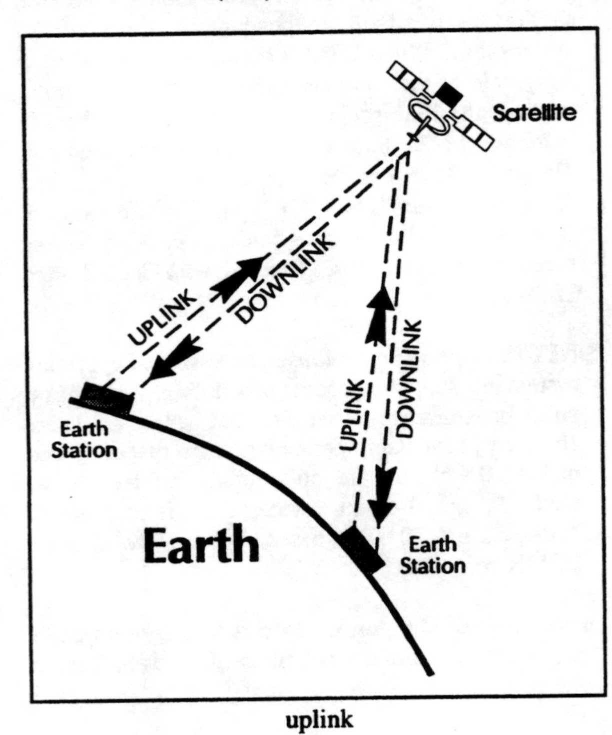

uplink

upright position: *Synonym* **erect position.**

UPS: *Acronym for* **uninterruptible power supply.**

upstream: **1.** The direction opposite the data flow. **2.** With respect to the flow of data in a communications path: at a specified point, the direction toward which data are received earlier than at the specified point.

UPT: *See* **Universal Personal Telecommunications service.**

UPT access code: In universal personal telecommunications service, the code that UPT users may need to dial, when using some terminals and networks, to enter the UPT environment before executing any UPT procedures.

UPT database: In universal personal telecommunications service, a repository for information, such as a service profile, that is related to a set of UPT customers and UPT users.

UPT environment: In universal personal telecommunications service, the environment within which the UPT service facilities are offered, consisting of combinations of networks and UPT service control facilities that, when combined, enable UPT users to make use of telecommunication services offered by these networks. *Note:* To the UPT user, the UPT environment appears as a single global network that provides personal mobility. However, when utilizing telecommunication services, the UPT network user may be limited by restrictions imposed by the network, by the capabilities of the terminal and network used, or by regulatory requirements.

uptime: The time during which a functional unit is fully operational. (188)

UPT indicator: In universal personal telecommunications, that portion (or portions) of the UPT number that identifies a call as a UPT call.

UPT number: In universal personal telecommunications, the number that uniquely identifies a UPT user and that is used to place a call to, or to forward a call to, that user. *Note:* A user may have multiple UPT numbers, *e.g.,* a business UPT number for business calls and a private UPT number for private calls. In the case of multiple numbers, each UPT number is considered, from a network vantage point, to identify a distinct UPT user, even if all such numbers identify the same person or entity. *Synonym* **universal personal telecommunications number.**

UPT routing address: In universal personal telecommunications, the number used by the network to direct a call according to the user's UPT service profile.

UPT service profile: In universal personal telecommunications, a record that contains all information related to a UPT user, which information is required to provide that user with UPT service such as subscriptions to basic and supplementing services and call-routing preferences. *Note:* Each UPT service profile is associated with a single UPT number. *Synonym* **service profile.**

UPT service profile management: In universal personal telecommunications, authorized access to and manipulation of the UPT service profile. *Note:* UPT service profile management can be performed by the UPT user, by the UPT customer (subscriber), or by the UPT service provider. *Synonym* **service profile management.**

UPT subscriber: In universal personal telecommunications, a person who (or entity that) obtains a UPT service from a UPT service provider on behalf of one or more UPT users.

UPT user: In universal personal telecommunications, a person who (or entity that) has access to universal personal telecommunication (UPT) services and who has been assigned a UPT number.

UPT user group: A specific set of universal personal telecommunication users.

usable line length: *See* **available line.**

usage: *Synonym* **occupancy.**

useful line: *Synonym* **available line.**

user: **1.** A person, organization, or other entity (including a computer or computer system), that employs the services provided by a telecommunication system, or by an information processing system, for transfer of information. (188) *Note:* A user functions as a source or final destination of user information, or both. *Synonym* **subscriber.** **2.** [A] person or process accessing an AIS by direct connections (*e.g.*, via terminals) or indirect connections. *Note:* "Indirect connection" relates to persons who prepare input data or receive output that is not reviewed for content or classification by a responsible individual. [NIS]

User Datagram Protocol (UDP): In the Internet Protocol suite, a standard, low-overhead, connectionless, host-to-host protocol that is used over packet-switched computer communications networks, and that allows an application program on one computer to send a datagram to an application program on another computer. *Note:* The main difference between UDP and TCP is that UDP provides connectionless service, whereas TCP does not.

user information: Information transferred across the functional interface between a source user and a telecommunications system for delivery to a destination user. *Note:* In telecommunications systems, user information includes user overhead information.

user information bit: A bit transferred from a source user to a telecommunications system for delivery to a destination user. *Note 1:* User information bits do not include the overhead bits originated by, or having their primary functional effect within, the telecommunications system. *Note 2:* User information bits are encoded to form channel bits.

user information block: A block that contains at least one user information bit. (188)

user line: *Synonym* **loop (def. #2).**

user overhead information: *See* **overhead information.**

user service class: *Synonym* **class of service.**

UT: *Abbreviation for* **Universal Time.**

UTC: *Abbreviation for* **Coordinated Universal Time.**

UTC(i): Coordinated Universal Time (UTC), as kept by the *"i"* laboratory, where *i* is any laboratory cooperating in the determination of UTC. *Note:* In the United States, the official UTC is kept by the U.S. Naval Observatory and is referred to as UTC (USNO).

utility load: *Synonym* **nonoperational load.**

utility program: A computer program that is in general support of the operations and processes of a computer. *Note:* Examples of utility programs include diagnostic programs, trace programs, input routines, and programs used to perform routine tasks, *i.e.,* perform everyday tasks, such as copying data from one storage location to another. *Synonyms* **service program, service routine, tool, utility routine.**

utility routine: *Synonym* **utility program.**

uv: *Abbreviation for* **ultraviolet.**

validation: **1.** Tests to determine whether an implemented system fulfills its requirements. **2.** The checking of data for correctness or for compliance with applicable standards, rules, and conventions. (188) **3.** [The] process of applying specialized security test and evaluation procedures, tools, and equipment needed to establish acceptance for joint usage of an AIS by one or more departments or agencies and their contractors. *Note:* This action will include, as necessary, final development, evaluation, and testing, preparatory to acceptance by senior security test and evaluation staff specialists. [NIS] **4.** In universal personal telecommunications, the process of verifying that a user or terminal is authorized to access UPT services.

value-added carrier: A company that sells the services of a value-added network.

value-added network (VAN): A network using the communication services of other commercial carriers, using hardware and software that permit enhanced telecommunication services to be offered.

VAN: *Acronym for* **value-added network.**

variable length buffer: A buffer into which data may be entered at one rate and removed at another rate without changing the data sequence. (188) *Note:* Most first-in first-out (FIFO) storage devices are variable-length buffers in that the input rate may be variable while the output rate is constant or the output rate may be variable while the input rate is constant. Various clocking and control systems are used to allow control of underflow or overflow conditions.

variable slope delta modulation: *See* **continuously variable slope delta modulation.**

variance: In statistics, in a population of samples, the mean of the squares of the differences between the respective samples and their mean, expressed mathematically as:

$$\sigma^2 = \frac{1}{n}\sum_{i=1}^{n}(x_i - \bar{x})^2 \quad ,$$

where n is the number of samples, x_i is the value of sample i, $\bar{x}$ is the mean of the samples, and σ^2 is the variance. *Note:* The square root of the variance, σ, is the standard deviation.

variant: **1.** One of two or more code symbols which have the same plain text equivalent. [NIS] **2.** One of several plain text meanings that are represented by a single code group. [JP1]

variation monitor: In ac power distribution, a device for sensing deviations of any measured variable, such as voltage, current, or frequency, and capable of initiating a programmed action, such as transfer to other power sources, when programmed limits of voltage, current, frequency, or time are exceeded, or providing an alarm, or both. (188)

vars: *Abbreviation for* **volt-amperes reactive.**

VC: *Abbreviation for* **virtual circuit.**

VDU: *Abbreviation for* **visual display unit.** *See* **monitor.**

vector processor: *Synonym* **array processor.**

verified off-hook: In telephone systems, a service provided by a unit that is inserted on each end of a transmission circuit for the purpose of verifying supervisory signals on the circuit. (188) *See* **automatic ringdown circuit.**

vertex angle: In an optical fiber, the angle formed by the extreme bound meridional rays accepted by the fiber, or emerging from it, equal to twice the acceptance angle; the angle formed by the largest cone of light accepted by the fiber or emitted from it. [FAA]

vertical redundancy check (VRC): *Synonym* **transverse redundancy check.**

very high frequency (VHF): Frequencies from 30 MHz to 300 MHz. (188)

very low frequency (VLF): Frequencies from 3 kHz to 30 kHz. (188)

vestigial sideband (VSB) transmission: Modified AM transmission in which one sideband, the carrier, and only a portion of the other sideband are transmitted. (188)

VF: *Abbreviation for* **voice frequency.**

VFCT: *Abbreviation for* **voice frequency carrier telegraph.** *See* **voice-frequency telegraph.**

VFCTG: *Abbreviation for* **voice-frequency carrier telegraph.** *See* **voice-frequency telegraph.**

VF patch bay: *See* **voice frequency primary patch bay.**

VFTG: *Abbreviation for* **voice-frequency telegraph.**

VHF: *Abbreviation for* **very high frequency.** *See* **electromagnetic spectrum.**

via net loss (VNL): Pertaining to circuit performance prediction and description that allows circuit parameters to be predetermined and the circuit to be designed to meet established criteria by analyzing actual, theoretical, and calculated losses.

video: **1.** An electrical signal containing timing (synchronization), luminance (intensity), and often chrominance (color) information that, when displayed on an appropriate device, gives a visual image or representation of the original image sequences. **2.** Pertaining to the sections of a television system that carry television signals, either in unmodulated or modulated form. (188) **3.** Pertaining to the demodulated radar signal that is applied to a radar display device. (188) **4.** Pertaining to the bandwidth or data rate necessary for the transmission of real-time television pictures. [After FAA] *Note:* In practice, the baseband bandwidth required for the transmission of NTSC television pictures (not including the audio carriers) is approximately 5 MHz.

video codec: *See* **codec.**

videoconference: *See* **video teleconference.**

video display terminal: *Synonym* **visual display unit.** *See* **monitor.**

video display unit: *Synonym* **visual display terminal.** *See* **monitor.**

video frame: *See* **frame (def. #6).**

videophone: **1.** A telephone that is coupled to an imaging device that enables the call receiver or the call originator, or both, to view one another as on television, if they so desire. [From Weik '89] **2.** A military communications terminal that (a) has video teleconference capability, (b) is usually configured as a small desktop unit, designed for one operator, and (c) is a single, integrated unit. (188) [From Weik '89]

video teleconference: **1.** A teleconference that includes video communications. (188) **2.** Pertaining to a two-way electronic communications system that permits two or more persons in different locations to engage in the equivalent of face-to-face audio and video communications. *Note*: Video teleconferences may be conducted as if all of the participants are in the same room. (188)

video teleconferencing unit (VTU): Equipment that performs video teleconference functions, such as coding and decoding of audio and video signals and multiplexing of video, audio, data, and control signals, and that usually does not include I/O devices, cryptographic devices, network interface equipment, network connections, or the communications network to which the unit is connected. (188)

view: In satellite communications, the quality or degree of visibility of a satellite to a ground station; *i.e.,* the degree to which the satellite is sufficiently above the horizon and clear of obstructions so that it is within a clear line of sight by an Earth terminal. (188) *Note:* A pair of satellite Earth terminals has a satellite in mutual view when both have unobstructed line-of-sight contact with the satellite simultaneously.

viewdata: A type of information-retrieval service in which a subscriber can (a) access a remote database via a common carrier channel, (b) request data, and (c) receive requested data on a video display over a separate channel. *Note:* The access, request, and reception are usually via common carrier broadcast channels. *Contrast with* **teletext.**

violation: *See* **AMI violation.**

virtual call: A call, established over a network, that uses the capabilities of either a real or virtual circuit by sharing all or any part of the resources of the circuit for the duration of the call.

virtual call capability: A service feature in which (a) a call set-up procedure and a call disengagement procedure determine the period of communication between two data terminal equipments (DTEs) in which user data are transferred by the network in the packet mode of operation, (b) end-to-end transfer control of packets within the network is required, (c) data may be delivered to the network by the call originator before the call access phase is completed, but the data are not delivered to the call receiver if the call attempt is unsuccessful, (d) the network delivers all the user data to the call receiver in the same sequence in which the data are received by the network, and (e) multi-access DTEs may have several virtual calls in progress at the same time. *Synonym* **virtual call facility.**

virtual call facility: *Synonym* **virtual call capability.**

virtual carrier frequency: In radio or carrier systems in which no carrier is transmitted, *e.g.*, single sideband or double sideband with suppressed carrier, the location in the frequency spectrum that the carrier would occupy if it were present. (188)

virtual circuit (VC): A communications arrangement in which data from a source user may be passed to a destination user over various real circuit configurations during a single period of communication. (188) *Note:* Virtual circuits are generally set up on a per-call basis and are disconnected when the call is terminated; however, a permanent virtual circuit can be established as an option to provide a dedicated link between two facilities. *Synonyms* **logical circuit, logical route.**

virtual circuit capability: A network-provided service feature in which a user is provided with a virtual circuit. *Note:* Virtual circuit capability is not necessarily limited to packet mode transmission. For example, an analog signal may be converted to a digital signal and then be routed over the network via any available route.

virtual connection: A logical connection that is made to a virtual circuit.

virtual height: The apparent height of an ionized layer, as determined from the time interval between the transmitted signal and the ionospheric echo at vertical incidence. (188)

virtual instruction: *Synonym* **teletraining.**

virtual memory: In computer systems, the memory as it appears to, *i.e.*, as it is available to, the operating programs running in the central processing unit (CPU). *Note:* The virtual memory may be smaller, equal to, or larger than the real memory present in the system.

virtual network: A network that provides virtual circuits and that is established by using the facilities of a real network.

virtual path: *See* **virtual circuit.**

virtual reality: An interactive, computer-generated simulated environment with which users can interact using specialized peripherals such as data gloves and head-mounted computer-graphic displays.

virtual storage: The storage space that may be regarded as addressable main storage by the user of a computer system in which virtual addresses are mapped into real addresses. *Note:* The size of virtual storage is limited by the addressing scheme of the computer system and by the amount of auxiliary storage available, and not by the actual number of main storage locations.

virtual terminal (VT): In open systems, an application service that (a) allows host terminals on a multi-user network to interact with other hosts regardless of terminal type and characteristics, (b) allows remote log-on by local-area-network managers for the purpose of management, (c) allows users to access information from another host processor for transaction processing, and (d) serves as a backup facility.

virus: An unwanted program which places itself into other programs which are shared among computer systems, and replicates itself. *Note:* A virus is usually manifested by a destructive or disruptive effect on the executable program that it affects.

visible spectrum: The region of the electromagnetic spectrum that can be perceived by human vision, approximately the wavelength range of 0.4 μm to 0.7 μm.

visual display unit (VDU): *See* **monitor.**

vitreous silica: Glass consisting of almost pure silicon dioxide (SiO_2). *Synonym* **fused silica.**

VLF: *Abbreviation for* **very low frequency.** *See* **electromagnetic spectrum.**

VNL: *Abbreviation for* **via net loss.**

V number: *Synonym* **normalized frequency (def. #1).**

vocoder: *Abbreviation for* **voice-coder.** A device that usually consists of a speech analyzer, which converts analog speech waveforms into narrowband digital signals, and a speech synthesizer, which converts the digital signals into artificial speech sounds. (188) *Note 1:* For COMSEC purposes, a vocoder may be used in conjunction with a key generator and a modulator-demodulator to transmit digitally encrypted speech signals over narrowband voice communications channels. These devices are used to reduce the bandwidth requirements for transmitting digitized speech signals. *Note 2:* Some analog vocoders move incoming signals from one portion of the spectrum to another portion.

vodas: *Acronym for* **voice-operated device anti-sing.** A device used to prevent overall voice-frequency singing in a two-way telephone circuit by ensuring that transmission can occur in only one direction at any given instant. (188)

vogad: *Acronym for* **voice-operated gain-adjusting device.** A device that has a substantially constant output amplitude over a wide range of input amplitudes. (188)

voice band: *Synonym* **voice frequency.**

voice coder: *See* **vocoder.**

voice-data signal: *See* **quasi-analog signal.**

voice frequency (VF): Pertaining to those frequencies within that part of the audio range that is used for the transmission of speech. (188) *Note 1:* In telephony, the usable voice-frequency band ranges from approximately 300 Hz to 3400 Hz. *Note 2:* In telephony, the bandwidth allocated for a single voice-frequency transmission channel is usually 4 kHz, including guard bands. *Synonym* **voice band.**

voice frequency carrier telegraph (VFCT): *Synonym* **voice-frequency telegraph.**

voice-frequency (VF) channel: A channel capable of carrying analog and quasi-analog signals. (188)

voice frequency (VF) primary patch bay: A patching facility that provides the first appearance of local-user VF circuits in the technical control facility (TCF). (188) *Note:* The VF primary patch bay provides patching, monitoring, and testing for all VF circuits. Signals will have various levels and signaling schemes depending on the user terminal equipment.

voice-frequency telegraph (VFTG): A method of multiplexing one or more dc telegraph channels onto a nominal 4-kHz voice frequency channel. (188) *Synonym* **voice frequency carrier telegraph.**

voice grade: In the public regulated telecommunications services, a service grade that is described in part 68, Title 47 of the *Code of Federal Regulations [CFR]. Note:* Voice-grade service does not imply any specific signaling or supervisory scheme.

voice-operated device anti-sing: *See* **vodas.**

voice-operated gain-adjusting device: *See* **vogad.**

voice operated relay circuit: *Synonym* **vox.**

voice operated transmit: *Synonym* **vox.**

voice-plus circuit: *Synonym* **composited circuit.**

volatile storage: A storage device in which the contents are lost when power is removed.

volatility: *See* **data volatility.**

voltage standing wave ratio (VSWR): In a transmission line, the ratio of maximum to minimum voltage in a standing wave pattern. *Note:* The VSWR is a measure of impedance mismatch between the transmission line and its load. The higher the VSWR, the greater the mismatch. The minimum VSWR, *i.e.,* that which corresponds to a perfect impedance match, is unity. (188)

volt-amperes reactive (vars): In alternating-current power transmission and distribution, the product of the rms voltage and amperage, *i.e.,* the apparent power, multiplied by the sine of the phase angle between the voltage and the current. *Note 1:* Vars represents the power not consumed by a reactive load, *i.e.,* when there is a phase difference between the applied voltage and the current. *Note 2:* Only effective power, *i.e.,* the actual power delivered to or consumed by the load, is expressed in watts. Volt-amperes reactive is properly expressed only in volt-amperes, never watts. *Note 3:* To maximize transmission efficiency, vars must be minimized by balancing capacitive and inductive loads, or by the addition of an appropriate capacitive or inductive reactance to the load.

volume: A portion of data, with its physical storage medium, that can be handled conveniently as a unit. *Note:* An example of a volume is a "floppy" diskette.

volume unit (vu): A unit of measurement of the power of an audio-frequency signal, as measured by a vu meter. (188) *Note 1:* The vu meter is built and used in accordance with American National Standard C16.5-1942. *Note 2:* When using the vu meter to measure sine wave test tone power, 0 vu equals 0 dBm.

vox: An acoustoelectric transducer and a keying relay connected so that the keying relay is actuated when sound, or voice, energy above a certain threshold is sensed by the transducer. *Note:* A vox is used to eliminate the need for push-to-talk operation of a transmitter by using voice energy to turn on the transmitter. (188) *Synonyms* **voice operated relay circuit, voice operated transmit.**

V reference point: The interface point in an ISDN environment between the line termination and the exchange termination.

VSB: *Abbreviation for* **vestigial sideband.** *See* **vestigial sideband transmission.**

V-series Recommendations: Sets of telecommunications protocols and interfaces defined by CCITT (now ITU-T) Recommendations. *Note:* Some of the more common V.-series Recommendations are:

V.21: A CCITT Recommendation for modem communications over standard commercially available lines at 300 b/s. This protocol is generally not used in the United States.

V.22bis: A CCITT Recommendation for modem communications over standard commercially available voice-grade channels at 2,400 b/s and below.

V.32: A CCITT Recommendation for modem communications over standard commercially available voice-grade channels at 9.6 kb/s and below.

V.32bis: A CCITT Recommendation for modem communication over standard commercially available voice-grade channels at 14.4 kb/s and below.

V.34: An ITU-T Recommendation for modem communication over standard commercially available voice-grade channels at 28.8 kb/s and below.

V.42: A CCITT Recommendation for error correction on modem communications.

V.42bis: A CCITT Recommendation for data compression on a modem circuit.

V.FAST: A new CCITT Recommendation for high-speed modems currently under development.

VSWR: *Abbreviation for* **voltage standing wave ratio.**

vu: *Abbreviation for* **volume unit.**

wafer: A thin slice of semiconducting material, such as a silicon crystal, upon which microcircuits are constructed by diffusion and deposition of various materials. *Note:* Millions of individual circuit elements, constituting hundreds of microcircuits, may be constructed on a single wafer. The individual microcircuits are separated by scoring and breaking the wafer into individual chips ("dice").

WAIS: *Acronym for* **Wide Area Information Servers.** A distributed text searching system that uses the protocol standard ANS Z39.50 to search index databases on remote computers. *Note 1:* WAIS libraries are most often found on the Internet. *Note 2:* WAIS allows users to discover and access information resources on the network without regard to their physical location. *Note 3:* WAIS software uses the client-server model.

WAN: *Acronym for* **wide area network.**

wander: Relative to jitter and swim, long-term random variations of the significant instants of a digital signal from their ideal positions. *Note 1:* Wander variations are those that occur over a period greater than 1 s (second). *Note 2:* Jitter, swim, wander, and drift have increasing periods of variation in that order.

warm boot: *Synonym* **warm restart (def. #2).**

warm restart: 1. A sequence of operations that is performed to reset a previously running system, after an unintentional shutdown. *Synonym* **warm start. 2.** In computer operations, the restarting of equipment, after a sudden shutdown, that allows reuse of previously retained initialized input data, retained programs, and retained output queues. *Note 1:* A warm restart may be needed after a program failure. *Note 2:* A warm start or restart cannot occur if initial data, programs, and files are not retained after closedown. *Synonyms* **hot boot, warm boot.** [From Weik '89]

warm start: *Synonym* **warm restart (def. #2).**

Warner exemption: A statutory exemption pertaining to the acquisition of telecommunications systems that meet the exclusionary criteria of the Warner Amendment, Public Law 97-86, 1 December 1981, which is also known as the Brooks Bill. (188) *Note:* Use of FTS2000 by U.S. Government agencies is mandatory when telecommunications are required. However, the Warner Amendment excludes the mandatory use of FTS2000 in instances related to maximum security.

WATS: *Acronym for* **Wide Area Telephone Service.**

wave equation: *See* **Maxwell's equations.**

waveform: The representation of a signal as a plot of amplitude versus time.

wavefront: The surface defined by the locus of points that have the same phase, *i.e.,* have the same path length from the source. [After 2196] *Note 1:* The wavefront is perpendicular to the ray that represents an electromagnetic wave. *Note 2:* The plane in

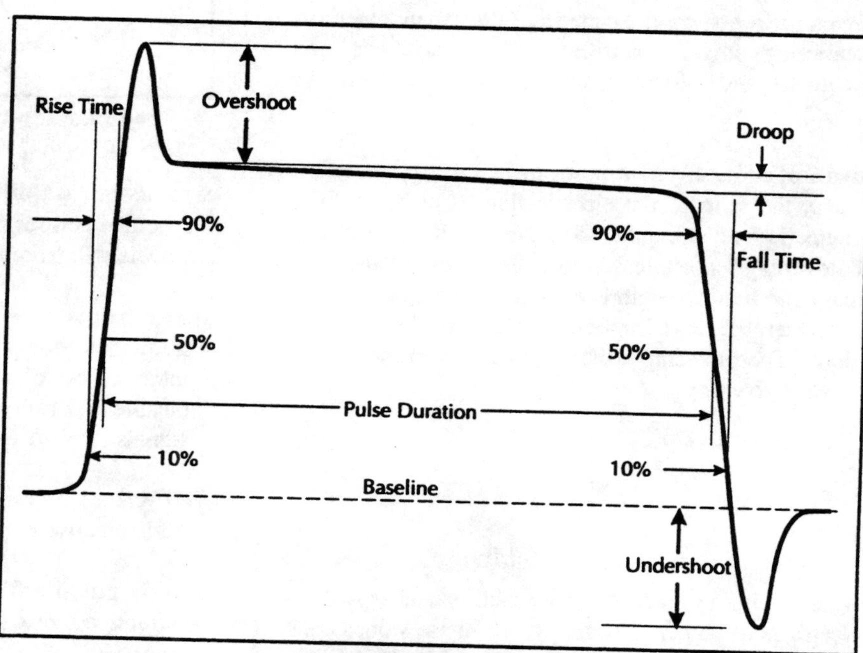

representative pulse waveform

which the electric and magnetic field vectors lie is tangential to the wavefront at every point. *Note 3:* The vector that represents the wavefront indicates the direction of propagation. *Note 4:* For parallel, *i.e.*, collimated, rays, the wavefront is plane. For rays diverging from a point, or converging toward a point, the wavefront is spherical. For rays with varying divergence or convergence, the wavefront has other shapes, such as ellipsoidal and paraboloidal, depending on the nature of the source.

waveguide: A material medium that confines and guides a propagating electromagnetic wave. (188) *Note 1:* In the microwave regime, a waveguide normally consists of a hollow metallic conductor, usually rectangular, elliptical, or circular in cross section. This type of waveguide may, under certain conditions, contain a solid or gaseous dielectric material. *Note 2:* In the optical regime, a waveguide used as a long transmission line consists of a solid dielectric filament (optical fiber), usually circular in cross section. In integrated optical circuits an optical waveguide may consist of a thin dielectric film. *Note 3:* In the rf regime, ionized layers of the stratosphere and refractive surfaces of the troposphere may also act as a waveguide.

waveguide dispersion: *See* **dispersion.**

waveguide scattering: Scattering (other than material scattering) that is attributable to variations of geometry and refractive index profile of an optical fiber.

wave impedance: At a point in an electromagnetic wave, the ratio of the electric field strength to the magnetic field strength. (188) *Note 1:* If the electric field strength is expressed in volts per meter and the magnetic field strength is expressed in ampere-turns per meter, the wave impedance will have the units of ohms. The wave impedance, Z, of an electromagnetic wave is given by

$$Z = \sqrt{\frac{\mu}{\epsilon}} ,$$

where μ is the magnetic permeability and ϵ is the electric permittivity. For free space, these values are $4\pi \times 10^{-7}$ H/m (henries per meter) and $(1/36\pi)$ F/m (farads per meter), from which 120π, *i.e.*, 377, ohms is obtained. In dielectric materials, the wave impedance is 377/n, where *n* is the refractive index. *Note 2:* Although the ratio is called the wave impedance, it is also the impedance of the free space or the material medium.

wavelength: The distance between points of corresponding phase of two consecutive cycles of a wave. (188) *Note:* The wavelength, λ, is related to the propagation velocity, *v*, and the frequency, *f*, by $\lambda = v/f$.

wavelength-division multiplexing (WDM): In optical fiber communications, any technique by which two or more optical signals having different wavelengths may be simultaneously transmitted in the same direction over one fiber, and then be separated by wavelength at the distant end. (188)

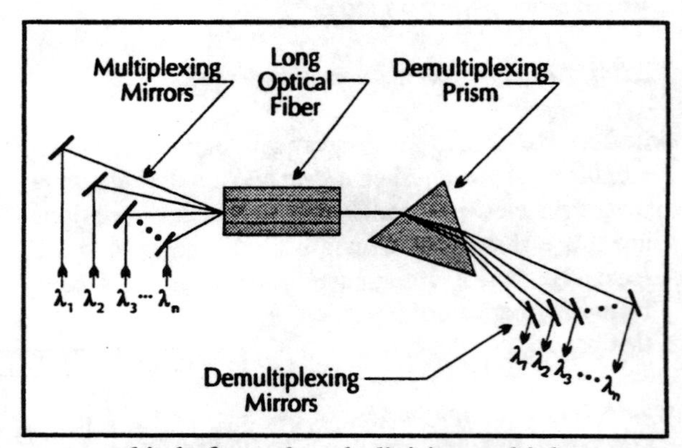

one kind of wavelength-division multiplexer

wavelength stability: Of an optical source during a specified period, the maximum deviation of the peak wavelength from its mean value.

wave trap: A device used to exclude unwanted frequency components, such as noise or other interference, of a wave. *Note:* Traps are usually tunable to permit selection of unwanted or interfering signals. [From Weik '89]

WDM: *Abbreviation for* **wavelength-division multiplexing**.

weakly guiding fiber: An optical fiber in which the refractive index contrast is small (substantially less than 1%).

web browser: A user interface (usually graphical) to hypertext information on the World Wide Web.

weighting network: A network having a loss that varies with frequency in a predetermined manner, and which network is used for improving or correcting transmission characteristics, or for characterizing noise measurements. (188)

whip antenna: A flexible rod antenna, usually between 1/10 and 5/8 wavelength long, supported on a base insulator.

white area: The area or population which does not receive interference-free primary service from an authorized AM station or does not receive a signal strength of at least 1 mV/m from an authorized FM station. [47CFR]

white facsimile transmission: **1.** In an amplitude-modulated facsimile system, transmission in which the maximum transmitted power corresponds to the minimum density, *i.e.,* the white area, of the object. (188) **2.** In a frequency-modulated facsimile system, transmission in which the lowest transmitted frequency corresponds to the minimum density *i.e.,* the white area, of the object. (188)

white noise: Noise having a frequency spectrum that is continuous and uniform over a specified frequency band. (188) *Note:* White noise has equal power per hertz over the specified frequency band. *Synonym* **additive white gaussian noise.**

white pages: **1.** A hard copy telephone directory listing of subscriber names, addresses, and telephone numbers. *Note: White pages* is associated with the residential subscriber listings in the standard directories distributed by the Bell System before divestiture. **2.** An electronic information database that contains user names and their associated network addresses, in the manner of a telephone directory. *Note*: Electronic white pages usually contain additional information, such as office location, phone number, and mailstop.

white signal: In facsimile systems, the signal resulting from scanning a minimum-density area, *i.e.,* the white area, of the object. (188)

who-are-you (WRU) character: A transmission-control character used for (a) switching on an answer-back unit in the station with which the connection has been established, (b) triggering the receiving unit to transmit an answer-back code to the terminal that transmitted the WRU signal, and (c) initiating a response that might include station identification, an indication of the type of equipment that is in service, and the status of the station. *Note 1:* The WRU signal corresponds to the 7-bit code assigned to the WRU. *Note 2:* The receiving unit may be a telegraph unit, data terminal equipment (DTE), or other unit. *Synonym* **WRU signal.** [From Weik '89]

Wide Area Information Servers (WAIS): *See* **WAIS.**

wide area network (WAN): A physical or logical network that provides data communications to a larger number of independent users than are usually served by a local area network (LAN) and is usually spread over a larger geographic area than that of a LAN. *Note 1:* WANs may include physical networks, such as Integrated Services Digital Networks (ISDNs), X.25 networks, and T1 networks. *Note 2:* A metropolitan area network (MAN) is a WAN that serves all the users in a metropolitan area. WANs may be nationwide or worldwide.

Wide Area Telephone Service (WATS): A toll service offering for customer dial-type telecommunications between a given customer [user] station and stations within specified geographic rate areas employing a single access line between the customer [user] location and the serving central office. Each access line may be arranged for either outward (OUT-WATS) or inward (IN-WATS) service, or both. [47CFR] *Note:* The offering is for fixed-rate inter- and intra-LATA services measured by zones and hours.

wideband: **1.** The property of any communications facility, equipment, channel, or system in which the range of frequencies used for transmission is greater than 0.1 % of the midband frequency. (188) *Note: "Wideband"* has many meanings depending upon application. *"Wideband"* is often used to distinguish it from *"narrowband,"* where both terms are subjectively defined relative to the implied context. **2.** In communications security systems, a bandwidth exceeding that of a nominal 4-kHz telephone channel. (188) [From Weik '89] **3.** The property of a circuit that has a bandwidth wider than normal for the type of circuit, frequency of operation, or type of modulation. **4.** In telephony, the property of a circuit that has a bandwidth greater than 4 kHz. (188) **5.** Pertaining to a signal that occupies a broad frequency spectrum. *Synonym* **broadband.** [From Weik '89]

wideband channel: A communication channel of a bandwidth equivalent to twelve or more voice-grade channels. [47CFR]

wideband modem: **1.** A modem whose modulated output signal can have an essential frequency spectrum that is broader than that which can be wholly contained within, and faithfully transmitted through, a voice channel with a nominal 4-kHz bandwidth. (188) **2.** A modem whose bandwidth capability is greater than that of a narrowband modem.

wildcard character: **1.** A character that may be substituted for any of a defined subset of all possible characters. *Note 1:* In high-frequency (HF) radio automatic link establishment, the wildcard character "?" may be substituted for any one of the 36 characters, "A" through "Z" and "0" through "9." *Note 2:* Whether the wildcard character represents a single character or a string of characters must be specified. (188) **2.** In computer (software) technology, a character that can be used to substitute for any other character or characters in a string. *Note:* The asterisk (*) usually substitutes as a wildcard character for any one or more of the ASCII characters, and the question mark (?) usually substitutes as a wildcard character for any one ASCII character.

wild-point detection: *Synonym* **reasonableness check.**

WIN: *Abbreviation for* **WWMCCS Intercomputer Network.**

window: **1.** In fiber optics, a band of wavelengths at which an optical fiber is sufficiently transparent for practical use in communications applications. [After FAA] *Synonyms* **spectral window, transmission window.** *See* **first window, second window, third window.** **2.** In imagery, a portion of a display surface in which display images pertaining to a particular application can be presented. *Note:* Different applications can be displayed simultaneously in different windows. **3.** A period during which an event can occur, can be expected to occur, or is allowed to occur.

windowing: Sectioning of a video display area into two or more separate regions for the purpose of displaying images from different sources. (188) *Note:* In windowing, one window could display data, another motion video from a remote site, and another, graphics.

wink: In telephone switching systems, a single supervisory pulse, *i.e.*, the momentary presence of, or interruption of, a supervisory signal. (188) *Note:* An example of a wink is the momentary flash of a supervisory light on an attendant's switchboard.

wink pulsing: In telephone switching systems, recurring pulsing in which the off-condition is relatively short compared to the on-condition. *Note:* On key-operated telephone instruments, the hold position, *i.e.*, the hold condition, of a line is often indicated by wink pulsing the associated lamp at 120 pulses per minute. During 6% of the pulse period the lamp is off and 94% of the period the lamp is on, *i.e.*, 30 ms (milliseconds) off and 470 ms on. (188)

wired radio frequency systems: Systems employing restricted radiation devices in which the radio frequency energy is conducted or guided along wires or in cables, including electric power and telephone lines. [NTIA]

wireless access mode: In personal communications service, interfacing with a network access point by means of a standardized air interface protocol without the use of a hardwired connection to the network.

wireless mobility management: In Personal Communications Service (PCS), the assigning and controlling of wireless links for terminal network connections. *Note:* Wireless mobility management provides an "alerting" function for call completion to a wireless terminal, monitors wireless link performance to determine when an automatic link transfer is required, and coordinates link transfers between wireless access interfaces.

wireless terminal: Any mobile terminal, mobile station, personal station, or personal terminal using non-fixed access to the network.

wireline common carrier: Common carriers [that] are in the business of providing landline local exchange telephone service. [47CFR]

word: A character string or a bit string considered to be an entity for some purpose. (188) *Note:* In telegraph communications, six character intervals are defined as a word when computing traffic capacity in words per minute, which is computed by multiplying the data signaling rate in baud by 10 and dividing the resulting product by the number of unit intervals per character.

word length: The number of characters or bits in a word.

word processing: The use of a computer system to manipulate text. *Note:* Examples of word processing functions include entering, editing, rearranging, sorting, storing, retrieving, displaying, and printing text. *Synonym* **text processing.**

work space: In computers and data processing systems, the portion of main storage that is used by a computer program for the temporary storage of data.

work station: **1.** In automated systems, such as computer, communications, and control systems, the input, output, display, and processing equipment that provides the operator-system interface. **2.** A configuration of input, output, display, and processing equipment that constitutes a stand-alone system not requiring external access.

World Time: *Synonym* **Coordinated Universal Time.**

World Wide Web (WWW): An international, virtual-network-based information service composed of Internet host computers that provide on-line information in a specific hypertext format. *Note 1:* WWW servers provide hypertext metalanguage (HTML) formatted documents using the hypertext transfer protocol (HTTP). *Note 2:* Information on the WWW is accessed with a hypertext browser such as Mosaic, Viola, or Lynx. *Note 3:* No hierarchy exists in the WWW, and the same information may be found by many different approaches.

worst hour of the year: That hour of the year during which the median noise over any radio path is at a maximum. (188) *Note:* This hour is considered to coincide with the hour during which the greatest transmission loss occurs.

W-profile fiber: *Synonym* **doubly clad fiber.**

wrapping: **1.** In a network using dual counter-rotating ring architecture, reconfiguration to circumvent a failed link or node. **2.** In open systems architecture, the use of a network to connect two other networks, thus providing an increased interaction capability between the two connected networks. *Note:* Recurring application of wrapping usually results in a hierarchical structure. [From Weik '89]

write: To make a permanent or transient recording of data in a storage device or on a data medium.

write cycle time: The minimum time interval between the starts of successive write cycles of a storage device that has separate reading and writing cycles.

write head: A magnetic head capable of writing only.

write protection label: *See* **write-protect tab.**

write-protect tab: A movable or removable tab, label, or other device, the presence or absence of which on the casing of a recording medium prevents writing on the medium. *Note:* An example of a write-protect tab is the sliding tab on a 3 ½-inch (8.85-cm) magnetic diskette of the type used in conjunction with desktop computers.

WRU signal: *Synonym* **who-are-you (WRU) character.**

WWMCCS: *Abbreviation for* **Worldwide Military Command and Control System.**

(this page intentionally left blank)

X-dimension of recorded spot: In facsimile systems, the effective recorded spot dimension measured in the direction of the recorded line. (188) *Note 1:* By "effective recorded spot dimension" is meant the largest center-to-center spacing between recorded spots, which gives minimum peak-to-peak variation of density of the recorded line. *Note 2:* "X-dimension of recorded spot" implies that the facsimile equipment response to a constant density in the object (original) is a succession of discrete recorded spots.

X-dimension of scanning spot: In facsimile systems, the distance between the centers of adjacent scanning spots measured in the direction of the scanning line on the object. (188) *Note:* The numerical value of the X-dimension of scanning spot depends upon the type of system.

xerographic recording: Recording enabled by the scanning action of an optical beam on a photoconducting surface on which an electrostatically charged latent image is developed with a resinous powder (toner).

XO: *Abbreviation for* **crystal oscillator**.

XOFF: An abbreviation for the ASCII transmission-control character meaning "Transmitter off." (188)

XON: An abbreviation for the ASCII transmission-control character meaning "Transmitter on."(188)

X.-series Recommendations: Sets of data telecommunications protocols and interfaces defined by CCITT Recommendations. *Note:* Some of the more common X.-series Recommendations are:

X.25: A CCITT Recommendation for public packet switched communications between a network user and the network itself.

X.75: A CCITT Recommendation for public packet switched communications between network hubs.

X.400: An addressing scheme for use with E-mail.

X.500: An addressing scheme for directory services.

XT: *Abbreviation for* **crosstalk.**

x-y mount: *Synonym* **altazimuth mount.**

(this page intentionally left blank)

Yagi antenna: A linear end-fire antenna, consisting of three or more half-wave elements (one driven, one reflector, and one or more directors). *Note 1:* A Yagi antenna offers very high directivity and gain. *Note 2:* The formal name for a *"Yagi antenna"* is *"Yagi-Uda array."*

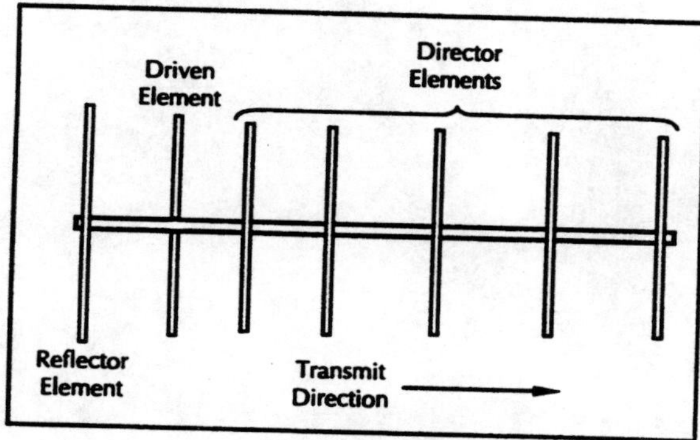

Yagi antenna

Y-dimension of recorded spot: In facsimile systems, the distance between the centers of adjacent recorded spots on adjacent lines measured perpendicular to the recorded line. (188)

Y-dimension of scanning spot: In facsimile systems, the distance between the centers of adjacent scanning spots on adjacent lines measured perpendicular to the scanning line on the object. (188) *Note:* The numerical value of the Y-dimension of scanning spot depends upon the type of system.

(this page intentionally left blank)

Z

Z: *Abbreviation for* **Zulu time.** *See* **Coordinated Universal Time.**

zero-bit insertion: A bit-stuffing technique used with bit-oriented protocols to ensure that six consecutive "1" bits never appear between the two flags that define the beginning and the ending of a transmission frame. *Note:* When five consecutive "1" bits occur in any part of the frame other than the beginning and ending flag, the sending station inserts an extra "0" bit. When the receiving station detects five "1" bits followed by a "0" bit, it removes the extra "0" bit, thereby restoring the bit stream to its original value.

zero dBm transmission level point (0 dBm TLP): In a communication system, a point at which the reference level is 1 mW, *i.e.,* 0 dBm. *Note:* The actual power level of the communications traffic is not necessarily 0 dBm. It is usually below the reference level. The reference is for system design and test purposes. *Synonym* **zero transmission level point.**

zero-dispersion slope: In a single-mode optical fiber, the rate of change of dispersion, with respect to wavelength, at the fiber's zero-dispersion wavelength. *Note 1:* In silica-based optical fibers, the zero-dispersion wavelength occurs at approximately 1.3 µm, but this wavelength may be shifted toward the minimum-loss window by the addition of dopants to the fiber material during manufacture. *Note 2:* Doubly and quadruply clad single-mode fibers have two zero-dispersion points, and thus two zero-dispersion slopes.

zero-dispersion wavelength: **1.** In a single-mode optical fiber, the wavelength or wavelengths at which material dispersion and waveguide dispersion cancel one another. *Note:* In all silica-based optical fibers, minimum material dispersion occurs naturally at a wavelength of approximately 1.3 µm. Single-mode fibers may be made of silica-based glasses containing

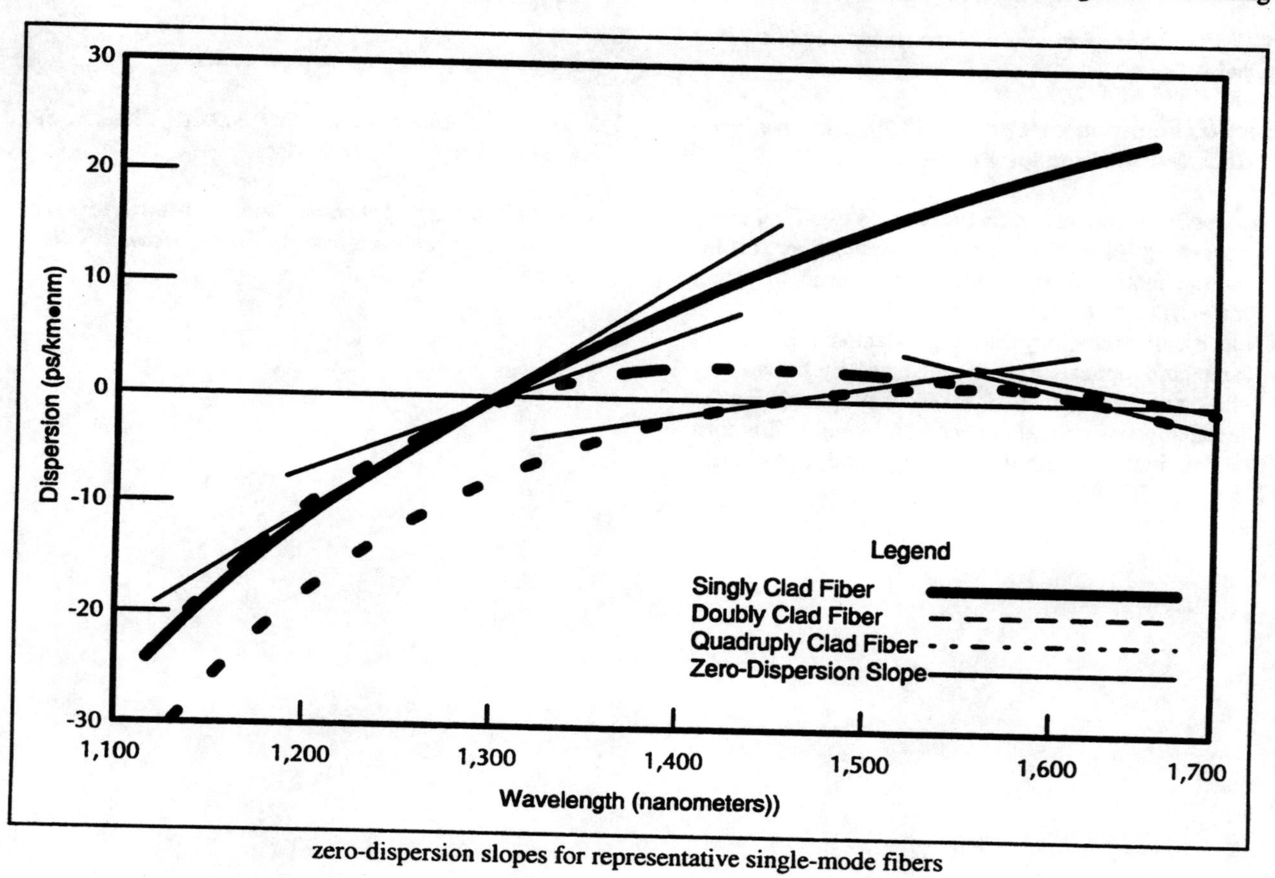

zero-dispersion slopes for representative single-mode fibers

dopants that shift the material-dispersion wavelength, and thus, the zero-dispersion wavelength, toward the minimum-loss window at approximately 1.55 µm. The engineering tradeoff is a slight increase in the minimum attenuation coefficient. **2.** Loosely, in a multimode optical fiber, the wavelength at which material dispersion is minimum, *i.e.*, essentially zero. *Synonym* **minimum-dispersion wavelength.**

zero dispersion window: *Synonym* **minimum dispersion window.**

zerofill: To fill unused storage locations with the representation of the character denoting "0".

zero-level decoder: A decoder that yields an analog level of 0 dBm at its output when the input is the digital mailed signal. [47CFR] *Note:* The signal is a 1-kHz sine wave.

zero suppression: The elimination of nonsignificant zeros from a numeral.

0TLP: *Abbreviation for* **zero transmission level point.**

zero transmission level point (0TLP): *Synonym* **zero dBm transmission level point.**

zip-cord: In optical communications, a two-fiber cable consisting essentially of two single-fiber cables having their jackets conjoined by a strip of jacket material. *Note 1:* This name is borrowed from electrical terminology referring to lamp cord. As with lamp cord, optical zip-cord may be easily furcated by slitting or tearing the two jackets apart, permitting the installation of optical connectors. *Note 2:* Zip-cord cables include both loose-buffer and tight-buffer designs. [FAA]

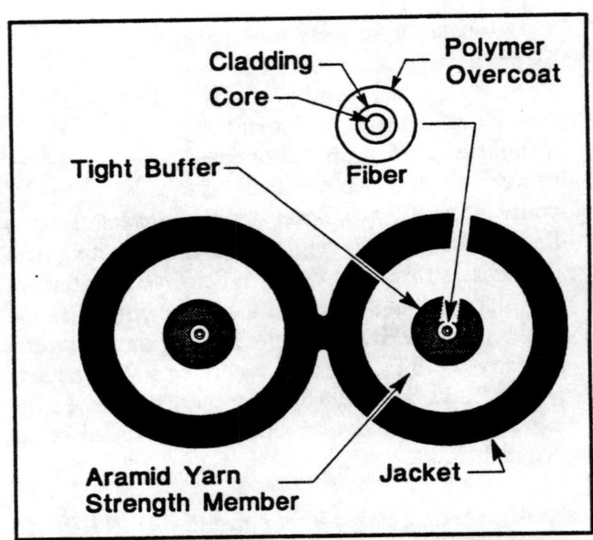

cross section of a two-fiber tight-buffered zip-cord optical cable

zone: *See* **communications zone, Fresnel zone, skip zone.**

zone of silence: *Synonym* **skip zone.**

Z Time: *Abbreviation for* **Zulu Time.** *See* **Coordinated Universal Time.**

Zulu Time (Z): *Synonym* **Coordinated Universal Time.** *Formerly a synonym for* **Greenwich Mean Time.**

APPENDIX A: ABBREVIATIONS and ACRONYMS

[Items shown in **bold** *represent term names or abbreviations that are defined in the glossary.]*

a	atto (10^{-18})
Å	**angstrom**
A	ampere
AAR	automatic alternate routing
AARTS	automatic audio remote test set
ABCA	American, British, Canadian, Australian [armies]
ac	alternating current
ACA	automatic circuit assurance
ACC	automatic callback calling
ACD	**automatic call distributor**
ac-dc	alternating current - direct current
ACK	**acknowledge character**
ACTS	Advanced Communications Technology Satellite
ACU	**automatic calling unit**
A-D	analog-to-digital
ADC	**analog-to-digital converter**
	analog-to-digital conversion
ADCCP	**Advanced Data Communication Control Procedures**
ADH	**automatic data handling**
ADP	**automatic data processing**
ADPCM	**adaptive differential pulse-code modulation**
ADPE	**automatic data processing equipment**
ADPSSO	automatic data processing system security officer
ADU	automatic dialing unit
ADX	automatic data exchange
AECS	**Aeronautical Emergency Communications System [Plan]**
AF	**audio frequency**
AFC	area frequency coordinator
	automatic frequency control
AFNOR	Association Français Normal
AFRS	**Armed Forces Radio Service**
AGC	**automatic gain control**
AGE	aerospace ground equipment
AI	**artificial intelligence**
AIG	address indicator group
	address indicating group
AIM	amplitude intensity modulation
AIN	**advanced intelligent network**
AIOD	**automatic identified outward dialing**
AIS	**automated information system**
ASIC	application-specific integrated circuits
AJ	anti-jamming
ALC	automatic level control
	automatic load control
ALE	**automatic link establishment**
ALU	**arithmetic and logic unit**

A/m	amperes per meter
AM	**amplitude modulation**
AMA	**automatic message accounting**
AMC	administrative management complex
AME	amplitude modulation equivalent
	automatic message exchange
AMI	**alternate mark inversion [signal]**
AM/PM/VSB	amplitude modulation/phase modulation/vestigial sideband
AMPS	**automatic message processing system**
AMPSSO	automated message processing system security officer
AMSC	American Mobile Satellite Corporation
AMTS	**automated maritime telecommunications system**
ANI	**automatic number identification**
ANL	automatic noise limiter
ANMCC	Alternate National Military Command Center
ANS	American National Standard
ANSI	**American National Standards Institute**
AP	**anomalous propagation**
APC	**adaptive predictive coding**
APD, apd, a.p.d.	**avalanche photodiode**
API	**application program interface**
APL	**average picture level**
APK	amplitude phase-shift keying
ARP	**address resolution protocol**
ARPA	Advanced Research Projects Agency [now DARPA]
ARPANET	Advanced Research Projects Agency Network
ARQ	automatic repeat-request
ARS	**automatic route selection**
ARSR	air route surveillance radar
ARU	**audio response unit**
ASC	AUTODIN Switching Center
ASCII	American Standard Code for Information Interchange
ASN.1	**abstract syntax notation one**
ASP	Aggregated Switch Procurement
	adjunct service point
ASR	automatic send and receive
	airport surveillance radar
AT	access tandem
ATACS	Army Tactical Communications System
ATB	**all trunks busy**
ATCRBS	air traffic control radar beacon system
ATDM	**asynchronous time-division multiplexing**
ATE	automatic test equipment
ATM	**asynchronous transfer mode**
ATV	**advanced television**
au	astronomical unit
AUI	**attachment unit interface**
AUTODIN	**Automatic Digital Network**
AUTOSEVOCOM	**Automatic Secure Voice Communications Network**
AUTOVON	**Automatic Voice Network**

AVD	alternate voice/data
AWG	American wire gauge
AWGN	additive white Gaussian noise
AZ	azimuth
b	**bit**
B	**bel**
	byte
balun	balanced to unbalanced
basecom	**base communications**
BASIC	beginners' all-purpose symbolic instruction code
BCC	**block check character**
BCD	**binary coded decimal**
	binary-coded decimal notation
BCI	**bit-count integrity**
Bd	**baud**
B8ZS	bipolar with eight-zero substitution
bell	BEL character
BER	**bit error ratio**
BERT	**bit error ratio tester**
BETRS	**basic exchange telecommunications radio service**
BEX	**broadband exchange**
BIH	**International Time Bureau**
B-ISDN	**broadband ISDN**
bi-sync	**binary synchronous [communication]**
bit	**binary digit**
BIT	built-in test
BITE	built-in test equipment
BIU	bus interface unit
BNF	Backus Naur form
BOC	**Bell Operating Company**
BPSK	binary phase-shift keying
bpi	**bits per inch**
	bytes per inch
b/s	**bits per second**
b/in	**bits per inch**
BPSK	binary phase-shift keying
BR	**bit rate**
BRI	**basic rate interface**
BSA	**basic serving arrangement**
BSE	**basic service element**
BSI	British Standards Institution
B6ZS	bipolar with six-zero substitution
B3ZS	bipolar with three-zero substitution
BTN	billing telephone number
BW	**bandwidth**
c	centi (10^{-2})
CACS	centralized alarm control system
CAM	computer-aided manufacturing

CAMA	**centralized automatic message accounting**
CAN	**cancel character**
CAP	**competitive access provider**
	customer administration panel
CARS	**cable television relay service [station]**
CAS	**centralized attendant services**
CASE	computer-aided software engineering
	computer aided system engineering
	computer-assisted software engineering
CATV	**cable TV**, cable television
	community antenna television
CBX	computer branch exchange
C^2	**command and control**
C^3	**command, control, and communications**
C^3CM	C^3 countermeasures
C^3I	command, control, communications and intelligence
CCA	carrier-controlled approach
CCD	charge-coupled device
CCH	**connections per circuit hour**
CCIF	International Telephone Consultative Committee *[a predecessor to the CCITT]*
CCIR	International Radio Consultative Committee
CCIS	**common-channel interoffice signaling**
CCIT	International Telegraph Consultative Committee *[a predecessor to the CCITT]*
CCITT	International Telegraph and Telephone Consultative Committee *{now ITU-T}*
CCL	continuous communications link
CCS	hundred call-seconds
CCSA	**common control switching arrangement**
CCTV	closed-circuit television
CCW	**cable cutoff wavelength**
cd	candela
CD	collision detection, compact disk
CDF	**combined distribution frame**
	cumulative distribution function
CDMA	**code-division multiple access**
CDPSK	**coherent differential phase-shift keying**
CDR	**call detail recording**
CD ROM	compact disk read-only memory
CDT	control data terminal
CDU	central display unit
C-E	**communications-electronics**
CEI	**comparably efficient interconnection**
CELP	**code-excited linear prediction**
CEP	circular error probable
CFE	contractor-furnished equipment
cgs	centimeter-gram-second
ChR	channel reliability
CIAS	circuit inventory and analysis system
CIC	content indicator code
CIF	common intermediate format
CIFAX	ciphered facsimile

CINC	commander-in-chief
ciphony	ciphered telephony
CiR	**circuit reliability**
C/kT	**carrier-to-receiver noise density**
CLASS	**custom local area signaling service**
cm	centimeter
CMI	coded mark inversion
CMIP	**Common Management Information Protocol**
CMIS	**common management information service**
CMOS	complementary metal oxide substrate
CMRR	**common-mode rejection ratio**
CNR	**carrier-to-noise ratio**
	combat-net radio
CNS	**complementary network service**
C.O.	**central office**
COAM	**customer owned and maintained equipment**
coax	**coaxial cable**
COBOL	common business oriented language
codec	coder-decoder
COG	**centralized ordering group**
COMINT	**communications intelligence**
COMJAM	**communications jamming**
compandor	compressor-expander
COMPUSEC	**computer security**
COMSAT	Communications Satellite Corporation
COMSEC	**communications security**
CONEX	**connectivity exchange**
CONUS	Continental United States
COP	Committee of Principals
COR	Council of Representatives
COT	**customer office terminal**
CPAS	cellular priority access services
CPE	**customer premises equipment**
cpi	**characters per inch**
cpm	counts per minute
cps	**characters per second**
	cycles per second *[deprecated]*
CPU	central processing unit
	communications processor unit
CR	**channel reliability**
	circuit reliability
CRC	**cyclic redundancy check**
CRITICOM	Critical Intelligence Communications
CROM	control read-only memory
CRT	cathode ray tube
c/s	cycles per second *[deprecated]*
CSA	**Canadian Standards Association**
CSC	circuit-switching center
	common signaling channel
CSMA	**carrier sense multiple access**

CSMA/CA	**carrier sense multiple access with collision avoidance**
CSMA/CD	**carrier sense multiple access with collision detection**
CSU	**channel service unit**
	circuit switching unit
	customer service unit
CTS	clear to send
CTX	**Centrex® [service]**
	clear to transmit
CVD	chemical vapor deposition
CVSD	**continuously variable slope delta [modulation]**
cw	**carrier wave**
	composite wave
	continuous wave
CX	**composite signaling**
cxr	**carrier**
D*	**specific detectivity**
d	deci (10^{-1})
D(λ)	dispersion coefficient
da	deka (10)
D-A	digital-to-analog
	digital-to-analog converter
D/L	**downlink**
DACS	**digital access and cross-connect system**
DAMA	**demand assignment multiple access**
DARPA	Defense Advanced Research Projects Agency *[formerly ARPA]*
dB	decibel
dBa	decibels adjusted
dBa(F1A)	noise power measured by a set with F1A-line weighting
dBa(HA1)	noise power measured by a set with HA1-receiver weighting
dBa0	noise power measured at zero transmission level point
dBc	dB relative to carrier power
dBm	dB referred to 1 milliwatt
dBm(psoph)	noise power in dBm measured by a set with psophometric weighting
DBMS	**database management system**
dBmV	dB referred to 1 millivolt across 75 ohms
dBm0	noise power in dBm referred to or measured at 0TLP
dBm0p	noise power in dBm0 measured by a psophometric or noise measuring set having psophometric weighting
dBr	power difference in dB between any point and a reference point
dBrn	dB above reference noise
dBrnC	noise power in dBrn measured by a set with C-message weighting
dBrnC0	noise power in dBrnC referred to or measured at 0TLP
dBrn(f_1-f_2)	flat noise power in dBrn
dBrn(144-line)	noise power in dBrn measured by a set with 144-line weighting
dBv	dB relative to 1 V (volt) peak-to-peak
dBW	dB referred to 1 W (watt)
dBx	dB above reference coupling
dc	direct current
DCA	Defense Communications Agency *(now DISA)*

DCE	data circuit-terminating equipment
DCL	direct communications link
DCPSK	differentially coherent phase-shift keying
DCS	**Defense Communications System**
DCTN	Defense Commercial Telecommunications Network
DCWV	direct-current working volts
DDD	**direct distance dialing**
DDN	**Defense Data Network**
DDS	digital data service
DEL	delete character
demarc	**demarcation point**
demux	demultiplex
	demultiplexer
	demultiplexing
dequeue	**double-ended queue**
DES	**Data Encryption Standard**
detem	detector/emitter
DFSK	**double-frequency shift keying**
DIA	Defense Intelligence Agency
DID	**direct inward dialing**
DIN	**Deutsches Institut für Normung**
DIP	**dual in-line package**
DISA	Defense Information Systems Agency *{formerly DCA}*
DISC	**disconnect command**
DISN	Defense Information System Network
DISNET	Defense Integrated Secure Network
DLA	Defense Logistic Agency
DLC	**digital loop carrier**
DLE	**data link escape character**
DM	**delta modulation**
DMA	Defense Mapping Agency
	direct memory access
DME	**distance measuring equipment**
DMS	Defense Message System
DNA	Defense Nuclear Agency
DNIC	**data network identification code**
DNPA	**data numbering plan area**
DNS	**Domain Name System**
DO	**design objective**
DoC	Department of Commerce
DOD	Department of Defense
	direct outward dialing
DODD	Department of Defense Directive
DODISS	Department of Defense Index of Specifications and Standards
DOD-STD	Department of Defense Standard
DOS	Department of State
DPCM	**differential pulse-code modulation**
DPSK	**differential phase-shift keying**
DQDB	**distributed-queue dual-bus [network]**
DRAM	dynamic random access memory

DRSI	destination station routing indicator
DS	**digital signal**
	direct support
DS0	**digital signal 0**
DS1	**digital signal 1**
DS1C	**digital signal 1C**
DS2	**digital signal 2**
DS3	**digital signal 3**
DS4	**digital signal 4**
DSA	**dial service assistance**
DSB	**double sideband [transmission]**
	Defense Science Board
DSB-RC	**double-sideband reduced carrier transmission**
DSB-SC	**double-sideband suppressed-carrier transmission**
DSC	**digital selective calling**
DSCS	Defense Satellite Communications System
DSE	**data switching exchange**
DSI	**digital speech interpolation**
DSL	**digital subscriber line**
DSN	**Defense Switched Network**
DSR	**data signaling rate**
DSS	direct station selection
DSSCS	Defense Special Service Communications System
DSTE	data subscriber terminal equipment
DSU	**data service unit**
DTE	data terminal equipment
DTMF	**dual-tone multifrequency [signaling]**
DTG	**date-time group**
DTN	data transmission network
DTS	Diplomatic Telecommunications Service
DTU	data transfer unit
	data tape unit
	digital transmission unit
	direct to user
DVL	direct voice link
DX signaling	**direct current signaling**
	duplex signaling
E	exa (10^{18})
E-MAIL, e-mail	**electronic mail**
EAS	**extended area service**
EBCDIC	extended binary coded decimal interchange code
E_b/N_0	signal energy per bit per hertz of thermal noise
EBO	embedded base organization
EBS	**Emergency Broadcast System**
EBX	electronic branch exchange

EC	**Earth coverage**
	Earth curvature
ECC	**electronically controlled coupling**
	enhance call completion
ECCM	**electronic counter-countermeasures**
ECM	**electronic countermeasures**
EDC	error detection and correction
EDI	electronic data interchange
EDTV	**extended-definition television**
EHF	**extremely high frequency**
EIA	Electronic Industries Association
eirp	**effective isotropically radiated power**
	equivalent isotopically radiated power
EIS	Emergency Information System
el	elevation
ELECTRO-OPTINT	**electro-optical intelligence**
ELF	**extremely low frequency**
ELINT	**electronics intelligence**
	electromagnetic intelligence
ELSEC	**electronics security**
ELT	**emergency locator transmitter**
EMC	**electromagnetic compatibility**
EMCON	**emission control**
EMD	**equilibrium mode distribution**
EME	**electromagnetic environment**
emf	electromotive force
EMI	**electromagnetic interference**
	electromagnetic interference control
EMP	**electromagnetic pulse**
EMR	**electromagnetic radiation**
EMR hazards	**electromagnetic radiation hazards**
e.m.r.p.	**effective monopole radiated power**
	equivalent monopole radiated power
EMS	**electronic message system**
EMSEC	**emanations security**
emu	electromagnetic unit
EMV	**electromagnetic vulnerability**
EMW	electromagnetic warfare
	electromagnetic wave
ENQ	**enquiry character**
EO	**end office**
E.O.	Executive Order
EOD	end of data
EOF	end of file
EOL	end of line
EOM	end of message
EOP	end of program
	end output
EOS	**end-of-selection character**

EOT	**end-of-transmission character**
	end of tape
EOW	engineering orderwire
EPROM	erasable programmable read-only memory
EPSCS	enhanced private switched communications system
ERL	echo return loss
ERLINK	emergency response link
ERP, e.r.p.	**effective radiated power**
ES	**end system**
	expert system
ESC	**escape character**
	enhanced satellite capability
ESF	**extended superframe**
ESM	**electronic warfare support measures**
ESP	enhanced service provider
ESS	**electronic switching system**
ETB	**end-of-transmission-block character**
ETX	**end-of-text character**
EW	**electronic warfare**
EXCSA	Exchange Carriers Standards Association
f	femto (10^{-15})
f	**frequency**
FAA	Federal Aviation Administration
FAQ file	Frequently Asked Questions file
FAX	**facsimile**
FC	**functional component**
FCC	Federal Communications Commission
FCS	frame check sequence
FDDI	**fiber distributed data interface**
FDDI-2	fiber distributed data interface–2
FDHM	**full duration at half maximum**
FDM	**frequency-division multiplexing**
FDMA	**frequency-division multiple access**
FDX	full duplex
FEC	**forward error correction**
FECC	Federal Emergency Communications Coordinators
FED-STD	Federal Standard
FEMA	Federal Emergency Management Agency
FEP	**front-end processor**
FET	field effect transistor
FIFO	**first-in first-out**
FIP	Federal Information Processing
FIPS	Federal Information Processing Standards
FIR	finite impulse response
FIRMR	Federal Information Resources Management Regulations
FISINT	**foreign instrumentation signals intelligence**
FLTSATCOM	Fleet Satellite Communications
flops	floating-point operations per second
FM	**frequency modulation**

FO	**fiber optics**
FOC	final operational capability
	full operational capability
FOT	frequency of optimum traffic
	frequency of optimum transmission
FPIS	forward propagation ionospheric scatter
ft/min	feet per minute
fps	foot-pound-second
FPS	frames per second
	focus projection and scanning
FRP	Federal Response Plan
FSDPSK	**filtered symmetric differential phase-shift keying**
FSK	**frequency-shift keying**
FSS	fully separate subsidiary
FSTS	Federal Secure Telephone Service *(no longer in operation)*
FT	fiber optic T-carrier
FTAM	**file transfer, access, and management**
FTF	Federal Telecommunications Fund
ft/s	feet per second
FTS	**Federal Telecommunications System**
FTS2000	Federal Telecommunications System 2000
FTSC	Federal Telecommunications Standards Committee
FTP	file transfer protocol
FWHM	**full width at half maximum**
FX	**fixed service**
	foreign exchange service
FYDP	Five Year Defense Plan
g	**profile parameter**
G	giga (10^9)
GBH	**group busy hour**
GCT	Greenwich Civil Time
GDF	**group distribution frame**
GETS	Government Emergency Telecommunications Service
GFE	Government-furnished equipment
GGCL	government-to-government communications link
GHz	**gigahertz**
GII	Global Information Infrastructure
GMT	**Greenwich Mean Time**
GOS	**grade of service**
GOSIP	Government Open Systems Interconnection Profile
GSA	General Services Administration
GSTN	general switched telephone network
G/T	**antenna gain-to-noise-temperature**
GTP	Government Telecommunications Program
GTS	Government Telecommunications System
GUI	**graphical user interface**

h	hecto (10^2)
	hour
	Planck's constant
HCS	hard clad silica [fiber]
HDLC	**high-level data link control**
HDTV	**high-definition television**
HDX	**half-duplex [operation]**
HE_{11} mode	the fundamental hybrid mode [of an optical fiber]
HEMP	**high-altitude electromagnetic pulse**
HERF	**hazards of electromagnetic radiation to fuel**
HERO	**hazards of electromagnetic radiation to ordnance**
HERP	**hazards of electromagnetic radiation to personnel**
HF	**high frequency**
HFDF	**high-frequency distribution frame**
HLL	**high-level language**
HPC	high probability of completion
HV	high voltage
Hz	**hertz**
IA	International Alphabet *[CCITT]*
I&C	installation and checkout
IC	**integrated circuit**
ICI	**incoming call identification**
ICNI	Integrated Communications, Navigation, and Identification
ICW	**interrupted continuous wave**
IDDD	International Direct Distance Dialing
IDF	**intermediate distribution frame**
IDN	**integrated digital network**
IDTV	**improved-definition television**
IEC	International Electrotechnical Commission
IEEE	Institute of Electrical and Electronics Engineers
IES	Industry Executive Subcommittee
IF	**intermediate frequency**
I/F	**interface**
IFF	**identification, friend or foe**
IFRB	**International Frequency Registration Board (*now* Radio Regulations Board)**
IFS	**ionospheric forward scatter**
IIR	infinite impulse response
IITF	Information Infrastructure Task Force
ILD	**injection laser diode**
ILS	**instrument landing system**
IM	**intensity modulation**
	intermodulation
IMA	individual mobilization augmentee
I&M	installation and maintenance
IMD	**intermodulation distortion**
IMP	**interface message processor**
IN	**intelligent network**
IN/1, IN/1+, IN/2	**intelligent network concepts**
INFOSEC	**information systems security**

INS	inertial navigation system
INTELSAT	International Telecommunications Satellite Consortium
INWATS	Inward Wide-Area Telephone Service
I/O	**input/output [device]**
IOC	**integrated optical circuit**
	initial operational capability
	input-output controller
IP	**Internet protocol**
	intelligent peripheral
IPA	intermediate power amplifier
IPC	**information processing center**
IPM	impulses per minute
	interference prediction model
	internal polarization modulation
	interruptions per minute
in/s	inches per second
ips	interruptions per second
IPX	Internet Packet Exchange
IQF	intrinsic quality factor
IR	**infrared**
IRAC	Interdepartment Radio Advisory Committee
IRC	international record carrier
	Interagency Radio Committee
ISB	**independent-sideband [transmission]**
ISDN	Integrated Services Digital Network
ISM	**industrial, scientific, and medical [applications]**
ISO	International Organization for Standardization
ITA	International Telegraph Alphabet
ITA-5	**International Telegraph Alphabet Number 5**
ITC	International Teletraffic Congress
ITS	Institute for Telecommunication Sciences
ITSO	International Telecommunications Satellite Organization
ITU	**International Telecommunication Union**
IVDT	integrated voice data terminal
IXC	**interexchange carrier**
JANAP	Joint Army-Navy-Air Force Publication(s)
JCS	Joint Chiefs of Staff
JPL	Jet Propulsion Laboratory
JSC	Joint Steering Committee *[now JTSSG]*
	Joint Spectrum Center
JTC^3A	Joint Tactical Command, Control and Communications Agency
JTIDS	**Joint Tactical Information Distribution System**
JTRB	**Joint Telecommunications Resources Board**
JTSSG	Joint Telecommunications Standards Steering Group *[formerly JSC]*
JWID	Joint Warrior Interoperability Demonstration

k	kilo (10^3)
	Boltzmann's constant
K	coefficient of absorption
	kelvin
KDC	**key distribution center**
KDR	keyboard data recorder
KDT	keyboard display terminal
kg	kilogram
kg•m•s	kilogram-meter-second
kHz	**kilohertz**
km	**kilometer**
kΩ, k	kilohm
KSR	keyboard send/receive device
kT	**noise power density**
KTS	**key telephone system**
KTU	key telephone unit
λ	wavelength
λ_{co}	cutoff wavelength
LAN	**local area network**
LANTFLT	Atlantic Fleet
LAP-B	Data Link Layer protocol [CCITT Recommendation X.25 (1989)]
LAP-D	link access procedure D
laser	light amplification by stimulated emission of radiation
LASINT	**laser intelligence**
LATA	**local access and transport area**
LBO	line buildout
LC	limited capability
LCD	**liquid crystal display**
LD	long distance
LDM	limited distance modem
LEC	**local exchange carrier**
LED	**light-emitting diode**
LF	**low frequency**
LFB	**look-ahead-for-busy [information]**
LIFO	**last-in first-out**
LLC	**logical link control [sublayer]**
l/m	lines per minute
LMF	language media format
LMR	land mobile radio
LNA	**launch numerical aperture**
LOF	lowest operating frequency
loran	long-range aid to navigation system
	long-range radio navigation
	long-range radio aid to navigation system
LOS	**line of sight,** loss of signal

LP	**linearly polarized [mode]**
	linear programming
	linking protection
	log-periodic [antenna]
	log-periodic [array]
LPA	linear power amplifier
LPC	**linear predictive coding**
LPD	low probability of detection
LPI	low probability of interception
lpi	lines per inch
lpm	lines per minute
LP_{01}	the **fundamental mode** [of an optical fiber]
LQA	**link quality analysis**
LRC	**longitudinal redundancy check**
LSB	lower sideband, least significant bit
LSI	large scale integrated [circuit]
	large scale integration
	line status indication
LTC	**line traffic coordinator**
LUF	**lowest usable high frequency**
LULT	line-unit-line termination
LUNT	line-unit-network termination
LV	low voltage
m	meter
	milli (10^{-3})
M	mega (10^{6})
MAC	**medium access control [sublayer]**
MACOM	major command
MAN	**metropolitan area network**
MAP	manufacturers' automation protocol
maser	microwave amplification by the stimulated emission of radiation
M[λ]	**material dispersion coefficient**
MAU	**medium access unit**
MCC	**maintenance control circuit**
MCEB	Military Communications-Electronics Board
MCM	**multicarrier modulation**
MCS	Master Control System
MCW	modulated continuous wave
MCXO	microcomputer compensated crystal oscillator
MDF	**main distribution frame**
MDT	mean downtime
MEECN	Minimum Essential Emergency Communications Network
MERCAST	**merchant-ship broadcast system**
MF	**medium frequency**
	multifrequency [signaling]
MFD	**mode field diameter**
MFJ	**Modification of Final Judgment**
MFSK	**multiple frequency-shift keying**
MHF	medium high frequency

MHS	message handling service
	message handling system
MHz	**megahertz**
mi	mile
MIC	**medium interface connector**
	microphone
	microwave integrated circuit
	minimum ignition current
	monolithic integrated circuit
	mutual interface chart
MILNET	military network
MIL-STD	Military Standard
min	minute
MIP	**medium interface point**
MIPS, mips	million instructions per second
MIS	**management information system**
MKS	meter-kilogram-second
MLPP	**multilevel precedence and preemption**
MMW	**millimeter wave**
modem	modulator-demodulator
modified AMI code	modified alternate mark inversion code
mol	mole
ms	millisecond (10^{-3} second)
MSB	most significant bit
MSK	minimum-shift keying
MTBF	**mean time between failures**
MTBM	mean time between maintenance
MTBO	**mean time between outages**
MTBPM	mean time between preventive maintenance
MTF	modulation transfer function
MTSO	mobile telephone switching office
MTSR	**mean time to service restoral**
MTTR	**mean time to repair**
μ	micro (10^{-6})
μs	microsecond
MUF	**maximum usable frequency**
muldem	**multiplexer/demultiplexer**
MUX	multiplex
	multiplexer
MUXing	**multiplexing**
mw	**microwave**
MWI	message waiting indicator
MWV	maximum working voltage
n	nano (10^{-9})
	refractive index
N_0	sea level refractivity
	spectral noise density
NA	**numerical aperture**
NACSEM	National Communications Security Emanation Memorandum

NACSIM	National Communications Security Information Memorandum
NAK	**negative-acknowledge character**
NASA	National Aeronautics and Space Administration
NATA	North American Telecommunications Association
NATO	North Atlantic Treaty Organization
NAVSTAR-GPS	Navigational Satellite Timing and Ranging—Global Positioning System
NBFM	narrowband frequency modulation
NBH	network busy hour
NBRVF	**narrowband radio voice frequency**
NBS	National Bureau of Standards *[now NIST]*
NBSV	narrowband secure voice
NCA	**National Command Authorities**
NCC	**National Coordinating Center for Telecommunications**
NCS	**National Communications System**
	net control station
NCSC	National Communications Security Committee
NDCS	network data control system
NDER	National Defense Executive Reserve
NEAÇP	National Emergency Airborne Command Post
NEC	**National Electric Code®**
NEP	**noise equivalent power**
NES	noise equivalent signal
NF	**noise figure**
NFS	Network File System
NIC	**network interface card**
NICS	NATO Integrated Communications System
NID	**network interface device**
	network inward dialing
	network information database
NII	**National Information Infrastructure**
NIOD	network inward/outward dialing
NIST	National Institute of Standards and Technology *[formerly NBS]*
NIU	**network interface unit**
NLP	National Level Program
nm	nanometer
NMCS	National Military Command System
nmi	**nautical mile**
NOD	**network outward dialing**
Np	**neper**
NPA	numbering plan area
NPR	noise power ratio
NRI	**net radio interface**
NRM	network resource manager
NRRC	Nuclear Risk Reduction Center
NRZ	**non-return-to-zero**
NRZI	non-return-to-zero inverted
NRZ-M	**non-return-to-zero mark**
NRZ-S	**non-return-to-zero space**
NRZ1	**non-return-to-zero, change on ones**
NRZ-1	non-return-to-zero mark

ns	nanosecond
NSA	National Security Agency
NSC	National Security Council
NS/EP telecommunications	National Security or Emergency Preparedness telecommunications
NSTAC	National Security Telecommunications Advisory Committee
NTCN	National Telecommunications Coordinating Network
NTDS	Naval Tactical Data System
NTMS	National Telecommunications Management Structure
NT1	**Network termination 1**
NT2	**Network termination 2**
NTI	network terminating interface
NTIA	National Telecommunications and Information Administration
NTN	**network terminal number**
NTSC	National Television Standards Committee
	National Television Standards Committee [standard]
NUL	**null character**
NVIS	**near vertical incidence skywave**
Ω	ohm
O&M	operations and maintenance
OC	operations center
OCC	**other common carrier**
OCR	**optical character reader**
	optical character recognition
OCU	orderwire control unit
OCVCXO	oven controlled-voltage controlled crystal oscillator
OCXO	oven controlled crystal oscillator
OD	**optical density**
	outside diameter
OFC	optical fiber, conductive
OFCP	optical fiber, conductive, plenum
OFCR	optical fiber, conductive, riser
OFN	optical fiber, nonconductive
OFNP	optical fiber, nonconductive, plenum
OFNR	optical fiber, nonconductive, riser
OMB	Office of Management and Budget
ONA	**open network architecture**
opm	operations per minute
OPMODEL	operations model
OPSEC	**operations security**
OPX	**off-premises extension**
OR	off-route service
	off-route aeronautical mobile service
OSHA	Occupational Safety and Health Administration
OSI	open switching interval
	Open Systems Interconnection
OSI-RM	**Open Systems Interconnection—Reference Model**
OSRI	originating stations routing indicator
OSSN	originating stations serial number
OTAM	over-the-air management of automated HF network nodes

OTAR	**over-the-air rekeying**
OTDR	**optical time domain reflectometer**
	optical time domain reflectometry
OW	**orderwire [circuit]**
p	pico (10^{-12})
P	peta (10^{15})
PABX	private automatic branch exchange
PAD	**packet assembler/disassembler**
PAL	phase alternation by line
PAL-M	phase alternation by line—modified
PAM	**pulse-amplitude modulation**
PAMA	**pulse-address multiple access**
p/a r	**peak-to-average ratio**
PAR	performance analysis and review
PARAMP	**parametric amplifier**
par meter	peak-to-average ratio meter
PAX	private automatic exchange
PBER	pseudo-bit-error-ratio
PBX	private branch exchange
PC	**carrier power (of a radio transmitter)**
	personal computer
PCB	**power circuit breaker**
	printed circuit board
PCM	**pulse-code modulation**
	plug compatible module
	process control module
PCS	Personal Communications Services
	personal communications system
	plastic-clad silica [fiber]
PCSR	parallel channels signaling rate
PD	**photodetector**
PDM	pulse delta modulation
	pulse-duration modulation
PDN	**public data network**
PDS	**protected distribution system**
	power distribution system
	program data source
PDT	programmable data terminal
PDU	**protocol data unit**
PE	**phase-encoded [recording]**
PEP	**peak envelope power (of a radio transmitter)**
pF	picofarad
PF	**power factor**
PFM	**pulse-frequency modulation**
PI	**protection interval**
PIC	plastic insulated cable
ping	packet Internet groper
PIV	peak inverse voltage
PLA	**programmable logic array**

PLL	**phase-locked loop**
PLN	private line network
PL/I	programming language 1
PLR	**pulse link repeater**
PLS	**physical signaling sublayer**
pm	**phase modulation**
PM	mean power
	polarization-maintaining [optical fiber]
	preventive maintenance
	pulse modulation
PMB	**pilot-make-busy [circuit]**
PMO	program management office
POI	**point of interface**
POP	**point of presence**
POSIX	portable operating system interface for computer environments
POTS	plain old telephone service
PP	polarization-preserving [optical fiber]
P-P	**peak-to-peak [value]**
P/P	point-to-point
PPM	**pulse-position modulation**
pps	pulses per second
PR	pulse rate
PRF	**pulse-repetition frequency**
PRI	**primary rate interface**
PRM	pulse-rate modulation
PROM	**programmable read-only memory**
PRR	**pulse repetition rate**
PRSL	primary area switch locator
PS	**permanent signal**
psi	pounds [force] per square inch
PSK	**phase-shift keying**
PSN	**public switched network**
p-static	**precipitation static**
PSTN	**public switched telephone network**
PTF	**patch and test facility**
PTM	**pulse-time modulation**
PTT	postal, telephone, and telegraph
	push-to-talk [operation]
PTTC	paper tape transmission code
PTTI	precise time and time interval
PU	power unit
PUC	public utility commission
	public utilities commission
PVC	**permanent virtual circuit**
	polyvinyl chloride [insulation]
pW	picowatt
PWM	**pulse-width modulation** *[deprecated]*
pWp	picowatt, psophometrically weighted
pWp0	picowatts, relative to a 0TLP
pX	**peak envelope power** *[in dB]*

PX	peak envelope power
	private exchange
	peak envelope power *[in watts]*
pY	mean power *[in dB]*
PY	mean power *[in watts]*
pZ	carrier power *[in dB]*
PZ	carrier power *[in watts]*
QA	**quality assurance**
QAM	**quadrature amplitude modulation**
QC	**quality control**
QCIF	**quarter common intermediate format**
QMR	qualitative material requirement
QOS	**quality of service**
QPSK	**quadrature phase-shift keying**
QRC	quick reaction capability
QSTAG	Quadripartite Standardization Agreement
R	route service
racon	**radar beacon**
rad	**radian**
	radiation absorbed dose
radar	radio detection and ranging
RADHAZ	**electromagnetic radiation hazards**
RADINT	**radar intelligence**
RAM	**random access memory**
	reliability, availability, and maintainability
R&D	research and development
RATT	radio teletypewriter system
RBOC	**Regional Bell Operating Company**
RbXO	rubidium-crystal oscillator
RC	**reflection coefficient**
	resource controller
RCC	**radio common carrier**
RCVR	receiver
RDF	**radio-direction finding**
REA	Rural Electrification Administration
REN	**ringer equivalency number**
RGB	red-green-blue
rf	**radio frequency**
	range finder
RFI	radio frequency interference
RFP	request for proposal
RFQ	request for quotation
RH	relative humidity
RHR	**radio horizon range**
RI	**routing indicator**
RISC	reduced instruction set chip
RJ	**registered jack**
RJE	**remote job entry**

rms	**root-mean-square [deviation]**
RO	read only
	receive only
ROA	**recognized operating agency**
ROC	required operational capability
ROM	**read-only memory**
ROSE	**remote operations service element protocol**
rpm	revolutions per minute
RPM	rate per minute
RPOA	recognized private operating agency
rps	revolutions per second
RQ	repeat-request
RR	repetition rate
RSL	**received signal level**
rss	root-sum-square
R/T	**real time**
RTA	**remote trunk arrangement**
RTS	request to send
RTTY	**radio teletypewriter**
RTU	remote terminal unit
RTX	request to transmit
RVA	reactive volt-ampere
RVWG	Reliability and Vulnerability Working Group
RWI	**radio and wire integration**
RX	receive
	receiver
RZ	**return-to-zero**
s	**second**
SCC	**specialized common carrier**
SCE	service creation environment
SCF	service control facility
SCP	service control point
SCPC	single channel per carrier
SCR	semiconductor-controlled rectifier
	silicon-controlled rectifier
SCSR	single channel signaling rate
SDLC	**synchronous data link control**
SDM	**space-division multiplexing** *[deprecated]*
SDN	software-defined network
SECAM	système electronique couleur avec memoire
SECDEF	Secretary of Defense
SECORD	secure voice cord board
SECTEL	secure telephone
SETAMS	systems engineering, technical assistance, and management services
SEVAS	Secure Voice Access System
S-F	**store-and-forward**
SF	**single-frequency [signaling]**
SGDF	**supergroup distribution frame**
S/H	sample and hold

SHA	sidereal hour angle
SHARES	Shared Resources (SHARES) HF Radio Program
SHF	super high frequency
SI	**International System of Units**
SID	**sudden ionospheric disturbance**
SIGINT	**signals intelligence**
SINAD	signal-plus-noise-plus-distortion to noise-plus-distortion ratio
S(λ)	chromatic dispersion slope
SLD	superluminescent diode [*see* **superluminescent LED**]
SLI	service logic interpreter
SLP	service logic program
SMDR	**station message-detail recording**
SMSA	standard metropolitan statistical area
SNR	**signal-to-noise ratio**
SOH	**start-of-heading character**
SOM	start of message
sonar	sound navigation and ranging
SONET	synchronous optical network
SOP	standard operating procedure
SOR	start of record
SOW	statement of work
(S+N)/N	**signal-plus-noise-to-noise ratio**
sr	**steradian**
S/R	send and receive
SSB	**single-sideband [transmission]**
SSB-SC	**single-sideband suppressed carrier [transmission]**
SSN	station serial number
SSP	service switching point
SSUPS	solid-state uninterruptible power system
STALO	stabilized local oscillator
STANAG	Standardization Agreement *[NATO]*
STD	subscriber trunk dialing
STFS	**standard time and frequency signal [service]**
	standard time and frequency service
STL	**standard telegraph level**
	studio-to-transmitter link
STP	standard temperature and pressure
	signal transfer point
STU	secure telephone unit
STX	**start-of-text character**
SUB	**substitute character**
SWR	**standing wave ratio**
SX	**simplex signaling**
SXS	**step-by-step switching system**
SYN	**synchronous idle character**
SYSGEN	**system generation**
T	tera (10^{12})
TADIL	**tactical data information link**
TADIL-A	**tactical data information link-A**

TADIL-B	**tactical data information link-B**
TADS	teletypewriter automatic dispatch system
TADSS	**Tactical Automatic Digital Switching System**
TAI	**International Atomic Time**
TASI	**time-assignment speech interpolation**
TAT	trans-Atlantic telecommunication (cable)
TC	toll center
TCB	**trusted computing base**
TCC	telecommunications center
TCCF	Tactical Communications Control Facility
TCF	**technical control facility**
TCP	**transmission control protocol**
TCS	**trusted computer system**
TCU	**teletypewriter control unit**
TCVXO	temperature compensated-voltage controlled crystal oscillator
TCXO	temperature controlled crystal oscillator
TD	time delay
	transmitter distributor
TDD	**Telecommunications Device for the Deaf**
TDM	**time-division multiplexing**
TDMA	**time-division multiple access**
TE	**transverse electric [mode]**
TED	**trunk encryption device**
TEK	**traffic encryption key**
TEM	**transverse electric and magnetic [mode]**
TEMPEST	compromising emanations
TEMS	telecommunications management system
TGM	**trunk group multiplexer**
THD	**total harmonic distortion**
THF	**tremendously high frequency**
THz	**terahertz**
TIA	Telecommunications Industry Association
TIE	**time interval error**
TIFF	**tag image file format**
TIP	terminal interface processor
T_K	**response timer**
TLP	**transmission level point**
TM	**transverse magnetic [mode]**
TP	toll point
TRANSEC	transmission security
TRC	**transverse redundancy check**
TRF	tuned radio frequency
TRI-TAC	tri-services tactical [equipment]
TSK	**transmission security key**
TSP	**Telecommunications Service Priority [system]**
TSPS	**traffic service position system**
TSR	telecommunications service request
	terminate and stay resident
TTL	transistor-transistor logic
TTTN	**tandem tie trunk network**

TTY	**teletypewriter**
TTY/TDD	**Telecommunications Device for the Deaf**
TV	**television**
TW	**traveling wave**
TWT	traveling wave tube
TWTA	traveling wave tube amplifier
TWX®	**teletypewriter exchange service**
TX	transmit
	transmitter
UDP	**User Datagram Protocol**
UHF	**ultra high frequency**
ULF	**ultra low frequency**
U/L	**uplink**
UPS	**uninterruptible power supply**
UPT	**Universal Personal Telecommunications service**
USB	upper sideband
USDA	U.S. Department of Agriculture
USFJ	U.S. Forces, Japan
USFK	U.S. Forces, Korea
USNO	U.S. Naval observatory
USTA	U.S. Telephone Association
UT	**Universal Time**
UTC	**Coordinated Universal Time**
uv	**ultraviolet**
V	volt
	normalized frequency
VA	value-added *[network service]*
	volt-ampere
VAN	**value-added network**
VAR	value added reseller
VARISTAR	variable resistor
vars	**volt-amperes reactive**
VC	**virtual circuit**
VCO	voltage-controlled oscillator
VCXO	voltage-controlled crystal oscillator
V/D	voice/data
Vdc	volts direct current
VDU	video display unit
	visual display unit
VF	**voice frequency**
VFCT	voice frequency carrier telegraph
VFCTG	voice frequency carrier telegraph
VFDF	voice frequency distribution frame
VFO	variable-frequency oscillator
VFTG	**voice-frequency telegraph**
VHF	**very high frequency**
VLF	**very low frequency**
V/m	volts per meter

VNL	**via net loss**
VNLF	via net loss factor
vocoder	voice-coder
vodas	voice-operated device anti-sing
vogad	voice-operated gain-adjusting device
volcas	voice-operated loss control and echo/signaling suppression
vox	voice-operated relay circuit
	voice operated transmit
VRC	vertical redundancy check
VSAT	very small aperture terminal
VSB	**vestigial sideband [transmission]**
VSM	vestigial sideband modulation
VSWR	**voltage standing wave ratio**
VT	**virtual terminal**
VTU	**video teleconferencing unit**
vu	**volume unit**
WADS	wide area data service
WAIS	Wide Area Information Servers
WAN	**wide area network**
WARC	World Administrative Radio Conference
WATS	Wide Area Telecommunications Service
	Wide Area Telephone Service
WAWS	Washington Area Wideband System
WDM	**wavelength-division multiplexing**
WHSR	White House Situation Room
WIN	WWMCCS Intercomputer Network
WITS	Washington Integrated Telecommunications System
WORM	write once, read many times
wpm	words per minute
wps	words per second
WRU	**who-are-you [character]**
wv	working voltage
WVDC	working voltage direct current
WWDSA	worldwide digital system architecture
WWMCCS	Worldwide Military Command and Control System
WWW	**World Wide Web**
XMIT	transmit
XMSN	transmission
XMTD	transmitted
XMTR	transmitter
XO	**crystal oscillator**
XOFF	transmitter off
XON	transmitter on
XT	**crosstalk**
XTAL	crystal
Z	**Zulu time**
ZD	zero defects
Z_o	**characteristic impedance**
0TLP	zero transmission level point

APPENDIX B

ABBREVIATED INDEX

This index consists of groupings of related principal term names of terms defined in this glossary. The groupings are arranged by technological families to provide the user the opportunity to browse among related definitions. For brevity, where numerous alphabetically consecutive term names of defined terms begin with the same word, only the first defined term name is included in this index. For example, eight defined terms beginning with the word *"facsimile"* follow the entry for *facsimile* in the body of the glossary. In this mini index, only the first, parent, term is listed because the additional entries are easily found after locating the parent term in the body of the glossary.

radiation pattern
radiation resistance
radio beam
radio horizon
radio horizon range (RHR)
radio range
reference antenna
reflective array antenna
rhombic antenna
shadow loss
side lobe
single-polarized antenna
skip zone
sky wave
slewing
slot antenna
space diversity
spillover
spot beam
test antenna
uniform linear array
whip antenna
Yagi antenna

Codes / Coding:

adaptive predictive coding (APC)
alphabetic code
alphanumeric code
analog decoding
analog encoding
analog-to-digital converter (ADC)
balanced code
bar code
Baudot code
BCH code
B8ZS
binary code
binary-coded decimal (BCD)
binary-coded decimal (BCD) notation
binary element
binary notation
bipolar signal
bit configuration
bit pairing
B6ZS
B3ZS
code
codec
code character
coded character set
code-division multiple access (CDMA)
comma-free code
data network identification code (DNIC)
delay encoding
dense binary code
differential encoding
differential Manchester encoding
digital alphabet
dipulse coding
direct-sequence spread spectrum
duobinary signal
EBCDIC
error-correcting code
error-detecting code
error-detecting system
escape character (ESC)
filtered symmetric differential phase-shift keying (FSDPSK)
gold code
Gray code
Hagelbarger code
Hamming code
Huffman coding
hypertext
linear predictive coding (LPC)
line code
Manchester code
mark
microcode
modified AMI code
motion compensation
multiple frequency-shift keying (MFSK)
multiprogramming
n-ary code
non-return-to-zero change-on-ones (NRZ1)
non-return-to-zero (NRZ)
non-return-to-zero mark (NRZ-M)
non-return-to-zero space (NRZ-S)
n-unit code
paired disparity code
parity
reconstructed sample
redundant code
return-to-zero (RZ)
run-length encoding
scrambler
secure transmission
segmented encoding law
self-synchronizing code
shannon
synchronization code
time code
time-gated direct-sequence spread spectrum
transcoding
two-out-of-five code
uniform encoding
unit-distance code
variant
zero-level decoder

Computer Graphics:

addressability
addressable point
computer graphics
hard copy
joy stick
landscape mode
pointer
portrait mode
regeneration
stroke

Computer Hardware:

communications processor unit (CPU)
comparator
compatibility
computer
computer architecture
computer system
CPU
digital computer
direct connect
file server
general purpose computer
hardware
host computer
hybrid computer
input-output channel
inverter
mainframe
microcomputer
micro-mainframe link
microprocessor
monitor

mouse
multiprocessor
operating system
parallel computer
parallel port
POSIX
process computer system
process control
processing unit
process interface system
processor
register
security filter
security kernel
serial computer
serial port
special purpose computer
standby
stored-program computer
touch-sensitive
trusted computing base (TCB)
trusted computer system (TCS)
warm restart
work station

Computer Programming / Languages / Software:

abstract syntax notation one (ASN.1)
Ada®
application program interface (API)
assemble
assembler
assembly language
assembly time
binding
branch
bug
call
CASE
CASE technology
C-language
COBOL
compartmentation
compatibility
compile
compiler
computer language
computer-oriented language
computer program
computer word
core dump
cross assembler
diagnostic program
firmware
global
graphical user interface (GUI)
graphic character
high-level language (HLL)
HTML
hypermedia
hypertext
icon
identifier (ID)
illegal character
instruction
intermediate language
intermediate-level language
interpret
job
kernel
label
LAN application (software)
language
language processor
link
load
loop
machine-independent
machine instruction
machine language
microinstruction
microprogram
monitor
multitasking
network control program (NCP)
network operating system
operand
operating system
overlay
path
pointer
procedure-oriented language
program
program architecture
programmable
programmable read-only memory (PROM)
programmer
programming language
prompt
reserved word
resident
restart
retrieval service
routine
run
screen
security filter
security kernel
SGML
shell
software
source language
source program
statement
structured programming
subroutine
supervisory program
switch
symbolic language
syntax
system software
target language
time-sharing
trace program
translator
trusted computing base (TCB)
trusted computer system (TCS)
underflow
UNIX™
utility program
virus
wildcard character
word processing

Control / Control Characters / Command / Error Correction / Handshaking:

acknowledge character (ACK)
ARQ
backward channel
backward signal
backward supervision
BCH code
block check
block parity
call control character
call control signal
cancel character (CAN)

carrier sense multiple access (CSMA)
channel bits
character
character check
character interval
check bit
check character
check digit
checksum
clear confirmation signal
command
command frame
command menu
command protocol data unit
common-channel signaling
connection-in-progress signal
control character
control communications
control operation
convolutional code
cyclic redundancy check (CRC)
data link escape character (DLE)
data transfer request signal
DCE clear signal
DCE waiting signal
delimiter
disabling tone
end-of-medium character
end-of-selection character
end-of-text character (ETX)
end-of-transmission-block character (ETB)
end-of-transmission character (EOT)
enquiry character (ENQ)
error control
error-correcting code
error-correcting system
error-detecting code
error-detecting system
flag
forward error correction (FEC)
Hagelbarger code
Hamming code
handshaking
idle character
longitudinal redundancy check (LRC)
negative-acknowledge character (NAK)
nontransparent mode
overhead information
password
pilot
process computer system
process control
process control equipment
process control system
ready-for-data signal
redundancy
redundancy check
residual error ratio
respond opportunity
response
Shannon's law
source quench
start signal
substitute (SUB) character
summation check
synchronous idle character (SYN)
transverse parity check
undetected error ratio
who-are-you (WRU) character

Distortion / Dispersion / Diffraction:

amplitude distortion
amplitude-vs.-frequency distortion
bias distortion
characteristic distortion
cyclic distortion
degree of isochronous distortion
degree of start-stop distortion
delay distortion
diffraction
dispersion
dispersion-limited operation
distortion
distortion-limited operation
end distortion
feeder echo noise
fortuitous distortion
harmonic distortion
intermodulation distortion
intramodal distortion
isochronous distortion
knife-edge effect
margin
multimode distortion
nonlinear distortion
phase-frequency distortion
quantizing distortion
signal droop
single-harmonic distortion
start-stop distortion
total harmonic distortion (THD)

Encryption / Decryption:

bulk encryption
cipher
cipher system
cipher text
ciphony
civision
codress message
compromise
COMSEC equipment
cover
cryptanalysis
CRYPTO
Data Encryption Standard (DES)
decipher
decode
decrypt
descrambler
encipher
encode
encoding law
encrypt
end-to-end encryption
garble
group
key
key distribution center (KDC)
key management
key stream
link encryption
multiplex link encryption
net control station (NCS)
null
OTAR
plain text
protected distribution system (PDS)
protection interval (PI)
pseudorandom number generator
public key cryptography
RED / BLACK concept
RED signal
rekeying
remote rekeying

security management
spoofing
superencryption
synchronous crypto-operation
transmission security key (TSK)
trunk encryption device (TED)
type 1 product
type 2 product
type 3 algorithm
type 4 algorithm
unique key
variant
vocoder

Facsimile:
analog facsimile equipment
aperture distortion
aspect ratio
black facsimile transmission
black recording
black signal
center frequency
confirmation to receive
continuous tone copy
contouring
cooperation factor
density
diametral index of cooperation
digital facsimile equipment
direction of scanning
direct recording
drum factor
drum speed
electrochemical recording
electrolytic recording
electronic line scanning
elemental area
facsimile (FAX)
flood projection
frame
graphics
halftone characteristic
hard copy
horizontal resolution
index of cooperation
Kendall effect
landscape mode
line-to-line correlation
maximum keying frequency
maximum modulating frequency
multipath
multiple-spot scanning
noisy black
noisy white
nominal linewidth
object
pel
phasing
photosensitive recording
picture black
picture frequency
picture white
portrait mode
record communication
recording spot
record medium
reproduction speed
run-length encoding
scanning
scanning direction
scanning line
signal contrast
simple scanning
skew
solid-state scanning
spot projection
spot size
spot speed
stagger
start-record signal
stop-record signal
stroke speed
synchronizing
synchronizing pilot
tailing
total line length
transmission time
transverse redundancy check (TRC)
underlap
white facsimile transmission
white signal
X-dimension of recorded spot
X-dimension of scanning spot
Y-dimension of recorded spot
Y-dimension of scanning spot

Fiber Optics:
abrasive
acceptance angle
acceptance cone
aligned bundle
all-silica fiber
angle of deviation
angle of incidence
aramid yarn
armor
axial propagation constant
bandwidth•distance product
bound mode
braid
break out
breakout cable
buffer
bundle
cable
cable cutoff wavelength (λ_{cc})
cladding
cladding diameter
cladding mode
cladding mode stripper
cleave
concentricity error
core
core area
core diameter
coupled modes
coupling efficiency
critical angle
cutback technique
cutoff mode
cutoff wavelength
deeply depressed cladding fiber
dielectric waveguide
differential quantum efficiency
dispersion
dispersion-shifted fiber
dispersion-unshifted fiber
dopant
doubly clad fiber
duplex cable
effective mode volume
electrical length
end finish
equilibrium length
equilibrium mode distribution (EMD)
equilibrium mode simulator
evanescent field
evanescent mode
extrinsic joint loss
fiber amplifier

fiber optic cable
fiber optic link
fiber optics (FO)
first window
flooding compound
fundamental mode
fusion splice
gap loss
gap-loss attenuator
gel
glass
graded-index fiber
graded-index profile
group delay
group index
group velocity
guided ray
hockey puck
homogeneous cladding
hybrid cable
hybrid mode
hydroxyl ion absorption
inclusion
index dip
index-matching material
insertion loss
intensity modulation (IM)
intramodal distortion
intrinsic joint loss
joint
lateral offset loss
launch angle
launching fiber
launch numerical aperture (LNA)
lay length
leaky mode
leaky ray
long wavelength
macrobend
macrobend loss
mandrel wrapping
material dispersion coefficient [M(λ)]
material scattering
mechanical splice
meridional ray
microbend
minimum bend radius
minimum-dispersion window
minimum-loss window
modal distribution
modal loss
modal noise
mode
mode coupling
mode field diameter (MFD)
mode partition noise
mode scrambler
mode volume
multifiber cable
multifiber joint
multimode distortion
multimode optical fiber
near-field region
near-field scanning
nonlinear scattering
numerical aperture loss
numerical aperture (*NA*)
OFC
OFN
OFNP
OFNR
open waveguide
optical connector
optical fiber
optical isolator
optical junction
ovality
overfill
parabolic profile
phase term
pigtail
plastic-clad silica (PCS) fiber
polarization-maintaining (PM) optical fiber
power-law index profile
precision-sleeve splicing
primary coating
profile parameter (*g*)
quadruply clad fiber
radiation angle
radiation-hardened fiber
radiation mode
radiation pattern
reference surface
reflection loss
refracted ray
refractive index contrast
refractive index (n, η)
refractive index profile
ribbon cable
rip cord
scattering center
second window
short wavelength
silica
single-mode optical fiber
skew ray
slab-dielectric waveguide
soliton
SONET
spectral loss curve
splice loss
splice organizer
star coupler
step-index fiber
step-index profile
strength member
substitution method
surface wave
tap
tapered fiber
T-coupler
terminus
thin-film optical waveguide
third window
tolerance field
transmitter central wavelength range (λ_{tmax} - λ_{tmin})
transverse electric and magnetic (TEM) mode
transverse electric (TE) mode
transverse magnetic (TM) mode
traveling wave
vertex angle
waveguide
weakly guiding fiber
window
zero-dispersion slope
zero-dispersion wavelength
zip-cord

Filters:

bandpass filter
bandpass limiter
band-stop filter
character filter
dichroic filter
digital filter
filter
high-pass filter
interference filter

Kalman filter
loop filter
low-pass filter
mode filter
optical filter
roofing filter
security filter

Frequency:
aircraft emergency frequency
automatic frequency control (AFC)
band
bandwidth (BW)
barrage jamming
baseband
calling frequency
carrier frequency
channel offset
characteristic frequency
cutoff frequency
extremely high frequency (EHF)
extremely low frequency (ELF)
frequency
guarded frequency
heterodyne
high frequency
hiss
hold-in frequency range
image frequency
image rejection ratio
insertion-loss-vs.-frequency characteristic
intermediate frequency (IF)
intermodulation product
lock-in frequency
lock-in range
lowest usable high frequency (LUF)
low frequency (LF)
maximum keying frequency
maximum modulating frequency
maximum usable frequency (MUF)
medium frequency (MF)
military common emergency frequency
natural frequency
nominal bandwidth
normalized frequency (V)
notch
occupied bandwidth
precise frequency
preemphasis
primary frequency
primary frequency standard
primary service area
priority
protected frequency
pull-in frequency range
pulse repetition rate
pump frequency
quartz oscillator
radio channel
radio frequency (rf)
radio wave
Rayleigh distribution
reference frequency
secondary frequency standard
sideband
single-frequency (SF) signaling
single-tone interference
slewing
spectral density
spectrum signature
spread spectrum
spurious emission
standard test tone
standard time and frequency signal (STFS) service
stereophonic sound subcarrier
stopband
syntonization
threshold frequency
transition frequency
ultra high frequency (UHF)
ultra low frequency (ULF)
very high frequency (VHF)
very low frequency (VLF)
virtual carrier frequency
voice frequency (VF)
wideband

Frequency Allocation / Assignment:
allocation (of a frequency band)
allotment (of a radio frequency or radio frequency channel)
assigned frequency
assigned frequency band
assignment (of a radio frequency or radio frequency channel)
authorized bandwidth
authorized frequency
C-band
frequency assignment
frequency assignment authority

Gain:
absolute gain
antenna gain
antenna gain-to-noise-temperature (G / T)
automatic gain control (AGC)
directive gain
diversity gain
equal gain combiner
fade margin
gain
gain medium
height gain
insertion gain
loop gain
net gain
process gain
signal processing gain
singing margin
singing point

Interfacing:
cell relay
data station
data transmission circuit
demarcation point (demarc)
dial mode
DTE clear signal
EIA interface
framed interface
gateway
interchange circuit
interexchange carrier (IXC)
interface
interface functionality
interface message processor (IMP)
interface payload
interface standard
medium interface point (MIP)
net radio interface (NRI)
network interface
network interface device (NID)
network termination 1 (NT1)
network termination 2 (NT2)
optical interface

point of interface (POI)
point of presence (POP)
POSIX
primary rate interface (PRI)
protocol translator
R interface
router
service
service access point (SAP)
service data unit (SDU)
S interface
terminal adapter
T-interface
transparent interface
T reference point
U interface
V reference point

Interference:
adjacent-channel interference
anti-interference
blanketing
co-channel interference
common-mode interference
common-mode rejection ratio (CMRR)
compatibility
conducted interference
differential mode interference
electromagnetic interference (EMI)
eye pattern
far-end crosstalk
FM blanketing
harmful interference
interaction crosstalk
interference
interference emission
interference filter
intersymbol interference
jamming margin
jamming to signal ratio (J / S)
limits of interference
optical interface
permissible interference
phase interference fading
precipitation static (p-static)
reradiation
single-frequency interference
single-tone interference
sunspot
white area

Internet:
Archie
browser
directed broadcast address
Domain Name System (DNS)
Gopher
ICMP
[The] Internet
Internet protocol (IP)
Internet protocol (IP) spoofing
lynx
Mosaic
SLIP
TCP
TCP / IP
TCP / IP Suite
Terminal Access Controller (TAC)
User Datagram Protocol (UDP)
web browser
World Wide Web (WWW)

ISDN:
access contention
B channel
bearer service
broadband ISDN (B-ISDN)
circuit transfer mode
contribution
customer access
D channel
demand service
digital subscriber line (DSL)
distribution
distribution service
event
frame
functional signaling
H-channel
hybrid interface structure
interactive service
interface functionality
interface payload
ISDN
labeled channel
labeled interface structure
labeled multiplexing
labeled statistical channel
LAP-D
messaging service
network element (NE)
network termination
network termination 1 (NT1)
network termination 2 (NT2)
positioned channel
post-production processing
primary rate interface (PRI)
reference configuration
reference point
reserved circuit service
retrieval service
R interface
Signaling System No. 7 (SS7)
S interface
synchronous transfer mode
teleaction service
terminal adapter
T-interface
transfer mode
transit delay
T reference point
U interface
V reference point
wide area network (WAN)

LANs / MANs / WANs:
baseband local area network
bridge
brouter
carrier sense
domain name server
Ethernet
fiber distributed data interface (FDDI)
head-end
jabber
LAN application (software)
lobe attaching unit
local area network (LAN)
logical link control (LLC) sublayer
metropolitan area network (MAN)
network control program (NCP)
network interface device (NID)
network topology
physical signaling sublayer (PLS)
ring latency
ring network
server
S interface

slotted-ring network
switched multimegabit data services (SMDS)
tail circuit
token
token-bus network
token passing
token ring adapter
truncated binary exponential backoff
wide area network (WAN)

Layered Systems / Open Architecture:

abstract syntax
encapsulation
expedited data unit
GOSIP
high-level control
intermediate system
LAP-B
layer
layered system
logical link control (LLC) sublayer
medium access control (MAC) sublayer
n-entity
network termination 1 (NT1)
network termination 2 (NT2)
n-user
open system
open systems architecture
Open Systems Interconnection (OSI)
Open Systems Interconnection (OSI)—architecture
Open Systems Interconnection (OSI)—Protocol Specification
Open Systems Interconnection—Reference Model (OSI—RM)
Open Systems Interconnection (OSI)—Service Definitions
Open Systems Interconnection (OSI)—Systems Management
peer entity
peer group
physical signaling sublayer (PLS)
protocol
protocol-control information
protocol data unit (PDU)
protocol hierarchy
service
service access point (SAP)
service data unit (SDU)
SONET
standardized profile
TCP
TCP / IP
TCP / IP Suite
terminal endpoint (TE) functional group
virtual terminal (VT)
wrapping

Loss / Attenuation:

absorption
absorption band
absorption coefficient
absorption index
absorption loss
angular misalignment loss
antenna dissipative loss
area loss
attenuation
attenuation coefficient
attenuation constant
attenuation-limited operation
attenuator
bridging loss
cloud attenuation
coupling loss
cutoff attenuator
differential mode attenuation
flat fading
free-space loss
gap loss
gap-loss attenuator
insertion loss
insertion-loss-vs.-frequency characteristic
lateral offset loss
loss
lossy medium
macrobend loss
modal loss
net loss
net loss variation
numerical aperture loss
path loss
precipitation attenuation
preemphasis
proration
reflection loss
return loss
scattering loss
selective fading
shadow loss
splice loss
transmission loss
via net loss (VNL)

Modems:

AT Commands
modem
modem patch
narrowband modem
wideband modem

Modulation:

absorption modulation
adaptive differential pulse-code modulation (ADPCM)
amplitude modulation (AM)
angle modulation
bandwidth (BW)
baseband
baseband modulation
binary modulation
bit-by-bit asynchronous operation
carrier shift
center frequency
chip rate
chirping
code-division multiple access (CDMA)
coherent differential phase-shift keying (CDPSK)
conditioned diphase modulation
continuously variable slope delta (CVSD) modulation
Costas loop
cross modulation
delta modulation (DM)
delta-sigma modulation
demand assignment multiple access (DAMA)
demodulation
differential modulation

differential phase-shift keying (DPSK)
differential pulse-code modulation (DPCM)
digital frequency modulation
digital modulation
digital phase modulation
direct-sequence modulation
disparity
double modulation
electro-optic modulator
exalted-carrier reception
fixed-reference modulation
frequency modulation (FM)
frequency-shift keying (FSK)
full modulation
incremental phase modulation (IPM)
indirect control
intensity modulation (IM)
isochronous demodulation
isochronous modulation
keying
load capacity
low-level modulation
maximum modulating frequency
mechanically induced modulation
modulation
multicarrier modulation (MCM)
overmodulation
percentage modulation
phase deviation
phase modulation (PM)
phase-shift keying (PSK)
preemphasis improvement
preemphasis network
pseudorandom noise
pulse-amplitude modulation (PAM)
pulse-code modulation (PCM)
pulse-duration modulation (PDM)
pulse-frequency modulation (PFM)
pulse-position modulation (PPM)
quadrature amplitude modulation (QAM)
quadrature modulation
quadrature phase-shift keying (QPSK)
quantization
quasi-analog transmission
radio
reduced carrier single-sideband emission
reduced carrier transmission
signal transition
significant condition
single-sideband (SSB) emission
single-sideband (SSB) equipment reference level
single-sideband suppressed carrier (SSB-SC) transmission
start-stop margin
start-stop modulation
stereophonic crosstalk
sub-band adaptive differential pulse code modulation (SB-ADPCM)
subcarrier
supergroup distribution frame (SGDF)
suppressed carrier transmission
thin-film optical modulator
unbalanced modulator
vestigial sideband (VSB) transmission
white facsimile transmission

Multiplexing:

asynchronous time-division multiplexing (ATDM)
asynchronous transfer mode (ATM)
carrier system
cell relay
channel bank
common-channel interoffice signaling (CCIS)
demultiplexing
D-4
digital multiplexer
digital multiplex hierarchy
digital signal (DS)
digroup
drop and insert
erect position
fractional T1
frame
framing
group
group distribution frame (GDF)
heterogeneous multiplexing
homogeneous multiplexing
inverse multiplexer
inverted position
labeled multiplexing
master frequency generator
multicarrier modulation (MCM)
multichannel
multiframe
multiplex aggregate bit rate
multiplex baseband
multiplexer / demultiplexer (muldem)
multiplexer (MUX)
multiplex hierarchy
multiplexing (MUXing)
multiplex link encryption
orthogonal multiplex
precombining
pregroup combining
quadruple diversity
space-division multiplexing
statistical multiplexing
statistical time-division multiplexing
supergroup distribution frame (SGDF)
synchronous TDM
synchronous transfer mode
system loading
T-carrier
thin-film optical multiplexer
time-division multiple access (TDMA)
time-division multiplexing (TDM)
transmultiplexer
trunk group multiplexer (TGM)
two-pilot regulation
wavelength-division multiplexing (WDM)

Networking:

access group
address field
asynchronous network
backbone
branching network
client
closed user group
commercial refile
control station
delayed-delivery facility
democratically synchronized network

despotically synchronized network
differentiating network
directed net
distributed control
distributed network
distributed-queue dual-bus (DQDB) [network]
endpoint node
equivalent network
free net
gateway
global address
GOSIP
group address
hierarchical computer network
hierarchically synchronized network
hotline
hybrid communications network
hypermedia
intelligent network (IN)
intelligent peripheral (IP)
interconnection
intermediate element
internetworking
interworking functions
leg
link
logical topology
loop transmission
maritime broadcast communications net
master station
mediation function
multicast
multichannel
multilevel precedence and preemption (MLPP)
multilink operation
multiple media
multipoint access
net operation
network
network control program (NCP)
network engineering
network interface
network interface card
network topology
noncentralized operation
nonsynchronous network
open network architecture (ONA)
overhead information
packet switching
path
PCS
physical topology
point-to-point link
polling
port
portability
POSIX
precedence
primary channel
primary station
public data network (PDN
public switched network (PSN)
radio net
relay
remote access
remote logon
route
routine message
seizing
server
service termination point
signal center
signal transfer point (STP)
sliding window
slot
slotted-ring network
slot time
sublayer
subnetwork
switched circuit
symmetrical channel
synchronous data link control (SDLC)
synchronous network
synchronous transmission
system failure transfer
system follow-up
system signaling and supervision
tandem
teleconference
Terminal Access Controller (TAC)
terminal adapter
terminal endpoint (TE) functional group
through group
through supergroup
T-interface
token-bus network
token passing
token-ring network
trace packet
unidirectional operation
U interface
Universal Personal Telecommunications (UPT) service
virtual network
virus
wrapping

Network Management:

accounting management
administrative management complex (AMC)
attribute
customer management complex
fault management
managed object
network administration
network management
network manager
operations system
performance management
security management
system management
telecommunications management network (TMN)

Networks:

ARPANet
Automatic Digital Network (AUTODIN)
Automatic Secure Voice Communications Network (AUTOSEVOCOM)
Automatic Voice Network (AUTOVON)
command net
common user network
communications net
communications network
computer network
Defense Communications System (DCS)
Defense Data Network (DDN)
Defense Switched Network (DSN)
Emergency Broadcast System (EBS)
integrated digital network (IDN)

[The] Internet
joint multichannel trunking and switching system
Joint Tactical Information Distribution System (JTIDS)
National Communications System (NCS)
National Information Infrastructure (NII)
value-added network (VAN)

Network Topology / Architecture:
client-server architecture
network architecture
network connectivity
network topology
node
physical topology
ring network

Noise:
ambient noise level
antenna noise temperature
atmospheric noise
background noise
blue noise
carrier noise level
carrier-to-noise ratio (CNR)
carrier-to-receiver noise density (*C/kT*)
channel noise level
circuit noise level
closed-loop noise bandwidth
C-message weighting
cosmic noise
effective input noise temperature
equipment intermodulation noise
equivalent noise resistance
equivalent noise temperature
equivalent pulse code modulation (PCM) noise
equivalent satellite link noise temperature
feeder echo noise
flat weighting
FM improvement factor
FM improvement threshold
front-end noise temperature
HA1-receiver weighting
idle-channel noise
impulse noise
in-band noise power ratio
intermodulation noise
intrinsic noise
loop noise
modal noise
mode partition noise
noise
notched noise
144-line weighting
144-receiver weighting
phase noise
photon noise
process gain
proration
pseudorandom noise
psophometric weighting
quantizing noise
quantum noise
quantum-noise-limited operation
random noise
received noise power
reference noise
shot noise
signal-plus-noise-to-noise ratio ((*S*+*N*) / *N*)
signal-to-noise ratio (SNR)
signal-to-noise ratio per bit
SINAD
snow
stereophonic crosstalk
thermal noise
total channel noise
white noise
worst hour of the year

NS / EP:
assignment
communications survivability
interoperability
National Communications System (NCS)
National Coordinating Center (NCC) for Telecommunications
NS / EP telecommunications
priority level
priority level assignment
private NS / EP telecommunications services
provisioning
public switched NS / EP telecommunications services
service identification
Telecommunications Service Priority (TSP) service
Telecommunications Service Priority (TSP) system

Organizations:
administration
AFNOR
American National Standards Institute (ANSI)
Bell Operating Company (BOC)
Canadian Standards Association (CSA)
CCIR
CCITT
common carrier
Deutsches Institut für Normung (DIN)
FCC
International Telecommunication Union (ITU)
ISO
Joint Telecommunications Resources Board (JTRB)
National Command Authorities (NCA)
National Communications System (NCS)
National Coordinating Center (NCC) for Telecommunications
public utilities commission (PUC)
Radio Regulations Board (*formerly* IFRB)
recognized operating agency (ROA)
Regional Bell Operating Company (RBOC)
specialized common carrier (SCC)

PCS / UPT / cellular mobile:
base station
call associated signaling (CAS)
call management
cell
cellular mobile
fixed access
interworking functions

mobile services switching center (MSC)
non-call associated signaling (NCAS)
non-fixed access
PCS
PCS switching center
PCS System
personal mobility
personal registration
personal terminal
public land mobile network (PLMN)
radio interface
service access
terminal mobility
terminal mobility management
Universal Personal Telecommunications (UPT) service
UPT access code
validation
wireless access mode
wireless mobility management
wireless terminal.

Power:
bel (B)
carrier power (of a radio transmitter)
carrier-to-receiver noise density (*C/kT*)
dB
power
radiance
radiant emittance
radiant intensity
radiant power
radiation efficiency
rated output power
reflectance
relative transmission level
rf power margin
sensitivity
signal level
signal-to-crosstalk ratio
SINAD
singing margin
single-polarized antenna
single-sideband (SSB) equipment reference level
spectral density
susceptibility threshold
system loading

Power (Electrical):
apparent power
auxiliary power
demand factor
demand load
disconnect switch
effective power
inverter
level
load
load factor
nonoperational load
operational load
power budget
power factor
power failure transfer
primary distribution system
primary power
primary substation
protector
station load
technical load
uninterruptible power supply
volt-amperes reactive (vars)

Protocols:
acknowledgement
Advanced Data Communication Control Procedures (ADCCP)
binary synchronous (bi-sync) communication
carrier sense multiple access with collision avoidance (CSMA / CA)
carrier sense multiple access with collision detection (CSMA / CD)
Common Management Information Protocol (CMIP)
connection-oriented data transfer protocol
disconnect command (DISC)
facility
frame relay
FTP
GOSIP
high-level data link control (HDLC)
HTTP
ICMP
internet protocol (IP)
LAP-B
LAP-D
link protocol
mode
network interface device (NID)
NFS
Open Systems Interconnection (OSI)— Protocol Specification
ping
protocol
protocol-control information
protocol converter
protocol data unit (PDU)
protocol hierarchy
protocol translator
remote operations service element (ROSE) protocol
response PDU
route
segment
serial
Simple Mail Transfer Protocol (SMTP)
Simple Network Management Protocol (SNMP)
SLIP
slot
slot time
synchronous data link control (SDLC)
TCP
TCP / IP
TCP / IP Suite
Telnet
transmission control protocol
User Datagram Protocol (UDP)
zero-bit insertion

Radio:
adaptive radio
airborne radio relay
air-ground radiotelephone service
allcall
amateur service
anycall
atmospheric noise

automated radio
automatic link establishment (ALE)
automatic message exchange (AME)
automatic sounding
brick
broadcasting-satellite service
broadcasting satellite space station
broadcasting service
broadcasting station
broadcast operation
capture effect
carrier synchronization
cellular mobile
clear channel
closed circuit
combat-net radio (CNR)
connectivity exchange (CONEX)
controller
cross-band radiotelegraph procedure
coverage
dead space
deemphasis
detector
deviation ratio
diffraction region
digital selective calling (DSC)
directed net
direction finding
discriminator
distance measuring equipment (DME)
diversity reception
diversity transmission
duplex operation
emergency locator transmitter (ELT)
emergency position-indicating radiobeacon station
emphasis
feeder link
FM blanketing
FM improvement factor
FM improvement threshold
Fresnel zone
harmful interference
heterodyne repeater
homing
horn
indefinite call sign
inside plant
k-factor
link
linking protection (LP)
link quality analysis (LQA)
maximum usable frequency (MUF)
meaconing
mean power [of a radio transmitter]
medium
message heading
message part
meteor burst communications
multi-satellite link
narrowband radio voice frequency (NBRVF)
near-vertical-incidence skywave
net control station (NCS)
net operation
net radio interface (NRI)
off-the-air
off-the-air monitoring
on-the-air
path
PBER
PCS
peak envelope power (of a radio transmitter) [PEP, pX, PX]
pilot
precipitation static (p-static)
preemphasis
preemphasis improvement
preemphasis network
preliminary call
primary service area
propagation mode
propagation path obstruction
protection interval (PI)
push-to-talk operation
quadruple diversity
query call
radiation
radio
radio and wire integration (RWI)
radio channel
radio common carrier (RCC)
radiocommunication
random access discrete address (RADA)
reflecting layer
rural radio service
rural subscriber station
scan-stop lockup
ship Earth station
shortwave
smooth Earth
sounding
sporadic E propagation
spurious response
stereophonic sound subcarrier
stressed environment
sudden ionospheric disturbance (SID)
sunspot
surface wave
tail circuit
terrestrial radiocommunication
terrestrial station
transmitter power output rating
traveling wave
TRI-TAC equipment
vestigial sideband (VSB) transmission
white area
wildcard character

Routing:

adaptive routing
alternate routing
avoidance routing
call tracing
circuit routing
circuit switching unit (CSU)
collective routing
destination routing
deterministic routing
dynamically adaptive routing
flood search routing
free routing
heuristic routing
hierarchical routing
hybrid routing
mediation function
multicast address
network administration
primary route
rerouting
route
route diversity
route matrix
routing
sender
spill-forward feature

spur

Satellites:

acquisition
active satellite
aeronautical mobile-satellite service
aeronautical mobile-satellite (OR) [off-route] service
aeronautical mobile-satellite (R) [route] service
amateur-satellite service
apogee
broadcasting-satellite service
broadcasting satellite space station
carrier-to-receiver noise density (*C/kT*)
C-band
chip
coast Earth station
communications satellite
community reception [in the broadcasting-satellite service]
coordination area
cross-site link
direct orbit
downlink (D / L)
dual access
Earth coverage
Earth exploration-satellite service
Earth station
Earth terminal
Earth terminal complex
equatorial orbit
equivalent satellite link noise temperature
fixed-satellite service
footprint
geostationary orbit
geosynchronous orbit
handoff
inclination of an orbit (of an Earth satellite)
inclined orbit
individual reception [in the broadcasting-satellite service]
inter-satellite service
land mobile-satellite service
maritime mobile-satellite service
meteorological-satellite service
mobile Earth station
multiple access
multi-satellite link
orbit
orbit determination
passive satellite
periapsis
perigee
period [of a satellite]
polar orbit
preassignment access plan
pulse-address multiple access (PAMA)
radiodetermination-satellite service
radionavigation-satellite service
resource controller (RC)
retrograde orbit
satellite
ship Earth station
space subsystem
spacecraft
space operation service
space radiocommunication
space research service
space system
space telecommand
space tracking
spot beam
standard frequency and time signal-satellite service
synchronous orbit
synchronous satellite
tail circuit
uplink (U / L)
view

Security:

access
anti-jam
anti-spoof
authenticate
authentication
authenticator
automated information systems security
between-the-lines entry
BLACK
BLACK signal
browsing
bug
certification
clipper chip
code
communications deception
communications-electronics (C-E)
communications intelligence (COMINT)
communications protection
communications security [COMSEC]
compromise
computer security (COMPUSEC)
COMSEC equipment
configuration management
controlled area
controlled security operation
controlled space
data security
electronic emission security
emanations security (EMSEC)
emission security
end-to-end security
environmental security
information security
information systems security (INFOSEC)
label
limited protection
monitoring
NAK attack
operations security
password
penetration
privacy
scrambler
secure communications
secure transmission
security
security filter
security kernel
security management
self-authentication
sensitive information
server
signals security
transmission security key (TSK)
trusted computing base (TCB)
trusted computer system (TCS)
Warner exemption

Services:

aeronautical fixed service
aeronautical mobile (OR) [off-route] service
aeronautical mobile (R) [route] service
aeronautical mobile-satellite service
aeronautical mobile-satellite (OR) [off-route] service
aeronautical mobile-satellite (R) [route] service
aeronautical mobile service
aeronautical multicom service
aeronautical radionavigation-satellite service
aeronautical radionavigation service
amateur-satellite service
amateur service
Archie
Armed Forces Radio Service (AFRS)
attendant position
base communications (basecom)
basic exchange telecommunications radio service (BETRS)
basic service
basic service element (BSE)
basic serving arrangement (BSA)
Centrex® (CTX) service
class of service
common management information service (CMIS)
comparably efficient interconnection (CEI)
complementary network service (CNS)
consolidated local telecommunications service
conversational service
custom local area signaling service (CLASS)
dedicated service
domestic fixed public service
domestic public radio services
eight-hundred (800) service
enhanced service
facility

Federal Telecommunications System (FTS)
five-hundred (500) service
flat rate service
foreign exchange (FX) service
FTS2000
interactive service
intercom
inter-satellite service
leased circuit
maritime mobile-satellite service
maritime mobile service
maritime radionavigation-satellite service
maritime radionavigation service
message service
meteorological aids service
meteorological-satellite service
mobile-satellite service
mobile service
multipoint distribution service
nine-hundred (900) service
other common carrier (OCC)
paging
PCS
port operations service
private NS / EP telecommunications services
provisioning
public data transmission service
public switched network (PSN)
public switched NS / EP telecommunications services
public switched telephone network (PSTN)
radio channel
radio common carrier (RCC)
radiodetermination-satellite service
radiolocation service
radionavigation-satellite service
resale carrier
resale service
reserved circuit service
restricted access
retrieval service
rural radio service
rural subscriber station
safety service
service access point (SAP)
service bit
seven-hundred (700) service
space research service
special grade of service
standard frequency and time signal-satellite service
teleaction service
telecommunications service
Telecommunications Service Priority (TSP) service
teletex
teletext
Telex®
unbundling
Universal Personal Telecommunications (UPT) service
universal service
viewdata
voice grade
white pages
Wide Area Telephone Service (WATS)
World Wide Web (WWW)

Service Features:

abbreviated dialing
access control
add-on conference
automatic answering
automatic identified outward dialing (AIOD)
automatic message accounting (AMA)
automatic number identification (ANI)
automatic redial
automatic sequential connection
busy verification
call detail recording (CDR)
called-line identification facility
called-party camp-on
caller ID
call forwarding
call hold
calling-line identification facility
calling-party camp-on
call pickup
call restriction
calls-barred facility
call splitting
call transfer
call waiting
classmark
conference call
consultation hold

dial service assistance (DSA)
dial-up
direct dialing service
direct distance dialing (DDD)
direct inward dialing (DID)
direct outward dialing (DOD)
essential service
extended area service (EAS)
fixed loop
group alerting and dispatching
system
line load control
multiaddress calling
network inward dialing (NID)
network outward dialing (NOD)
remote access
remote call forwarding
service feature
speed calling
spill-forward feature
three-way calling
transit network identification
verified off-hook
virtual call
virtual circuit capability

Standards:
ANSI / EIA / TIA-568
ASCII
basic mode link control
basic rate interface (BRI)
de facto standard
frequency standard
functional profile
GOSIP)
group
industry standard
interface standard
International Telegraph Alphabet
Number 5 (ITA-5)
interoperability standard
message handling system (MHS)
National Electric Code® (NEC)
network terminal number (NTN)
NTSC standard
Open Systems Interconnection
(OSI)
Open Systems Interconnection
(OSI)—architecture
Open Systems Interconnection
(OSI)—Protocol Specification
Open Systems
Interconnection—Reference
Model (OSI—RM)
Open Systems Interconnection
(OSI)—Service Definitions
Open Systems Interconnection
(OSI)—Systems Management
POSIX
precise frequency
primary frequency standard
proprietary standard
p x 64
quarter common intermediate
format (QCIF)
R interface
rubidium clock
rubidium standard
secondary frequency standard
secondary time standard
serial
Signaling System No. 7 (SS7)
SONET
standard
standard optical source
standard time and frequency signal
(STFS) service
system standard
terminal adapter
time standard
V-series Recommendations
X.-series Recommendations

Storage:
accumulator
archiving
arithmetic overflow
arithmetic register
arithmetic underflow
associative storage
auxiliary storage
buffer
cache memory
CD ROM
channel
clear
computer program origin
direct access
direct address
diskette
disk pack
elastic buffer
erase
field
file
file transfer, access, and
management (FTAM)
hard sectoring
head
interblock gap
internal memory
loader
machine-readable medium
magnetic card
magnetic core storage
magnetic drum
magnetic tape
main storage
medium
memory
off-line storage
optical disk
overlay
overload
packing density
permanent storage
phase-encoded (PE) recording
phase flux reversal
print-through
programmable read-only memory
(PROM)
queue
queue traffic
queuing
random access memory (RAM)
read head
reading
read-only memory (ROM)
read-only storage
record
record communication
record information
record medium
regeneration
register
rotational position sensing
sector
sectoring
seek
seek time

shift register
simple buffering
skew
soft sectoring
spooling
storage
storage cell
store-and-forward (S-F)
store-and-forward switching center
streaming tape drive
streaming tape recording
tag image file format (TIFF)
tape relay
track
track density
tracking error
variable length buffer
virtual memory
virtual storage
volatile storage
volume
work space
write

Switching:
analog switch
broadband exchange (BEX)
burst switching
call progress signal
circuit-switched data transmission service
circuit switching
class of office
common control
congestion
crosspoint
digital switching
distributed switching
electronic switching system (ESS)
fast packet switching
first-in last-out (FIFO)
interrupted continuous wave (ICW)
joint multichannel trunking and switching system
message switching
multiple access
multiple homing
nonblocking switch
off-hook signal
one-way trunk
pilot-make-busy (PMB) circuit
private exchange (PX)
public switched network (PSN)
public switched NS / EP telecommunications services
public switched telephone network (PSTN)
queuing delay
rotary hunting
seizing
semiautomatic switching system
signal transfer point (STP)
space-division switching
spill forward
split homing
step-by-step (SXS) switching system
switch
switchboard
switch busy hour
switching
synchronous transfer mode
Tactical Automatic Digital Switching System (TADSS)
thin-film optical switch
three-way calling
time-division switching
trunk
wink pulsing

Synchronization:
bilateral synchronization
carrier synchronization
digital synchronization
double-ended synchronization
frame synchronization
linear analog synchronization
lip synchronization
mutually synchronized network
mutual synchronization
nonsynchronous data transmission channel
pilot
single-ended synchronization
synchronism
synchronization
synchronization bit
synchronization code
synchronization pulse
synchronous
timing tracking accuracy
unilateral synchronization system

Telegraphy:
A-condition
Baudot code
cable
chad tape
characteristic distortion
character-stepped
double-frequency shift keying (DFSK)
end distortion
equal-length code
frequency-shift telegraphy
internal bias
local battery
mark
narrative traffic
neutral direct-current telegraph system
neutral operation
polar direct-current telegraph transmission
polarential telegraph system
polar operation
push-to-type operation
radiotelegram
radio telegraphy
record communication
reperforator
sending-end crossfire
space
tape relay
telegram
telegraphy
teleprinter
teletypewriter (TTY)
terminal equipment
tone diversity
TTY / TDD
twinplex
two-tone keying
voice-frequency telegraph (VFTG)
word

Telephony Hardware:
acoustic coupler
aerial cable
aerial insert

applique
central office (C.O.)
combined distribution frame (CDF)
common battery
conditioned loop
connecting arrangement
cord circuit
cord lamp
cordless switchboard
crossbar switch
cross-connection
cross-office trunk
customer office terminal (COT)
customer service unit (CSU)
dial switching equipment
digital access and cross-connect system (DACS)
digital circuit patch bay
digital loop carrier (DLC)
digital primary patch bay
digital switch
distribution frame
drop
DSA board
D-type patch bay
electrically powered telephone
embedded base equipment
embedded customer-premises equipment
end instrument
end office
entrance facility
entrance point
entrance room
equal-level patch bay
equipment room
equipment side
exchange
exchange facilities
extension bell
four-wire repeater
grandfathered systems
grandfathered terminal equipment
group distribution frame (GDF)
high-frequency distribution frame (HFDF)
individual line
inside plant
integrated station
intercom
intermediate distribution frame (IDF)
interoffice trunk
interposition trunk
interswitch trunk
intertoll trunk
intraoffice trunk
key set
key telephone system (KTS)
K-type patch bay
limited-protection voice equipment
line side
local battery
local exchange loop
local side
main distribution frame (MDF)
main station
master frequency generator
MM patch bay
new customer premises equipment
office classification
off-premises extension (OPX)
on-premises extension
orderwire multiplex
outside plant
patch and test facility (PTF)
plant
private exchange (PX)
private line
proration
protected distribution system (PDS)
public switched telephone network (PSTN)
rack
repertory dialer
rotary dial
SECORD
semiautomatic switching system
signal center
sound-powered telephone
special grade access line
splice closure
station battery
STU
supergroup distribution frame (SGDF)
switchboard
switched circuit
switched loop
tandem center
tandem office
technical control facility (TCF)
Telecommunications Device for the Deaf (TDD)
telecommunications facilities
telephone
terminal equipment
through-group equipment
through-supergroup equipment
tie trunk
translator
tributary office
videophone
vodas
vogad
voice frequency (VF) primary patch bay

Television:

advanced television (ATV)
aspect ratio
cable TV (CATV)
cable television relay service (CARS) station
chroma keying
chrominance signal
closed captioning
closed circuit
composite video
distribution-quality television
extended-definition television (EDTV)
fixed microwave auxiliary station
freeze frame television
full-motion operation
ghost
head-end
high-definition television (HDTV)
improved-definition television (IDTV)
lip synchronization
minimum picture interval
PAL
PAL-M
picture black
picture white
raster
reference black level
reference white level
RGB
scanning

SECAM
studio-to-transmitter link (STL)
television (TV)
video

Time / Timing:
absolute delay
cesium clock
channel time slot
clock
coasting mode
coherence time
coordinated clock
coordinated time scale
Coordinated Universal Time (UTC)
data transfer time
date
date-time group (DTG)
dating format
DoD master clock
epoch date
equipment clock
external timing reference
false clock
Greenwich Mean Time (GMT)
heterochronous
hierarchically synchronized network
homochronous
independent clocks
indirect control
International Atomic Time (TAI)
isochronous
isochronous demodulation
isochronous modulation
isochronous signal
Julian date
leap second
local clock
master clock
master-slave timing
mesochronous
mutually synchronized network
mutual synchronization
near real time
nodal clock
nonsynchronous network
oligarchically synchronized network
plesiochronous
precise time
precise time and time interval (PTTI)
primary time standard
principal clock
protection interval (PI)
pull-in frequency range
quartz clock
real time
reference clock
release time
remote clock
rise time
rubidium clock
rubidium standard
secondary time standard
serial
slave clock
standard frequency and time signal-satellite service
standard frequency and time signal service
standard frequency and time signal station
standard time and frequency signal (STFS) service
station clock
synchronism
synchronization
synchronous network
synchronous TDM
time
time ambiguity
time instability
time interval error (TIE)
time jitter
time marker
time of occurrence
time-out
time scale
time standard
timing recovery
timing signal
timing tracking accuracy
transmit-after-receive time delay
transmitter attack-time delay
transmitter release-time delay
uniform time scale
unit interval
Universal Time (UT)
UTC(*i*)

Traffic:
activity factor
adaptive channel allocation
busy hour
busy season
calling rate
call-second
connections per circuit hour (CCH)
directionalization
FOT
frequency prediction
grade of service (GOS)
group busy hour (GBH)
high-usage trunk group
holding time
interconnect facility
line traffic coordinator (LTC)
loading
minimize
narrative traffic
net control station (NCS)
off-line recovery
overflow
queue
queue traffic
queuing
record traffic
scanning
short haul toll traffic
switch busy hour
traffic

Transmission / Propagation:
absolute delay
analog transmission
anisochronous
anomalous propagation (AP)
asynchronous transmission
atmospheric duct
babble
backscattering
batched transmission
bit-stream transmission
burst transmission
channel
coupled modes
critical frequency
critical wavelength
cross-polarized operation
Dellinger effect

digital transmission group
digital transmission system
digital voice transmission
diurnal phase shift
double-sideband (DSB) transmission
double-sideband reduced carrier (DSB-RC) transmission
double-sideband suppressed-carrier (DSB-SC) transmission
D region
ducting
duplex operation
echo
electronic line of sight
elevated duct
E region
erlang
flutter
forward scatter
FOT
F region
Fresnel zone
ground wave
group velocity
homogeneous cladding
intermediate-field region
ionosphere
ionospheric absorption
ionospheric disturbance
ionospheric forward scatter
ionospheric reflection
ionospheric scatter
ionospheric turbulence
isochronous burst transmission
k-factor
knife-edge effect
line-of-sight (LOS) propagation
magnetic storm
magneto-ionic double refraction
material scattering
maximum usable frequency (MUF)
Mie scattering
multipath
near-field region
near-vertical-incidence skywave
parallel transmission
path clearance
path profile
primary service area
propagation
propagation constant
propagation mode
propagation path obstruction
propagation time delay
pulse broadening
quasi-analog transmission
radiation scattering
ray
Rayleigh fading
Rayleigh scattering
reflecting layer
reflection
refracted ray
refraction
refractive index (n, η)
relative transmission level
right-hand [clockwise] polarized wave
ring latency
scattering
scintillation
serial transmission
shadow loss
simplex operation
skip distance
skip zone
sky wave
smooth Earth
space diversity
sporadic E
sunspot
surface wave
synchronous transmission
time block
trapped electromagnetic wave
troposphere
tropospheric scatter
tropospheric wave
undisturbed day
virtual height
waveguide

Transmission Lines:

artificial transmission line
balanced line
building out
bundle
cable
Category 3
Category 4
Category 5
coaxial cable (coax)
composite cable
direct-buried cable
directional coupler
electrical length
fiber optic cable
field wire
filled cable
group delay
hybrid cable
land line
lay length
line
long line
loop
medium
nonloaded twisted pair
on-premises wiring
open wire
paired cable
plenum
quad
quadded cable
reflection
reflection coefficient (RC)
reflection loss
return loss
ribbon cable
sheath
sheath miles
shielded pair
shielded twisted pair
skin effect
slope
slope equalizer
spiral-four cable
strength member
symmetrical pair
terminal impedance
termination
transmission line
transverse electric and magnetic (TEM) mode
transverse electric (TE) mode
twin cable
unbalanced line
underground cable
uniform transmission line
voltage standing wave ratio (VSWR)

zip-cord

Video:

average picture level (APL)
block distortion
blurring
color errors
continuous presence
contrast
control field
edge busyness
error blocks
field
frame
freeze frame
image
interlaced scanning
jerkiness
mosquito noise
motion compensation
motion response degradation
motion video
object
object persistence
object retention
panning
p x 64
quarter common intermediate format (QCIF)
scene cut
scene cut response
smearing
snow
spatial application
spatial edge noise
still image
still video
temporal application
temporal edge noise
transmission service channel
video
videophone
windowing

Waves:

carrier (cxr)
coherence area
coherence length
coherent
continuous wave (cw)
depolarization
doppler effect
electromagnetic wave (EMW)
elliptical polarization
envelope
group delay
group delay time
guided wave
indirect wave
left-hand (anti-clockwise) polarized wave
linear polarization
material scattering
microwave
millimeter wave
plane wave
polarization
polarization diversity
Poynting vector
propagation
pulse carrier
radiant energy
radiation
radio
radio beam
radiocommunication
radio wave
reflection loss
reflection
reflection coefficient (RC)
right-hand [clockwise] polarized wave
scattering
signature
spurious response
square wave
standing wave
standing wave ratio (SWR)
tropospheric wave
waveform
wavefront
waveguide scattering
wave impedance
wavelength
wavelength stability
wave trap

Telecommunications Act Handbook

From the Foreword...

"[The Telecommunications Act Handbook] is not simply a dry recitation of the provisions of the new telecommunications law. Instead, it puts the new law in context-helping telecommunications professionals and laymen alike understand why it's important, and how it will affect their lives."

Congressman Jack Fields,
Chairman, Subcommittee on Telecommunications and Finance
U.S. House of Representatives

The Telecommunications Act of 1996 will dramatically change the way you do business in the telecommunications industry. Its impact on the market will result in huge transeferals of wealth.

Is your company prepared to claim its market share?

Now is the time for your to position your company to take advantage of its enormous opportunity, but to do so you need to understand the legal landscape with which your business operates.

The new **Telecommunications Act Handbook:** *A Complete Reference for Business* provides a practical, non-legalese explanations of this legal framework, and delves into the challenges, risks, and benefits confronting this market.

This comprehensive handbook also summarizes the history and development of each sector of the telecommunications industry, and then clearly explains how the new legislation will change each industry sector.

Don't allow your company to miss out on this $1 trillion dollar opportunity. With the **Telecommunications Act Handbook:** *A Complete Reference to Business,* you'll quickly learn the enormous impact of this law that establishes the framework that will govern telecommunications activities for years to come.

Telecommunications Act Handbook: ***A Complete Reference for Business***
Hardcover, Index, 640 pages, 1996, ISBN: 0-86587-545-6 **$89**

Government Institutes

Contact us for a complete catalog of our books.

4 Research Place, Rockville, Maryland, 20850
tel. **(301) 921-2355** fax **(301) 921-0373**

Website- **http://www.govinst.com**
E-mail- **giinfo@govinst.com**

GOVERNMENT INSTITUTES MINI-CATALOG

PC #	TELECOMMUNICATIONS TITLES	Pub Date	Price
545	Telecommunications Act Handbook: A Complete Reference for Business	1996	$89
675	Beyond the Telecommunications Act: A Domestic & International Perspective	1998	$89
564	Official Telecommunications Dictionary: Legal and Regulatory Definitions	1997	$49
586	Regulation of Wireless Communications Systems	1997	$89
606	Official Internet Dictionary: A Comprehensive Reference for Professionals	1998	$49
583	Telecommunications : Key Contacts and Information Sources	1998	$59

PC #	ENVIRONMENTAL TITLES	Pub Date	Price
585	Book of Lists for Regulated Hazardous Substances, 8th Edition	1997	$79
🖫 4088	CFR Chemical Lists on CD ROM, 1997 Edition	1997	$125
🖫 4089	Chemical Data for Workplace Sampling & Analysis, Single User	1997	$125
512	Clean Water Handbook, 2nd Edition	1996	$89
581	EH&S Auditing Made Easy	1997	$79
587	E H & S CFR Training Requirements, 3rd Edition	1997	$89
525	Environmental Audits, 7th Edition	1996	$79
548	Environmental Engineering and Science: An Introduction	1997	$79
578	Environmental Guide to the Internet, 4th Edition	1998	$59
560	Environmental Law Handbook, 14th Edition	1997	$79
625	Environmental Statutes, 1998 Edition	1998	$69
🖫 4000	Environmental Statutes on Disk for Windows-Single User	1998	$139
536	ESAs Made Easy	1996	$59
515	Industrial Environmental Management: A Practical Approach	1996	$79
🖫 4078	IRIS Database-Single User	1997	$495
510	ISO 14000: Understanding Environmental Standards	1996	$69
551	ISO 14001: An Executive Repoert	1996	$55
518	Lead Regulation Handbook	1996	$79
478	Principles of EH&S Management	1995	$69
554	Property Rights: Understanding Government Takings	1997	$79
582	Recycling & Waste Mgmt Guide to the Internet	1997	$49
603	Superfund Manual, 6th Edition	1997	$115
566	TSCA Handbook, 3rd Edition	1997	$95
534	Wetland Mitigation: Mitigation Banking and Other Strategies	1997	$75

PC #	SAFETY AND HEALTH TITLES	Pub Date	Price
547	Construction Safety Handbook	1996	$79
553	Cumulative Trauma Disorders	1997	$59
559	Forklift Safety	1997	$65
539	Fundamentals of Occupational Safety & Health	1996	$49
535	Making Sense of OSHA Compliance	1997	$59
563	Managing Change for Safety and Health Professionals	1997	$59
589	Managing Fatigue in Transportation, *ATA Conference*	1997	$75
598	Project Mgmt for E H & S Professionals	1997	$59
552	Safety & Health in Agriculture, Forestry and Fisheries	1997	$125
613	Safety & Health on the Internet, 2nd Edition	1998	$49
463	Safety Made Easy	1995	$49
590	Your Company Safety and Health Manual	1997	$79

🖫 Electronic Product available on CD-ROM or Floppy Disk

PLEASE CALL OUR CUSTOMER SERVICE DEPARTMENT AT (301) 921-2323 FOR A FREE PUBLICATIONS CATALOG.

Government Institutes
4 Research Place, Suite 200 • Rockville, MD 20850-3226
Tel. (301) 921-2323 • FAX (301) 921-0264
E mail: giinfo@govinst.com • Internet: http://www.govinst.com